Biodiesel Production and Processing Technology

Biodiesel Production and Processing Technology

Ava White
Connor Stone
Adam Wood

Kruger Brentt
Publishers
2026

Kruger Brentt Publishers UK. LTD.
Company Number 9728962

Regd. Office: 68 St Margarets Road, Edgware, Middlesex HA8 9UU

ISBN: 978-1-78715-602-9

Disclaimer:

Every possible effort has been made to ensure that the information contained in this book is accurate at the time of going to press, and the publisher and author cannot accept responsibility for any errors or omissions, however caused. No responsibility for loss or damage occasioned to any person acting, or refraining from action, as a result of the material in this publication can be accepted by the editor, the publisher or the author. The Publisher is not associated with any product or vendor mentioned in the book. The contents of this work are intended to further general scientific research, understanding and discussion only. Readers should consult with a specialist where appropriate.

Every effort has been made to trace the owners of copyright material used in this book, if any. The author and the publisher will be grateful for any omission brought to their notice for acknowledgement in the future editions of the book.

For information on all our publications visit our website at http://krugerbrentt.com/

PREFACE

The global search for sustainable and renewable energy alternatives has placed biodiesel at the forefront of biofuel research and industrial application. Derived primarily from vegetable oils, animal fats, and waste cooking oils, biodiesel offers a promising substitute for conventional petroleum-based diesel due to its biodegradability, low toxicity, and reduced greenhouse gas emissions. As the world faces challenges of energy security, rising fossil fuel prices, and the urgent need to mitigate climate change, biodiesel production and utilization have emerged as vital components of the renewable energy portfolio.

This book, Biodiesel Production and Processing Technology, provides a comprehensive and structured overview of the principles, processes, and applications involved in biodiesel technology. It is designed to serve as a reference for students, researchers, engineers, and industry professionals seeking both fundamental knowledge and practical insights into biodiesel production systems.

The book begins with an introduction to the energy crisis and the role of biodiesel as a sustainable fuel alternative. It discusses feedstock options, including edible and non-edible oils, algal sources, and waste-based raw materials, emphasizing the importance of feedstock selection in determining fuel quality and production costs. Detailed chapters explore the chemistry of transesterification, catalyst types, process optimization, and technological innovations such as supercritical fluid methods, enzymatic catalysis, and microwave-assisted reactions.

Attention is given to biodiesel purification and quality standards, addressing issues of fuel properties, blending behavior, storage stability, and compliance with international specifications such as ASTM D6751 and EN 14214. The integration of biodiesel with existing diesel infrastructure, as well as its environmental, economic, and social implications, are critically examined.

Special sections highlight advances in biodiesel from second-generation and third-generation feedstocks, including lignocellulosic biomass and microalgae, which offer solutions to the food-versus-fuel debate. The role of process engineering, reactor design, and life-cycle assessment (LCA) in enhancing the sustainability of biodiesel production is also discussed in depth.

Through a combination of theoretical frameworks, process flow diagrams, experimental approaches, and case studies from global biodiesel projects, this book presents a balanced perspective on current challenges and future opportunities in the field. It underscores the need for innovative technologies, supportive policies, and collaborative research to scale up biodiesel production and make it a viable contributor to global energy needs.

By blending scientific principles with technological applications, this book equips readers with the knowledge required to understand, design, and optimize biodiesel production systems for a sustainable energy future.

Ava White
Connor Stone
Adam Wood

CONTENTS

PREFACE V

1. **INEDIBLE OILS: FEEDSTOCKS FOR SUSTAINABLE BIODIESEL 1**

1. Introduction: An Overview 1

2. Agronomical aspects 3

3. Chemical aspects: Standardization of the raw materials and biodiesel production 8

4. Concise Summary 18

References 20

2. **PRODUCTION OF BIODIESEL FROM WASTE COOKING OIL 25**

1. Introduction: An Overview 25

2. Types of cooking oil 28

3. Biodiesel production process 32

4. Biodiesel production from used cooking oils 38

5. Concise Summary 49

References 49

3. **UTILIZATION OF ANIMAL FAT WASTE FOR BIODIESEL PRODUCTION 51**

1. Introduction: An Overview 51

2. Meat production around the world 54

3. Characterization and generation of animal fat wastes 57

4. Biodiesel production from animal fat wastes: technical challenges 61

5. Concise Summary 72

References 73

4. **EXTRACTION OF LIPIDS FOR BIODIESEL PRODUCTION FROM OLEAGINOUS FUNGI 83**

1. Introduction: An Overview 83

2. Ecology of oleaginous fungi 84

3. Biochemistry of lipid accumulation 87

4. Metabolic engineering of oleaginous yeasts 93

5. Cultivation condition of oleaginous yeasts 94
6. Batch, fed-batch and fermentation processes 95
7. Substrates and raw material 97
8. Concise Summary 100
References 100

5. MICROBIAL BIODIESEL PRODUCTION - OIL FEEDSTOCKS DERIVED FROM MICROBIAL CELL CULTIVATIONS 109

1. Introduction: An Overview 109
2. Biodiesel and current feedstocks 110
3. Microorganisms for microbial lipid production 111
4. Feedstocks for microbial lipid production 115
5. Conversion processes for microbial lipid production 117
6. Cell harvest and lipid extraction 121
7. Techno-economic analysis and life cycle assessment 125
8. Concise Summary 126
References 126

6. PRODUCTION OF ALGAL BIOMASS AND BIODIESEL 133

1. Introduction: An Overview 133
2. Algae as biological material 133
3. Diesel production problems 135
4. Biodiesel instead of diesel 136
5. Algae as potentials for biodiesel production 138
6. Biodiesel from algae 139
7. Environmental advantages of algal biofuels 144
8. Comparison between biodiesel production from algae and vegetables 145
9. The physical and chemical properties of biodiesel produced from algal cell 146
10. Enhancement the biodiesel production from algae 147
11. Wastewater nitrogen and phosphorous as microalgae nutrients 150
12. Economic importance 153
13. The issues associated with algae 153
14. Additional applications of algae 154
References 154

7. MICROALGAE AS FEEDSTOCKS FOR BIODIESEL SYNTHESIS 159

1. Introduction: An Overview 159
2. Potential of using microalgae as biodiesel feedstocks 162

3. Biodiesel production from microalgae 171

4. Genetic engineering of microalgae 178

5. Concise Summary 182

6. References 182

8. ECO-PHYSIOLOGICAL CONSTRAI TS AND TECHNOLOGICAL INNOVATIONS FOR BIODIESEL PRODUCTION FROM MICROALGAE................................195

1. Introduction: An Overview 195

2. Microalgae biomass generation 196

3. Thermodynamic efficiency of photosynthesis in microalgae 198

4. Limitations and improvements of photosynthetic biomass production of microalgae 201

5. Microalgal biomass harvesting 204

6. Biomass conversion to biodiesel 204

7. Concise Summary 208

References 208

9. BENEFITS AND OBSTACLES OF MICROALGAE AS A SOURCE OF OIL FOR BIODIESEL...213

1. Introduction: An Overview 213

2. Microalgae 213

3. Biodiesel from microalgae 215

4. Microalgal lipids 215

5. Cultivation of microalgae 217

6. Harvesting and processing 222

7. Economic and environmental feasibility 227

8. Optimizing lipid productivity 229

9. Concise Summary 234

10. References 236

10. AN INTEGRATED WASTE-FREE BIOMASS UTILIZATION SYSTEM FOR AN INCREASED PRODUCTIVITY OF BIOFUEL & BIOENERGY 243

1. Introduction: An Overview 243

2. Energy aspects of biomass utilization 244

3. New trends in biodiesel technologies 246

4. Butanol as raw material in biodiesel production 252

5. Other aspects of the integrated biomass utilization system 260

6. Concise Summary 269

References 269

11. PRODUCTION OF BIODIESEL VIA IN-SITU SUPERCRITICAL METHANOL TRANSESTERIFICATION 273

1. Introduction: An Overview 273

2. Materials and methods 275

3. Results and discussion 277

4. Concise summary 287

5. References 288

12. TRANSESTERIFICATION IN SUPERCRITICAL CONDITIONS. 295

1. Introduction: An Overview 295

2. Process overview 302

3. Process improvements 309

4. Process prospective 313

5. Concise Summary 314

References 315

13. ALTERNATIVE TECHNIQUES FOR THE PRODUCTION OF FATTY ACID ALKYL ESTERS: MICROWAVES, RADIO FREQUENCY, AND ULTRASOUND.... 323

1. Introduction: An Overview 323

2. Fatty acid alkylesters production assisted by microwaves 324

3. Fatty acid alkylesters production assisted by radio frequency 333

4. Fatty acid alkylesters production assisted by ultrasound 335

5. Future Directions 339

6. References 340

14. REACTIVE DISTILLATION FOR THE SYNTHESIS AND CHARACTERIZATION OF BIODIESEL VIA TRANSESTERIFICATION 347

1. Introduction: An Overview 347

2. Biodiesel scenarios worldwide 348

3. Reactive distillation 348

4. Case study - Transesterification by reactive distillation for synthesis and characterization of biodiesel 354

5. Experimental setup 357

6. Concise Summary 372

References 375

15. GAS-LIQUID PROCESS, THERMODYNAMIC PROPERTIES (19 BLENDS), EFFICIENCY AND ENVIRONMENTAL IMPLICATIONS, SEM PARTICULATE MATTER ASSESSMENT, AND ON-ROAD BUS TRIAL OF AN ESTABLISHED NOX-REDUCING BIODIESEL 377

1. Introduction: An Overview 377
2. Process description 378
3. Engine testing and exhaust gas analysis 383
4. Results and discussion 387
5. Concise Summary 396
References 398

16. PRODUCTION OF BIODIESEL UTILIZING SOLID CATALYST 401

1. Introduction: An Overview 401
2. Heterogeneous solid catalysts 401
3. Catalytic mechanism 412
4. Other methods or technologies 416
5. Concise Summary 419
References 420

17. AN ECO-FRIENDLY ALTERNATIVE FOR CASTOR OIL BIODIESEL: UTILIZATION OF SOLID SUPPORTED ACIDIC SALT CATALYST 427

1. Introduction: An Overview 427
2. Feedstock of biodiesel 429
3. Transesterification of vegetable oils 431
4. Materials and methods 434
5. Results and discussion 438
6. Mechanism 440
7. Influence of reaction parameters 441
8. Concise Summary 444
References 444

18. THE IMMOBILIZED LIPASES IN BIODIESEL PRODUCTION 449

1. Introduction: An Overview 449
2. Lipases mode of action and classification 450
3. The immobilized lipases in biodiesel production 452
4. Concise Summary 458
References 459

INDEX 467

1
Chapter

INEDIBLE OILS: FEEDSTOCKS FOR SUSTAINABLE BIODIESEL

1. INTRODUCTION: AN OVERVIEW

The EU directive 2003/30/EC defines biodiesel as "methyl ester derived from vegetable or animal oil, of diesel quality, intended for use as biofuel." The latest EU directive 2009/28/EC establishes targets for 2020, aiming for a 20% contribution of renewable energy sources to the EU's total energy consumption and a 10% contribution from renewable sources in the transport energy consumption of each member state. This context emphasizes the significance of developing sustainable and responsible biofuels production that does not affect the food chain.

Nowadays most biodiesel is produced through triglycerides transesterification of edible oils with methanol, in the presence of an alkaline catalyst (Lotero et al., 2005). The so obtained product has low viscosity and is a biofuel (fatty methyl ester) that can replace petroleum-based diesel fuel with no need of engine modifications (Suwannakarn et al., 2005). Furthermore, if compared to fossil fuel, the formed ester fuels are non-toxic, safe to handle, and biodegradable (Krawczyk, 1996). Glycerine is also obtained as by-product as shown in Fig. 1.

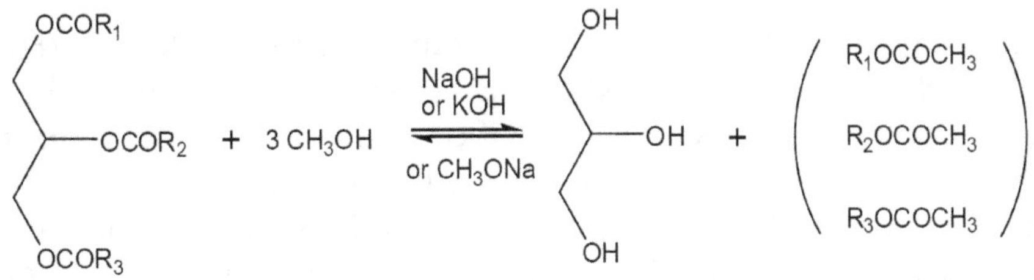

Triglyceride Methanol Glycerine FAME (Fatty alkyl methyl ester) BIODIESEL

Fig. 1. Transesterification of a trygliceride.

Refined, low acidity oilseeds (e.g. those derived from sunflower, soybean, rapeseed, tobacco etc.) may be easily converted into biodiesel, but their exploitation significantly raises the production costs, resulting in a biofuel that is not competitive with the petroleum-based diesel (Loreto et al., 2005). Presumably, as the market increases and technology is improved, costs will be driven down. In any case, the raw materials constitute a large portion of the manufacturing cost of biodiesel (up to 80%) (Bender, 1999).

Current oilseeds production systems raise environmental concerns because lands are intensively cultivated requiring high fertilizer and water inputs. These practices, aiming to increase yield, must be reduced or carefully regulated to prevent emissions of greenhouse gases or other environmental impacts. To do this, improved agronomic practices as the use of mixed species or crop rotation undoubtedly play a key role in mitigating negative impacts and enhancing biodiversity. A deep understanding of the microbial diversity of soils, its impacts on nutrient uptake and therefore on yield is crucial for sustainable cropping systems (The Royal Society, 2008).

Energy crops for industrial destination may represent a strategic opportunity in land use and income generation. However, in addition to the environmental aspects, economical concerns exist regarding the subtraction of lands for food cultivation. In a high market tension, it could have major impact on food/feed prices, increasing inequality, especially in developing countries. In addition, increased demand for food can result in the slow- down in biodiesel production due to reduced raw material availability. This was noticed in 2007 with industrial plants exploiting only 50% of their production capacity (Carvoli et al., 2008).

For all these reasons, it is highly desirable to produce biodiesel from crops specifically selected for their high productivity and characterized by low input requirements, or from low-cost feedstock such as waste cooking oil (WCO), animal fats and greases (Canakci et al., 2005; Zhang et. al, 2003). While edible crops available for biodiesel production are restricted to few species (mainly palm/ soybean in the U.S. and palm/ rapeseed in the E.U.), the intent of using dedicated alternative feedstock opens a wide choice for new species that may be more suitable for specific conditions resulting on high yields.

The high WCO potential is recognized also by the EU directive 2009/28/EC, where waste vegetable or animal oil biodiesel is reported to save about the 88% of greenhouse emissions, a quite high value if compared to biodiesel from common vegetable oils, whose greenhouse emission savings range from 36 to 62%. The main issue posed by such a raw material is the need of its standardization, especially with regard to acidity decrease. Several methods have been proposed to solve this problem. Among them it is worth mentioning, besides the cited alkali refining method, addition of excess catalyst (Ono & Yoshiharu, 1979), extraction with a solvent (Rao et al., 2009), distillation refining process (Xingzhong et al., 2008) and pre- esterification method (Loreto et al., 2005; Pirola et al., 2010; Bianchi et. al, 2010; Parodi and Martini, 2008). This last seems to be the most attractive approach and has recently received much attention.

The authors elucidate how waste materials or oils from non-food crops might be utilized as viable raw materials for biodiesel manufacturing.

2. AGRONOMICAL ASPECTS

The authors offer preliminary results from a three-year study on the viability of utilizing novel oilseed species for biodiesel production in Italy[1]. The objective is to present an innovative agronomic strategy that could influence the energy balance and enhance sustainability in oilseed production.

2.1 Non edible oil crops in the Mediterranean basin

A considerable amount of studies are available on mainstream and alternative crops for biodiesel feedstock. The authors made a selection of the most promising crops to be introduced in the Mediterranean zone, taking into account that currently the Mediterranean basin comprises not only temperate climate but also slightly-arid lands. Some of these are being effectively tested under the mentioned project as part of a unique rotation program.

Among oil crops the Brassicaceae family has an outstanding position. Rapeseed (*Brassica napus*) is the third largest oil crop with 12% of the world plant oil market with best yields when cultivated in cold-temperate regions (Carlsson, 2009). Yet, the large biodiversity of Brassicaceae reveal incipient species, among which *Brassica juncea*, *Brassica nigra*, *Brassica rapa*, *Brassica carinata*, *Sinapis alba*, *Camelina sativa*, *Eruca sativa* ssp. *oleifera*, etc. Besides the potential as raw material for biodiesel, their high content of glucosinolates (GSL) make them able to recover soils made marginal by soil-borne pests as nematodes (e.g. galling nematodes from the *Meloidogyne* genus and cist nematodes from *Heterodera* and *Globodera* genera) (Romero et al., 2009; Curto & Lazzeri, 2006). Many researchers also report weed- suppressive effects of Brassicaceae (e.g. Al-Khatib, 1997; Krishnan, 1998) as well as filtering- buffering effects against heavy metals pollution (Palmer et al., 2001).

On the other side an unexpected source of oilseed seems to arise from the tobacco culture. In anticipation of changes in tobacco market, selections of new varieties destined for energy production are coming out. Tobacco, as drought resistant species, seems a good option to face the shift of some previously fertile into arid lands caused by climate change.

2.1.1 Brassica carinata

The recent interest in *B. carinata* (also known as Ethiopian or Abyssinian mustard) is mainly a result of its high resistance to biotic and abiotic stresses such as drought tolerance. *Brassica carinata,* is an annual crop noted to be highly resistant to many rapeseed pests: blackleg (*Leptosphaeria maculans)*, white rust (*Albugo candida)*, *Sclerotinia* sp. and *Phyllotreta cruciferae* (Pan, 2009). According to Razon (2009), *B.carinata*, together with *sativa* ssp. *oleifera*, is the

1 SUSBIOFUEL project ("Studio di fattibilità per la produzione di biocarburanti da semi oleosi di nuove specie e da sottoprodotti o materiali di scarto" – D.M. 27800/7303/09), financially supported by the Ministry of Agricultural, Food and Forestry Policies – Italy.

most promising oilseed for biodiesel purpose in temperate zones, not just for the yield but also for its adaptability to hard pedo-climatic conditions. It may be used in a crop rotation system with cereals and on low nutrient soils. Best results are achieved sowing on autumn (IENICA, 2004). Harvesting may be done with same equipment used for rapeseed with the advantage that *B. carinata* shows a good resistance to the dehiscence of mature siliquae. The vegetable oil obtained from *B.carinata* is characterized by the p resence of erucic acid, making it unsuitable for human consumption. On the other hand, its physico-chemical properties meet the European specifications defined for biodiesel destination by the normative EN 14214:2002. Beyond its oil production capabilities, it was pointed out that the *B. carinata*'s lignocellulosic biomass can also be used to generate power and especially heat (Gasola et al., 2007), revealing an even greater potential.

2.1.2 Brassica juncea

Brassica juncea (also known as wild mustard or Indian mustard) varieties are grown for edible leaves or for condiment mustard only in some countries, while its use as an oilseed crop is increasingly growing. Canadian plant breeders have developed *B. juncea* cultivars with canola characteristics (Potts et al., 1999). As a result, canola varieties of *B. napus* and canola-type *B. juncea* have similar compositional characteristics. The key differences between *B. napus* and canola-type *B. juncea* lie in their agronomic characteristics. *Brassica juncea* tolerates high temperatures and drought better than *B. napus*, and thus it is better suited for the warmer, drier climates as the Upper Plains of the U.S. or the Mediterranean area. Green manure of *B.juncea* is a current practice in some countries (e.g. Italy and U.S.) making use of the GSL-Myrosinase system as a natural biofumigant. At the same time, this practice supplies organic matter to soil. To make the most of its biocidal activity against soil- borne pests and diseases, the mulching and incorporation to soil must be done at flowering time (Curto & Lazzeri, 2006).

2.1.3 *Nicotiana tabacum*

The tobacco (*N. tabacum*) is an annual herbaceous plant belonging to the Solanaceae family, widespread in North and South America, commonly grown for the collection of leaves. The seeds are very small (up to 10,000/g) and contain 36 to 39% of oil having a high percentage of linoleic acid (Giannelos et al., 2002). Currently, the common varieties directed to leaf production reach the modest order of 1 to 1.2 t seeds/ha (Patel, 1998, as cited in Usta, 2005) as a result of selection to reduce the amount of seed produced. Recently researchers were able to over express, through genetic engineering, genes responsible for the oil production in the leaves (Andrianov et al., 2010). However, the seeds potential for oil production is much higher. In this sense, another recent outcome on tobacco improvement is a variety that can at least triple seed (up to 5 t/ha) and oil production. The energy tobacco varieties exist both in the non GMO and the GMO version for resistance factors against herbicides and insects (Fogher, 2008). Its high oil yield makes it very competitive in front of mainstream oil crops as rapeseed, sunflower and soybean. The remaining meal revealed to be relevant for combustion or to be used as a protein source for livestock. Tests with pigs demonstrated its palatability to animals, a good conversion rate and therefore its equivalence to the soybean meal (Fogher, 2002). In addition, the presence of consolidate agricultural practices and know-how make clear the advantage of using a well-known species as tobacco as alternative feedstock for biodiesel. The research on Energy Tobacco has also found new economies for the transplant management as well as direct sowing techniques are currently under test. Combine- harvesters for the harvest of the whole inflorescences are available.

2.1.4 *Ricinus communis*

Ricinus communis (castor bean) is an oilseed crop that belongs to the Euphorbiaceae family, which includes other energy crops as cassava (*Manihot esculenta*), rubber tree (*Hevea brasiliensis*) and physic nut (*Jatropha curcas*). Among non-edible oils, the one extracted from castor bean is the most used for a wide variety of industrial purposes. Its oil is primarily of economic interest having cosmetic, medical and chemical applications. The presence of a high proportion of ricin oleic acid makes it suitable for the production of high-quality lubricants (Sanzone & Sortino, 2010). The use of castor oil is particularly supported in Brazil, with attempts to extract the ethyl esters using ethanol from sugarcane fermentation (although less reactive than methanol), making it a complete natural and renewable product (Pinto et al., 2005). Albeit the actual productivity is not very high, between 600 and 1,000 kg seeds/ha year, this value could triplicate with genetic improvement (Holanda, 2004). With the recent report on the draft genome sequence of castor bean revealing some key genes involved in oil synthesis (Chan et al., 2010), this possibility becomes even more palpable. In addition to this, the ease with which it can be cultivated in unfavorable environments contributes to its appeal as a raw material for sustainable biodiesel. In agreement to this, a two years field experiment conducted in south Italy using local ecotypes yielded around 2.3 t/ha of seeds, with up to 38% oil content, a quite high number for the dry conditions of the region (Sanzone & Sortino, 2010). The main limitation is the hand harvest, the current practice in the biggest producer countries as India, Brazil and China. However mechanization of harvesting is recently available for the collection of dwarf hybrid plants (Clixoo, 2010).

2.1.5 *Cynara cardunculus*

Among the species of interest for the production of biodiesel, the cardoon or artichoke thistle (*C. cardunculus*) is an important resource to be exploited, particularly in light of its adaptability to different soils. *Cinara cardunculus* is a perennial herbaceous species belonging to the family Asteraceae. Its deep root system allows the plant to extract water and nutrients from very deep soil zones revealing a plant with a small demand for fertilization and extremely resistant to drought. This characteristic makes it suitable to be grown on dry marginal or abandoned lands in the Mediterranean basin. Production reaches 30-35 t/ha per year, with about 2 tons of seeds; the seeds contain up to 25 % oil, with a similar composition to sunflower oil (Pasqualino, 2006). Recently, studies have been conducted within the EU project "Biocard - Global Process to Improve *C. cardunculus*". In the framework of this project, a research on the harvesting procedures, i.e. a crucial point of the cultivation of the thistle has also been conducted. As an example, a combine prototype designed to separate and thresh the capitula and to drop the biomass proved to be feasible, with a good cost/working capacity relation (Pari et al., 2008).

2.2 A new proposal for biodiesel production

The rationale of this proposal consists in the use of non-edible crops on soils no longer suitable for food production due to infestation by nematodes. The authors tested the possibility to rescue marginal soil fertility in consequence of the cultivation and the green manure of a naturally biocidal crop (*B. juncea* and *B. carinata*). Thanks to this practice the soil could be quickly good enough to produce oilseeds with satisfying yields for industrial destination. Furthermore a reduction in inputs of fertilizers is also expected due to preservation of organic matter content of soil. This practice offers the possibility to rescue soils availability for food production. Indeed, after some cycles of this rotation, the pest control and the progressive increase of organic matter should make the soil eligible again for quality productions.

2.2.1 Experimental details

The agronomic rotation was tested under a wide range of situations. Three field trial locations were chosen taking into account Italy's wide latitudinal distribution[2]. Experimental design was thought to produce oilseed from N. tabacum and from traditional oilseed crops (sunflower, soybean, and rapeseed), used as comparison to validate the methodology. Each field was divided into two parts and B.juncea was sown only in one half of the field. To maximize the biofumigant effect, green manuring with B.juncea biomass was carried out when the crop reached flowering. After this, sowing of soybean, sunflower and rapeseed as well as the transplant of tobacco plantlets took place in both parts of the field. In order to make the proposal as flexible as possible, four different fertilization treatments were used: low input (30 kg/ha of chemical fertilizer[3]), medium input (90 kg/ha of chemical

2 Altedo (BO), Vaccolino (FE) and Santa Margherita di Savoia (FG).

3 Urea (Nitrogen 46%)

fertilizer), high input (140 kg/ha of chemical fertilizer) or organic input (10000 kg/ha of poultry manure). Untreated plots were set up as control. All field tests were conducted under Good Experimental Practices (GEP).

To evaluate the effect of the green manure of *B.juncea* on nematode infection, countings of *Meloidogyne* spp. were carried out on soil samples taken from both sides of the field while effects on yield of crops grown in succession were monitored recording the fresh weight per hectare (kg/ha) of plant biomass from both sides of the field. Since the green manure of *B.juncea* supplies organic matter to soil, possibly increasing also its sulphur content, it's relevant to ensure that crops grown after this agronomical practice are not enriched in sulphur and therefore less suitable for biodiesel production[4].To check this, sulphur quantification in sunflower seeds and oil were done. Seed samples were taken from the unfertilized plots of both sides of the field, and sulphur content detected by ICP-MS (Inductively Coupled Plasma Mass Spectrometry).

2.2.2 Results and discussion on agronomical aspects

Research on alternative biofuel aims to face the increasing demand for energy requirements by means of a more sustainable energy supply. From this point of view, greenhouse gases saving is expected from biofuels.

The first year of experimentation makes clear that plants grown in succession of *B. juncea* resulted in higher biomass. This could be due either to the increase in the organic matter content or to the pest control. Indeed, counting of nematodes revealed a strong effect of the green manure of *B.juncea* on nematode control. The average number of larvae found was almost four times lower in the presence of the biofumigant crop. The use of *B. juncea* as green manure does not influence the sulphur content in sunflower seeds and oil, suggesting no sulphur accumulation occurs in succeeding crops.

In order to assess the chemical properties of *B. juncea* oil for biodiesel destination, the authors quantified the total sulphur, nitrogen and phosphorus content in oil from commercial seeds of *B. juncea*. In table 1 data of the quantifications are reported.

Table 1. Nitrogen, sulphur and phosphor content in *B. juncea* oil.

Element	Unit	Value	Standard Test Method
sulphur	mg/kg	112	UNI EN 20846:2005
nitrogen	% (mass)	0,35	ASTM D5291-09
phosphorus	mg/kg	< 4	UNI EN 14107:2003

In table 2 the mean percentage increasing of biomass of *B. napus*, *H. annus*, *G. max*, and *N. tabacum* produced after green manuring of *B. juncea* is summarized.

4 The contents of this element in the final product must be under 10 ppm (UNI EN 14214 - Automotive fuels. Fatty acid methyl esters (FAME) for diesel engines. Requirements and test methods).

Table 2. Increasing of biomass of oilseed crops produced after green manuring of B. *juncea.*

Crop	Unit	Biomass increasing
N. tabacum	%	21
B. napus	%	15
H. annus	%	26
G. max	%	28

3.CHEMICALASPECTS:STANDARDIZATIONOFTHERAWMATERIALS AND BIODIESEL PRODUCTION

3.1 Oil characterization

Oil characterization before proceeding with the standardization of the raw material is a very important issue. Some properties remain in fact unchanged from the starting material to the finished biodiesel, or they are anyway predetermined. It is so important to check that the values of such chemical and physical oil properties are in range with those required by the standard regulations (see Table 3). The experimental procedures to get the values of such properties are also standardized and are indicated in the regulations. The following are parameters for starting oil that can affect the quality of the final biodiesel.

Sulfur and phosphorous content:

High sulphur and phosphorous content in the fuels cause greater engine wear and in particular shorten the life of the catalyst. Biodiesel derived from soybean, rapeseed, sunflower and tobacco oils are known to contain virtually no sulphur (Radich, 2004; Zhiyuan et al., 2008).

The authors have nevertheless found that the oil obtained from *B.juncea* seeds may contain high concentrations of sulphur due to the presence in the plant's tissues of glucosinolates, the molecules responsible for the biofumigation effect.

Linoleic acid methyl ester, iodine value and viscosity

Soybean, sunflower, peanut and rapeseed oils contain a high proportion of linoleic fatty acids, so affecting the properties of the derived ester with a low melting point and cetane number. Quantitative determination of linoleic acid methyl ester is accomplished by gas chromatography with the use of an internal standard after the substrate has been transesterificated and allows also the quantification of the other acid methyl esters (Environment Australia, 2003). The super-critical chromatography is another useful analytical technique, suitable for the direct analysis of the oils.

Table 3. European Standard specifications for biodiesel (automotive fuels).

Specification	Units	limits Min	Max	Method
Ester content	% (m/m)	96.5		EN 14103
Density 15°C	kg/m3	860	900	EN ISO 3675 EN ISO 12185
Viscosity 40°C	mm2/s	3.50	5.00	EN ISO 3104
Sulphur	mg/kg	-	10.0	preEN ISO 20846 preEN ISO 20884
Carbon residue (10% dist.residue)	% (m/m)	-	0.30	EN ISO 10370
Cetane number		51.0	-	EN ISO 5165
Sulphated ash	% (m/m)	-	0.02	ISO 3987
Water	mg/kg	-	500	EN ISO 12937
Total contamination	mg/kg	-	24	EN 12662
Cu corrosion max		-		EN ISO 2160
Oxidation stability, 110°C	h (hours)	6.0	-	EN 14112
Acid value	mg KOH/g	-	0.5	EN 14104
Iodine value	gr I2/100 gr	-	120	EN 14111
Linoleic acid ME	% (m/m)	-	12.0	EN 14103
Methanol	% (m/m)	-	0.20	EN 14110
Monoglyceride	% (m/m)	-	0.80	EN 14105
Diglyceride	% (m/m)	-	0.20	EN 14105
Triglyceride	% (m/m)	-	0.20	EN 14105
Free glycerol	% (m/m)	-	0.02	EN 14105
Total glycerol	% (m/m)	-	0.25	EN 14105
Gp I metals (Na+K)	mg/kg	-	5.0	EN 14108 EN14109
Gp II metals (Ca+Mg)	mg/kg	-	5.0	EN14538
Phosphorous	mg/kg	-	5.0	EN 14538

An indicative fatty acid methyl esters composition of the raw oils typically used for biodiesel production and of the ones adopted by the authors, is given in Table 4 (Velasco et al., 1998; Tyson, 2002; Winayanuwattikun at al. 2008, Zheng & Hanna, 1996).

Table 4. Indicative acidic composition of some raw materials for biodiesel production.

Oil	Comon Name	Fatty acid composition, wt%
Arachis hypogea	Peanut	11.9 (16:0), 3.0 (18:0), 40.0 (18:1), 40.7 (18:2), 1.2 (20:0), 3.2 (22:0)
Brassica juncea	Indian mustard	3.6 (16:0), 1.1 (18:0), 13.9 (18:1), 21.5 (18:2), 13.7 (18.3), 8.7 (20:1), 33.5 (22:1)
Brassica napus	Canola	4.7 (16:0), 0.1 (16:1), 1.6 (18:0), 66.0 (18:1), 21.2 (18:2), 5.2 (18:3), 0.9 (20:0), 0.3 (22:0)
Carthamus tinctorius	Safflower	0.1 (14:0), 6.4 (16:0), 2.2 (18:0), 14.1 (18:1), 76.6 (18:2), 0.2 (18:3), 0.2 (20:0) 0.2 (22:0)
Elaeis guineensis	Palm	0.5 (12:0), 1.0 (14:0), 38.7 (16:0), 3.3 (18:0), 45.5 (18:1), 10.8 (18:2), 0.1 (18:3), 0.1 (20:0)
Glycine max	Soybean	10.7 (16:0), 3.0 (18:0), 24.0 (18:1), 56.6 (18:2), 5.3 (18:3), 0.2 (20:0), 0.2 (22:0)
Helianthus annus	Sunflower	6.6 (16:0), 3.1 (18:0), 22.4 (18:1), 66.2 (18:2), 1.0 (18:3), 0.3 (20:0), 0.4 (22:0)
Jatropha curcas	Physic nut	0.1 (12:0), 0.2 (14:0), 14.8 (16:0), 0.8 (16:1), 4.2 (18:0), 41.0 (18:1), 38.6 (18:2), 0.3 (18:3)
Nicotiana tabacum	Tobacco	6.6 (16:0), 3.1 (18:0), 22.4 (18:1), 66.2 (18:2), 1.0 (18:3), 0.3 (20:0), 0.4 (22:0)
Lard	-	4.8 (14:0), 28.4 (16:0), 4.7 (16:1) 14.8 (18:0), 44.6 (18:1), 2.7 (18:2)
Yellow grease	-	1.0 (14:0), 23.0 (16:0), 1.0 (16:1) 10.0 (18:0), 50.0 (18:1), 15.0 (18:2)
Brown grease	-	1.7 (14:0), 23.0 (16:0), 3.1 (16:1) 12.5 (18:0), 42.5 (18:1), 12.2 (18:2), 0.8 (18:3)

Iodine value, viscosity and density

The iodine value (IV) is an index of the number of double bonds in biodiesel, and therefore is a parameter that quantifies the degree of unsaturation of biodiesel. Both EN and ASTM standard methods measure the IV by addition of an iodine/chlorine reagent. Biodiesel viscosity is directly correlated to the IV of biodiesel for biodiesel with iodine numbers of between 107 and 150 (Environment Australia, 2003).

One of the main reasons for processing vegetable oils for use in engines is to reduce the viscosity thereby improving fuel flow characteristics. High viscosities can cause injector spray pattern problems that lead to excessive coking and oil dilution. These problems are associated with reduced engine life. Nevertheless, the necessary characteristics depend also on the end use; the engines for the production of energetic power in fact allow the use of fuels with higher viscosity (i.e. from palm oil).

Density dictates the energy content of fuel where high densities indicate more thermal energy for the same amount of fuel and therefore better fuel economy.

The authors have already published the results of the measurement of the IV obtained for some oils selected as potential raw materials for BD production (Pirola et al., 2011). In Table 5 the values of IV, viscosity and density found by the authors for waste cooking oil and its mixture with raw rapeseed oil are shown, demonstrating that the properties of the feedstock can be improved by the use of blends of different oils. The values reported in the Table 5 evidences that with the dilution with rapeseed oil it is possible to decrease the viscosity of WCO but increasing the number of IV. Nevertheless also in the case of most diluited sample the IV value is lower than those of rapeseed oil.

Table 5. IV, viscosities and densities of some potential raw materials for biodiesel production.

Oil	Iodine value (gI2/100g oil)	Viscosity (mm2/s 40 °C)	Density (kg/m3 15° C)
WCO	54	82.2	918
WCO:rapeseed oil 1:1	85	52.8	914
WCO:rapeseed oil 3:1	100	40.5	926
Rapeseed	115	n.d.	n.d.

It has to be taken into account that after the transesterification process the IV of the feedstock remain unchanged, the viscosity is reduced from 10 to 15 times, whereas density has been found to remain almost the same or to be reduced in some cases (Zheng & Hanna, 1996).

3.2 Oil standardization: Free fatty acids esterification reaction

As already mentioned in the introduction paragraph, the use of raw, non edible oils poses the problem of standardization before the transesterification process, especially with regard to acidity decrease. In fact oils, besides triglycerides contain also free fatty acids (FFA). These lasts are able to react with the alkaline catalyst used for the transesterification reaction yielding soaps which prevent the contact between the reagents. A FFA content lower than 0.5% wt is also required by the EN 14214.

Among the different deacidification methods listed in the introduction, the authors have recently paid attention to the pre-esterification process (Loreto et al., 2005; Pirola et al., 2010; Bianchi et al., 2010). This method is particularly convenient as it is not only able to lower the acidity content of the oils but also provides methyl esters already at this stage, so increasing the final yield in biodiesel. A scheme of the FFA esterification reaction is given in Fig.2.

$$RCOOH + CH_3OH \xrightleftharpoons{\text{acid catalysis}} RCOOCH_3 + H_2O$$

Fig. 2. Scheme of the Free Fatty Acid Esterification Reaction.

The use of heterogeneous catalysts (Sharma & Singh, 2011) is usually preferred to the use of homogeneous ones (Alsalme et al., 2008) as it prevents neutralization and separation costs, besides being not corrosive, so avoiding the use of expensive construction materials. Another important advantage is that the recovered catalysts can be potentially used for a long time and/or multiple reaction cycles.

In the recent years the authors have deepened the study of the pre-esterification process investigating the effect of the use of different kinds of oils, different types of reactors and catalysts and different operating conditions (Pirola et al., 2010; Bianchi et al., 2010; Pirola et al. 2011)

In the following paragraphs, the most relevant aspects of the experimental work and the results obtained by the authors for what concerns the pre-esterification process are reported.

3.2.1 Experimental details

A remarkable aspect of the proposed process is represented by the mild operative conditions, i.e. low temperature (between 303 and 338 K) and atmospheric pressure. Moreover, the adopted working temperature is the same of the following transesterification reaction and of the methanol recovery by distillation. Each single reaction has been carried out for six hours withdrawing samples from the reactor at pre-established times and analysing them through titration with KOH 0.1 M. The percentage of FFA content per weight was calculated as otherwise reported (Marchetti & Errazu, 2007, Pirola et al. 2010).

All the esterification experiments have been conducted using a slurry reactor as the one already described elsewhere (Bianchi et al., 2010). A slurry reactor is the simplest type of catalytic reactor, in which the catalyst is suspended in the mass of the regents thanks to the agitation.

Much attention has been paid by the authors to the use of acid ion exchange resins. Amberlyst ®46 (named A46 in this chapter), i.e. a commercial product by Dow

Advanced Materials, and D5081, a catalyst at the laboratory development stage by Purolite® have been successfully applied in this reaction. The main features of the employed catalysts are reported in Tab. 6.

Table 6. Main features of the ion exchange resins adopted as catalysts in the FFA esterification reaction.

Resin	Matrix	Functional Group	Ionic form	Acid capacity (meq H+/g)	Max. operating Temp (°C)
D5081	Styrene-divinylbenzene	R-SO3-	H+	1,0	130
A46	Styrene-divinylbenzene	R-SO3-	H+	0,60	120

The acid capacity of the catalysts, corresponding to the number of the active sites per gram of catalyst was also experimentally determined by the authors by ion exchange with a NaCl- saturated solution and successive titration with NaOH (López et al., 2007). The values were found to be in agreement with the ones provided by technical sheets.

A distinguishing feature of A46 and D5081 is represented by the location of the active acid sites: these catalysts are in fact sulphonated only on their surface and not inside the pores. Consequently, A46 and D5081 are characterized by a smaller number of acid sites per gram if compared to other Amberlysts®, which are also internally sulphonated (Bianchi et al., 2010).

3.2.2 Deacidification results

In Fig. 3 the results from the esterification reaction performed on different raw oils are shown.

From the graph it can be noticed that in almost all the cases it is possible to obtain a FFA concentration lower than 0.5% wt after 6 hours of reaction. The differences in the acidic composition seem not to affect the final yield of the reaction. What seems to influence the FFA conversion is the refinement degree of the oil. Waste cooking oil (WCO) is in fact more hardly processable with the esterification in comparison to refined oils, probably due to its higher viscosity which results in limitations to the mass transfer of the reagents towards catalysts. Indeed, the required acidity limit is not achieved within 6 hours of reaction. Adding rapeseed oil, less viscous, to the WCO in different ratios it is possible to increase the final FFA conversion and reaching a FFA content lower than 0.5% wt. The blend of a raw oil characterized by high viscosity with a less viscous one is also effective in shortening the time to reach the plateau of conversion, as displayed in Fig. 4.

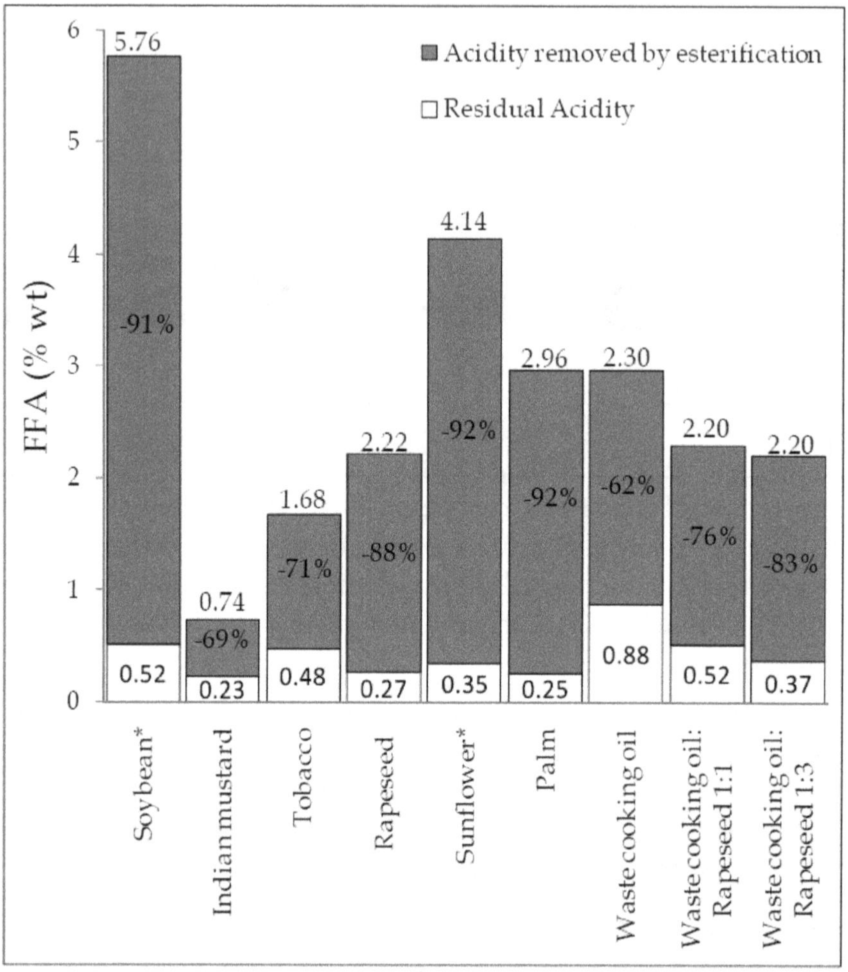

Fig. 3. Acidity removed by esterification (6 hr) and residual acidity of different oils used as raw material: slurry reactor, T=338K, catalyst: Amberlyst® 46 weight ratio methanol/oil= 16:100, weight ratio catalyst/oil=1:10; *commercial, refined oils with the addition of pure oleic acid.

In Fig. 4 the conversion curves concerning the recycles of the use of the catalyst A46 in the case of WCO are also shown. The catalyst does not show a drastic drop in its activity notwithstanding the used substrate is not refined. This decrease in the catalytic performance might be ascribable to the catalyst's settling in the reaction environment (Pirola et al., 2011) or to the presence of cations inside the oil. This aspect is still under investigation.

It is convenient to use an excess of methanol respect the stoichiometric amount in order to shift the equilibrium towards the product. Nevertheless, when adding methanol a double phase system is formed (the maximum solubility of methanol in oil is in the interval 6- 8%) and therefore it is not convenient to increase further this parameter.

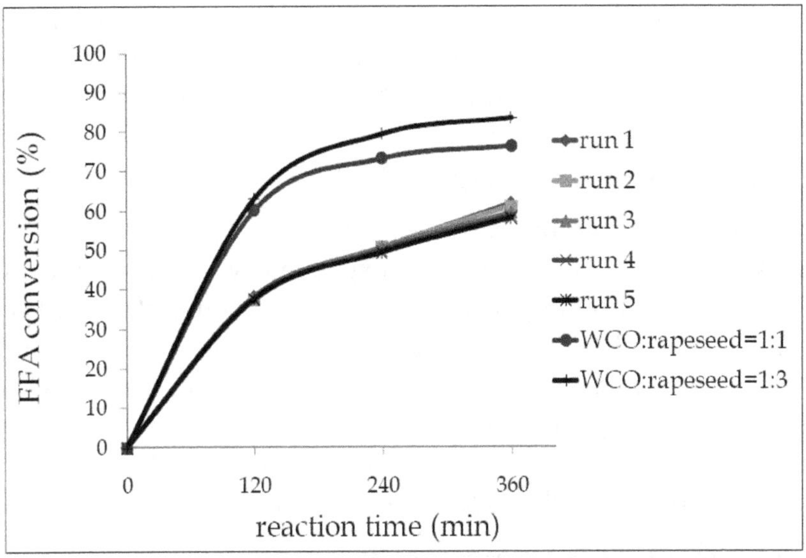

Fig. 4. FFA conversion (%) vs reaction time of waste cooking oil (WCO) and its blends with rapeseed oils: slurry reactor, T=338K, catalyst: Amberlyst® 46 weight ratio methanol /oil= 16:100, weight ratio catalyst/oil=1:10.

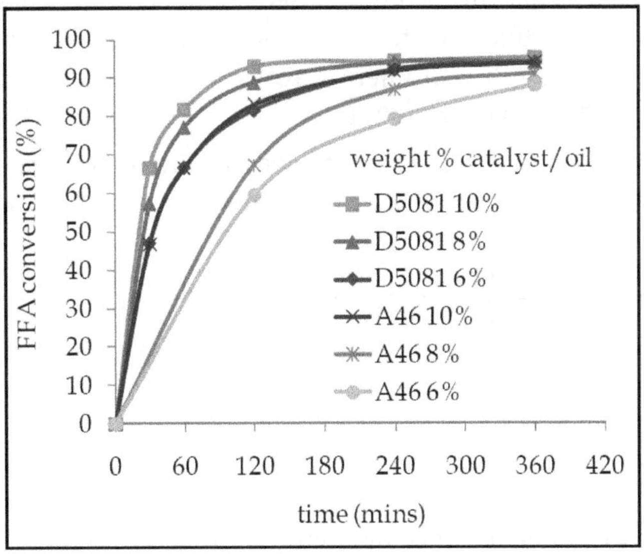

Fig. 5. FFA conversion (%) vs reaction time for different amounts of catalysts A46 and D5081, rapeseed oil with initial acidity=5%, slurry reactor, weight ratio methanol/oil= 16:100, T=338K. Dots are experimentally obtained.

Continue lines are simulated (see paragraph 3.2.3)

The lifetime of the catalyst is a very important issue from an industrial standpoint. The authors have already performed a deep study on the ion exchange resins endurance in the FFA esterification reaction (Pirola et al., 2010). The most important outcome of this study is that resins like A46 (Dow Advanced Materials) and D5081 (Purolite), which are functionalized only on their surface are very stable in the reaction conditions and can guarantee long operating times without being replaced.

A comparison between these two resins is displayed in Fig. 5.

As can be seen from the graph, catalyst D5081 shows better results than A46 at lower catalyst's loading. This can be easily explained by the higher number of acid sites located on its surface. In particular, the use of a ratio of 10 %wt of catalyst D5081 vs. oil allows reaching the maximum conversion in 2 hours. From the graph can be seen how the curves for 6% of D5081 and 10% wt catalyst/oil of A46 perfectly overlap. This outcome suggested that a fixed amount of acid active sites per gram of FFA was required to reach the maximum of conversion in 4 hours. Based on the experimental data obtained, this amount was found to be equal to 1.2 meq of H+.

3.2.3 Simulation of the catalytic results

The considered reaction system turns out to be an highly non-ideal system, being formed by a mixture of oil, methylester, methanol, FFA and water. Indeed, activity coefficients instead of concentrations are used not only for the phase and chemical equilibria calculations, but also for the kinetic expressions. Modified UNIFAC model was used adopting the parameters available in literature and published by Gmehling et al., 2002 (Pirola et al., 2011).

A pseudohomogeneous model was used for describing the kinetic behavior of the reaction (Pöpken et al., 2000). The adopted model is displayed in the following equation:

$$r = \frac{1}{m_{cat}} \frac{1}{\upsilon_i} \frac{dn_i}{dt} = k_1 a_{FFA} a_{methanol} - k_{-1} a_{methylester} a_{water}$$

where:

r= reaction rate

m_{cat}= dry mass of catalyst, gr

υ_i= stoichiometric coefficients of component i n_1= moles of component i

t = reaction time

k_1= kinetic constant of direct reaction

k_{-1}= kinetic constant of indirect reaction a_i= activity of component i

The temperature dependence of the rate constant is expressed by the Arrhenius law:

$$k_i = k_i^0 \exp\left(\frac{-E_{A,i}}{RT}\right)$$

where k_i^0 and $E_{A,i}$ are the pre-exponential factor and the activation energy of the reaction i, respectively (i=1 for the direct reaction, i=-1 for the indirect reaction), T is the absolute temperature and R the Universal Gas Constant. The adopted parameters set is the same reported by Steinigeweg (Steinigeweg & Gmehling, 2003).

All the simulations were carried using Batch Reactor of PRO II by Simsci – Esscor. The model turned out to be able to reproduce qualitatively the behavior of different systems, characterized by different catalyst type and content.

In the previous Figure 5, continue lines represent simulated behaviors using the same parameters, but considering a different catalyst mass due to different catalyst acidity and concentration.

3.3 Oil transformation: The transesterification reaction

The transesterification reaction has been performed by the authors on the rapeseed and *B.juncea* (Indian mustard) oilseeds deacidified with the esterification process described in the previous paragraph.

Sodium Methoxide (MeONa) was employed as catalyst. MeONa is known to be the most active catalyst for triglycerides transesterification reaction, but it requires the total absence of water (Schuchardt, 1996). For this reason, the unreacted methanol and the reaction water were evaporated from the deacidified oils before processing them with the transesterification reaction.

The employed experimental setup was the same employed for the slurry esterification. Being the transesterification an equilibrium reaction, it was performed in two steps, removing the formed glycerine after the first step. The adopted conditions were the following:

- 1st step: weight ratio methanol/oil=20:100, weight ratio MeONa/oil=1:100, 233 K, 1,5 h;
- 2nd step: weight ratio methanol/oil=5:100, weight ratio MeONa/oil=0.5:100, 233 K, 1 h.

The total ester content is a measure of the completeness of the transesterification reaction. Many are the factors affecting ester yield in the transesterification reaction: molar ratios of glycerides to alcohol, type of catalyst(s) used, reaction conditions, water content, FFA concentration, etc.

The European prEN14214 biodiesel standard sets a minimum limit for ester content of >96.5% mass, whereas the US ASTM D 6751 biodiesel standard does not set a specification for ester content.

Mono- and di-glycerides as well as tri-glycerides can remain in the final product in small quantities. Most are generally reacted or concentrated in the glycerine phase and separated from the ester.

Both in the case of rapeseed oil and *B.juncea* oilseed, the final yield in methylester was higher than 98%.

The analyses of methyl esters and unreacted mono-, di- and triglycerides are accomplished through gas chromatography.

The detailed requirements for biodiesel according to both EN 14214 and US ASTM D 6751 are listed in paragraph 1.

In the US a standard for biodiesel (ASTM D 6751 – Standard Specification for Biodiesel Fuel (B100) does not include the same number of parameters as prEN 14214 but the parameters that coincide have similar limits. The US specification covers sulfur biodiesel (B100) content much higher if compared to the one of European Standard. For use as a blend component with diesel fuel oils defined by ASTM D 975 Grades 1-D, 2-D, and low sulfur 1-D and 2-D. (Environment Australia, 2003).

4. CONCISE SUMMARY

Utilizing oilseed from alternative crops or waste oils as a feedstock for biodiesel production is a highly effective method to reduce the production costs of this biofuel..

From the agronomic point of view the authors verified that the green manure of *B.juncea* resulted in nematode infestation drastically decreased and improved soil quality, reflected in higher yield of crops in agronomic succession. In the first year of experimentation *B. juncea* was preferred to *B.carinata* because of its suitability to spring planting (starting period of the project SUSBIOFUEL). Further work will be necessary to improve the setting up of the agronomic proposal. Winter sowing of *B.carinata* will be done in the next years and alternative promising patented variety of tobacco (selected for seed production)[5] are currently under test. The authors are also evaluating the proposed rotation in comparison with commercial pellets[6] of defatted Brassicaceae meal. In addition, more outcomes are attended: yield grains[7], evaluation of the weed control potential of *B.juncea* and survival rate of transplanted *N.tabacum* plantlets following the green manuring or not.

The flexibility of Brassicaceae (efficient green manure and/or oil crop) allows using these species with a dual aim according to the situation, thus increasing the sustainability of the system. On the other hand new tobacco varieties promise yields above the best rape harvests around Europe. Under this light tobacco is a really interesting alternative oil crop especially in countries like Italy where it has

5 Kindly supplied by Sunchem Holding S.r.l.

6 Biofence by Triumph Italia S.p.a.

7 This kind of data is necessary to express results in terms of functional unit as required by a life cycle thinking approach.

been cultivated since a long time and Good Agricultural Practices (GAP) for this crop have long been known: all points in favour to the conversion of tobacco cultivation toward oil seeds production. To give a more comprehensive evaluation of innovations introduced in the whole biodiesel production chain, the authors aim to develop a method able to assess biodiesel sustainability.

The authors are aware that their proposal alone does not solve the overall sustainability problem of biodiesel production, but it contributes significantly to a wider portfolio of land- use strategy, stimulating the call for innovations both in technology and emissions reduction measures. Food production from marginal soils would worsen soil depletion and nematodes infestation. The restoring of soil fertility avoiding the chemicals usage, and in the mean time the generation of income from vegetable oils, assure the ethical, economical and environmental sustainability of the solution. Policy strategies will be needed to increasingly shift abandoned or low biodiversity value marginal lands to this kind of ecologically-friendly practices.

From the chemical point of view, the high concentration of FFA contained in these raw materials (waste or alternative crops) leading to the formation of soaps during the final transesterification step can be easily overcome by performing a pre-esterification reaction. This treatment allows lowering the acid content of the raw material below the limit required by the biodiesel standard, so avoiding also the formation of soaps during the transesterification stage. The FFA esterification is also helpful in increasing the final yield in biodiesel as it produces methyl esters.

The oilseeds of Brassica juncea, Nicotiana tabacum, rapeseed, palm, soybean, and sunflower have been effectively deacidified using an esterification reaction. Waste cooking oil (WCO) is not an ideal raw material for biodiesel production since its qualities do not adequately meet the necessary standards. Nonetheless, it is feasible to utilize this type of feedstock by blending it with other oils that possess a lower viscosity. The authors have effectively deacidified a blend of waste cooking oil and rapeseed oil, resulting in an enhanced reaction rate.

Two acid ion exchange resins have been used as catalysts: Amberlyst®46 (Dow Advanced Materials) and Purolite® D5081 (Purolite). Both resins yielded excellent outcomes in the examined response. D5081 shown to be more active than A46, achieving maximal conversion in shorter durations under same conditions.

A process simulation of the FFA esterification, capable of predicting reaction progression via thermodynamic and kinetic analysis, was successfully executed using the software PRO II (SimSci). A pseudohomogeneous model was employed to elucidate the kinetic behavior of the reaction, utilizing a modified UNIFAC model for the computation of activity coefficients, which were applied for both phase and chemical equilibrium calculations as well as kinetic expressions. The data derived from this model exhibited a strong correlation with the experimental results.

REFERENCES

Al-Khatib, K., Libbey, C. & Boydston, R. (1997). Weed Suppression with Brassica Green Manure Crops in Green Pea. *Weed Science,* Vol. 45, No. 3, pp. 439-445, ISSN 1939- 7291.

Alsalme, A. Kozhevnikova, E.F. & Kozhevnikov, I. (2008) Heteropoly acids as catalysts for liquid-phase esterification and transesterification. *Applied Catalysis A*, Vol. 349, pp.170-176.

Andrianov, V., Borisjuk, N., Pogrebnyak, N., Brinker, A., Dixon, J., Spitsin, S., Flynn, J., Matyszczuk, P., Andryszak, K., Laurelli, M., Golovkin, M. & Koprowski, H. (2010). Tobacco as a Production Platform for Biofuel: Overexpression of *Arabidopsis DGAT* and *LEC2* genes Increases Accumulation and Shifts the Composition of Lipids in Green Biomass. *Plant Biotechnology Journal,* Vol. 8, (April 2010), pp. 277–287, ISSN 1467-7644.

Bender, M. (1999). Economic Feasibility Review for Community-Scale Farmer Cooperatives for Biodiesel. *Bioresource Technology*, Vol. 70, (October 1999), pp. 81-87, ISSN 0960- 8524.

Bianchi, C.L., Boffito, D.C., Pirola, C. & Ragaini, V. (2010) Low temperature de-acidification process of animal fat as a pre-step to biodiesel production. *Catalysis Letters*, Vol. 134, (November, 2009), pp. 179-183.

Canakci, M. & Sanli, H. (2008) Biodiesel production from various feedstocks and their effects on the fuel properties. *Journal of Industrial Microbiology Biotechnology*, Vol. 35, pp. 431-441.

Carlsson, A. S. (2009). Plant Oils as Feedstock Alternatives to Petroleum - A Short Survey of Potential Oil Crop Platforms. *Biochimie*, Vol. 91, (June 2009), pp. 665–670, ISSN 0300-9084.

Carvoli, G., Ragaini, V. & Soave, C. (2008) Il futuro è dietro l'angolo. *La Chimica e l'Industria*, Vol. 1, pp. 38-41.

Chan, A. P., Crabtree,J., Zhao, Q., Lorenzi, H., Orvis, J., Puiu, D., Melake-Berhan, A., Jones, K. M., Redman, J., Chen, G., Cahoon, E. B., Gedil, M., Stanke, M., Haas, B. J., Wortman, J. R., Fraser-Liggett, C. M., Ravel, J. & Rabinowicz, P. D. (2010). Draft Genome Sequence of the Oilseed Species. *Ricinus communis. Nature Biotechnology,* Vol. 28, (August 2010), pp. 951-956, ISSN 1087-0156.

Clixoo (2010). Preview of Comprehensive Castor Oil Report, In: *Castor Oil Industry Reference & Resources*, 30.05.2011, Available from http://www.castoroil.in

Curto, G. & Lazzeri, L. (2006). Brassicacee, un Baluardo Sotterraneo Contro i Nematodi. *Agricoltura*, Vol. 34, No. 5, (May 2006), pp.110-112.

Environment Australia (2003). National Standards for Biodiesel - Discussion Paper, In *Setting National Fuels Quality Standards*, 0 642 54908 7 pp. 1-196, Retrieved from <www.ea.gov.au/atmosphere/transport/biodiesel/index.html>.

Fogher, C. (2008). Sviluppo di un Ideotipo di Tabacco per la Produzione di Seme da Usare a Fini Energetici. Presented at the workshop "*La Filiera del Tabacco in Campania: Ristrutturazione e/o Riconversione. La ricerca come motore per l'innovazione tecnologica, la sostenibilità e la competitività della filiera*", Portici, Italy, February, 2008

Gasol, C. M., Gabarrella, X., Antonc, A., Rigolad, M., Carrascoe, J., Ciriae, P., Solanoe, M. L. & Rieradeva, J. (2007). Life Cycle Assessment of a *Brassica carinata* Bioenergy Cropping System in Southern Europe. *Biomass and Bioenergy*, Vol. 31, No. 8, (August 2007), pp. 543–555, ISSN 0961-9534.

Giannelos, P. N., Zannikos, F., Stournas, S., Lois, E. & Anastopoulos, G. (2002). Tobacco Seed Oil as an Alternative Diesel Fuel: Physical And Chemical Properties. *Industrial Crops and Products*, Vol.16, (July 2002), pp. 1–9, ISSN 0926-6690.

Holanda, A. (2004). *Biodiesel e Inclusão Social*, Câmara dos Deputados, Coordenação de Publicações, Brasília, Brazil, Retrieved from http://www2.camara.gov.br/a- camara/altosestudos/pdf/biodiesel-e-inclusao-social/biodiesel-e-inclusao-social

IENICA (2004). Agronomy Guide, Generic guidelines on the agronomy of selected industrial crops. *IENICA - Interactive European Network for Industrial Crops and their Applications*, Retrieved from http://www.ienica.net/agronomyguide/agronomyguide05.pdf

Krawczyk, T. (1996). Biodiesel - Alternative Fuel Makes Inroads but Hurdles Remain. *INFORM*, Vol. 7, No. 8, (August 1996), pp. 801-829.

Krishnan, G., Holshauser, D. L. & Nissen, S. J. (1998). Weed Control in Soybean (*Glycine max*) with Green Manure Crops. *Weed Technology*, Vol. 12, No. 1, (Jan.-Mar. 1998), pp. 97-102.

López, D.E., Suwannakarn, K., Bruce, D.A., & Goodwin, J.G. (2007) Esterification and transesterification on tungstated zirconia: Effect of calcination temperature. *Journal of Catalysis*, Vol. 247, pp. 43-50.

Lotero, E., Liu, Y., Lopez, D.E., Suwannakara, K. Bruce, D.A. & Goodwin J.G. (2005). Synthesis of Biodiesel Via Acid Catalysis. *Industrial& Engineering Chemistry Research*, Vol. 44 (14), (January 2005) pp. 5353-5363.

Palmer, C.E., Warwick, S. & Keller, W. (2001). Brassicaceae (Cruciferae) Family, Plant Biotechnology, and Phytoremediation. *International Journal of Phytoremediation*, Vol. 3, No. 3, pp. 245–287, ISSN: 1549-7879.

Pan, X. (2009). A Two Year Agronomic Evaluation of Camelina sativa and Brassica carinata in NS, PEI and SK (Master's thesis). Dalhousie University, Halifax, Canada, Retrieved from Digital Repository Unimib database of Dalhousie University at http://dalspace.library.dal.ca/handle/10222/12370

Pari, L., Fedrizzi, M. & Gallucci, F. (2008). *Cynara cardunculus* Exploitation for Energy Applications: Development of a Combine Head for Theshing and Concurrent

Residues Collecting and Utilization. Proceedings of 16th European Biomass Conference & Exibition, ISBN 978-88-89407-58-1, Valencia, Spain, 2-6 June 2008

Parodi, A., Marini, L. (2008) Process for the production of biodiesel. Patent WO 2008/007231.

Pasqualino, J.C. (2006). *Cynara cardunculus as an Alternative Crop for Biodiesel Production* (Ph.D. Thesis). Universitat Rovira I Virgili, Tarragona, Spain, Retrieved from http://tdx.cat/bitstream/handle/10803/8545/PhDThesisJPasqualino.pdf?sequenc e=1

Pinto A. C., Guarierio L. L. N., Rezende M. J. C., Ribeiro N. M., Torres E. A., Lopes W. A., Pereira, P. A. de P. & de Andrade J. B. (2005). Biodiesel: an Overview. *Journal of the Brazilian Chemical Society*, Vol. 16, No.6b, pp. 1313-1330, ISSN 0103-5053

Pirola, C., Bianchi, C.L., Boffito, D.C., Carvoli, G. & Ragaini, V. (2010) Vegetable oil deacidification by Amberlyst : study of catalyst lifetime and a suitable reactor configuration. *Industrial & Engineering Chemistry Research*, Vol. 49 (2010), pp. 4601- 4606.

Pirola, C. Boffito, D.C. Carvoli, G., Di Fronzo, A. Ragaini, V. & Bianchi, C.L. (2011) Soybean oil de-acidification as a first step towards biodiesel production. In *Soybean/Book 2*, ISBN 978-953-307-533-4

Pöpken, T. Götze, L. Gmehling, J. (2000). *Reaction Kinetics and Chemical Equilibrium of Homogeneously and Heterogeneously Catalyzed Acetic Acid Esterification with Methanol and Methyl Acetate Hydrolysis.* Industrial Engineering Chemistry Research, Vol. 39, pp. 2601-2611.

Potts, D. A., Rakow, G. W. & Males, D. R. (1999). Canola-Quality Brassica juncea, a New Oilseed Crop for the Canadian Prairies. Proceedings of the 10th International Rapeseed Congress, Canberra, Australia, September 1999

Radich, A. (1998). Biodiesel Performance, Costs, and Use. Energy Information Administration Available from: http://www.eia.doe.gov/oiaf/analysispaper/biodiesel/pdf/biodiesel.pdf

Rao, K.V., Krishnasamy, S., Penumarthy, V. (2009) WO 2009047793

Razon, L. F. (2009). Alternative Crops for Biodiesel Feedstock. *CAB Reviews: Perspectives in Agriculture, Veterinary Science, Nutrition and Natural Resources*, Vol. 4, No. 56, (October 2009), pp. 1-15, ISSN 1749-8848

Romero, E., Barrau, C. & Romero, F. (2009). Plant Metabolites Derived from *Brassica* spp. Tissues as Biofumigant to Control Soil Borne Fungi Pathogens. *Proceedings of the 7th International Symposium on Chemical and non- Chemical Soil and Substrate Disinfestation*, ISBN 9789066056237, Leuven, Belgium, September 2009.

Sanzone, E. & Sortino, O. (2010). Ricino, Buone Prospettive Negli Ambienti Caldo-Aridi. *Terra e Vita*, No. 7, pp. 28-29, ISSN 0040-3776.

Sharma, Y.S. & Singh, B. (2011). Advancements in solid acid catalysts for ecofriendly and economically viable synthesis of biodiesel. *Biofuels, Bioproducts & Biorefining*, Vol. 5, pp. 69-92.

Steinigeweg, S. & Gmehling, J. (2003). Esterification of a Fatty Acid by Reactive Distillation. *Industrial Engineering Chemistry Research*, Vol. 42, pp. 3612-3619

Suwannakarn, K., Loreto, E., Goodwin J.G.Jr. & Lu, C. (2008). Stability of sulfated zirconia and the nature of the catalytically active species in the transesterification of tryglicerides. *Journal of Catalysis*, Vol 255, pp. 279-286.

T. Ono, K. Yoshiharu, US Patent 4,164, 506, 1979.

The Royal Society (2008). Sustainable Biofuels: Prospects and Challenges. *The Clyvedon Press Ltd*, ISBN 978 0 85403 662 2, Retrieved from http://royalsociety.org/Sustainable- biofuels-prospects-and-challenges/

Tyson, K.S. (2002) Brown grease feedstocks for biodiesel, In : *National Renewable Energy Laboratory*, January 3, 2008 available from http://www.nrbp.org/pdfs/pub32.pdfS.

Usta, N. (2005). Use of Tobacco Seed Oil Methyl Ester in Turbocharged Indirect Injection Diesel Engine. *Biomass & Bioenergy*, Vol. 28, (January 2005), pp. 77-86, ISSN 0961-9534.

Velasco, L., Goffman, F.D. & Becker H.C. (1998). Variability for the fatty acid composition of the seed oil in a germplasm collection of the genus Brassica. *Genteic Resources and Crop Evolution*, Vol. 45, pp. 371-382.

Winayanuwattikun, P., Kaewpiboon, C., Piriyakananon, K., Tantong, S., Thakernkarnkit, W., Chulalaksananukul, W. & Yongvanich, T. (2008). Potential plant oil feedstock for lipase-catalyzed biodiesel production in Thailand. *Biomass and Bioenergy*, Vol. 32, pp. 1279-1286, ISSN 0961-9534.

Xingzhong, Y., Jia, L., Guanming, Z., Jingang, S., Jingyi, T. & Guohe, H. (2009). Optimization of conversion of waste rapeseed oil with high FFA to biodiesel using response surface methodology. *Renewable Energy*, Vol. 33, pp. 1678-1684.

Zhang, Y., Dube, M.A., McLean, D.D. & Kates, M. (2003). Biodiesel production from waste cooking oil: 1. Process design and technological assessment. *Bioresource Technology*, Vol. 90, pp. 229-240.

Zheng, D. & Hanna, M.A. (1996) Preparation and properties of methyl esters of beef tallow. *Bioresource Technology*, Vol. 57, pp. 137-142, ISSN 0960-8524.

Zhiyuan, H., Piqiang, T., Xiaoyu & Y. Diming, L. (2008). Life cycle energy, environment and economic assessment of soybean-based biodiesel as an alternative automotive fuel in China. *Energy*, Vol. 33, pp. 1654-1658, ISSN 0360-5442.

2 Chapter

PRODUCTION OF BIODIESEL FROM WASTE COOKING OIL

1. INTRODUCTION: AN OVERVIEW

Biodiesel encompasses all forms of alternative fuels produced from vegetable oils or animal fats. The prefix bio denotes a renewable and organic essence, as opposed to conventional diesel sourced from petroleum; meanwhile, diesel fuel pertains to its application in diesel engines. Biodiesel is generated through the transesterification of triglycerides in oils such as palm oil, soybean, rapeseed, sunflower, and castor oil, resulting in methyl or ethyl esters. In this process, the three fatty acid chains of each triglyceride molecule react with an alcohol in the presence of a catalyst to yield ethyl or methyl esters.

The ASTM (American Society for Testing and Materials Standard) describes the biodiesel as esters monoalkyl of fatty acids of long chain that are produced from vegetable oil, animal fat or waste cooking oils in a chemical reaction known as transesterification.

Biodiesel has the same properties of diesel used as fuel for cars, trucks, etc. This may be mixed in any proportion with the diesel from the oil refined. It is not necessary to make any modifications to the engines in order to use this fuel.

"The use of pure biodiesel can be designated as B100 or blended with fuel diesel, designated as BXX, where XX represents the percentage of biodiesel in the blend. The most common ratio is B20 which represents a 20% biodiesel and 80% diesel"(Arbeláez & Rivera, 2007 pp 4). Colombia in South America, is taking advantage of the opportunities that biofuels will open to the agriculture. With more than a million liters a day, Colombia is the second largest producer of ethanol in Latin America, after Brazil. This has decongested the domestic market of sugar at more than 500 thousand tons. The result is strong revenue for the 300,000 people who derive their livelihood from the production of panela (from sugar cane).

In Colombia the biodiesel is produced from the palm oil and methanol, "being the last imported to meet the demand in the biodiesel production". In the past two years, the biodiesel production from Palm was between 300000 liters/day to 965000 liters per day, distributed in four plants located in the Atlantic coast and in the country center.

In the biodiesel production is technically possible to use methanol and ethanol alcohol (Cujia & Bula, 2010. pp 106).

The palm oil is one of oilseeds trade more productive on the planet; it is removed between six and ten times more oil than the other as soy, rapeseed and sunflower. Colombia has more than 300,000 hectares planted in Palm oil, generating permanent and stable employment for more than 90,000 people.

The biodiesel advantages are that it is a renewable and biodegradable biofuel; it produces less harmful emissions to the environment than those that produce fossil fuels. Specifically the Palm biodiesel pure or mixed with diesel fuel reduces the emissions of CO_2, nitrogen oxides (NOx) and particulate material. Table 1, shows the world production of vegetable oils.

Table 1. World production of vegetable oils, 2008/2009. (Source: "Oilseeds: World markets and trade". FAS-USDA, October 2008)

OILS	MILLION TONS
Palm oil (fruit)	43.20
Soy oil	38.11
Rapeseed oil	19.38
Sunflower oil	11.45
Cotton oil	4.94
Palm oil (seed)	5.10
Peanut oil	4.93
Coconut oil	3.62
Olive oil	2.97

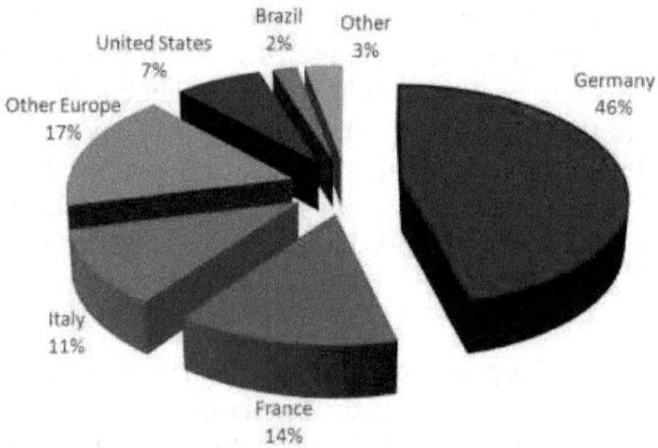

Fig. 1. World production of biodiesel (Source: National Federation of Oil Palm Growers (FEDEPALMA)).

The estimated consumption of diesel in the world at the end of the year 2005 was 960 billion liters. On the other hand, the production of biodiesel during the same year was 4.2 billion liters (Figure 1). For example, assuming that 2% of diesel was replaced with biodiesel, it would mean an increase of 15 billion liters in the biodiesel global production. This amount of biodiesel has other impacts, including overproduction of glycerin, the use of more land, etc.

Table 2. ASTM Features

FEATURES	UNIT	LIMITS Minimum	Maximum	TEST METHOD
Ester content	%(m/m)	96.5	-	EN 14103
Density a 15°C	Kg/m2	860	900	EN ISO 3675 EN ISO 12185
Viscosity a 40°C	Mm2/g	3.50	5.00	EN ISO 3104
Flash point	°C	120	-	Pr EN ISO 3679
Sulfur content	mg/kg	-	10.0	PrEN ISO 20846 pr EN ISO 20884
Carbon residue (in 10% of distilled residue)	% (m/m)	-	0.30	EN ISO 10370
Cetane index		51.0		EN ISO 5165
Sulphated ash content	% (m/m)	-	0.02	ISO 3987
Water content	mg/kg	-	500	EN ISO 12937
Total contamination	mg/kg	-	24	EN 12662
Cooper band corrosion (3 h at 50°C)	Classification	Class 1		EN ISO 2160
Oxidation stability 110°C	Hours	6.0		EN 14122
Acid index	mg KOH/g		0.50	EN 14111
Iodine index	g de iodine/100 g		140	EN 14103
Methyl ester of linoleic acid	%(m/m)		12.0	EN 14103
Methyl esters of methylpoli-unsaturated (> = 4 double bonds)	%(m/m)		1	
Methanol content	%(m/m)		0.20	EN 14110
Monoglycerides content	%(m/m)		0.80	EN 14105

FEATURES	UNIT	LIMITS		TEST METHOD
		Minimum	Maximum	
Diglycerides content	%(m/m)		0.20	EN 14105
Triglycerides content	%(m/m)		0.20	EN 14105
Free glycerin	%(m/m)		0.20	EN 14105 EN 14105
Total glycerin	%(m/m)		0.25	EN 14105
Metals of group 1 (Na+K)	mg/kg		5.0	EN 14108 EN 14109
Metals of group 2 (Ca+Mg)	mg/kg		5.0	PrEN 14538
Phosphorus content	mg/kg		10.0	EN 14107

ASTM has specified different fuel tests needed to ensure their proper functioning. Table 2, lists the specifications established for biodiesel and the corresponding test method.

1.1 Environmental problems for disposing used cooking oil

Used cooking oil causes severe environmental problems, "a liter of oil poured into a water course can pollute up to 1000 tanks of 500 liters". It's feasible to demonstrate the contamination with the dumping of these oils to the main water sources.

The oil which reaches the water sources increases its organic pollution load, to form layers on the water surface to prevent the oxygen exchange and alters the ecosystem. The dumping of the oil also causes problems in the pipes drain obstructing them and creating odors and increasing the cost of wastewater treatment. For this reason, has been necessary to create a way to recover this oil and reuse it. Also due to the wear and tear resulting in sewer pipes may cause overflows of the system, "generating diseases that can cause mild stomach cramps to diseases potentially fatal, such as cholera, infectious hepatitis and gastroenteritis, due to the sewage contains water which can transport bacteria, viruses, parasites, intestinal worms and molds" (Peisch. Consulted: http://www.seagrantpr.org/catalog/files/fact_sheets/54-aguas-usadas-de-PR.PDF).The dangerous odors generate impact negatively on health, "is formed hydrogen sulfide (H_2S), which can cause irritation of the respiratory tract, skin infections, headaches and eye irritation" (Peisch. Consulted: http://www.seagrantpr.org/ catalog/files/fact_sheets/54- aguas-usadas-de-PR.PDF).

2. TYPES OF COOKING OIL

Among the alternatives as a vegetal raw material to extract the oil are: oil palm, soybean, sesame, cotton, corn, canola, sunflower and olives.

2.1 Palm oil

Palm oil is retrieved from the mesocarp of the Palm fruit, this oil is regarded as the second most widely produced only surpassed by the soybean oil. The oil palm is a tropical plant characteristic of warmer climates that grows below 500 meters above sea level. "Its origin is located in the Guinea Gulf in West Africa." "Hence its scientific name, Elaeis guineensis Jacq and its popular name: African oil palm" (FEDEPALMA. Consulted: http://www.fedepalma.org/palma.htm).

Colombia is the largest producer of palm oil in Latin America and the fourth in the world. "The extracted oil from the palm contains a relationship 1:1 between saturated and unsaturated fatty acids, is also a major source of natural antioxidants as tocopherols, tocotrienols and carotenes"(FEDEPALMA. Consulted: http://www.fedepalma.org/ palma.htm). It has been proven that Palm oil is natural source vitamin E, in the form of tocopherols and tocotrienols. The tocotrienol act as protectors against cells aging, arthrosclerosis, cancer and some neurodegenerative diseases such as Alzheimer's disease. Unrefined palm oil is the richest in beta-carotene natural source; its consumption has proved to be very useful for preventing and treating the deficiency of vitamin A in risk populations.

2.1.1 Characteristics of plant

The oil palm presents fruit by thousands, spherical, ovoid or elongates, to form compact clusters of between 10 and 40 kilograms of weight. Inside, they kept a single seed, almonds or palmist, to protect with the fart, a woody endocarp, surrounded in turn by a fleshy pulp. Both, pulp and almond oil generously provide. The productive life of the oil palm can be most of fifty years, but from the twentieth or twenty-five the stem reaches a height that hinders the work of harvest and marks the beginning of the renewal in commercial plantations. 25 to 28 °C on average monthly temperatures are favorable, if the minimum average temperature is below 21 °C. Temperatures of 15 °C stop the growth of the seedlings from greenhouse and decrease the performance of adult palms. Between 1,800 and 2,200 mm precipitation is optimal, if it is well distributed in every month. Like the coconut palm, the palm oil is favored by deep, loose and well drained soils. A superficial phreatic level limits the development and nutrition of roots. In general, the physical characteristics good, texture and structure, are preferable to the level of fertility, as it can be corrected with mineral fertilization. The palm oil resists low acidity levels, up to pH 4. Too alkaline soils are harmful. Although you can plant with success on land of hills with slopes above of 20°, are preferred levels or slightly wavy, with no more than 15 ° gradients.

2.1.2 Pests

The major pest of palm oil and its damage are:

- Acaro: They are located on the underside of the leaves, mainly in vivarium palms. The damages are identified by the discoloration of the leaves, which reduces the photosynthetic area. We can fight it with Tedión.
- Arriera ant: it is common in tropical areas. This animal can cause serious defoliations in palms of all ages. We can fight it with bait poisoned as Mirex, applied to the nest mouths.
- Estrategus: Is a beetle of 50 to 60 mm long, black, with two horns. This animal drills in the ground, at the foot of the Palm, a gallery of even 80 cm; penetrates the tissues of the trunk base and destroys it. It is controlled with 200 g of heptachlor powdered 5%, slightly buried around the Palm.
- Rats: This animal can cause damage at the trunk base of young palms. Controlled with baits of coumarine, which must be changed regularly.
- Yellow beetle or alurnus: attacks the young leaves of the plant heart as well as on the coconut tree. It is controlled with sprayings of Thiodan 35 EC, solution of 800 cc in 200 liters of water. Apply 2 to 4 liters in palm.
- Beetles or black palm weevil: In Palm oil causes the same damage to the coconut palm.
- Lace bug: is 2.5 mm long. It is an insect of transparent grey color. It is located in the underside of the leaves. Their stings favor infections by various fungi, which may cause draining of the leaves.

2.2 Rapeseed or canola oil

Rapeseed is a "specie oilseed in the cruciferous family. Many of the species of this family have been cultivated since long time ago that their roots, stems, flowers and seeds are edible" (Iriarte, Consulted: http://www.inta.gov.ar/ediciones/idia/oleaginosa/colza01.pdf). Ideally grows in climates that go from temperate to slightly cold and wet (minimum of 0 °C and maximum of 40 °C). When the seeds of rapeseed are crushed we can obtain oil and a kind of pulp or prized residue from always to feed livestock, since that gives a 34% protein and 15% crude fiber. The biodegradable properties of rapeseed or canola oil make it ideal to be used on the basis of paints, herbicides, lubricants, food packaging, etc.

2.2.1 Characteristics of plant

Oilseed rape (Brassica napus) is a crucifer of deep and pivoting root. The stem has a size of 1.5 m approximately. The lower leaves are petiolate but the superiors entire and lanceolate. The flowers are small, yellow, and are grouped in terminal racemes. The fruits have a number of grains by pod around 20-25, depending on the variety. The rapeseed composition is showed in the table 3:

Table 3. Rapeseed composition.

COMPOSITION	%
Proteins	21,08
Fat	48,55
Fiber	6,42
Ashes	4,54
Nitrogen-free extracts	19,41
TOTAL	100,00

The seeds are spherical of 2 to 2.5 mm in diameter and when are mature have a reddish or black brown color. Rapeseed has a proportion (39%) of oil where there are a large number of fatty acids of long-chain, which quantitatively the most important is the erucic acid. The cultivation of rapeseed has ability to grow in temperate climates to temperate cold with good humidity. It adapts to different soil types, the ideals are the franc soils of good fertility and permeable which is a very sensitive crop to the superficial flooding.

2.2.2 Pests

Rape stem weevil (Ceuthorrhynchus napi): the grub of this insect deforms the stem of the rape, which is curved and often indenting in a certain length.

- Terminal bud Weevil (Ceuthorrhynchus picitarsis): adults do not cause damage, but the larvae destroy the terminal bud and force the plant to produce side shoots. The treatments are made with endosulfan and Fosalón.
- The siliques weevil (Ceuthorrhynchus assimilis): adults bite the young siliques and the larvae gnaw seeds causing a significant decrease in the harvest. Endosulfan and Fosalón are used in treatments.
- Cecydomia (Dasyneura brassiceae): The larvae of this insect destroy the siliques totally. The endosulfan and fosalon control this plague.
- Meligetos of the cruciferous (*Meligethes sp*): adults are in charge of gnawing the buttons of the rapeseed; these attacks are more important younger are the buttons. When begin the flowering the damage decrease.
- Flea of rapeseed (Psyllodes chrysocephala): adults appear in autumn rape fields, generally shortly after birth gnawing the leaves and can destroy large number of plants. Karate to doses of 40-80 cc/hL is recommended for the treatment.

- Flea of the cabbage (*Phylotreta sp*): adult insects wintering in the soil in September and appear in April. Karate works very well against these insects.

2.3 Sunflower oil

The oil extracted from sunflower seeds is considered to be of high quality for a low percentage of saturated fatty acids and a high percentage of unsaturated fatty acids. It also contains essential fatty acids and a considerable amount of tocopherols that gives it stability. The acidic composition of the sunflower depends on the genotype and the environment. There are currently three groups of genotypes: traditional, oleic medium and oleic high.

2.3.1 Characteristics of plant

The sunflower belongs at the family "*Asteraceae*, whose scientific name is *Helianthus annuus*. It is an annual plant with a vigorous development in all its organs. Within this species there are many types or subspecies grown as ornamental plants, oilseeds and forage plants" (INFOAGRO, Consulted: http://www.infoagro.com/herbaceos/oleaginosas /girasol.htm). Average sunflower cycle includes between 100 and 150 days according to genotypes, dates of planting, latitude and availability of water and nutrients. The "temperature is the most important factor in the control of the seeds germination being the optimal near to 26 °C with maximum temperatures of 40 °C and minimum from 3 to 6 °C. The threshold for soil temperature (0 to 5 cm) from which normally starts sowing is between 8 and 10 °C" (Diaz- Zorita et al, Consulted: http://www.asagir.org.ar /Publicaciones/cuadernillo_web.pdf). The availability of water acts on the soaking of seeds, on the subsequent growth of the seedling. The water excess decreases the amount of air in the soil.

2.3.2 Pests

Pests of early-onset (e.g. cutting caterpillars, leafcutter ants, velvety larvae, worm wire, tenebrionido of the sunflower, underground grille, weevils, black beetle, slugs, etc.) produce damage in seeds and seedlings. Slugs cause great damage to the leaves. The control is convenient with treatments of seeds or specific toxic baits.

3. BIODIESEL PRODUCTION PROCESS

The biodiesel production is given by the transesterification reaction which consists of three consecutive and reversible reactions. First, the triglyceride is converted in diacylglycerol, and running at monoglyceride and glycerin. In each reaction one mole of methyl ester is released as shown in Figure 2.

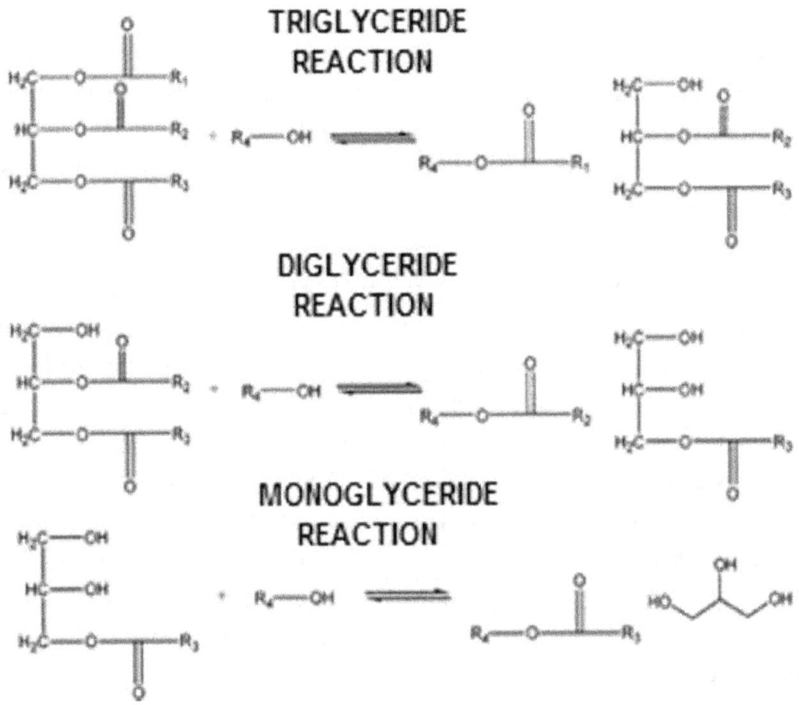

Fig. 2. Stages of the transesterification reaction (Arbeláez & Rivera, 2007. pp 13)

Figures 3 and 4 show the secondary reactions that may occur: the saponification reaction and the neutralization reaction of free fatty acids.

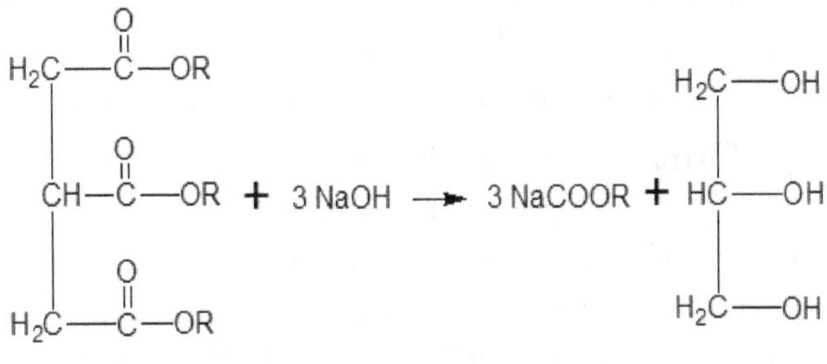

Fig. 3. Saponification reaction (Arbeláez & Rivera, 2007. pp 13)

$$\underset{\textit{Fatty acid}}{RCOOH} + NaOH \rightarrow \underset{\textit{Sodium carboxylate}}{\mathbf{R-COONa}} + H_2O$$

Fig. 4. Neutralization reactions of free fatty acids (Arbeláez & Rivera, 2007. pp 13)

3.1 Raw materials

The biodiesel production comes mostly from oils extracted oilseed plants especially sunflower, soy, rapeseed and animal fats. However, any material that contains triglycerides can be used for the biodiesel production. "In addition to the oil or fat is needed an alcohol and catalyst to convert oils and fats in alkyl esters". (Arbeláez & Rivera, 2007. pp7)

3.1.1 Alcohol

"Primary and secondary alcohols with string of 1-8 carbons are used for the biodiesel production, among the alcohols that can be used in this process are: methanol, ethanol" (Cujia & Bula, 2010. pp 106), propanol y butanol. "When are used alcohols such as ethanol is more complicated the recovery of pure alcohol in the process because the azeotrope that forms with water"(Cheng et al. 2008. pp 4) and the performance of ethyl esters is less compared to the methyl esters due methanol has a lower molecular weight (32.04 g/mole) compared to ethanol (46.07 g/mole)."On the other hand if you use methanol, not would contribute to environmental issues and sustainability, biodiesel would not be completely bio, by having a fossil component provided by the alcohol, because methanol is made from natural gas, which is fossil" "(Cheng et al. 2008. pp 4).To use methanol or ethanol is needed "a mechanical agitation to encourage the transfer of mass"(Arbeláez & Rivera, 2007. pp 10)."In the course of the reaction form emulsions, using methanol is easy and quickly dissolved, forming a glycerol-rich bottom layer and a higher layer in methyl esters, while using ethanol these emulsions are more stable making the process of separation and purification of ethyl esters more difficult"(Arbeláez & Rivera, 2007. pp 10).

Is preferred to use methanol in the biodiesel production because of their low viscosity (0.59 m * Pa * s at 20 °C), because using alcohols such as ethanol with high viscosity (1,074 m * Pa * s at 20 °C), the biodiesel viscosity increases and as a result a "fuel of high viscosity not will be pulverized properly by injection systems that have diesel engines. Also increase the opacity of fumes which limits their application in automotive engines"(Benjumea et al. 2007. pp 149).

In the reaction performance is feasible to reach "higher conversions with methanol, ethanol using the process is more complex, expensive, requires a higher consumption of energy and time"(EREN. 2003.pp 38). "We found that it requires less reaction time when using methanol rather than ethanol, either in acid or alkaline catalysis, reaching high yields"(Giron et al. 2009.pp 18).

With the above, the methanol is selected to be used in the biodiesel production due to its lower cost, better performance and less time and energy during the reaction.

3.1.2 Catalysts

Homogeneous, heterogeneous or enzyme catalysts are used in the biodiesel production. Homogeneous catalysts are soluble in the middle of reaction, i.e. they are in a single phase either liquid or gaseous. "One of the advantages of homogeneous catalysis is the high speed of reaction, and moderate temperature and pressure conditions" (EREN. 2003.pp 4). The catalysts can be acids or alkalis, the acid catalysts are effective but require a time interval extremely long and temperatures exceeding 100 °C for its action. "Getting conversions of 99% with a concentration of 1% sulfuric acid in relation to the amount of oil, it takes about 50 hours" "(EREN. 2003.pp 13). We can use this catalytic process when the oils have a high degree of acidity and "harm the action of alkali catalysts with acidity greater than 10 %"(EREN. 2003.pp 39).We can use sulfuric acid (H_2SO_4), phosphoric acid (H_3PO_4), among others. When is used "acid catalysts with alcohol excess is that the recovery of glycerin is more difficult as the quantities of alcohol are quite large compared to other type of catalyst" (Arbeláez & Rivera, 2007. pp13).

"Using HCl are achieved yields of 61% and with H2SO4 we can obtain 80%"(Liu et al. 2006a pp 186), but these "catalysts are more corrosive than alkali catalysts"(Errazu et al. 2005 pp 1305). In comparison with the acidic catalysts, the basic catalysts accelerate the reaction rate, the disadvantage of basic catalysts is that produces soaps due to the high amounts of free fatty acids and water by which we must add the appropriate amount of base to neutralize fatty acids free. The most commonly used are sodium hydroxide (NaOH), potassium hydroxide (KOH) and inappropriate for industrial application (CH3ONa) sodium methoxide since this is more expensive and "requires total absence of water" (EREN. 2003 pp 40). "The catalysts are dissolved in the reaction mixture alcohol-oil what does that not can be recovered at the end of the transesterification reaction" (Arbeláez & Rivera, 2007. pp13). "By using KOH as a catalyst we can produce potassium fertilizers such as potassium chloride, potassium sulphate and potassium nitrate if the product with phosphoric acid is neutralized" (Arbeláez & Rivera, 2007. pp14).

"The maximum yield found with NaOH is 85% at a sodium hydroxide concentration of 1,0%. Adding an excess in the amount of the catalyst, it gives rise to the formation of an emulsion which increases viscosity and leads to the gel formation" "(Cheng et al. 2008. pp 2210)."With regard to the use of catalyst as ($NaOCH_3$) sodium methoxide and ($KOCH_3$) potassium methoxide we can observe high efficiency compared with other alkali catalysts"(Cheng et al. 2008. pp 2210). The temperature of the transesterification reaction "should not exceed the boiling point of alcohol, because it vaporizes and forms bubbles which limit the reaction in the interfaces alcohol/oil/biodiesel"(Giron et al. 2009.pp 18) "To be used as catalyst NaOH with methanol, has been found that the optimum temperature to achieve high yields was 60 °C, while using KOH to this same temperature not achieved such high yields and higher catalyst concentrations should be used to using NaOH" (Liu et al. 2006b pp 110). "In an alkali catalyzed process is reached high purity and yields in short periods of time ranging between 30 - 60 minutes" (Liu et al. 2006a pp 186).

Heterogeneous catalysts are found in two phases and a contact area, "the use of these catalysts simplifies and makes more economical the purification process due the easy separation of the products and reactants. The disadvantage is the difficulty to temperature control for very exothermic reactions, limitations on mass transfer of reactants and products, as well as high mechanical resistance to the catalyst" (Arbeláez & Rivera, 2007. pp12). Among the most common catalysts are the metal oxides (MgO, CaO), acids of Lewis ($SnCl_2$), etc. For example, by using zinc oxide are obtained yields of 50.7%, when using Al_2O_3 is obtained 57.5% and using CaO yield of 65%"(Rojas & Torres. 2009 pp 15). "These catalysts have limitations on transfer of mass of reactants and products" (Arbeláez & Rivera, 2007. pp12), but they have the advantage that they are not corrosive to the reactor"(Guan et al. 2009 pp 520).The easy separation of the products generates a "simplification of the manufacturing process since the catalyst can be separate from the products of reaction with a simple filtration process"(Lles et al. 2008 pp 63). "Don't generate byproduct of soap by reaction with free fatty acids (AGL)". (Bournay et al. 2005. pp 191) "Using CaO is achieved a yield of 65% and by using MgO a yield of 64%"(Bournay et al. 2005. pp 192). To achieve high yields the reaction must be carried out "to a higher temperature increasing energy costs" (Bournay et al. 2005. pp 191).Reported high reaction times, because the "speed of transesterification reaction with these catalysts is lower in comparison with homogeneous catalysts, due to the mass transfer resistance" (Guan et al. 2009 pp 522).

Finally, the lipases being effective for the transesterification reaction can be used between the enzyme catalysts. "This type of catalysis has the advantage of allowing the use of alcohol with high content of water (more than 3%), low temperatures, which is an energy-saving and high degrees of acidity in oils" (EREN. 2003. pp 41).

3.1.3 Waste cooking oil

The waste cooking oil is generated from the fried food, which need large amounts of oil because it requires the full immersion of food at temperatures greater than 180 °C. Accordingly to the high temperatures are generated changes in its chemical and physical composition, as well as in its organoleptic properties which affect both the food and oil quality.

Reuse of domestic oil has a high risk to the health of consumers as depending on the type of food subjected to frying, "this absorbs between 5% and 20% of the used oil, which can increase significantly the amount of hazardous compounds that provide degraded oil to food" (EREN. 2003. pp 31)."In an alkali catalyzed process is reached high purity and high yields in short periods of time ranging between 30 - 60 minutes"(Liu et al. 2006 pp 186).

Used cooking oil is normally black, a strong odor and does not have large amount of solids because its collection is passed through a fine mesh. In Figure 5, we can see a sample of used oil from the hotel sector.

Fig. 5. Sample of waste cooking oil

3.1.3.1 Domestic waste oil treatment

Wastes containing these types of oils are products of decomposition that impair the oil quality causing reduction in productivity in the transesterification reaction and may also generate undesirable by-products which hurt the final product. For these reasons, it is important to refine the waste domestic oil for the biodiesel production. "This type of refinement has a right effect on the yield of the reaction from 67% to 87% after bleaching". (EREN. 2003. pp 36). For the treatment of adequacy of waste domestic oil, the operations that can be applied are filtration, de-acidification or neutralization and whitening. The processes of degumming and deodorization aren't needed because the oils have already been treated prior to use and although during degradation odors occur, the removal is not essential for the biodiesel production.

- *Filtration.* The operation is for removing solids, inorganic material, and other contaminants in the oil. It can be carried out at temperatures higher than 60 °C, where substances carbonaceous produced from burnt organic material, pieces of paper, waste food and other solids are removed or occur at low temperatures which depend on the physical condition of the oil. In addition, we can delete solid fats or products of low melting points from the frying process.
- *Desacidification* It is the process by which free oils fatty acids are removed, various methods are used:
  a. Neutralization with alkaline solution: in this process the acids are removed in the form of soaps.
  b. Esterification with glycerin: seeks to regenerate the triglyceride.
  c. Extraction by solvents: where it is used ethanol in proportions 1.3 times the amount of oil.
  d. The distillation of fatty acids, this method requires a high energy cost.
  e. Removal of fatty acids with ion-exchange: a resin of strongly basic char-

acter for the removal of free fatty acids and the color of the oil is used. Method that provides greater account of productivity in the removal of free fatty acids is the neutralization by caustic soda, since it not only are obtained high relations, but also helps in the bleaching of the oil, because made soaps help dragging the color generators. There are basically two procedures:

- Neutralization with dilute alkali: are used concentrations of 0.75 to 2 N.
- Neutralization with concentrated alkali, where the concentration of caustic soda vary between 2 and 5 N.

In each of the procedures mentioned above neutralization is carried out hot, with oil at a temperature between 50 - 60 °C and addition of caustic soda between 70-80 °C.

3.1.3.2 Chemical characteristics of the used oil

Chemical characterization of the used oil, is presented in table 4

Table 4. Characterization of cooking oil collected by the hotel sector (Source: Avalquímico Ltda.2010)

FEATURES	OIL COLLECTED BY THE HOTEL SECTOR
Acidity (%)	0.56
Moisture	0.25
Viscosity at 37ºC (centistokes)	44.78
Iodine index (Cgl2/g)	108.22
Peroxide index (meq. Oxygen active/Kg of sample)	16.61
Unsaponifiable material (%)	1.70
Saponification index (mg KOH/g)	195.87
Ash (%)	0.030
Refractive index 25°C	1.4700
Density 15°C (g/mL)	0.9216

4. BIODIESEL PRODUCTION FROM USED COOKING OILS

According to table 5, the catalyst with higher industrial scaling, economic cost, high yields and short reaction time, is the alternative of basic catalysis, using sodium hydroxide. Although soap can be formed using sodium hydroxide in the transesterification reaction, this occurs if the content of free fatty acids is greater than 1% and the type of oil collected from the hotel sector has a percentage of acidity of 0.54%, so it is not problem to use this type of catalyst for the biodiesel production.

Table 5. Comparison of alternatives for the biodiesel production.

SELECTION PARAMETERS	Comparison of alternatives		
	Alkaline	Heterogeneous	Acid
Catalyst	NaOH	CaO	H_2SO_4
Alcohol	Methanol	Methanol	Methanol
Scaling	High	Low	Low
Catalyst separation	Low	High	Low
By-products formation	Glycerin Soap, pasty	Glycerin	Glycerin
Environmental impact	High	Low	High
Cost	Economic	High	Moderate
Availability	Medium	Low	Medium
Catalyst concentration (%w/w)	Low	Medium	High
Molar ratio alcohol/oil	Medium	Low	High
Temperature (°C)	Low	High	Medium
Yield	High	Low	Medium
Reaction time (hours)	Low	Low	High
Safety level	High	Low	High

To select the best alternative for the biodiesel production are defined three ranges for the operation conditions to be used in the transesterification reaction (table 6).

Table 6. Ranges established for the operation conditions of the transesterification reaction (Rojas & Torres. 2009. pp 18).

SELECTION PARAMETERS	RANGES		
	Low	Middle	High
Catalyst concentration (% w/w)	0.2% - 1%	>1% - 3%	>3% - 15%
Molar ratio alcohol/oil	3:1 - 6:1	>6:1 - 12:1	>12:1 - 80:1
Temperature (°C)	50 - 60	>60 - 100	>100 – 200
Yield (%)	20 – 70	>70 - 90	> 90 – 100
Reaction time (hours)	0.16 – 1	>1 - 2	> 2 – 40

4.1 Experimental design

It is a key to make a design from which the most appropriate values for each of the design factors can be established. The selected factors were: molar ratio alcohol/oil, percentage of catalyst, temperature and washing agent, where the first three are design variables and the latter is a design condition. Before starting the design is important defines the ranges and levels for these factors, for this reason, we search the experimental phase in scientific articles related to the project (table 7).

Table 7. Ranges for the design factors

DESIGN FACTORS	RANGE
Molar ratio alcohol/oil	6:1- 15:1
Catalyst Percentage (% wt.)	0,4-1
Temperature (°C)	40-70
Washing agent	Water (40°C) - Acetic acid

Based on the ranges set out in table 7, we provide that the factorial design appropriate for the process is the factorial design 2^k, which can be solved by the technique of Yate contrasts which establish two levels for each of the design factors, these levels are high (+) and low (-) (see table 8)

Table 8. Levels for each design factor

Design Factor	High level (+)	Low level (-)
Molar ratio alcohol/oil	9:1	6:1
Catalyst Percentage(% wt)	0.7	0.5
Temperature(°C)	60	50
Washing agent	Acetic acid	Water (40°C)

For the molar ratio alcohol/oil is found a ratio of 6: 1 that is optimum for achieving high conversions, some articles display that lower ratio not is possible to reach a complete transesterification reaction. There are also good results with ratios ranging between 9:1 and 12:1, while if we use higher than 15:1 molar ratio there are difficulties in the separation of glycerin and methyl esters.

For the catalyst concentration the values vary in a range of 0.4 - 2% being the concentration 1% better but the reaction yield is not very high.

For the design factorial mentioned we can set the number of trials, having clear that 2 is the number of levels and k is the number of factors, i.e. which has a total of 2^4 treatments and is carried out a duplicated for each one, as should be taken into account the time limit and the project costs, which in total we have 16 experiments. Then is defined the signs matrix (table 9), where it is necessary to enumerate the trials and then is assigned a combination of treatments that aims to relate the design factors.

Consecutively is assigned the level for each combination, positive for the design factor that is being evaluated in the trial and negative for whose are not related. For this type of design the first trial has low levels and the final test has the higher levels.

Table 9. Matrix of signs

Trial	Combination of Treatments	Design Factors			
		A	B	C	D
1	1	-	-	-	-
2	A	+	-	-	-
3	B	-	+	-	-
4	C	-	-	+	-
5	D	-	-	-	+
6	AB	+	+	-	-
7	AC	+	-	+	-
8	AD	+	-	-	+
9	BC	-	+	+	-
10	BD	-	+	-	+
11	CD	-	-	+	+
12	ABC	+	+	+	-
13	ABD	+	+	-	+
14	ACD	+	-	+	+
15	BCD	-	+	+	+
16	ABCD	+	+	+	+

Performance is evaluated according to the treatments combination, considering that a duplicated is made by treatment. With data from the response variable, which is the yield, we carry out a statistical analysis by means of the ANOVA table or analysis of variance for data, which gives the more appropriate conditions for each factors of design and allows establishing that trials were the best.

For each main effect and interaction effect we have associated a single degree of freedom, so this is calculated using the following expression:

$$GL_i = N^0_{\text{niveles}} - 1 \quad (1)$$

Where:

i: is any combination of treatment

$N_{levels} = 2$, because we have a high level and one low

To determine that interactions are significant an f for each source of variation is calculated and compared to $f_{0,05}(1,16) = 4,49$, with this we can determine in which region (probable region RP or critical region RC) is each treatment and thus be able to establish that treatment is accepted or rejected with the help of figure 6.

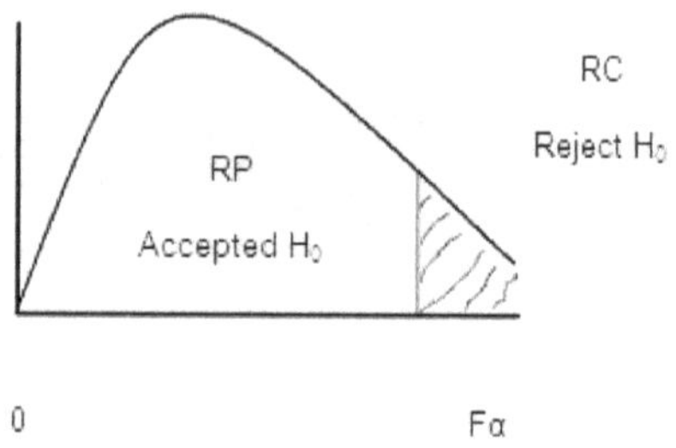

Fig. 6. Function f

4.2 Experimental development at the laboratory level

Table 10, shows the quantities determined at the laboratory level for each compound that is involved in the biodiesel production.

Table 10. Amount of reagents

AMOUNT OF REAGENTS	
Waste cooking oil (mL)	150
Amount of methanol (6:1) (mL)	40.63
Amount of methanol (9:1) (mL)	60.95
Amount of NaOH grams (0.5%p/p)	0.823
Amount of NaOH grams (0.7%p/p)	1.105
Amount of HCl grams (0.5% p/p)	1.466
Amount of HCl grams (0.7% p/p)	2.053

First is the filtration of used oil, then mixing alcohol/catalyst to add it to the reactor which contains the oil at the temperature of the transesterification reaction, then is the separation of biodiesel and glycerin, washes the biodiesel and finally is the distillation of the biodiesel. "The dissolution is agitated at low rpm because that at high revolutions sodium hydroxide can be oxidized" (Arbeláez & Rivera, 2007. pp45), should also be covered because amount of methanol due to its volatility can be lost. For the transesterification reaction are used reactors of four mouths with capacity of 500 mL and 1000 mL, magnetic stirrers, plates of agitation, spiral capacitors, mercury thermometers, thermostat bath and temperature controller. Figure 7, shows the setup for the biodiesel production.

To carry out the transesterification reaction is loaded oil at the reactors and heated up to reaction temperature, while it reaches the temperature is made the mixture of the catalyst with alcohol, then it is added to the reactor. At the end of the reaction time is added HCl to 37 per cent in order to neutralize the reaction.

Completed reaction, the product is poured into the separation funnels and let a minimum time of 8 hours, to ensure good separation of the phases (Figure 8). Separation times were not equal for all runs varied between 10-24 hours.

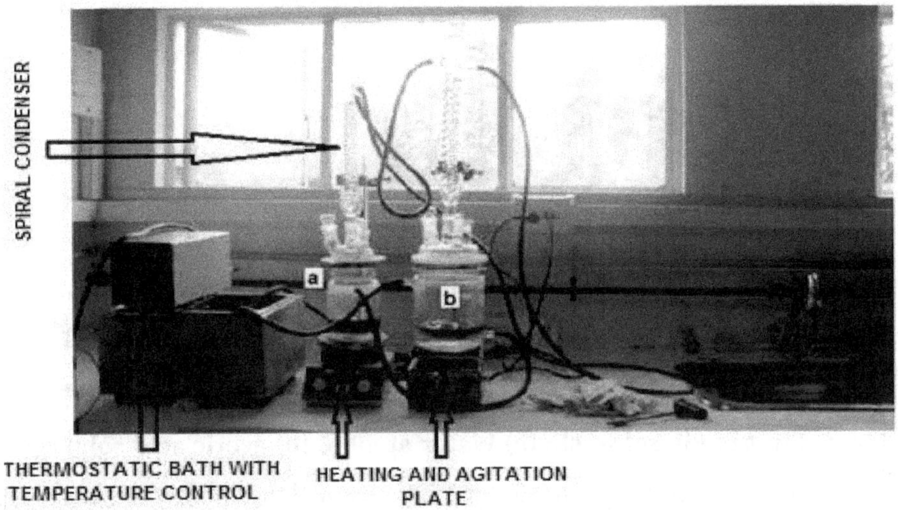

Fig. 7. Biodiesel production assembly: a) Reactor of 500 mL b) Reactor of 1L.

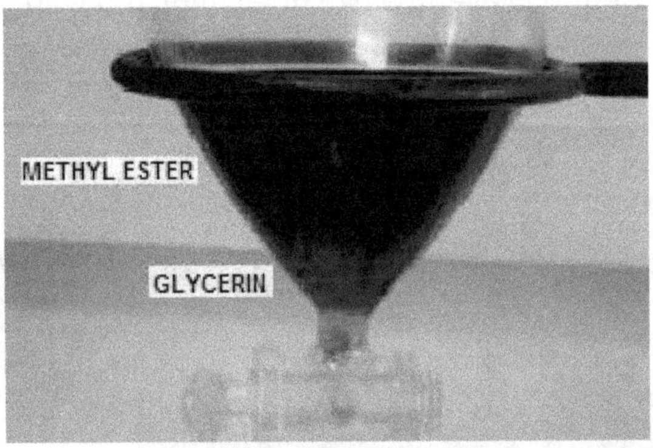

Fig. 8. Biodiesel-Glycerin separation

Once separate the glycerin from biodiesel, it is carried out the washing on the funnels. Separate the glycerin, the biodiesel must be washed because that may contain residues of catalyst, methanol, soaps and glycerides without reacting. Two types of washings are established according to the experimental design.

In one of the washing, water at 40 °C is used for three washes. Other tests of washing have been conducted with acetic acid 10% wt, where is used the same amount of used cooking oil; two washes are carried out with acid solution and the third is done with deionized water.

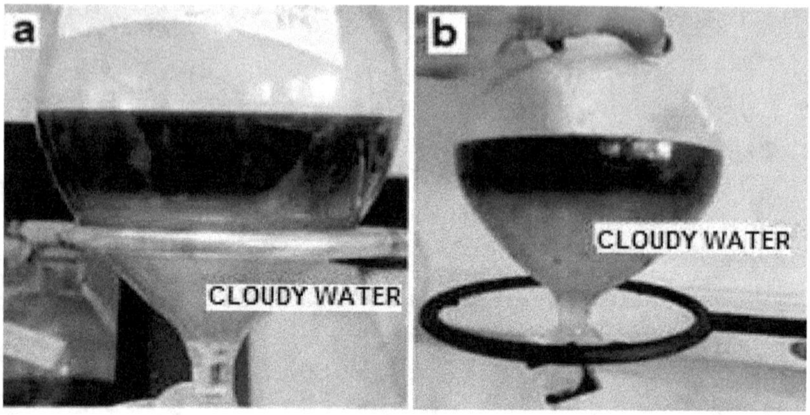

Fig. 9. First biodiesel wash: (a) Water at 40 °C (b) acetic acid solution

The distillation is carried out at 40 °C, temperature which is below the boiling point of methanol. The vacuum pump is used in order to minimize the time of distillation and vacuum trap is used to prevent waste of alcohol and water to the pump.

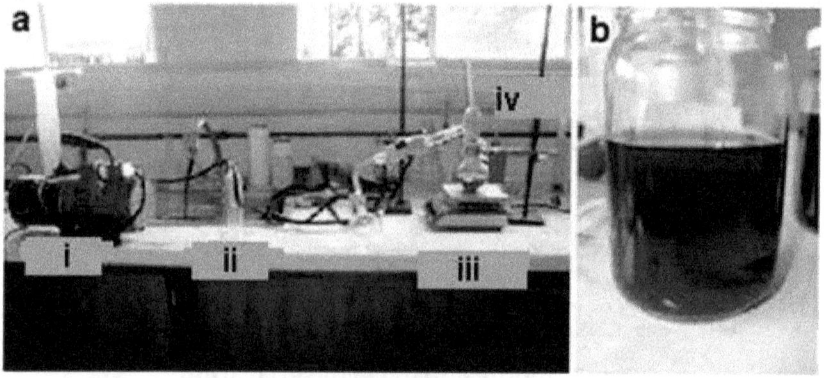

Fig. 10. Biodiesel distillation (a) Mounting: i. vacuum pump ii. Vacuum trap iii. Hot plate iv. Thermometer (b) Distilled biodiesel

4.3 Results and analysis

We show the results and analysis of tests conducted at the laboratory level, as its density and the analysis of variance. According to the literature retrieved biodiesel is "liquid, transparent and reddish color without any content of solids or gels" (Arbeláez & Rivera, 2007. pp37).

Taking into account the results of the biodiesel appearance, it can be concluded that the catalyst percentage influences in the biodiesel stability because that similar conditions were equal in appearance. The majority of samples that had contained solids, gels and their appearance was opaque, were the samples where used 0.5% NaOH.

Opaque samples as the 3, 4 and 7 are because have much water content, this might be for the washes with water at 40 °C and its water content was increased. Opacity is an indicator that the methyl esters have presence of water.

The table 11 shows the densities for each test sample.

Table 11. Density for each sample

TEST	DUPLICATE 1 Biodiesel density (g/ml)	DUPLICATE 2 Biodiesel density (g/ml)
1	0,902	0,910
2	0,902	0,877
3	0,867	0,887
4	0,893	0,881
5	0,897	0,903
6	0,885	0,887
7	0,883	0,906
8	0,906	0,860
9	0,871	0,887
10	0,885	0,893
11	0,914	0,906
12	0,897	0,891
13	0,900	0,895
14	0,885	0,875
15	0,883	0,862
16	0,873	0,891

The biodiesel density according to standard ASTM D-1298 must be in a range of (0.86 - 0.90 g/ml), the density does not guarantee that retrieved biodiesel is of good quality, but is taken to compare samples with each other, due there is evidence that satisfy with the provided density but its appearance is not adequate or are samples which do not comply with the permitted density value but its appearance is appropriated. As for example, the duplicate 2 from sample 1 exceeds the density range allowed for biodiesel, but the appearance fulfill with the stipulated features. For the sample 11, both the duplicate 1 and the 2 exceed the density level and its appearance is a little opaque.

Samples 11 and 12 are in the density range allowed but the appearance does not meet any features because have high solids content and are highly opaque.

Some samples contain solids and do not fulfill with the physical characteristics of biodiesel, have a density within the level agreed by the ASTM D-1298 standard for a biodiesel of good quality, which must be due to that the solids are smaller proportion than the liquid phase that is rich in methyl esters.

It can be concluded that the density and appearance analysis are not reliable parameters to determine the biodiesel quality. For this reason, based on the order of the table 12, we can determine the variable response (table 13), which is expressed as the ratio between the mass of the biodiesel produced and the mass of oil used for the production (productivity percent).

Table 12. Rearranged experimental matrix

TEST	COMBINATION OF TREATMENTS	DESIGN FACTORS			
		A	B	C	D
1	1	6:1	0,5	50	Water (40°C)
2	A	9:1	0,5	50	Water (40°C)
3	B	6:1	0,7	50	Water (40°C)
4	AB	9:1	0,7	50	Water (40°C)
5	C	6:1	0,5	60	Water (40°C)
6	AC	9:1	0,5	60	Water (40°C)
7	BC	6:1	0,7	60	Water (40°C)
8	ABC	9:1	0,7	60	Water (40°C)
9	D	6:1	0,5	50	Acetic Acid (T amb)
10	AD	9:1	0,5	50	Acetic Acid (T amb)
11	BD	6:1	0,7	50	Acetic Acid (T amb)
12	ABD	9:1	0,7	50	Acetic Acid (T amb)
13	CD	6:1	0,5	60	Acetic Acid (T amb)
14	ACD	9:1	0,5	60	Acetic Acid (T amb)
15	BCD	6:1	0,7	60	Acetic Acid (T amb)
16	ABCD	9:1	0,7	60	Acetic Acid (T amb)

Table 13. Reaction productivity

TEST	COMBINATION OF TREATMENTS	PRODUCTIVITY (%)	
		DUPLICATE 1	DUPLICATE 2
1	1	70	44,9
2	A	79,5	74,3
3	B	85,6	84,5
4	AB	93,7	93,9
5	C	69,4	68,4
6	AC	62,3	88,2
7	BC	64,5	62,6
8	ABC	88,7	85,5
9	D	62,1	63,1
10	AD	89,5	60,7
11	BD	76,2	80,7
12	ABD	88,9	88,5
13	CD	61,3	57,5
14	ACD	81,2	80,3
15	BCD	74,8	71,8
16	ABCD	86,2	81,8

We carry out the ANOVA statistical analysis (table 14) and the test of hypothesis (table 15).

Table 14. ANOVA TABLE

COMBINATION OF TREATMENTS	EFECTS	SQUARES SUM	FREEDOM DEGREES	MEAN SQUARE	f CALCULATED
A	14,1125	1593,30	1	1593,30	22,78
B	12,2	1190,72	1	1190,72	17,02
C	-3,225	83,205	1	83,21	1,19
D	-0,7125	4,06125	1	4,06	0,06
AB	-0,8	5,12	1	5,12	0,07
AC	1,375	15,125	1	15,12	0,22
AD	-0,4125	1,36125	1	1,36	0,02
BC	-6,2875	316,26125	1	316,26	4,52
BD	-0,55	2,42	1	2,42	0,03
CD	1,375	15,125	1	15,13	0,22
ABC	2,437500	47,531250	1	47,53	0,68
ABD	-2,425000	47,045000	1	47,05	0,67
ACD	0,950000	7,220000	1	7,22	0,10
BCD	3,212500	82,561250	1	82,56	1,1803
ABCD	-4,5375	164,71125	1	164,71	2,35
ERROR		1119,21	16	69,950625	
TOTAL		4694,97875	31		

Table 15. Hypothesis test

COMBINATION OF TREATMENTS	EFECTS	f CALCULATED	f alfa	DECISION
A	14,1125	22,77751271	4,49	NOT ACCEPTED
B	12,2	17,02229251	4,49	NOT ACCEPTED
C	-3,225	1,189481867	4,49	ACCEPTED
D	-0,7125	0,058058809	4,49	ACCEPTED
AB	-0,8	0,073194485	4,49	ACCEPTED
AC	1,375	0,216223944	4,49	ACCEPTED
AD	-0,4125	0,019460155	4,49	ACCEPTED
BC	-6,2875	4,521206923	4,49	NOT ACCEPTED
BD	-0,55	0,034595831	4,49	ACCEPTED

COMBINATION OF TREATMENTS	EFECTS	f CALCULATED	f alfa	DECISION
CD	1,375	0,216223944	4,49	ACCEPTED
ABC	2,437500	0,679497145	4,49	ACCEPTED
ABD	-2,425000	0,672545814	4,49	ACCEPTED
ACD	0,950000	0,103215661	4,49	ACCEPTED
BCD	3,212500	1,180278947	4,49	ACCEPTED
ABCD	-4,5375	2,354678747	4,49	ACCEPTED

The factor A, the molar ratio alcohol/oil, has a significant effect on the reaction productivity, since the effect is positive by increasing the molar ratio increases productivity, for this reason the highest 9:1 ratio is selected.

The factor B, the catalyst concentration has a significant effect on the reaction productivity so increasing the catalyst concentration increases the reaction productivity for this reason we select 0.7% w/w. The factor C, the reaction temperature has no significant effect, but is recommended to work with the lowest level, i.e. at a temperature 50 °C to prevent a further loss of methanol due to its volatility. The factor D, the washing agent has no significant effect because the effect is negative, we select the low level (-) water at 40 °C.

The combination of treatments AB, AC, AD, CD, ABC, ABD, ACD, BCD, ABCD, do not has significant effects on the reaction productivity. The treatment combination BC has significant effects on the reaction productivity, so it is advisable to work with levels higher for each factor, catalyst concentration of 0.7 %p/p and reaction temperature of 50 °C.

Table 16. Properties of the biodiesel sample to the best conditions

Properties	Unit	Results	Standard (ASTM D-6751)
Density at 15.6 °C	kg/m^3	889.9	860- 900
Grades API		27	Minimum 32
Cinematic Viscosity at 40°C	Cst	5.21	1.9 – 6
Cetane index		48	Minimum 47
Caloric value	J/g	40,873.00	37,216.00
Temperature 90% distilled	°C	332	Maximum 360

Table 16, shows the made characterization of the sample obtained to a molar ratio alcohol/oil 9:1, catalyst concentration 0.7% w/w, reaction temperature of 50 °C and water at 40 °C as washing agent that is the biodiesel with a best properties.

The API gravity of the retrieved biodiesel ranges from 32 to 34 degrees. The examined sample exhibited a result below the stated ranges, indicating that the biodiesel derived from this sample possesses a high density of 889.9 kg/m^3, considering that API gravity is inversely related to density. The other analyzed attributes align with the values documented in the literature, thereby ensuring the quality of biodiesel.

5. CONCISE SUMMARY

It can be concluded that employing acetic acid or water as a cleaning agent does not influence reaction productivity, just as the reaction temperature does not impact the variable response within the values utilized in the study. The distinct variables influencing biodiesel generation are the catalyst concentration and the molar ratio of alcohol to oil.

According with the above, the best conditions of operation are:

- Molar ratio alcohol/aceite: 9:1
- Catalyst concentration of: 0.7% w/w
- Reaction temperature: 50 °C
- Washing agent: water at 40 °C

REFERENCES

Arbeláez, M., Rivera, M. (2007). *Diseño conceptual de un proceso para la obtención de biodiesel a partir de algunos aceites vegetales colombianos.* Universidad Eafit. Medellín.

Benjumea P, Benavides A Y Pashova V. (2007). *El biodiesel de aceite de higuerilla como combustible alternativo para motores diesel,* Universidad Nacional De Colombia Sede Medellín.

Bournay L, Casanave D, Chodorge J, Delfort D Y Hillion G. (2005), *new heterogeneous process for biodiesel production: a way to improve the quality and the value of the crude glycerin produced by biodiesel plants,* France.

Chen G, Meng X Y Wang Y. (2008). *Biodiesel production from waste cooking oil via alkali catalyst and its engine test,* China Architecture Design & Research Group, Beijing, China.

Cujia G., Bula A. (2010). *Potencial obtención de gas de síntesis para la producción de metanol a partir de la gasificación de residuos de palma africana.* Asociación Interciencia Venezuela.

Diaz-Zorita, M., Et Al., El cultivo De Girasol, In: *ASAGIR,* 2003, <Http://Www.Asagir.Org.Ar/Publicaciones/Cuadernillo_Web.Pdf>

Ente Regional De La Energía De Castilla Y León (EREN) (2003). *Plan piloto de recogida de aceites de cocina usados para la producción de biodiesel.* España.

Errazu A, Marchetti J Y Miguel V.(2005). *Possible methods for biodiesel production,* Planta Piloto De Ingeniería Química, Argentina.

FEDEPALMA(n.d.), Biodiesel, In: *FEDEPALMA.* 20 of june of 2011, Available from: <Http://Www.Fedepalma.Org/>

Girón E, Rojas A Y Torres Harlen. (2009).*variables de operación en el proceso de transesterificación de aceites vegetales: una revisión. In: Catálisis Química,* Vol 29 No 3.

Guan G, Kusakabe K Y Yamasaki S.(2009). *Tri-potassium phosphate as a solid catalyst for biodiesel production from waste cooking oil,* Department Of Living Environmental Science, Fukuoka Women's University,

INFOAGRO, El Cultivo Del Girasol (1ª Parte) 20 of june of 2011, In: INFOAGRO, Available from: <Http://Www.Infoagro.Com/Herbaceos/Oleaginosas/Girasol.Htm>

Iriarte, L. El Cultivo De Colza En Argentina, In: *ANTI,* 20 of june of 2011, Available from: <Http://Www.Inta.Gov.Ar/Ediciones/Idia/Oleaginosa/Colza01.Pdf>

Liu P, Ou S, Wang, Y. Zhang Z. (2006a). *Preparation of biodiesel from waste cooking oil via two- step catalyzed process,* Department Of Food Science And Engineering.

Liu P, Ou S, Tang S, Wang Y Y Xue F.(2006b) *comparison of two different processes to synthesize biodiesel by waste cooking oil,* Department Of Food Science And Engineering, Jinan University.

Peisch, S. (n.d.). Aguas Usadas En Puerto Rico: Amenaza A La Salud Publica Y Al Ambiente, In: *Datos Marinos #54.* 20 of june of 2011, Available from: <Http://Www. Seagrantpr.Org/Catalog/Files/Fact_Sheets/54-Aguas-Usadas-De- Pr.Pdf>

Rojas A Y Torres H. (2009). *Variables de operación en el proceso de transesterificación de aceites vegetales: una revisión* In: *Catálisis Química.*

Lles M, Pires E, Royo C. (2008). *Estudio de catalizadores heterogéneos en la síntesis de biodiesel a partir de aceite de girasol.*

3 Chapter

UTILIZATION OF ANIMAL FAT WASTE FOR BIODIESEL PRODUCTION

1. INTRODUCTION: AN OVERVIEW

Our society is significantly reliant on petroleum for its operations. Nonetheless, petroleum is a limited resource and contributes to other environmental issues, including elevated carbon dioxide concentrations in the atmosphere. Approximately 90% is utilized as an energy source for transportation, heating, and electricity generation, while the remaining sources serve as feedstocks in the chemical industry. (Carlsson, 2009). As demands for energy are increasing and fossil fuels are limited, research is directed towards alternative renewable fuels (Bhatti et al., 2008). High petroleum prices and the scarcity of known petroleum reserves demand the study of other sources of energy. In this context, agroindustrial wastes (animal fats, wood, manure) play an important role as energetic materials. Oils and fats are basically triacylglycerols (TAG) composed of three long-chain fatty acids. These triacylglycerols have higher viscosity and therefore cannot be used as fuel in common diesel engines. In order to reduce viscosity, triacylglycerols are converted into esters by transesterification reaction. By this means, three smaller molecules of ester and one molecule of glycerin are obtained from one molecule of fat or oil. Glycerin is removed as by-product and esters are known as biodiesel (Fazal et al., 2011).

Biodiesel fuels are attracting increasing attention worldwide as a blending component or a direct replacement for diesel fuel in vehicle engines. Biodiesel consists of a mixture of fatty acid (chain length C_{14}-C_{22}) alkyl esters, derived from a renewable lipid feedstock, such as vegetable oil or animal fat. In the case when methanol or ethanol are used as reactants, it will be a mixture of fatty acid methyl esters (FAME) or fatty acid ethyl esters (FAEE), respectively. However, methanol is commonly and widely used in biodiesel production due to its low cost and availability. Other alcohols such as isopropanol and butyl may also be used. A key quality factor for the primary alcohol is the water content, which interferes with the transesterification reactions and can result in poor yields and high level of soap, free fatty acids (FFA) and TAG in the final fuel (Demirbas, 2009a; Lam et al., 2010).

Biodiesel is a low-emission diesel substitute fuel made from renewable resources and waste lipid. The most common way to produce biodiesel is through transesterification, especially alkali-catalyzed transesterification (Leung et al, 2010). The most commonly used catalysts for converting TAG to biodiesel are sodium hydroxide, potassium hydroxide and sodium methoxide. The alkaline catalysts are highly hygroscopic and form chemical water when dissolved in the alcohol reactant. They also absorb water from the air during storage. Acid catalysts include sulfuric and phosphoric acids, being more related to directly esterification of FFA, although they are considered to be slow for industrial processing (Demirbas, 2009a). When the raw materials (oils or fats) have a high percentage of FFA or water, the alkali catalyst will react with the FFA to form soaps (Leung et al, 2010).

An alternative fuel to petrodiesel must be technically feasible, economically competitive, environmentally acceptable and easy available (Demirbas, 2009a). FAME from vegetable oils and animal fats have shown promise as biodiesel, due to improved viscosity, volatility and combustion behaviour relative to triacylglycerols, and can be used in conventional diesel engines without significant modifications (Bhatti et al., 2008). The advantages of biodiesel over diesel fuel are its portability, ready availability, renewability, higher combustion efficiency, lower sulphur and aromatic content, higher cetane number, higher biodegradability, better emission profile, safer handling, besides being non-toxic (Lapuerta et al., 2008; Demirbas, 2009a, Balat & Balat, 2010). Besides the superb lubricating property of biodiesel and its similarities in physicochemical properties to diesel, makes it an excellent fuel for compression ignition engines, revealing its potentials and practical usability for the replacement of petrodiesel in the nearest future (Atadashi et al., 2010). Moreover, biodiesel offers advantages regarding the engine wear, cost, and availability. When burned, biodiesel produces pollutants that are less detrimental to human health (Fazal et al., 2011).

Biodiesel has superior emission profile than diesel, substantially reducing emissions of unburned hydrocarbons, carbon monoxide, sulfates, polycyclic aromatic hydrocarbons, nitrated polycyclic aromatic hydrocarbons, and particulate matter (Lapuerta et al., 2008). Diesel blends containing up to 20% biodiesel can be used in nearly all diesel-powered equipment, and higher level blends and pure biodiesel can be used in many engines with little or no modification. Lower-level blends are compatible with most storage and distribution equipments, but special handling is required for higher-level blends (Demirbas, 2009a).

Usage of biodiesel will allow a balance to be sought between agriculture, economic development and environment (Demirbas, 2009a). Lower cost feedstocks are needed since biodiesel from food-grade oils is not economically competitive with petroleum-based diesel fuel. Main animal fat sources are beef tallow, lard, poultry fat and fish oils. Yellow greases can be mixtures of vegetable oils and animal fats. The FFA content affects the type of biodiesel process used and the yield of fuel from that process. Other contamination present can affect the extent of feedstock preparation necessary to use a given reaction chemistry (Demirbas, 2009a). Tallow is beef fat produced by slaughterhouse, while lard is hog fat and chicken fat refers to poultry. Brown grease comes from restaurant grease traps, sewage plants, and "black grease" (sludge). The brown one is gelatinous at room temperature and has low overall oil content. Yellow and brown grease as well as tallow can be converted into biodiesel, although the costs of processing are higher and the per-gallon biodiesel yield is lower. According to the USDA, the United States produces over 1.4 billion gallons of used cooking oil and animal fat each year. In fact, around 74% of the inedible tallow and grease produced goes to animal feed, while the remainder is used to make soaps, lubricants and other products such as Biodiesel (Tickell, 2006).

Soybean oil is the major feedstock for biodiesel in the USA and in other parts of the world. Rapeseed oil is the major source of oil in Europe and it contributes about 85% of the oil for world biodiesel production, followed by sunflower seed oil, soybean oil and palm oil. Some sources for vegetable oil extraction to be use in biodiesel production are: castor berry, palm pulp, palm kernel oil, babassu kernel, sunflower seeds, coconut kernel, cotton seed, peanut grain, canola seed (Leung et al., 2010). According to European Biodiesel Board (EBB, 2008), European production of biodiesel reached 5.7 million tons compared to US production of 1.7 billion liters in 2007. Germany is the largest producer of biodiesel among EU countries, accounting for about half of the total European biodiesel production. In the East Asian countries, palm oil is the major feedstock for biodiesel, being the annual average production expected to be about 31.4 million tons/year over the period 2006-2010 (Shrestha & Gerpen, 2010).

In 2010, about 2.4 billion liters of biodiesel were produced in Brazil, corresponding to 14% of the global participation. The country has a wide variety of feedstocks to be used in the production of oil and fatty acids. However, it is important to find new sources that don't compete with food chains. Therefore, it is necessary to invest in finding residual oils and other products (Pacheco, 2006). Research is being conducted on sustainable options for biodiesel generation utilizing enzymes, which facilitate mild reaction conditions and simplify glycerol recovery, thereby circumventing the disadvantages associated with chemical synthesis. (Rodrigues & Ayub, 2011).

2. MEAT PRODUCTION AROUND THE WORLD

In the last years, meat production has increased significantly. World meat production reached 237.7 million tons in 2010, from which 42.7%, 33.4%, 23.9% corresponds to respectively pork, poultry and beef (USDA, 2010). Consequently, a larger amount of residues from animal processing-plants has been generated in countries with intensive livestock production. Within agroindustrial residues, lipid sources may be used as feedstock to biodiesel supply, helping to solve inappropriate environmental disposal, besides contributing to energy demand.

Animal protein consumption in the world is a great well-being indicator of corporations (excluding those who decide for several reasons do not consume animal protein). As can be seen in Figure 1, consumption growth is directly related to the population income level and tends to rise as income rises, because in rich countries energy consumption is 3,470 kcal, while in poorest countries this value is 2,660 kcal (FAO, 2010). In Brazil, the studies done by Hoffmann (2000), Schlindwein (2006) and Pintos-Payeras (2009) also demonstrate income great importance in meat consumption. The percentage of meat in the diet is approximately twice the richer countries. In Brazil, it is noted that meat national consumption is already, in proportional terms, similar to consumption in rich countries, although in absolute terms the energy consumption of 3,060 kcal was below the same period.

These data show that income growth in peripheral countries will have a pronounced impact on meat consumption. Thus, per capita income growth in underdeveloped countries (China, India, Brazil and Russia) in the last three decades will be determinant of consumption growth and meat production. The possibility of per capita income growth in Africa will undoubtedly be a propeller of meat consumption in the near future. The surprising economic growth in China over the past 10 years and its impact on animal protein production has turned this country into a major propeller of meat dynamics in the world. In this period the gross national per capita income has grown at rates of 13.44%, and following the same line, meat consumption grew 2.3% annually (12.7 million tons in the period). In India, increase in per capita income of 9.23% over the same period was responsible for an increase in meat consumption in the same order of 6.68% per year (2.2 million tons). In Brazil, although less intense, we can see economic growth of 5.22% from 2000 to 2009. However, it must be taken into account that the best income distribution of the Brazilian economy lead to higher meat consumption by the poorest population (+3.64% per year meaning 4.8 million tons). A similar phenomenon was observed in Russia where meat consumption grew 5.07% per year between 2000 and 2009 (3 million tons).

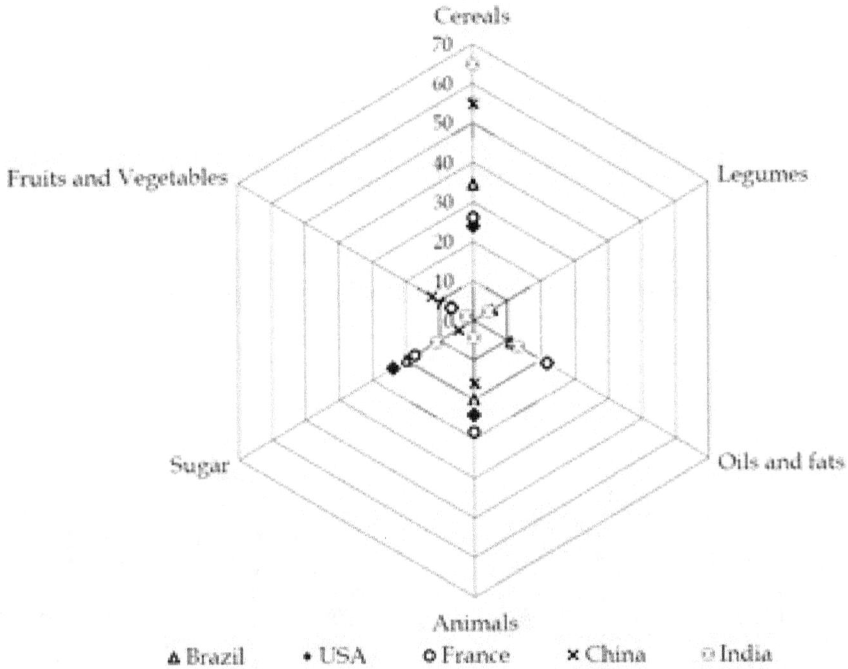

Fig. 1. Sources of per capita energy consumption from the diet as a percentage of the average period 2001-2003.

Source: FAO (2010).

The gradual effect shown in Figure 1 is a major driver of food economy, domestic and global levels, through increasing the middle class and the adoption by families who arrive there, the consumer behavior of those already there (Homem de Mello, 1990). As a result of this graduation, there would be huge demand in animal protein, legumes, fruits and vegetables. This dynamic evolution of food demand, as economic growth explains meat production growth in the world in the last two decades. Although it has been reported a large percentage increase in major meat consumption, the absolute volume consumed in India is still very low (3.8 kg) including beef, pork and poultry, when compared to 89.69, 49.9 and kg in Brazil, China and Russia, respectively. Thus, in this country is to be expected that the continuous increase in per capita income by more than a decade will boost meat consumption to a level closer to developing countries. Worldwide, source of animal protein (except milk) most produced and consumed is pork with 29.86% (Figure 2), followed by chicken meat (22.97%), eggs (18.05%) and beef (17.56%). These four groups of sources account for 88.44% of animal protein total consumption in the world.

In a second group of sources, the following four are responsible for more than 7.00% of animal protein consumption. This group comprises the consumption of sheep (2.39%), turkey (1.77%), eggs of other birds (1.42%) and goat (1.42%).

Rounding out the meat group: duck (1.09%), buffalo (0.97%), goose and guinea fowl (0.69%), rabbits (0.53%), hunt (0.49%), other meat (0.36%), horse (0.29%), camel (0.10%), other birds (0.03%) and ostrich (0.004%). At the same group, deserves attention the consumption of buffalo that occurs almost entirely in India, where beef consumption is forbidden by Hindu religion (83% of India population). India and Pakistan are also important producers of sheep and goats. Similarly horse consumption is concentrated in Asian countries. Rabbits are produced mostly in China, Venezuela and Italy. China also concentrates the production of geese, goats, ducks and sheep.

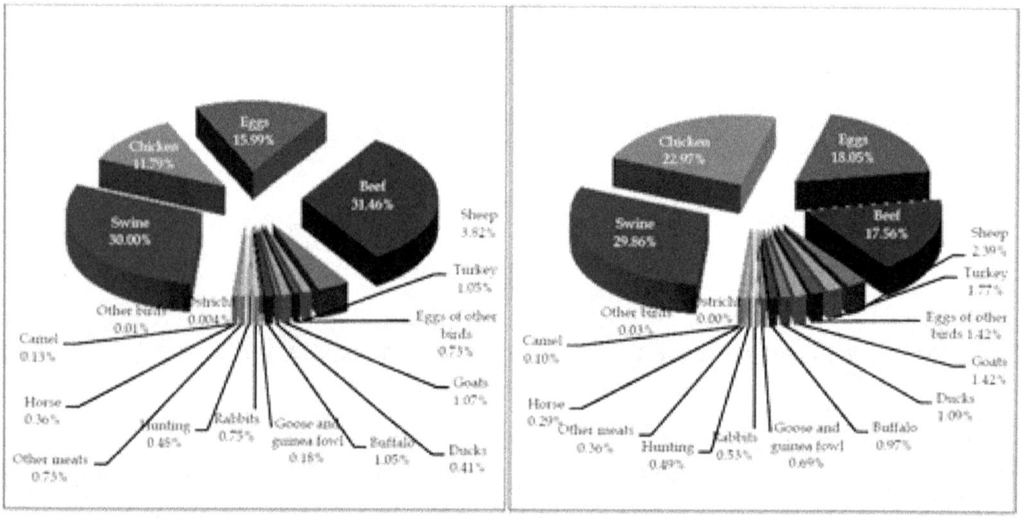

Fig. 2. Distribution of animal protein production in the world in 1975 (a) and 2008 (b) Source: FAO (2010) adapted by the authors.

Globally, a great dynamic in animal protein production can be noted. Even being less expressive regarding total production, the highest growth rates in animal protein production are focused on meat of birds. Among the most important sources, chicken may be highlighted showing an annually growth of 4.1% over the past 10 years. The negative highlight is related to beef which presented one of the lowest growth rates (1.1%). In an intermediate form, pork production increased in the order of 2.52%. Since 1975, year after year, poultry industry is consolidating itself as one of the most important animal protein sources for the population. According to data from the United States Department of Agriculture (USDA), world production of broilers grew consistently over the past 35 years, from 10.6 million tons in 1975 to 71 million tons by the end of the first decade of this century. Brazil has a different dynamic for meat production. Unlike the rest of the world, the main animal protein is chicken (41.31%), beef (36.49%), pork (12.19%), and eggs (7.38), which represent 97.37% of total produced in the country (Figure 3).

As a result of these factors one should expect a continued growth in production and consumption of meat, mainly chicken, followed by pork and beef (Figure 4).

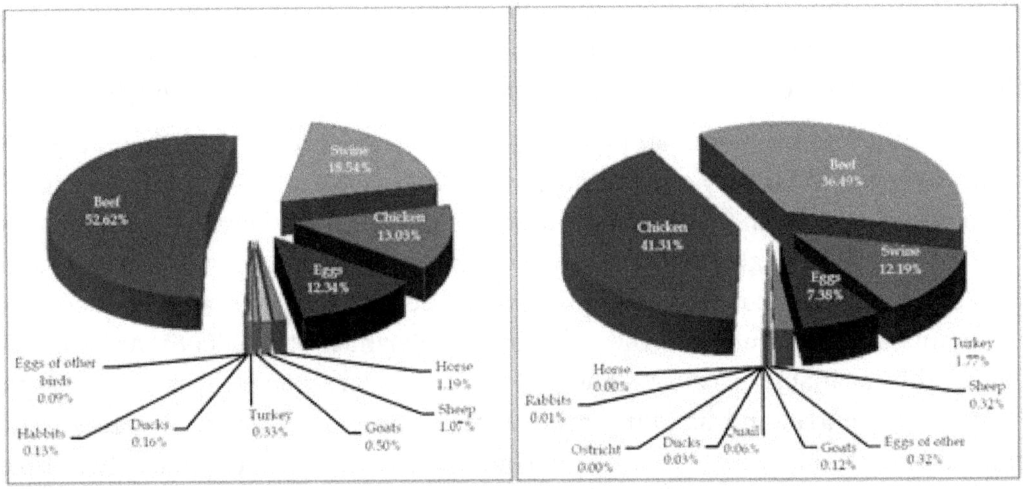

Fig. 3. Distribution of animal protein production in Brazil in 1975 (a) and 2008 (b) Source: FAO (2010) adapted by the authors.

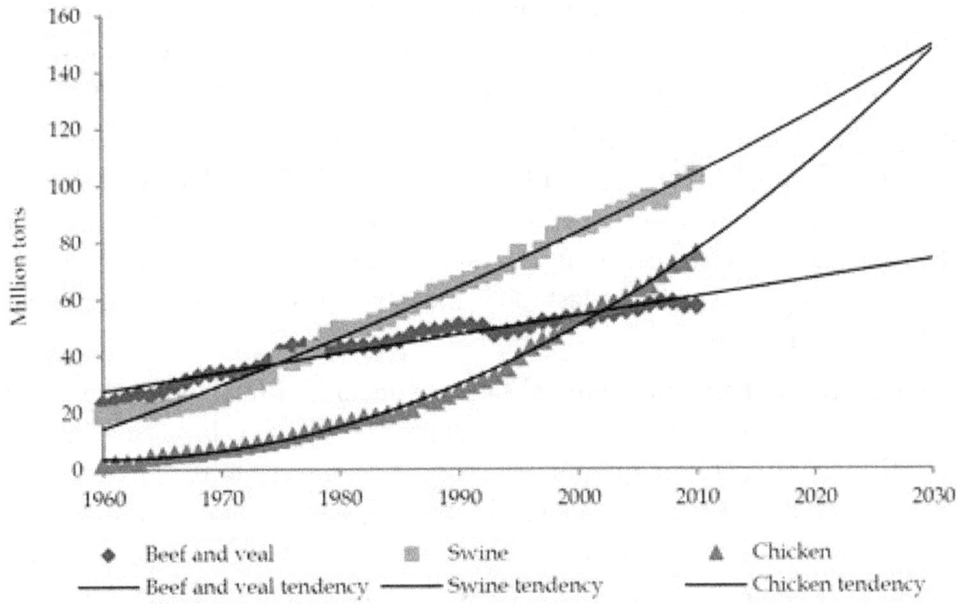

Fig. 4. Tendency of main animal protein production in the World Source: USDA (2010) adapted by the authors.

3. CHARACTERIZATION AND GENERATION OF ANIMAL FAT WASTES

Oils and fats are found in living organisms, consisting essentially of fatty acid esters and glycerin mixtures, and are known as triacylglycerols (commonly called triglycerides), which are hydrolyzed during extraction processes and storage, releasing fatty acids and glycerin.

Moreover, the use of oils or fats as fuel for internal combustion engines and their derivatives have been proposed for this intention over the past 100 years, when Rudolf Diesel applied in their assays crude petroleum and peanut oil. However, the problems of petroleum supply on the world market, generated by armed conflict that began in the 30s, led to the search for viable solutions for replacing fossil fuel.

The use of greases and animal fats eliminates the need to dispose them, besides contributing to the supply of biodiesel (Janaun & Ellis, 2010). Animal fats are highly viscous and mostly in solid form at ambient temperature because of their high content of saturated fatty acids. The high viscous fuels lead to poor atomization and result in incomplete combustion. The consequences are the increased emissions of pollutants and particulate in the exhaust gas (Kerihuel et al., 2006). Animal fats are readily available because slaughter industries are generally well managed for product control and handling procedures. However, there's a biosafety issue related to animal fats that could come from the contaminated animals. The future research to ensure biodiesel quality from animal waste (cradle to grave) has been highlighted (Janaun & Ellis, 2010). Biodiesel made from used cooking oil or from animal fat is less resistant to cold weather than biodiesel made from virgin soybean oil or most other virgin oils. As additives are developed specifically for the biodiesel industry, even this distinction could soon disappear (Tickell, 2006).

Black greases are defined loosely as greases resulting from sewage or other unconventional oil sources. It has a low conversion factor to biodiesel due to its high content in FFA. Brown greases are generally defined as a combination of greases and trappings from the slaughter industry. Yellow greases comprehend the oils and greases produced in the fast food industry and collected by the rendering industry (Tickell, 2006).

In Brazilian meat chain, most animal fats are generated in slaughterhouses and rendering plants. Products of rendering industry usually have lower market value. Materials that for aesthetic or sanitary reasons are not suitable for human food are are intended as feedstocks for rendering processes. Among these materials, there are fatty trimmings, bones, and offal, as well as entire carcasses of animals condemned at slaughterhouses, and those that have died on farms (deadstock) or in transit. The raw materials are collected in slaughterhouses, butcherhouses and supermarkets by trucks that take them to rendering plants. There are some industrial-scale slaughterhouses that process the residues within their own facilities. Once in the rendering plants the residues are chopped and heated in a steam-jacketed vessel to drive off the moisture and simultaneously release the fat from the fat cells using this so called "dry" method. The internal temperature of reactor reaches 1200 ºC under 5-6 kg/cm^2 of pressure during 2 h per batch.

In Brazil, does not exists an animal fat classification, only a general designation based on the animal from which the fat originates, such as chicken fat or fish oil, tallow and lard. The greases produced in Brazil are generally described as follows:

a. Tallow: extracted from residues of bovine slaughter and it can be filtered or not since it has guaranteed that the product contains minimum 90% total fatty acids, unsaponifiable impurities maximum 1.5% and no FFA or fat degradation products;
b. Lard: extracted from swine slaughter residues, being its specification and quality guarantees the same as for tallow;
c. Chicken fat: extracted from broiler slaughter residues and it can be filtered or not since it has guaranteed that the product contains minimum 90% total fatty acids, maximum 3% unsaponifiable impurities, without FFA or fat degradation products;
d. Animal fat mix: extracted from slaughter residues of mammals or birds. It can be filtered or not since it has guaranteed that the product contains total fatty acids minimum 90%, maximum 2% unsaponifiable impurities, without FFA or products of fat degradation unless the ones generated even with good production practices implemented.

The animal species from which the fat originates must be specified. Additions of antioxidants must be informed in any of these products. The main difference between animal fat and vegetable oil is their fatty acid composition. Vegetable oils have high content of unsaturated fatty acids, mainly oleic and linoleic acid, while animal fat composition has higher proportion of saturated fatty acids (Table 1).

Table 1. Average fatty acid composition of some vegetable oil and animal fat (Pearl, 2002; Rostagno et al., 2011)

Oil or Fat	12:0	14:0	16:0	16:1	18:0	18:1	18:2	18:3	20:4	≥ 20
Chicken	0.1	1-1.3	17-20.7	5.4	6-12	42.7	20.7	0.7-1.3	0.1	1.6
Lard	0.1	1-2	23.6-30	2.8	12-18	40-50	7-13	0-1	1.7	1.3
Tallow	0.1	3-6	23.3-32	4.4	19-25	37-43	2-3	0.6-0.9	0.2	1.8
Fish	0.2	6.1	14.3	10.0	3.0	15.1	1.4	0.7	0.7	56.5
Butter	-	7-10	24-26	-	10-13	1-2.5	2-5	-	-	-
Soybean	-	0.1	6-10.2	-	2-5	20-30	50-60	-	-	-
Rapeseed	0.2	0.1	3.9	0.2	1.7	60.0	18.8	9.5	-	4.0
Corn	-	1-2	8-12	0.1	2-5	19-49	34-62	0.7	-	2.0
Olive	-	-	9-10	-	2-3	73-84	10-12	Traces	-	-
Cotton	-	-	20-25	-	1-2	23-35	40-50	Traces	-	-

Traditionally in Brazil, cleaning and toilet products industries use part of animal fat residues to produce soaps and waxes while other parts are employed in the production of lubricants and leather preservatives. Nevertheless, in Brazil, the beginning of National Program of Biodiesel Production and Use (Law #11.097) has rapidly changed this scenario and between October/2008 and March/2009 biodiesel plants consumed 43% of total tallow, which corresponds to approximately 15% of whole biodiesel produced. Although Brazil is also a major producer of chicken and swine meat, fats from these species are still not being used for biodiesel production. According to UBABEF (2009) data, Brazil produced around 23 million tons of meat, from which 3, 9 and 11 million correspond to swine, cattle and poultry, respectively. Considering the amount of residues 45% (wt/wt) cattle and 25% (wt/wt) swine and poultry contain approximately 15% fat. Thus, feedstock potential amount is 607,500, 412,500 and 112,500 tons for cattle, poultry and swine, respectively.

Wastes from slaughterhouses are constituted by non-edible by-products and wastewater which pass through flocculation and flotation process. Non-edible animal by-products are sent for rendering plants where flours are processed and good-quality fats besides acid fats are originated. Good-quality fats are destined for drugs and cosmetics, while acidic fats (which don't attend industry acid requirements and flotation stage) have low or no commercial value, being their promising target energy or biodiesel production. Wastewater undergoes flocculation and flotation process with the aid of coagulants, being separated into solid (flotation stage) and liquids (liquid phase). The first one is destined to rendering plants, while the second one goes to treatment lagoons, as is shown in Figure 5. Animal fats are classified in three categories (low, medium and high-grade quality fat) according to the risk level, following Regulation (EC) 1774/2002 of the European Parliament and of the Council of 3 October 2002. High-grade quality fat has below 2% of FFA, which are mainly used for drugs and cosmetics, besides pet food. Medium-grade fat presents 3-5% FFA, while low-grade fat has above 5% FFA and are destined to biofuel production.

In Brazil, according to the National Petroleum Agency (ANP 2011), raw materials of animal origin used for biodiesel production account for 14.82% of authorized nominal capacity. This value is still low compared to the raw materials of vegetable origin that account for 84.45% of biodiesel production. However, volume of animal feedstocks tends to grow since Brazil has one of the largest animal herds in the world. Brazil is currently the second largest cattle producer (over 9.1 million tons), the fourth largest pork producer (more than 3.1 million tons), and the third largest chicken meat producer, with more than 11.4 million tons. In this context, lipid by-products from slaughterhouses should become attractive, especially for economical and environmental reasons.

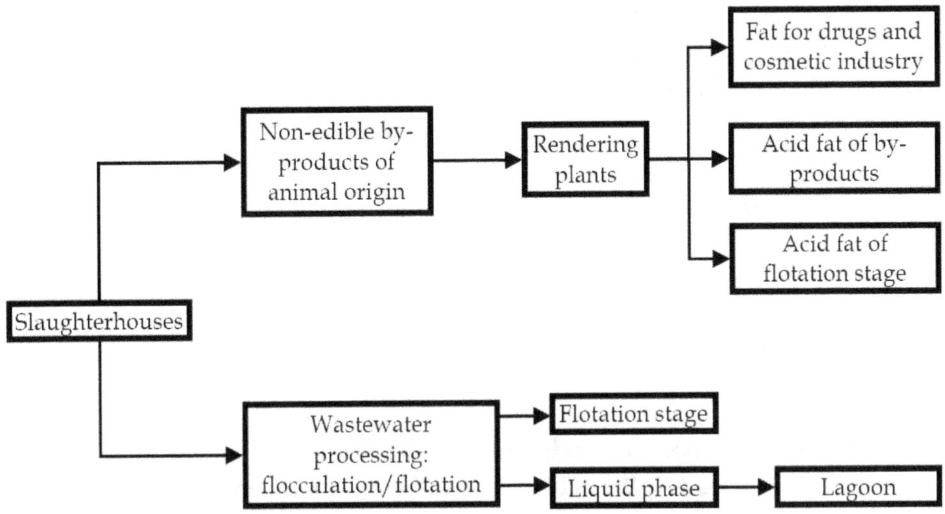

Fig. 5. Scheme of fat processing in slaughterhouses.

4.BIODIESELPRODUCTIONFROMANIMALFATWASTES:TECHNICAL CHALLENGES

The feedstocks issue is the critical point affecting the economic feasibility of biodiesel production, since accounts around 80% of the biofuel total cost. In this context, several efforts have been carried out in order to reduce biodiesel prices, essentially by altering lipid sources (Zhang et al., 2003a, 2003b; Canakci, 2007; Canakci & Sanli, 2008; Wang, 2009; Janaun & Ellis, 2010; Martins et al., 2011). Nowadays, edible vegetable oils are the major starting materials for biodiesel preparation. In consequence, prospection for novel feedstocks has been primarily attributed to investigations involving oleaginous species for inedible oil extraction (Nass et al., 2007). In recent, alternatively lipid residues as waste frying oil and inedible animal fats have also receiving considerable attention from biofuel sector. To take advantage of these low cost and low quality resources, a convenient action would be to reuse residues in order to integrate sustainable energy supply and waste management in food processing facilities.

To get a better understanding of challenges involved on biodiesel synthesis from animal fat wastes, a brief review regarding to fundamental reactions of carboxylic acids and esters is presented (Formo, 1954; Carey & Sundberg, 1983; Morrison & Boyd, 1996). As illustrated in figure 6a, carboxylic acids originate their salts (soaps) by treatment with aqueous alkaline solutions (hydroxides or carbonates). Additionally, carbonyl group confers an interesting synthetic versatility to carboxylic acids since they can be converted into derivatives by nucleophilic substitution. In fact, esters are directly obtained reacting carboxylic acids using alcohols as acyl-acceptors under acidic conditions, being this process usually referred as Fisher esterification (Figure 6b). The strategy frequently employed to shift equilibrium to the right includes the use of large amounts of alcohol and water removal from the reactional medium.

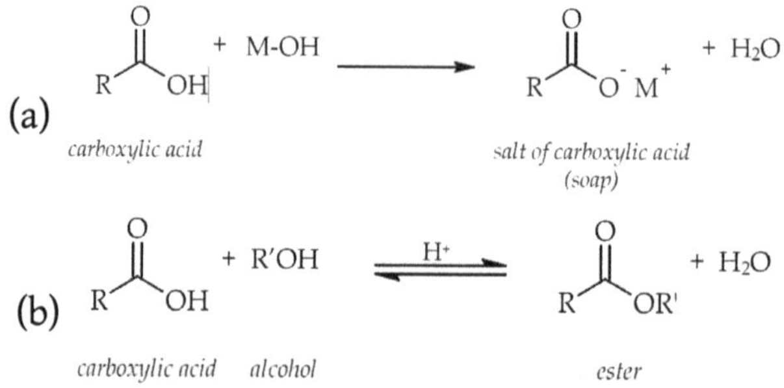

Fig. 6. Reactions of carboxylic acids: (a) acid-base neutralization (where M is Na^+ or K^+); (b) acid-catalyzed esterification

Esters are carboxylic acid derivatives that can be hydrolyzed either in acid or basic medium. The alkali-catalyzed process is essentially irreversible (Figure 7a). On the other hand, hydrolysis in acidic solution is an equilibrium reaction, being dependent on the relative alcohol and water concentrations (Figure 7b).

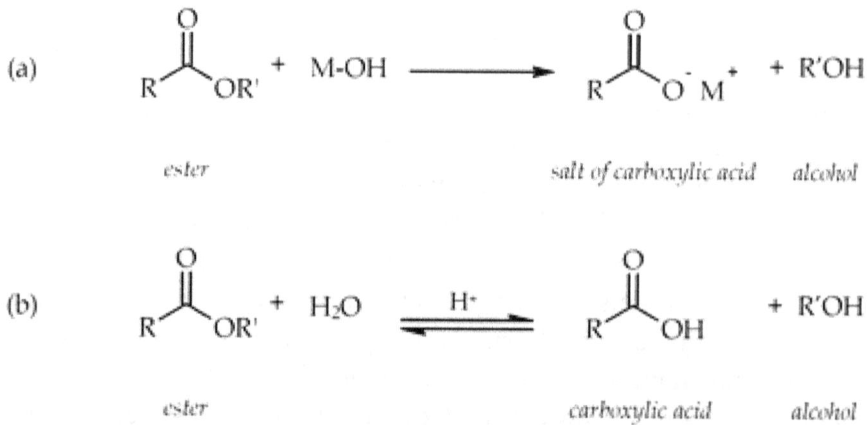

Fig. 7. Ester hydrolysis: (a) alkali-catalyzed (where M is Na^+ or K^+); (b) acid-catalyzed

According to Xu (2003), interesterification is a general term for the reactions between an ester and a fatty acid (acidolysis), an alcohol (alcoholysis), or another ester (transesterification). Esters are converted into another by alkoxy group exchanger as exemplified in figure 8. The displacement of —OR' molecular unit is carried out when the original ester react with an alcohol to provide a new carboxylic acid derivative. Alcoholysis is usually denominated by most authors as transesterification, general term that will be used from now on to describe biodiesel production reaction. The transesterification is an equilibrium process and addition of an excessive amount of alcohol can be used in favor of products' synthesis.

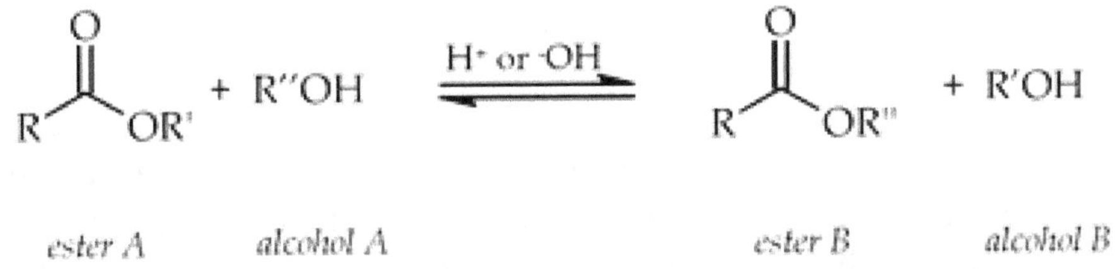

Fig. 8. Transesterification type alcoholysis

Oils and fats are complex lipids derived respectively from vegetable and animal sources. Their compositions are primarily based on triacylglycerols (TAG), which molecules consist of a glycerol backbone attached by ester bonds to three long-chain carboxylic acids (fatty acids). Reactions of ester linkages of oils and fats were recognized a long time by their technological importance (Formo, 1954). Nowadays, non-hydrolytic ester reactions (esterification and interesterification) play a fundamental role in the applied chemistry. For instance, biodiesel is a mixture of fatty acid mono-alkyl esters readily produced from TAG transesterification by using a short chain alcohol, as showed in figure 9.

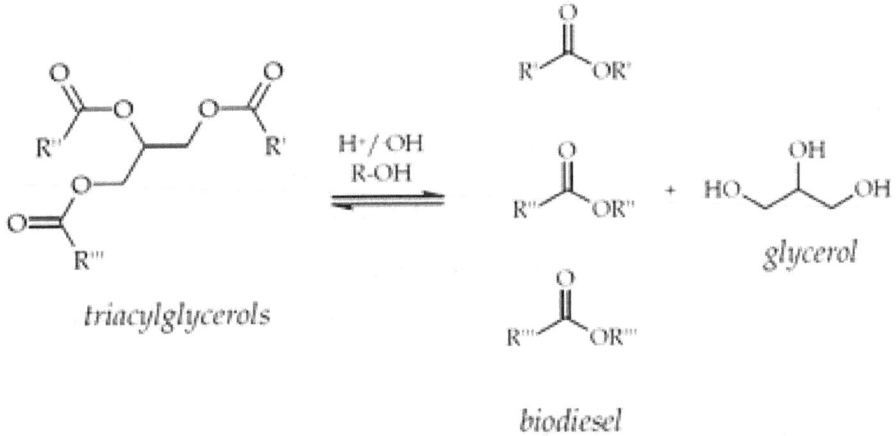

Fig. 9. Overall scheme of the TAG transesterification for biodiesel production (where R is $-CH_3$ or $-CH_3CH_2$)

Transesterification of TAG is a process of three consecutives and reversible acid- or basic- catalyzed reactions. Diacylglycerols (DAG) and monoacylglycerols (MAG) are intermediates. The stoichiometry of the overall reaction requires a molar ratio of 1:3 (TAG:alcohol) to give 3 mol of ester and 1 mol of glycerol. Its course involves stepwise conversions of TAG to DAG to MAG to glycerol (GL) (Figure 10).

Few studies were concerned with detailed kinetic aspects of the transesterification of vegetable oils (Freedman et al., 1986; Noureddini & Zhu, 1997; Darnoko & Cheryan, 2000; Komers et al., 2002). Freedman et al. (1986) investigated the kinetics of acid- and base- catalyzed transesterification of TAG with methanol and

1-buthanol at 6:1 and 30:1 molar ratio alcohol:oil. The authors proposed pseudo first-order kinetics at high molar ratio alcohol:oil and second-order kinetics combined with a shunt-reaction at low alcohol:oil ratio. According to Noureddini & Zhu (1997), alkali-catalyzed methanolysis of oils can be described as follow: a) initially reaction is characterized by a mass transfer controlled regime (slow) that results from low miscibility of reactants; b) ester produced at beginning can act as mutual solvent and favor a kinetic controlled regime (fast) characterized by a sudden surge in products formation; c) in the final, an equilibrium regime (slow) is approached. Figure 11 shows typical distribution of reactants, intermediates and products during the course of transesterification, where a sigmoid behavior for ester production is exemplified.

Reaction 1 $$\mathrm{TAG + alcohol} \underset{}{\overset{\text{catalyst}}{\rightleftharpoons}} \mathrm{DAG + RCOOR'}$$

Reaction 2 $$\mathrm{DAG + alcohol} \underset{}{\overset{\text{catalyst}}{\rightleftharpoons}} \mathrm{MAG + RCOOR'}$$

Reaction 3 $$\mathrm{MAG + alcohol} \underset{}{\overset{\text{catalyst}}{\rightleftharpoons}} \mathrm{GL + RCOOR'}$$

Overall reaction $$\mathrm{TAG + alcohol} \underset{}{\overset{\text{catalyst}}{\rightleftharpoons}} \mathrm{GL + 3\ RCOOR'}$$

Fig. 10. Chemistry of TAG transesterification

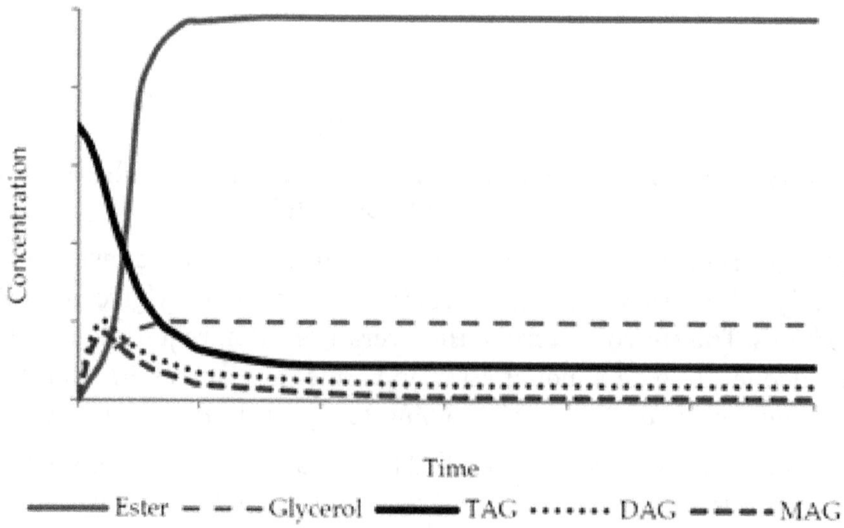

Fig. 11. Illustrative scheme of component concentration change during transesterification

Besides transesterification, reactions showed previously in this section can be involved during biodiesel preparation from lipid feedstocks depending on catalyst used. In fact, this reaction comprises a complex system. Komers and co-workers (2001b), in a fundamental research, were able to show that the reaction mixture of alkaline methanolysis of oils includes the following main components: TAG, DAG, MAG, methyl esters, methanol, soaps, KOH (in the form of OH-), CH_3OK (as CH_3O-), and water. Considering the system summarized in figure 12, the great issue is to establish appropriate conditions to minimize possible side reactions (hydrolysis and soaps formation) and, in consequence, drive the process toward ester production.

As is already well-know, transesterification may be influenced by several factors such as: feedstock composition; FFA content in raw materials, water concentration; alcohol to TAG molar ratio; catalyst type and concentration; type of alcohol; temperature; pressure; and mixing intensity. Researches have been intensively conducted to evaluate variables affecting ester yields and their respective interactions. Background about these parameters is detailed in several critical reviews (Schuchardt et al., 1998; Ma & Hanna, 1999; Fukuda et al., 2001; Van Gerpen & Knothe, 2005; Meher et al., 2006; Sharma et al., 2008; Vasudevan & Briggs, 2008; Demirbas, 2009b; Basha et al., 2009; Helwani et al., 2009; Vasudevan & Fu, 2010; Atadashi et al., 2010; Leung et al., 2010).

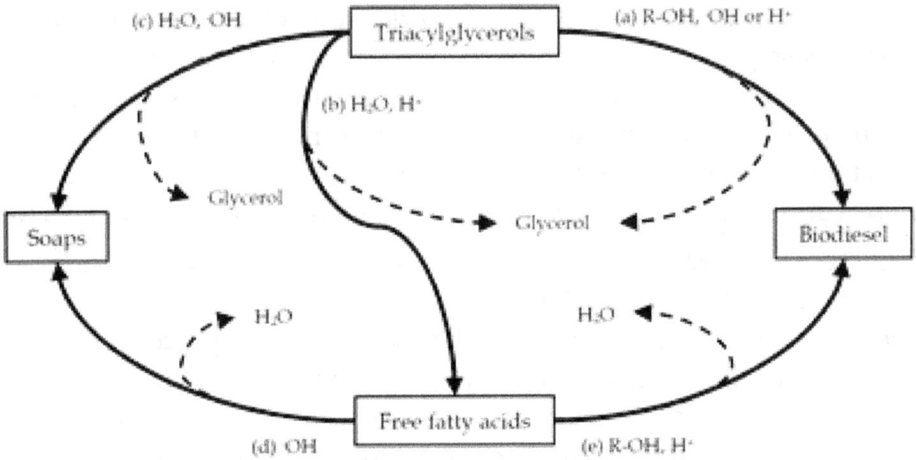

Fig. 12. Reactions involved in conventional biodiesel production: (a) alkali-catalyzed transesterification (expected route); (b) acid-catalyzed hydrolysis; (c) alkali-catalyzed hydrolysis; (d) acid-base neutralization; (e) acid-catalyzed esterification (expected route)

Homogeneous alkali-catalyzed transesterification is the most widely employed industrial process for biodiesel production (Helwani et al., 2009; Atadashi et al., 2010; Leung et al., 2010). This fact is because the base-catalyzed reation is faster than the acid one under mild conditions (Formo, 1954) resulting in a fuel-grade biodiesel. Alkaline catalysts are furthermore less corrosive than acidic compounds. Batch reactors are used for transesterification of refined vegetable oils with alcohol (molar ratio alcohol to oil 6:1) under anhydrous conditions. In summary, high esters conversion rates (>95%) are obtained in short times (after 1 h) at atmospheric pressure in temperatures ranged from 40 to 70°C.

Metal hydroxides (NaOH and KOH) and methoxides (NaOCH3 and KOCH3) are generally applied as catalysts in concentrations ranging from 0.5 to 2% wt/wt of oil (Vicent et al., 2004; Dias et al., 2008). The most common acyl-receptor is methanol owing to its low cost. However, ethanol can be successfully used as well (Feuge & Gros, 1949; Wu et al., 1999; Encinar et al., 2002; Ghassan et al., 2004; Ferrari et al., 2005; Bouaid et al. 2007). Ethylic route is particularly interesting in countries with consolidated sugarcane industry like Brazil (Nass et al., 2007), allowing biodiesel production entirely based on biomass resources. Afterwards reaction achievement, spontaneous separation of biodiesel and rich-glycerol phases occurs by gravitational settling. In some cases, a centrifugation step may be used to speed up the separation of phases. Then biodiesel is isolated and purified by removal alcohol excess, water washing, drying, and vacuum distillation.

Rendered animal fats are attractive raw materials for biodiesel industry once they are immediately available and found in huge amounts at relative low-prices in regions with intensive livestock. The mentioned lipid sources are generated in meat-processing plants with different quality degrees. Often, inedible residual fats don't present specific requirements for direct application in conventional biodiesel approach mediated by alkalis. According to system showed in figure 12, feedstocks and reactants necessarily should meet suitable quality with respect to FFA and moisture. For that reason, refined vegetable oils are favored instead of lipid wastes.

The main technical restrictions with processing animal fat wastes are their relative high FFA (ranging from 5% to 30%) and water content. These two factors are key parameters for determining viability of transesterification process, because they may cause catalyst effectiveness and promote soaps formation. In fact, alkaline catalysts are consumed by neutralization with FFA in the reactional medium, leading to soaps and water formation. As a result of catalyst deactivation, ester yield is significantly reduced. In addition, post-treatment of the final mixture is more difficult by the occurrence of soaps, which prevents phase separation between esters and glycerol, promoting stable emulsion establishment in washing operations. Kusdiana & Saka (2004) were able to demonstrate this effect on TAG methanolysis using 1.5% NaOH (wt/wt) as illustrated in figure 13a. Restrictive limits of FFA ranging from <1% to <3%, as recently reported (Atadashi et al., 2010). According to reports involving fat residues, starting materials for basic-catalyzed transesterification should not exceed values beyond 0.5% FFA, which corresponds to an acid number of 1 mg KOH/g of oil (Ma et al., 1998; Canakci & Van Gerpen, 2001). For vegetable oils, a FFA value lower than 3% (6 mg KOH/g of oil) is recommended for good conversion efficiency (Dorado et al., 2002; Tamasevic & Siler-Marinkovic, 2003; Phan & Phan, 2008). In both cases, transesterification rate can be enhanced with bases if FFA is around 5%, although further quantity of catalyst must be added to compensate higher acidity and loss due soap formation (Van Gerpen, 2005). Particularly, this procedure involving excessive amount of catalyst is not recommended since it gives rise to the formation of gels that interfere in the reaction, hinder glycerol separation, and contribute to emulsification during water washing.

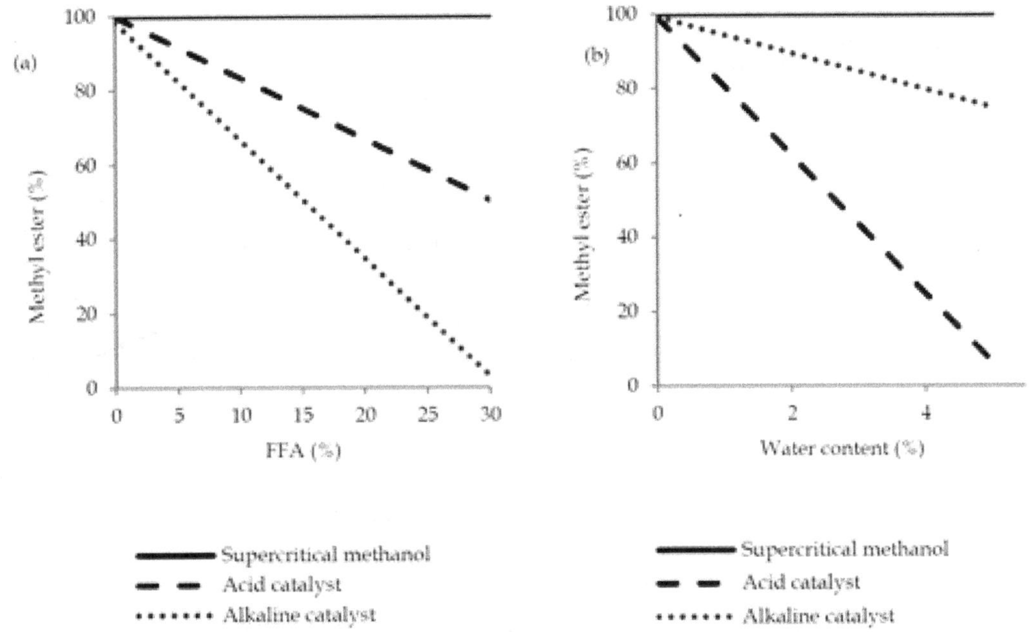

Fig. 13. Effects of FFA (a) and water (b) contents on the transesterification reaction of oils (adapted from Kusdiana & Saka, 2004).

It's well-established that TAG transesterification with basic catalysis is also sensitive to water content. Water is one of the main causes for side reactions besides alcoholysis. The effective catalyzing agents in the alkaline catalyzed transesterification are alkoxide ions (RO-). According to equilibrium study reported by Komers et al. (2001a), initial concentration of alkolate (RO–) decreases with an increasing amount of water in methanol and KOH. This effect can also occur by water presence in oils and fats. Then, transesterification doesn't occur without catalyst generation and besides hydrolysis may take place as competitive reaction follow-on to soaps production. For the alkaline-catalyzed methanolysis of oils, ester conversion was slightly reduced when water concentration increased in reaction system, as showed in figures 13b and 14a (Canakci & Van Gerpen, 1999; Kusdiana & Saka, 2004). The effects of FFA and water content on alkali-transesterification of beef tallow were investigated by Ma and co-workers (1998). A significant interaction between two factors was clearly observed, characterizing synergistic negative effect on the reaction, according to data showed in figure 15b. With respect to the single effect, the apparent yield of beef tallow methyl esters (BTME) was the highest without addition of FFA and water. The apparent yield decreased with the increase of the water amount without addition of FFA. A similar behavior was noted without water addition when FFA level increased. Water generally can be removed from raw materials by drying, gravitational settling or with desiccant agents before processing transesterification.

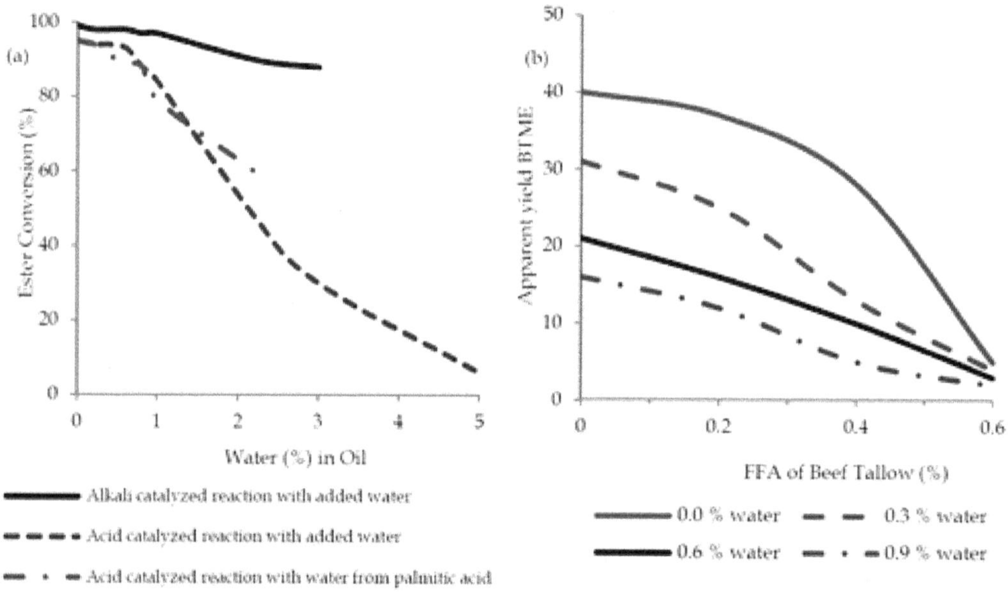

Fig. 14. (a) Water effect on transesterification of oils (adapted from Canakci & Van Gerpen, 2001); (b) FFA and water effects on alkali-catalyzed transesterification of beef tallow (adapted from Ma et al., 1998)

The FFA content turn waste lipids unsuitable for conventional biodiesel route. Transesterification via acid catalysis is an alternative process claimed as more tolerant to high FFA levels (Lotero et al., 2005). The homogeneous acid-catalyzed transesterification is slower than alkaline process. Generally, this reaction is performed at high molar rations of alcohol:oil (50:1) at 80°C, and high catalyst concentrations (3% by weight of lipid feedstock). Besides, strong mineral acids (HCl and H2SO4) are corrosives, causing damages to reactors. As can be see in figure 13b, water is the major obstacle to this reaction, being more critical than in base catalysis. According to Canakci and Van Gerpen (1999), in order to achieve good ester conversion, the acid catalyst also requires water content lower than 0.5%, which is around the same for alkaline reaction. Only 0.1% of water in reaction medium is enough to result in some reduction of the methyl ester yield (Kusdiana & Saka, 2004). In acid-catalyzed transesterification mechanism, the key-step is the protonation of the carbonyl oxygen. This increases the electrophilicity of the carbonyl carbon, making it more prone to nucleophilic attack. When present in reactional medium, water can form clusters around protons with less acid strength than alcohol-only proton complexes. Therefore, the catalytic species (H+) are deactivated by hydration, and don't allow TAG and their intermediates susceptible to alcohol attack (Helwani et al., 2009). On the other hand, acids are able to simultaneously catalyze both transesterification and esterification. Acid catalysts are effective at converting FFA to ester quickly. The integrated process is convenient to produce biodiesel from feedstocks having high FFA levels (Canakci & Van Gerpen, 2001; Zhang et al., 2003a, 2003b). The two-step approach includes an acid-catalyzed pre-treatment to esterification of FFA prior to alkali- catalyzed transesterification of TAG.

Nevertheless, as mentioned previously in this section, acid-catalyzed esterification is an equilibrium reaction, and hydrolysis occurs as inverse process. Water is produced in reactional medium when FFA react with alcohol to give esters. Canakci & Van Gerpen (1999), simulating FFA content in oil with palmitic acid, showed that water formed during acid-catalyzed esterification has similar negative effect on transesterification than when water was deliberatively added to reaction mixture. This fact is noted in figure 15a by coincident lines of acid-catalyzed transesterification (3% H2SO4, molar ratio 6:1, at 60°C) with water from reaction of palmitic acid and only with water addition. Then, water formed in the esterification limited FFA levels in the lipid source to 5%.

Table 2. Conditions of animal fats transesterification for biodiesel preparation

Feedstock	Catalyst (wt/wt of fat)	Alcohol	Molar ratio	T (°C)	Time (h)	Conv. (%)	Reference
Beef tallow	Step 1: NaOH 1% Step 2: NaOH 0.2%	MeOH	6:1 20%	70 60	0.5 1	-	Zheng & Hanna, 1996
Beef tallow	H2SO4 1%	MeOH	6:1	60	48	13.0	Alcantara et al., 2000
	NaOCH3 1%	MeOH	6:1	60	3	Quantitative	Alcantara et al., 2000
Beef tallow	KOH 2%	MeOH	–	65	1.5	>95	Moraes et al., 2008
Beef tallow	KOH 1.50%	MeOH	6:1	65	3	Quantitative	da Cunha et al., 2009
Beef tallow	Sulfonated polystyrene20 mol%	MeOH	100:1	64	18	70.0	Soldi et al., 2009

Feedstock	Catalyst (wt/wt of fat)	Alcohol	Molar ratio	T (°C)	Time (h)	Conv. (%)	Reference
Beef tallow: sunflower oil blends	NaOH 1%	MeOH	6:1	60	1	–	Taravus et al., 2009
Chicken tallow	H_2SO_4 25%	MeOH	30:1	50	24	Quantitative	Bhatti et al., 2008
	NaOH 1.5%	MeOH	–	30	1	88.1	
Feather meal fat	KOH 1%	MeOH	9:1	70	1/4	Quantitative	Kondamudi et al., 2009
Mutton tallow	H_2SO_4 25%	MeOH	30:1	60	24	93.2	Bhatti et al., 2008
	NaOH 1.5%	MeOH	–	30	1	78.3	
	KOH 1%	MeOH	6:1	65	3	79.7	
Duck tallow	NaOH 1%	MeOH	6:1	65	3	62.3	Chung et al., 2009
	$NaOCH_3$ 1%	MeOH	6:1	65	3	79.3	
Lard:soybean oil blends	NaOH 0.8%	MeOH	6:1	65	1	81.7-88.6	Dias et al., 2008
Lard:waste frying oil blends	NaOH 0.8%	MeOH	6:1	65	1	81.7-88.0	Dias et al., 2008
Lard	Step 1: H_2SO_4 2% Step 2: NaOH 1%	MeOH	6:1	65	Step 1: 5 Step 2: 1	66.2	Dias et al., 2009
Lard:soybean oil blend (25:75 wt/wt)	Step 1: H_2SO_4 2% Step 2: NaOH 1%	MeOH	6:1	65	Step 1: 5 Step 2: 1	64.4	Dias et al., 2009
Lard:soybean oil blend (25:75 wt/wt)	NaOH 1%	MeOH	6:1	65	1	77.8	Dias et al., 2009
Lard	Immobilized-lipase (*Candida* sp. 99-125)	MeOH	1:1(3x)	40	1/2	87.4	Lu et al., 2007
Lard	KOH 1.26%	MeOH	7.5:1	65	1/3	98.6	Jeong et al., 2009
Lard	KOH 0.9%	MeOH	6:1	60	1/3	89.2	Berrios et al., 2009
Leather	KOH 0.75%	MeOH	6:1	50	1/4	Quantitative	Ísler et al., 2010
Poultry fat	Mg-Al hydrocalcite 10%	MeOH	30:1	120	8	93.0	Liu et al., 2007
Tallow	NaOH 0.5%	MeOH	6:1	60	3	–	Öner & Altun, 2009
Waste animal fat	H_2SO_4 2.25 M	EtOH	–	50	2	78.0	Ghassan et al., 2004
Waste animal fat	Step 1: H_2SO_4 0.08% Step 2: NaOH 0.01%	MeOH	–	62	2	89.0	Gürü et al., 2009

Even with all the above mentioned details regarding to raw materials' properties, several researches have stated that animal fat wastes are really important sources for biodiesel production. In Table 2, reactional conditions for biodiesel preparation from different animal fats are summarized.

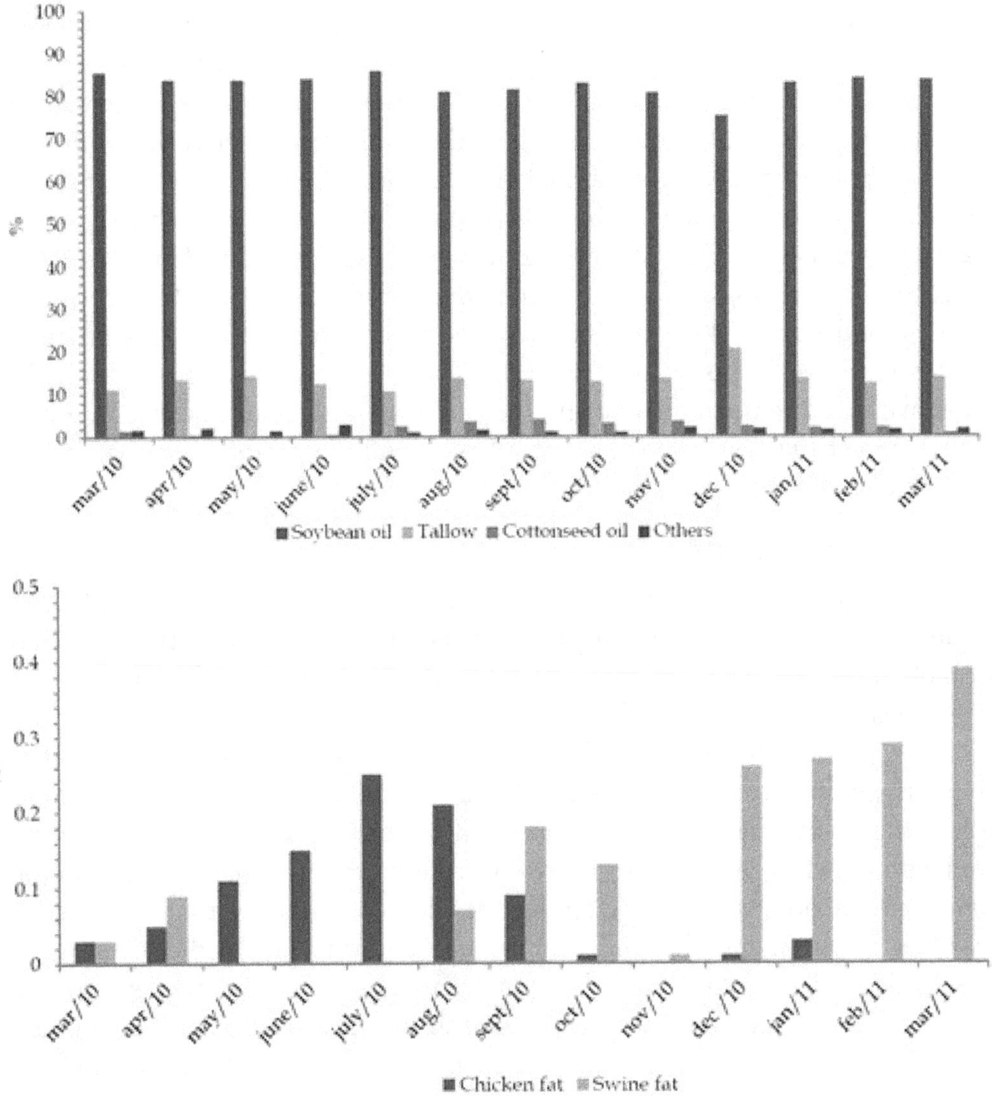

Fig. 15. Raw materials used for biodiesel production in Brazil Source: (ANP, 2011)

Figure 15 presents the raw materials employed for biodiesel production in Brazil from March 2010 until March 2011. As can be seen, soybean oil is the major feedstock. Additionally, beef tallow also plays an important role in this economic segment. The application of animal lipid sources in the Brazilian bioenergy sector is likely to increase because of accessibility to others profitable raw materials such as chicken and swine fat wastes. Recently, the simulation of investment in an industrial plant, made by Santos Filho et al. 2010, with processing capacity of 10,000 liters per day presented results that attest to the profitability of the enterprise.

The internal rate of return for an undertaking was 191%, the payback time was 1.51 years and the minimum price that enables the project was R$ 1,57, about US$ 1.00 (currency exchange August, 3rd, 2011), which is lower than the worst market since 2005. According to the authors, the results indicate that the use of acid fat from the slaughter of pigs and poultry for biodiesel production is technically and economically feasible, because there is high supply of raw material in different states of the country, facilitating logistics and providing a low cost transport of products. Increased demand for biofuels, especially for biodiesel, which year after year has been more requested for blending with diesel fuel, rising from 2% (2008) to 5% (2010) representing a consistent demand for production. Therefore, conversion of swine and poultry fats into biodiesel is advantageous for meat- processing industries that use this waste for burning and heat generation for boilers. Its use allows an increase in income and the chain will also be promoting an increase in the competitiveness of pork and poultry, turning a product with virtually no value into an income generator.

5. CONCISE SUMMARY

Globally, vegetable oils are favored as the primary lipid feedstocks for biodiesel production. Nevertheless, animal fats possess significant potential as feedstocks for the biofuel sector, as they are not commodities and have a diminished market value. Over the last years, meat production has increased significantly attaining 237.7 million tons in 2010, from which 42.7%, 33.4%, 23.9% corresponds respectively to pork, poultry and beef. Then a larger amount of residues from animal processing-plants has been generated in countries with intensive livestock production. Within agroindustrial residues, lipid sources may be used to solve inappropriate environmental disposal, besides contributing to energy supply. Brazilian government demands increasing addition of biodiesel into fossil diesel, taking place in 2010 a novel regulatory mark which raised the level up to 5%. Therefore, it has been encouraged the search for other renewable raw materials for application in the biofuel industry, such as non-edible oils and waste animal fats. Brazil is one of the main meat producers account to 9.1 beef, 3.2 pork, and 12.3 poultry million tons, dominating the world market together with the USA. In Brazil there is a broad range of residual lipid sources from slaughterhouse and rendering establishments ready available for application in biodiesel synthesis, including tallow, lard, poultry fat, mixed animal fat (mammal and poultry fat), and floating material from wastewater treatment plants. In a couple of years, researches focusing on fat residues should be accomplished mainly in order to improve feedstocks standardization process, because FFA and water content are decisive factors determining economic viability and biodiesel quality. Besides, researches in the field of prominent process such as heterogeneous catalysis (Di Serio et al., 2008; Cordeiro et al., 2011), enzyme-based process (Shimada et al., 2002), and supercritical fluids (Demirbas, 2006) should be carried out using animal fat wastes turning these raw materials more and more attractive.

REFERENCES

Alcantara, R., Amores, J., Canoira, L., Fidalgo, E., Franco, M. J., & Navarro, A. (2000). Catalytic production of biodiesel from soy-bean oil, used frying oil and tallow. *Biomass and Bioenergy*, Vol.18, No.6, (June 2000), pp. 515-527, ISSN 0961-9534

ANP, 2011. Boletim mensal do biodiesel. Agência Nacional do Petróleo, Gás Natural e Biocombustível. Superintendência de Refino e Processamento de Gás Natural – SRP. Ministério de Minas e Energia. April 2011. Available from: <http://www.udop.com.br/download/estatistica/anp______biodiesel/2011/0411_boletim_mensal.pdf>

Atadashi, I. M., Aroua, M. K., & Aziz, A. A. (2010). High quality biodiesel and its diesel engine application: A review. *Renewable and Sustainable Energy Reviews*, Vol.14, No.7, (September 2010), pp.1999-2008, ISSN 1364-0321

Balat, M. & Balat, H. (2010). Progress in biodiesel processing. *Applied Energy*, Vol.87, No.6, (June 2010), pp.1815-1835, ISSN 0306-2619

Basha, S. A.; Gopal, K. R., & Jebaraj, S. (2009). A review on biodiesel production, combustion, emissions and performance. *Renewable and Sustainable Energy Reviews*. Vol.13, No.6-7, (August /September 2009), pp. 1628-1634, ISSN 1364-0321

Berrios, M., Gutiérrez, M. C., Martín, M. A., & Martin, A. (2009). Application of the factorial design of experiments to biodiesel production from lard. *Fuel processing and Technology*, Vol.90, No.12, (December 2009), pp. 1447-1451, ISSN 0378-3820

Bhatti, H. N., Hanif, M. A., Qasim, M., & Rehman, A. U. (2008). Biodiesel production from waste tallow. *Fuel*, Vol.87, N.13-14, (October 2008), pp.2961–2966, ISSN 0016-2361

Bouaid, A., Martinez, M., & Aracil, J. (2007). A comparative study of the production of ethyl esters from vegetable oils as a biodiesel fuel optimization by factorial design. *Chemical Engineering Journal* , Vol.134, No.1-3 (November 2007) pp. 93–99, ISSN 1385-8947

Canakci, M. & Gerpen, J. V. (1999). Biodiesel production via acid catalysis. *Transactions of the ASABE*, Vol.42, No.5, pp. 1203-1210, ISSN 0001-2351

Canakci, M. & Gerpen, J. V. (2001). Biodiesel production from oils and fats with high free fatty acids. *Transactions of the ASAE,* Vol.44, No.6, (November/December 2001), pp. 1429-1436, ISSN 0001-2351

Canakci, M. (2007). The potential of restaurant waste lipids as biodiesel feedstocks. *Bioresource Technology*, Vol.98, No.1, (January 2007), pp. 183-190, ISSN 0960-8524 Canakci, M. & Sanli, H. (2008). Biodiesel production from various feedstocks and their effects on the fuel properties. *Journal of Industrial Microbiology and Biotechnology*, Vol.35, No.5, (May 2008), pp. 431-441, ISSN 1476-5535

Carey, F. A., Sundberg, R. J. (1983). *Advanced Organic Chemistry* (2nd edition), Plenum Press, ISBN 0-306-41088-5, New York

Carlsson, A. S. (2009). Plant oils as feedstock alternatives to petroleum – A short survey of potential oil crop platforms. *Biochimie*, Vol.91, No.6, (June 2009), pp. 665–670, ISSN 0300-9084

Chung, K.-H. & Kim, J. (2009). Biodiesel production by transesterification of duck tallow with methanol on alkali catalysts. *Biomass and Bioenergy,* Vol.33, No.1, (January 2009), pp. 155-158, ISSN 0961-9534

Cordeiro, C. S., da Silva, F. R., Wypych, F. & Ramos, L. P. (2011). Heterogeneous catalysts for biodiesel production. *Química Nova*, Vol.34, No.3, (March, 2011), pp. 477-486, ISSN 0100-4042

da Cunha, M. E., Krause L. C., Moraes, M. S. A., Faccini, C. S., Jacques, R. A., Almeida, S. R., Rodrigues, M. R. A., & Caramão, E. B. (2009). Beef tallow biodiesel produced in a pilot scale. *Fuel Processing Technology*, Vol.90, No.4, (April 2009), pp. 570-575, ISSN 0378-3820

Darnoko, D. & Cheryan, M. (2000). Kinetics of palm oil transesterification in a batch reactor. *Journal of the American Oil Chemists' Society,* Vol.77, No.12, (December 2000), pp. 1263-1267, ISSN 0003-021X

Demirbas, A. (2006). Biodiesel production via non-catalytic SCF method and biodiesel fuel characteristics. *Energy Conversion and Management*, Vol.47, No.15-16, (September 2006), pp. 2271-2282, ISSN 0196-8904

Demirbas, A. (2008). Comparison of transesterification methods for production of biodiesel from vegetable oils and fats. *Energy Conversion and Management*, Vol.49, N.1, (January 2008), pp. 125–130, ISSN 0196-8904

Demirbas, A. (2009a). Biofuels. In: Demirbas A, editor. Biofuels: Securing the planet's future energy needs. London (England): Springer; 336 p, e-ISBN 978-1-84882-011-1

Demirbas, A. (2009b). Progress and recent trends in biodiesel fuels. *Energy Conversion and Management*, Vol.50, No.1, (January 2009), pp. 14–34, ISSN 0196-8904

Dias, J. M., Alvim-Ferraz, M. C. M., & Almeida, M. F. (2008). Comparison of the performance of different homogeneous alkali catalysts during transesterification of waste and virgin oils and evaluation of biodiesel quality. *Fuel*, Vol.87, No.17-18, (December 2008), pp. 3572–3578, ISSN 0016-2361

Dias, J. M., Alvim-Ferraz, M. C., & Almeida, M. F. (2009). Production of biodiesel from acid waste lard. *Bioresource Technology*, Vol.100, No.24, (August 2009), pp. 6355-6361, ISSN 0960-8524

Di Serio, M., Tesser, R., Pengmei, L., & Santacesaria, Elio. (2008). Heterogeneous catalyst for biodiesel production. *Energy & Fuels*, Vol.22, No.1, (December 2007), pp. 207-217, ISSN 1520-5029

Dorado, M. P., Ballesteros, E., de Almeida, J. A., Schellert, C., Löhrlein, H. P., & Krause, R. (2002). An alkaline-catalyzed transesterification process for high free fatty acid waste oils. *Transactions of the ASABE*, Vol.45, No.3, pp. 525-529, ISSN 0001-2351

Encinar, J. M., González, J. F., Rodríguez, J. J., & Tejedor, A. (2002). Biodiesel fuels from vegetable oils: transesterification of *Cynara cardunculus* L. oils with ethanol. *Energy & Fuels*, Vol.16, No.2, (January 2002), pp. 443-450, ISSN 1520-5029

European Parliament (October 10 2002), 2011 Regulation (EC) 1774/2002 of the European Parliament and of the Council of 3 October 2002 (laying down health rules concerning animal by-products not intended for human consumption), Available from: <http://eur-lex.europa.eu/LexUriServ/LexUriServ do?uri=OJ:L:2002:273:0001:0095:EN:PDF>FAO, 2010. Faostat. March 2010. Available from <http://www.fao.org/corp/statistics/en/>

Fazal, M.A., Haseeb, A. S. M. A., & Masjuki, H. H. (2011). Biodiesel feasibility study: An evaluation of material compatibility; performance; emission and engine durability. *Renewable & Sustainable Energy Reviews,* Vol.15, No.2, (February 2011), pp. 1314- 1324, ISSN 1364-0321

Ferrari, R. A., Oliveira, V. S., & Scabio, A. (2005). Biodiesel de soja — Taxa de conversão em ésteres etílicos, caracterização físico-química e consumo em gerador de energia. *Química Nova*, Vol.28, No.1, (January/February 2005), pp 19-23, ISSN 1678- 7064

Feuge, R. O. & Gros, A. T. (1949). *Journal of the American Oil Chemists' Society*, Vol.26, No.3, (March 1949), pp. 97-102, ISSN 0003-021X

Formo, M. W. (1954). Ester reactions of fatty materials. *Journal of the American Oil Chemists' Society*, Vol.31, No.11, (November 1954), pp. 548-559, ISSN 0003-021X

Freedman, B., Pryde, E. H., & Mounts, T. L. (1984). Variables affecting the yields of fatty esters from transesterified vegetable oils. *Journal of the American Oil Chemists' Society*, Vol.61, No.10, (October 1984), pp. 1638-1643, ISSN 0003-021X

Freedman, B., Butterfield, R. O., & Pryde, E. H. (1986). Transesterification Kinetics of Soybean Oil. *Journal of the American Oil Chemists' Society*, Vol.63, No.10, (October 1986), pp. 1375-1380, ISSN 0003-021X

Fukuda, H., Kondo, A., & Noda, H. (2001). Biodiesel fuel production by transesterification of oils. *Journal of Bioscience and Bioengineering*, Vol.92, No.5, pp. 405-416, ISSN 1389- 1723

Gürü, M., Artukoglu, B. D., Keskin, A., & Koca, A. (2009). Biodiesel production from waste animal fat and improvement of its characteristics by synthesized nickel and magnesium additive. *Energy Conversion and Management* Vol.50, No.3, (March 2009), pp. 498-502, ISSN 0196-8904

Helwani, Z., Othman, M. R., Aziz, N., Fernando, W. J. N., & Kim, J. (2009). Technologies for production of biodiesel focusing on green catalytic techniques: A review. Fuel Processing Technology, Vol.90, No.12, (December 2009), pp. 1502-1514, ISSN 0378- 3820

Hoffmann, R. (2000). Elasticidade-renda das despesas com alimentos em regiões metropolitanas do Brasil em 1995-96. *Informações Econômicas*, Vol.30, No.2, (Fevereiro 2000), pp. 17-24, ISSN 1678-832X

Homem de Mello, F. (1990). O crescimento agrícola brasileiro dos anos 80 e as perspectivas para os anos 90. *Revista de Economia Política*, Vol. 10, No 3, (Julho-Setembro 1990), pp. 22-30, ISSN 0101-3157

Isler, A., Sundu, S., Tuter, M., Karaosmanoglu, F. (2010). Transesterification reaction of the fat originated from solid waste of the leather industry. Waste Management, Vol.30, No.12, (December 2010), pp. 2631-2635, ISSN 0956-053X

Janaun, J. & Ellis, N. (2010). Perspectives on biodiesel as a sustainable fuel. *Renewable and Sustainable Energy Reviews* Vol.14, No.4, (May 2010), pp.1312–1320, ISSN 1364-0321

Jeong, G.-T., Yang, H.-S., & Park, D.-H. (2009). Optimization of transesterification of animal fat ester using response surface methodology. *Bioresource Technology*, Vol.100, No.1, (January 2009), pp. 25-30, ISSN 0960-8524

Kerihuel, A, Kumar, M. S., Bellettre, J., & Tazerout, M. (2006). Ethanol animal fat emulsions as a diesel engine fuel – Part 1: Formulations and influential Parameters. *Fuel*, Vol.85, No.17-18, (December 2006), pp. 2371-2684, ISSN 0016-2361

Komers, K., Machek, J., & Stloukal, R. (2001a). Biodiesel from rapeseed oil, methanol and KOH 2. Composition of solution of KOH in methanol as reaction partner of oil.

European Journal of Lipid Science and Technology, Vol.103, No.6, (June 2001), pp. 359- 362, ISSN 1438-9312

Komers, K., Stloukal, R., Machek, J., & Skopal, F. (2001b). Biodiesel from rapeseed oil, methanol and KOH 3. Analysis of composition of actual reaction mixture. *European Journal of Lipid Science and Technology*, Vol.103, No.6, (June 2001), pp. 363-371, ISSN 1438-9312

Komers, K., Skopal, F., Stloukal, R., & Machek, J. (2002). Kinetics and mechanism of the KOH – catalyzed methanolysis of rapeseed oil for biodiesel production. *European Journal of Lipid Science and Technology*, Vol.104, No.11, (November 2002), pp. 728– 737, ISSN 1438-9312

Kondamudi, N., Strull, J., Misra, M., & Mohapatra, S. K. (2009). A green process for producing biodiesel from feather meal. *Journal of Agricultural and Food Chemistry*, Vol.57, No.14, (June 2009), pp. 6163-6166, ISSN 0021-8561

Kusdiana, D. & Saka, S. (2004). Effects of water on biodiesel fuel production by supercritical methanol treatment. *Bioresource Technology*, Vol.91, No.3, (February 2004), pp. 289- 295, ISSN 0960-8524

Lam, M. K., Lee, K. T., & Mohamed, A. R. (2010). Homogeneous, heterogeneous and enzymatic catalysis for transesterification of high free fatty acid oil (waste cooking oil) to biodiesel: A review. *Biotechnology Advances* Vol.28, No.4, (July-August 2010), pp.500–518, ISSN 0734-9750

Lapuerta, M., Armas, O., & Rodríguez-Fernández, J. (2008). Effect of biodiesel fuels on diesel engine emissions. *Progress in Energy and Combustion Science*, Vol.34, No.2, (April 2008), pp. 198-223, ISSN 0360-1285

Leung, D. Y C., Wu, X., & Leung, M. K. H. (2010). A review on biodiesel production using catalyzed transesterification. *Applied Energy,* Vol.87, N.4, (April 2010), pp. 1083– 1095, ISSN 0306-2619

Liu, Y., Lotero, E., Goodwin Jr, J. G., & Mo, X. (2007). Transesterification of poultry fat with methanol using Mg-Al hydrotalcite derived catalysts. *Applied Catalysis A: General,* Vol.331, pp. 138-148, ISSN 0926-860X

Lotero, E., Liu, Y., Lopez, D. E., Suwannakarn, K., Bruce, D. A., & Goodwin Jr., J. G. (2005). Synthesis of biodiesel via acid catalysis. *Industrial & Engineering Chemistry Research*, Vol.44, No.14, (January 2005), pp.5353-5363, ISSN 1520-5045

Lu, J., Nie, K., Xie, F., Wang, F., & Tan, T. (2007). Enzymatic synthesis of fatty acid methyl esters from lard with immobilized Candida sp. 99-125. *Process Biochemistry*, Vol.42, No.9, (September 2007), pp. 1367-1370, ISSN 0032-9592

Ma, F., Clements, L. D., & Hanna, M. A. (1998). The effects of catalyst, free fatty acids, and water on transesterification of beef tallow, *Transactions of the ASAE,* Vol.41, No.5, pp. 1261-1264, ISSN 0001-2351

Ma, F. & Hanna, M. A. (1999). Biodiesel production: a review. *Bioresource Technology*, Vol.70, No.1, (October 1999), pp. 1-15, ISSN 0960-8524

Martins, R., Nachiluk, K., & Bueno, C. R. F. & de Freitas, S. F. (2011). O biodiesel de sebo bovino no Brasil. *Informações Econômicas*, Vol.41, No.5, (Maio 2011), pp. 56-70, ISSN 1678-832X

Meher, L. C., Sagar, D. V., & Naik, S. N. Technical aspects of bidiesel production by transesterification – a review. (2006). *Renewable & Sustainable Energy Reviews,* Vol.10, No.3, (June 2006), pp. 248-268, ISSN 1364-0321

Monteiro, M. R., Ambrozin, A. R. P., Lião, L. M., & Ferreira, A. G. (2008). Critical reviews on analytical methods for biodiesel characterization. *Talanta,* Vol.77, No.2, (December 2008), pp. 593-605, ISSN 0039-9140

Moraes, M. S. A., Krause, L. C., da Cunha, M. E., Faccini, C. S., de Menezes, E. W., Veses, R. C., Rodrigues, M. R. A., & Caramão, E. B. (2008). Tallow Biodiesel: Properties Evaluation and Consumption Tests in a Diesel Engine. *Energy & Fuels*, Vol.22, No.3, (March 2008), pp. 1949-1954, ISSN 1520-5029

Morrison, R. & Boyd, R. (1996). *Química Orgânica* (13th edition), Fundação Calouste Gulbenkian, ISBN 972-31-0513-6, Lisboa

Nass, L. L., Pereira, P. A. A., & Ellis, D. (2007.) Biofuels in Brazil: An Overview. *Crop Science*, Vol.47, No.6 (November/December 2007), pp. 2228-2237, ISSN 1435-0653

Noureddini, H. & Zhu, D. (1997). Kinetics of transesterification of soybean oil. *Journal of the American Oil Chemists' Society*, Vol.74, No.11, (December 1997), pp. 1457-1463, ISSN 0003-021X

Öner, C. & Altun, S. (2009). Biodiesel production from inedible animal tallow and an experimental investigation of its use as alternative fuel in a direct injection diesel engine. *Applied Energy,* Vol. 86, No. 10, (October 2009), pp. 2114-2120, ISSN 0306-2619

Pacheco, J. W. (2006). *Guia técnico ambiental de graxarias - Série P+L*, CETESB, São Paulo.

Phan, A. N. & Phan, T. M. (2008). Biodiesel production from waste cooking oils. *Fuel,* Vol.87, No.17-18 (December 2008), pp. 3490-3496, ISSN 0016-2361

Pinto-Payeras, J. A. Estimação do sistema quase ideal de demanda para uma cesta ampliada de produtos empregando dados da pof de 2002-2003. Economia Aplicada, v. 13, n. 2, 2009, pp. 231-255

Rodrigues, R. C. & Ayub, M. A. Z. (2011). Effects of the combined use of *Thermomyces lanuginosus* and *Rhizomucor miehei* lipases for the transesterification and hydrolysis of soybean oil. *Process Biochemistry*, Vol.46, No.3, (March 2011), pp. 682-688, ISSN 1359-5113

Rostagno, H. S., Albino, L. F. T., Donzele, J. L., Gomes, P. C., Oliveira, R. F., Lopes, D.C., Ferreira, A. S., Barreto, S. L. T., & Euclides, R. F. (2011). *Brazilian Tables for poultry and swine: Food Composition and Nutritional Requirements.* Viçosa Federal University (UFV), CDD 22. ed. 636.085, 3rd ed, 252 p., Viçosa, MG, Brazil

Santos Filho, J. I. (2010). Viabilidade econômica da utilização de gordura ácida de frangos e suínos na produção de biodiesel, *Procedings of 4° Congresso da Rede Brasileira de Tecnologia de Biodiesel*, Belo Horizont (MG), October 2010

Schlindwein, M. M. Influência do custo de oportunidade do tempo da mulher sobre o padrão de consumo alimentar das famílias brasileiras. 2006. 118 p. Tese (Doutorado em Economia Aplicada) – Escola Superior de Agricultura "Luiz de Queiroz", Universidade de São Paulo, Piracicaba

Schuchardt, U., Sercheli, R., & Vargas, R. M. (1998). Transesterification of vegetable oils: a Review. *Journal of the Brazilian Chemical Society*, Vol.9, No.3, (May/June 1998), pp. 199-210, ISSN 0103-5053

Sharma, Y.C., Singh, B., & Upadhyay. (2008). Advancements in development and characterization of biodiesel: A review. *Fuel*, Vol.87, No.12 (September 2008), pp. 2355-2373, ISSN 0016-2361

Shimada, Y., Watanabe, Y., Sugihara, A., & Tominaga, Y. (2002). Enzymatic alcoholysis for biodiesel fuel production and application of the reaction to oil processing. *Journal of Molecular Catalysis B: Enzymatic*, Vol.17, No.3-5, (June 2002), PP. 133-142, ISSN 1381- 1177

Shrestha, D. & Gerpen, J. V. (2010). Biodiesel from Oilseed Crops, In: *Industrial Crops and Uses*, B.P. Singh, (Ed.), 140-156, ISBN-13 9781845936167, CABI, Wallingford, U.K. 512 p.

Soldi, R. A., Oliveira, A. R. S., Ramos, L. P., & César-Oliveira, M. A. F. (2009). Soybean oil and beef tallow alcoholysis by acid heterogeneous catalysis. *Applied Catalysis A: General*, Vol.361, No.1-2 (June 2009), pp. 42–48, ISSN 0926-860X

Taravus, S., Temur, H. & Yartasi, A. (2009). Alkali-Catalyzed Biodiesel production from Mixtures of Sunflower Oil and Beef Tallow, *Energy & Fuels*, Vol.23, No.8, (July 2009), pp. 4112–4115, ISSN 1520-5029

Tashtoush, G. M., Al-Widyan, M. I., Al-Jarrah, M. M. (2004). Experimental study on evaluation and optimization of conversion of waste animal fat into biodiesel. *Energy Conversion and Management* Vol.45, No.17, (October 2004), pp. 2697-2711, ISSN 0196-8904

Teixeira, R., Almeida, E., Oliveira Jr., L., & Fernandes, H. (2010). Sep 28. Produtividade e logística na produção de biodiesel. Ensaios FEE, Porto Alegre, Vol.31, No.1, (August 2010), pp. 7-30. Available at: http://revistas.fee.tche.br/index.php/ensaios/article/view/2213/2731

Tickell, J. (2006). *Biodiesel America: How to achieve energy security, Free America from middle-east oil dependence and make money growing fuel.* ISBN 9780970722744, Yorkshire Press, USA

Tomasevic, A. V. & Siler-Marinkovic, S. S. (2003). Methanolysis of used frying oil. *Fuel Processing Technology*, Vol.81, No.1, (April 2003), pp. 1-6, ISSN 0378-3820

UBABEF. Brazilian Poultry Union. Meat production in Brazil. May 27 2011, Available from: <http://www.brazilianchicken.com.br/english/the-poultry-industry/the-world- of-chicken.php>

USDA. United States Department of Agriculture. Livestock and poultry: World markets and trade. October 2010. May 23 2011, Available from:

<http://www.fas.usda.gov/psdonline/circulars/livestock_poultry.pdf> USDA. Foreign Agricultural Service, March 18 2010, Available from: <http://www.fas.usda.gov/psdonline/>van Gerpen, J. & Knothe, G. (2005). Basics of the transesterification reaction. In: Knothe G, Van Gerpen J. H. and Krahl, J., editors. The Biodiesel Handbook. Illinois: AOCS Press; 302 p, ISBN 0-000000-00-00

Van Gerpen, J. (2005). Biodiesel processing and production. *Fuel Processing Technology*, Vol. 86, No.10, (June 2005), pp. 1097-1107, ISSN 0378-3820

Vasudevan, P. T. & Briggs, M. (2008). Biodiesel production—current state of the art and challenges. *Journal of Industrial Microbiology and Biotechnology*, Vol.1, No.1, (March 2010), pp. 47–63, ISSN 1367-5435

Vasudevan, P. T. & Fu, B. (2010). Environmentally Sustainable Biofuels: Advances in Biodiesel Research. *Waste Biomass Valorization*, Vol.35, No.5, (May 2008), pp. 421- 430, ISSN 1877-2641

Vicente, G., Martínez, M., & Aracil, J. (2004). Integrated biodiesel production: a comparison of different homogeneous catalysts systems. *Biresource Technology*, Vol.92, No.3, (May 2004), pp. 297-305, ISSN 0960-8524

Wang, L. (2009). *Biodiesel production from waste oils and fats,* CRC Press, ISBN 978-1-4200-6339- 4

Wu, W. H., Foglia, T. A., Marmer, W. N., & Phillips, J. G. (1999). Optimizing production of ethyl esters of grease using 95% ethanol by response surface methodology. *Journal of the American Oil Chemists' Society*, Vol.76, No.4, (April 1999), pp. 517-521, ISSN 0003-021X

Xu, X. (2003). Engineering of enzymatic reactions and reactors for lipid modification and synthesis. *European Journal of Lipid Science and Technology*, Vol.105, No.6, (June 2003), pp. 289–304, ISSN 1438-7697

Zhang, Y. Dubé, M. A., McLean, D. D., & Kates, M. (2003a). Biodiesel production from waste cooking oil: 1. Process design and technological assessment. *Bioresource Technology*, Vol.89, No.1, (August 2003), pp. 1-16, ISSN 0960-8524

Zhang, Y. Dubé, M. A., McLean, D. D., & Kates, M. (2003b). Biodiesel production from waste cooking oil: 2. Economic assessment and sensitivity analysis. *Bioresource Technology*, Vol.90, No.3, (December 2003), pp. 229-240, ISSN 0960-8524

Zheng, D. & Hanna, M. A. (1996). Preparation and properties of methyl esters of beef tallow. *Bioresource Technology*, Vol.57, No.2, (August 1996), pp. 137-142, ISSN 0960-8524

4
Chapter

EXTRACTION OF LIPIDS FOR BIODIESEL PRODUCTION FROM OLEAGINOUS FUNGI

1. INTRODUCTION: AN OVERVIEW

The production of biomass-based biofuels is a crucial strategy to address elevated energy costs, mitigate the potential exhaustion of crude oil reserves, decrease greenhouse gas emissions, and promote a sustainable economy. (Zinoviev et al., 2010). Microbial lipids can represent a valuable alternative feedstock for biodiesel production, and a potential solution for a bio- based economy.

Nowadays, the production of biodiesel is based mostly on plant oils, even though animal fats, and algal oils can also be used. In particular, soybean, rapeseed, and palm oils are adopted as the major feedstock for biodiesel production. They are produced on agricultural land, opening the debate on the impact of the expansion of bioenergy crop cultures, which displace land from food production. Furthermore, their price restricts the large-scale development of biodiesel to some extent.

In order to meet the increasing demand of biodiesel production, other oil sources have been explored. Recently, the development of processes to produce single cell oil (SCO) by using heterotrophic oleaginous microorganisms has triggered significant attention (Azocar et al., 2010). These organisms accumulate lipids, mostly consisting of triacylglycerols (TAG), that form the storage fraction of the cell. The occurrence of TAG as reserve compounds is widespread among all eukaryotic organisms such as fungi, plants and animals, whereas it has only rarely been described in bacteria (Meng et al., 2009). In fact, bacteria generally accumulate polyhydroxyalkanoates as storage compound and only few bacterial species, belonging to the actinobacterial genera *Mycobacterium*, *Streptomyces*, *Rhodococcus* and *Nocardia* produce relevant amounts of lipids (Alvarez & Steinbuchel, 2002).

Among heterotrophic microorganisms, oleaginous fungi, encompassing both molds and yeasts, have been increasingly recognized as proficient makers of triacylglycerols (TAG). This chapter will concentrate on recent advancements in their metabolism, physiology, and the outcomes attained in strain enhancement, process engineering, and raw material utilization.

2. ECOLOGY OF OLEAGINOUS FUNGI

Oleaginous microorganisms are able to accumulate lipids above the 20% of their biomass, on dry basis. Several species of yeasts and filamentous fungi are regarded as oleaginous, since they have the capability to synthetize and accumulate high amounts of TAG within their cells, up to 70% of the biomass weight. These lipids have similar composition and energy value to plant and animal oils, but their production do not compete for food resources, in particular if it is based on inexpensive carbon sources, such as raw materials, by-products, and surplus. Furthermore, fungal SCO have a short process cycle, and their production is not subjected to seasonal and cyclical weather variations.

The study of oleaginous yeasts has a long history: their ability to accumulate lipids has been known from the 70s, but only in the last years the attention has been focused on exploitation of SCO for biodiesel production. The yeasts represent a part of the microbiota in all natural ecosystems, such as soils, freshwaters and marine waters, from the ocean surface to the deep sea. Widely distributed in the natural environment, they colonize also more extreme environments, such as low temperatures, low oxygen availabilities, and oceanic waters (Butinar et al., 2007). Approximately 1500 species of yeasts belonging to over 100 genera have been described so far (Satyanarayana & Kunze, 2010). Although the vast majority of yeasts are beneficial to human life, only a few are opportunistic human pathogens. As a whole, they play a pivoltal role in the food chain, and in the carbon, nitrogen and sulphur cycles. Among the huge number of species that have been described, only 30 are able to accumulate more than 25% of their dry weight as lipids (Beopoulos et al., 2009b).

Basidiomycetous yeasts strongly prevail among oleaginous yeasts, representing most of all the strains identified as lipid producers, even though some important oleaginous species have been identified among Ascomycota as well (e.g. *Yarrowia lipolytica*).

The most deeply investigated oleaginous yeasts belong to the genera *Yarrowia*, *Candida*, *Rhodotorula*, *Rhodosporodium*, *Cryptococcus*, and *Lypomyces* (Ageitos et al., 2011; Li et al., 2008; Rossi et al., 2009). *Yarrowia lipolytica*, previously referred to as *Candida lipolytica*, is a good candidate for single-cell oil production (Beopoulos et al., 2009a; Beopoulos et al., 2009b). *Yarrowia* are hemiascomycetous dimorphic fungi that belong to the order *Saccharomycetales*. They are able to degrade hydrophobic substrates such as n-paraffins and oils very efficiently and this physiological feature prompted the scientific community to explore several biotechnological applications (Bankar et al., 2009). The common habitats of these fungi are oil-polluted environments and foods such as cheese, yogurt, kefir, shoyu, meat, and poultry products. Despite *Y. Lipolytica* is distantly related to the conventional yeast

Saccharomyces cerevisiae, the genome displays an expansion of protein families and genes involved in hydrophobic substrate (such as alkanes and lipids) utilization. Wild-type strains accumulate up to 38% of dry weight (DW) as lipids. Albeit the levels are lower than those of other oleaginous yeasts, it became a model organism because it can be subjected to genetic and metabolic engineering, having be developed a reliable and versatile system for disruption, cloning and expression of target genes.

Within the *Candida* genus, *Candida curvata* (Holdsworth & Ratledge, 1991) also referred as *Apiotrichum curvatum* and *Candida freyschussii* (Amaretti et al., 2011) synthetize and store significant amount of lipids. *Candida* comprises an extremely heterogeneous group of Ascomycota that can all grow with yeast morphology, classified in 150 heterogeneous species, among which only a minority have been implicated in human diseases, since approximately 65% of *Candida* species are unable to grow at 37°C, then they can not be successful pathogens or commensals of humans (Calderone, 2002). Therefore, most of the species can be exploited for biotechnological applications, despite of unwarranted negative public perceptions.

Lipomyces spp. present a great propensity to accumulate triacylglycerols. This genus belongs to the *Saccharomycetales* order and represents a unique branch in the evolution of the ascomycetes (van der Walt, 1992). *Lipomyces* are true soil inhabitants and have a worldwide distribution. The oleaginous species *Lipomyces starkeyi* has the capability to accumulate over 70% of its cell biomass as lipid under defined culture conditions, and can produce lipid on xylose, ethanol, and L-arabinose, or using a mixture of glucose and xylose (Zhao et al., 2008), as well as other wastes (Angerbauer et al., 2008).

Cryptococcus curvatus is a yeast with industrial potential as single-cell oil because it can grow and accumulate lipid on a very broad range of substrates. It requires minimal nutrients for growth, accumulating up to 60% of its cellular dry weight (DW) as intracellular lipid (Meesters et al., 1996; Zhang et al., 2011). Yeasts of the *Cryptococcus* genus are widely distributed in nature and may be isolated from various substrates such as air, soil, bird excreta, water, animal surfaces and mucosae, leaves, flowers, and decomposing wood. Most species are considered as free-living (non-symbiotic) and only a few have medical importance being responsible for disease in man and animals (*C. neoformans* and *C. gattii*). *C. curvatus* is recognized as an opportunistic pathogen of animals, including humans (Findley et al., 2009).

Species belonging to the genus *Rhodosporidium*, and to its asexual counterpart *Rhodotorula*, have been claimed as oleaginous yeasts. They belong to one of the three main lineages of the *Basidiomycota*, the *Pucciniomycotina*. *Rhodotorula* is a common environmental inhabitant. The synthesis of different commercially important natural carotenoids by yeast species belonging to the genus *Rhodotorula* has led to consider these microorganisms as a potential pigment sources. Within this genus, the mesophilic red yeast *Rhodotorula glutinis* is able to synthetize and store lipids also growing on glycerol, whereas the psychrophilic species *Rhodotorula glacialis,* that are not red yeasts, accumulates lipids in a range of temperature between -3 and 20°C (Amaretti et al., 2010). The red yeast *Rhodosporidium toruloides* is an oleaginous mesophilic species. *Rhodosporidium* are able to carry out a number of diverse biochemical reactions such as biodegradation of epoxides, biphenyls and oxiranes (Smit, 2004), biosynthesis of carotenoids (de Miguel et al., 1997) and other types of biotransformations, but a major biotechnological exploitation is associated to their ability to convert glycerol and lignocellulosic biowastes into lipids (Hu et al., 2009; Yu et al., 2011). Among the oily yeasts, two novel species of the anamorphic basidiomycetous genus *Trichosporon* have been recently identified (*T. cacaoliposimilis* and *T. oleaginosus*) (Gujjari et al., 2011), despite lipid accumulation has not yet explored in the perspective of biodiesel production. *Trichosporon* are basidiomycetous yeasts widely distributed in nature, consisting of soil- and water-associated species, predominantly found in environmental substrates, such as decomposing wood. They present distinct morphological characteristics of budding yeast cells and true mycelia that disarticulate to form arthroconidia. Some species are causative agents of diseases in man and cattle. They can occasionally belong to the gastrointestinal microbiota of humans as well as transiently colonize the skin and respiratory tract.

Exploitation of oleaginous filamentous fungi for biodiesel production has a more recent history, which, with few exceptions, derives from studies focused to poly-unsaturated fatty acid production (PUFA), such as arachidonic acid and γ-linolenic acid. The most relevant example of this biotechnological application is represented by exploitation of *Mortierella alpina* to produce oils containing n-1, n-3, n-4, n-6, n-7, and n-9 PUFAs (Sakuradani et al., 2009). Among the major lipid producers there is *Mucor circinelloides,* a zygomycete fungus, which is emerging as opportunistic pathogen in immunocompromised patients (Li et al., 2011). *M. circinelloides* has been used for the first commercial production of microbial lipids (Ratledge, 2004). Lipid accumulation in *M. circinelloides* has been extensively studied (Wynn et al., 2001), and its TAG have been proposed as feedstock for producing biodiesel by direct transformation of its lipids (Vicente et al., 2009). *M. circinelloides* represents an outstanding model within the *Zygomycota* phylum, based on the availability of an efficient transformation procedure (Gutierrez et al., 2011) and on the whole sequence of genome (http://genome.jgi- psf.org/Mucci2/Mucci2.home.html). Also the phylogenetically related *Umbelopsis isabellina* has emerging as a promising species to convert biomass residues to biodiesel precursors (Meeuwse et al., 2011a). To the best of our knowledge, limited are the attempts to get lipids with *Aspergillus oryzae* that, conversely, is extensively studied as lipase producer to carry out transesterification of TAG (Adachi et al., 2011).

3. BIOCHEMISTRY OF LIPID ACCUMULATION

Lipid accumulation in oleaginous yeasts and molds has been demonstrated to occur when a nutrient in the medium (e.g. the nitrogen or the phosphorus source) becomes limited and the carbon source is present in excess. Nitrogen limitation is the most efficient condition for inducing lipogenesis. During the growth phase, nitrogen is necessary for the synthesis of proteins and nucleic acids, while the carbon flux is distributed among energetic and anabolic processes yielding carbohydrates, lipids, nucleic acids and proteins. When nitrogen gets limited, the growth rate slows down and the synthesis of proteins and nucleic acids tends to cease. In non-oleaginous species, the carbon excess remains unutilized or is converted into storage polysaccharides, while, in oleaginous species, it is preferentially channeled toward lipid synthesis, leading to the accumulation of TAG within intracellular lipid bodies (Ratledge & Wynn, 2002; Granger at al., 1993).

The biochemical pathway of lipid biosynthesis is not very different among eukaryotic organisms and does not differ in oleaginous and non-oleaginous fungi. The ability to accumulate high amounts of lipid depends mostly on the regulation the biosynthetic pathway and the supply of the precursors (i.e. acetyl-CoA, malonyl-CoA, and glycerol-3- phosphate) and the cofactor NADPH.

Most information were obtained from the model yeast *Saccharomyces cerevisiae* (Kohlwein, 2010), that does not accumulate lipids, and *Yarrowia lipolytica*, that represent a model for bio- oil production and is suitable for genetic manipulation (Beopoulos et al., 2009b).

3.1 Fatty acids biosynthesis and modifications

De novo synthesis of fatty acids (FA), the first step of lipid accumulation, is carried out in the cytosol by fatty acids synthetase (FAS) complex. In yeasts, FAS bears phosphopantheteine transferase activity to activate its acyl carrier protein (ACP) by loading the coenzyme pantothenate. FAS is a multimer of 6 α and 6 β subunits encoded by *fas2* and *fas1*, respectively, each subunit containing four functional domains. Therefore, FAS consists in a $\alpha6\beta6$ molecular complex of 2.6 MDa with 48 functional centers that catalyze all reactions required for synthesis of fatty acids through cycles of multistep reactions. FAS firstly loads acetyl-CoA on its β-ketoacyl-ACP synthase (KS), then it exherts β-ketoacyl-ACP reductase (KR), β-hydroxyacyl-ACP dehydratase (DH), and enoyl-ACP reductase (EAR) activities. This set of reactions is repeated cyclically seven times to yield palmitoyl-ACP (Fig. 1) (Tehlivets et al, 2007).

The biosynthesis of FA requires the constant supply of acetyl-CoA as initial biosynthetic unit and of malonyl-CoA as the elongation unit, supplying two carbons at each step. Non- oleaginous yeasts receive acetyl-CoA mostly from glycolysis. In oleaginous yeasts, acetyl- CoA is mostly provided by the cleavage in the cytosol of citrate, which accumulated as a consequence of nitrogen limitation (Ratledge, 2002) (Fig. 2). In fact, lipid accumulation by oleaginous fungi does not occur under balanced nutrient conditions.

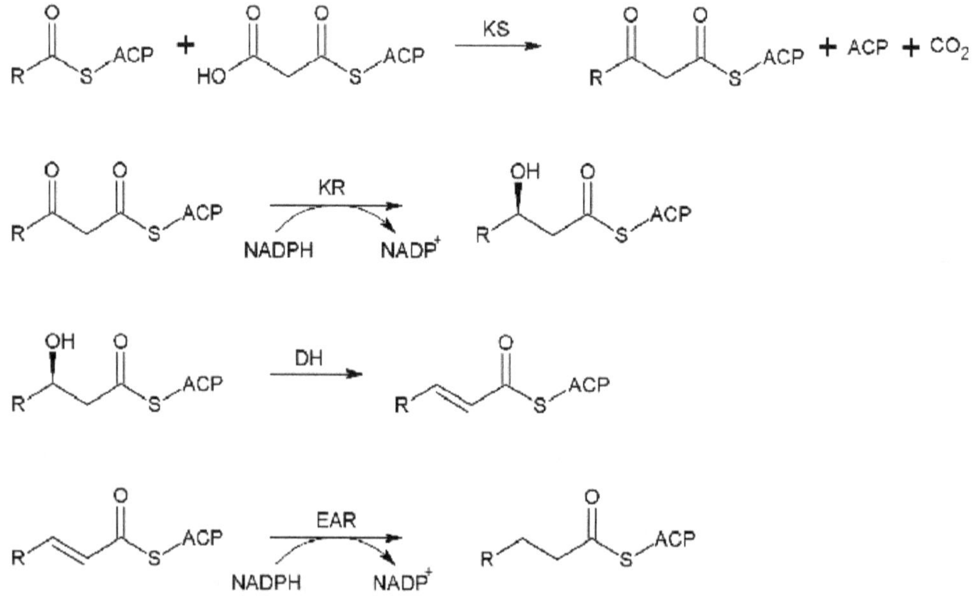

Fig. 1. Reactions occurring sequentially in fatty acid synthetase: condensation of acyl-ACP and malonyl-ACP mediated by KS, NADPH-dependent reduction of the keto group to a hydroxyl group by means of KR, dehydration to create a double bond with DH and reduction of the double bond by means of EAR. R = H, $CH_3(CH_2)_{2n}$; n_{max}=7.

In oleaginous yeasts, nitrogen limitation activates AMP-deaminase (Ratledge & Wynn, 2002), which supply ammonium to the nitrogen-starved cell. As a consequence, mitochondrial AMP concentration decreases, causing isocitrate dehydrogenase activity to drop. The TCA cycle is then blocked at the level of isocitrate, which accumulates and equilibrates with citrate through aconitase. Excess of citrate from TCA cycle is exported out of the mitochondrion via the malate/citrate antiport. Cytosolic ATP-citrate lyase (ACL) cleaves citrate to give oxaloacetate and acetyl-CoA (Fig. 2).

ACL represents one of the key enzymes that contribute to the oleaginous trait of yeasts, whereas its activity is negligible in non-oleaginous species. ACL is composed of two subunits, encoded by *ACL1* and *ACL2* and is negatively regulated by exogenous FA. Malonyl-CoA is produced from acetyl-CoA by acetyl-CoA carboxylase (ACC) that condensate an acetyl-CoA unit with bicarbonate:

$$\text{Acetyl-CoA} + HCO_3^- + \text{ATP} \leftrightarrows \text{malonyl-CoA} + \text{ADP} + \text{Pi}$$

ACC is also a key enzyme in *de novo* FA synthesis, since *ACC1* mutants became FA auxotrophs or maintain low levels of ACC activity (Tehlivets et al., 2007). ACC1 undergoes allosteric activation by citrate. Furthermore the transcription of *FAS1*, *FAS2*, and *ACC1* is coordinately regulated, being negatively regulated by FA.

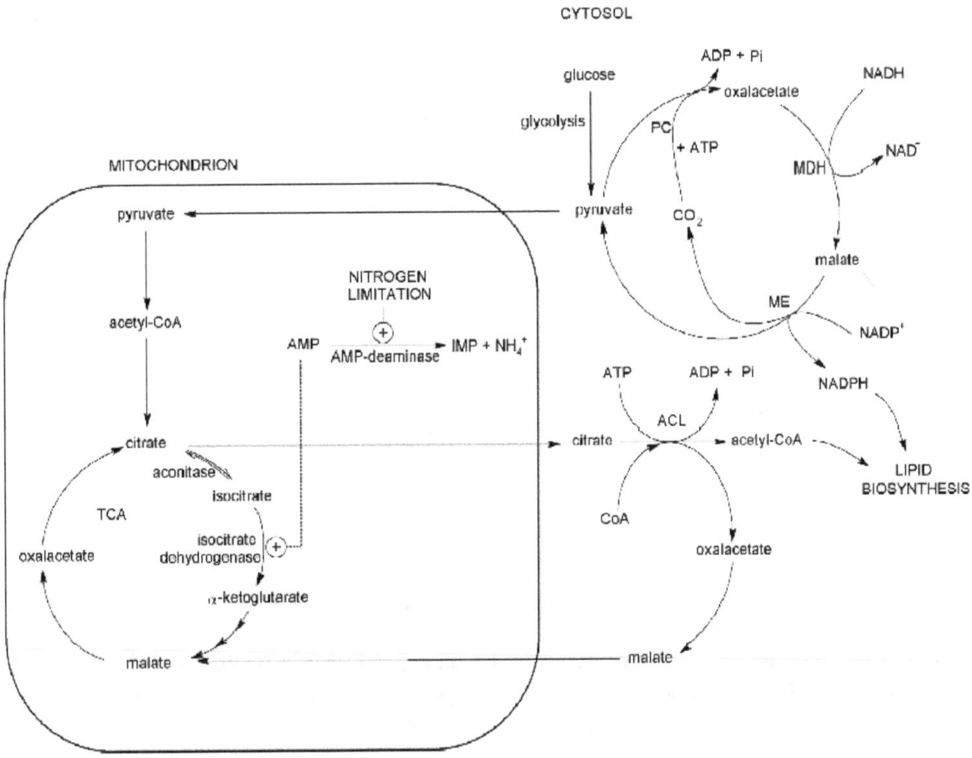

Fig. 2. Lipid biosynthesis from excess of citrate as a consequence of nitrogen limitation. Adapted from Ratledge, 2004.

Cytosolic NADPH is required for KR and EAR functions of FAS. For each elongation step of the acyl chain, two molecules of NADPH are required. One of the major sources of cytosolic NADPH are the pentose phosphate pathway and the transhydrogenase cycle, which transforms NADH into NADPH through the activity of pyruvate carboxylase (PC), malate dehydrogenase (MDH), and malic enzyme (ME), catalyzing the following reactions:

$$\text{Pyruvate} + CO_2 + \text{ATP} \leftrightarrows \text{oxaloacetate} + \text{ADP} + \text{Pi (PC)}$$

$$\text{Oxaloacetate} + \text{NADH} \leftrightarrows \text{Malate} + \text{NAD}^{+} \text{ (MDH)}$$

$$\text{Malate} + \text{NADP}^{+} \leftrightarrows \text{pyruvate} + CO_2 + \text{NADPH (ME)}$$

$$\text{NADH} + \text{NADP}^{+} \leftrightarrows \text{NADPH} + \text{NAD}^{+}$$

ME has been found in several oleaginous fungi and it has been regarded as a key enzyme involved in lipid accumulation (Ratledge, 2002). In *Mortierella circinelloides*, overexpression of ME enhanced lipid accumulation (Zhang et al., 2011), whereas overexpression of the ME homologous in *Yarrowia lipolytica* did not result in yield improvement.

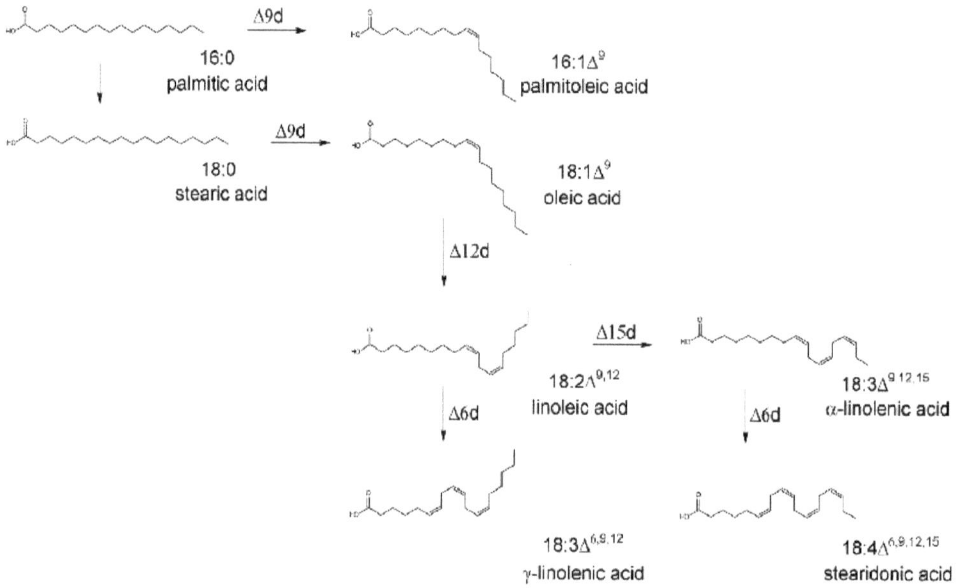

Fig. 3. Biosynthesis of poly-unsaturated fatty acid. Δ9d, Δ12d and Δ15d are the most common desaturases which are present in the endoplasmic reticulum (Ratledge 2004).

The final products of FAS are myristic or palmitic acids, depending on the yeast species. Reactions resulting in further elongation or desaturation occur in the endoplasmic reticulum (ER). Elongation reactions are catalyzed by elongases (such as malonyl-palmitoil transacylase, MPT) organized in a complex that requires malonyl-CoA provided by ACC. Desaturations are introduced by ER desaturases, hydrophobic membrane-bound proteins. The most common desaturases are Δ9, which inserts the first double bond onto palmitic and/or stearic acids, and Δ12, which catalyzes the insertion of the second unsaturation into oleic acid to produce linoleic acid. Δ6 and Δ15 desaturase activities have been recently described in in psychrophilic oleaginous yeasts, based on production of γ and α-linolenic acids, respectively (Fig. 3).

3.2 Biosynthesis of triacyl-glycerol

The fatty acyl-CoA produced by *de novo* synthesis are esterified with glycerol or sterols to produce triacyl-glycerol (TAG) and steryl-esters (SE), respectively. In oleaginous fungi, the neutral lipids SE and TAG are store inside the lipid bodies (LB).

TAG are mostly formed by consecutive acylation of glycerol-3-phosphate (G3P), carried out by diverse acyl transferases. G3P is formed from glycerol by glycerol kinase or can be synthesized from dihydroxyacetone phosphate (DHAP) by G3P dehydrogenase, in a reversible reaction. *S. cerevisiae* can use both G3P and DHAP as acyl-group acceptor. The addition of the first acyl group leads to 1-acyl G3P, also named lysophosphatidic acid (LPA). LPA can also be formed by reduction of acyl-DHAP, carried out by a NADPH dependent reductase. A second acyltransferase loads an other acyl group, producing 1,2-diacyl G3P (phosphatidic acid, PA). Phosphate is removed from PA by phosphatidate phosphatase isoenzymes, generating diacylglycerol (DAG). DAG can be the direct precursor of TAG, or can be channeled toward phospholipids biosynthesis (Fig. 4).

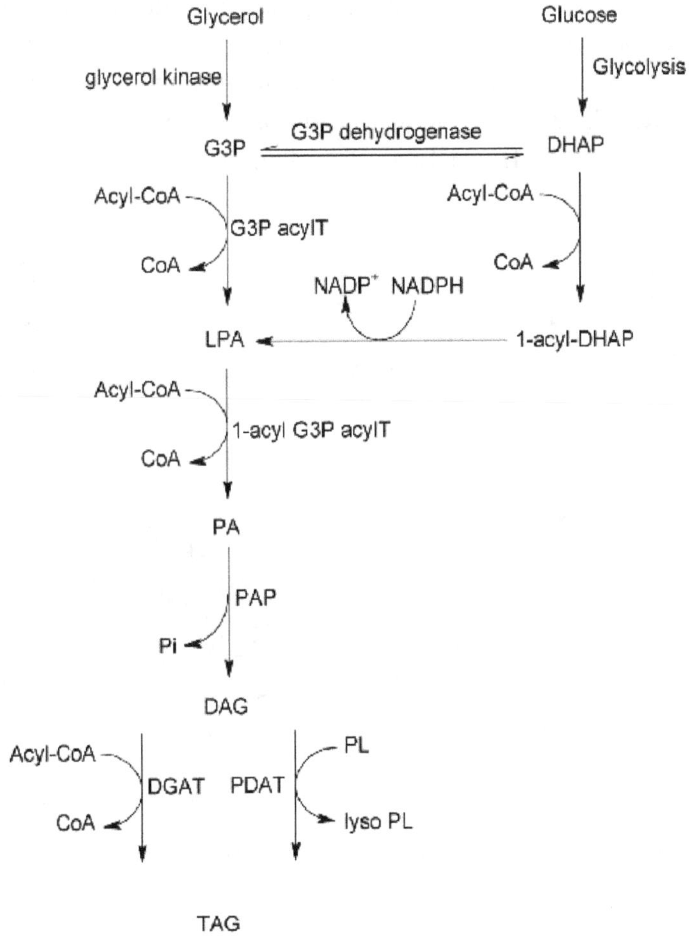

Fig. 4. ***de novo*** **synthesis of TAG (adapted from Czabany et al., 2007)**

The last step of *de novo* synthesis of TAG can be carried either by using diverse acyl donors, such as acyl-CoA or with phospholipids. In the former case, DAG acyl transferases (DGAT), which are integral proteins of the ER, can directly load the third Acyl-CoA. A DGAT enzyme is present in *S. cerevisiae* and *Y. lipolytica* and is mostly active during the stationary phase, although it is expressed also during the exponential phase. A second DGAT, more active during the exponential growth phase, has been identified in *Y. lipolytica.* In *S. cerevisiae* the phospholipid:DAG acyltransferase (PDAT) is localized in the ER, whereas in *Y. lipolytica* it is present both in the ER and in the surface of LB (Fig. 4).

3.3 Biogenesis of lipid bodies

In eukaryotes, neutral lipids are stored in specialized compartments known as lipids bodies (LB). They are assembled at a specialized subdomain of the ER where most biosynthetic enzymes and structural proteins are located (Waltermann et al., 2005). The neutral lipids do not fit among phospholipids and are thus deposited between the two leaflets of the membrane bilayer. However, substantial amounts of neutral lipids cannot be incorporated into membrane bilayer of ER and ongoing neutral lipid synthesis leads to the formation of a bud which buds off of the ER as a mature LB after reaching the critical size (Fig. 5).

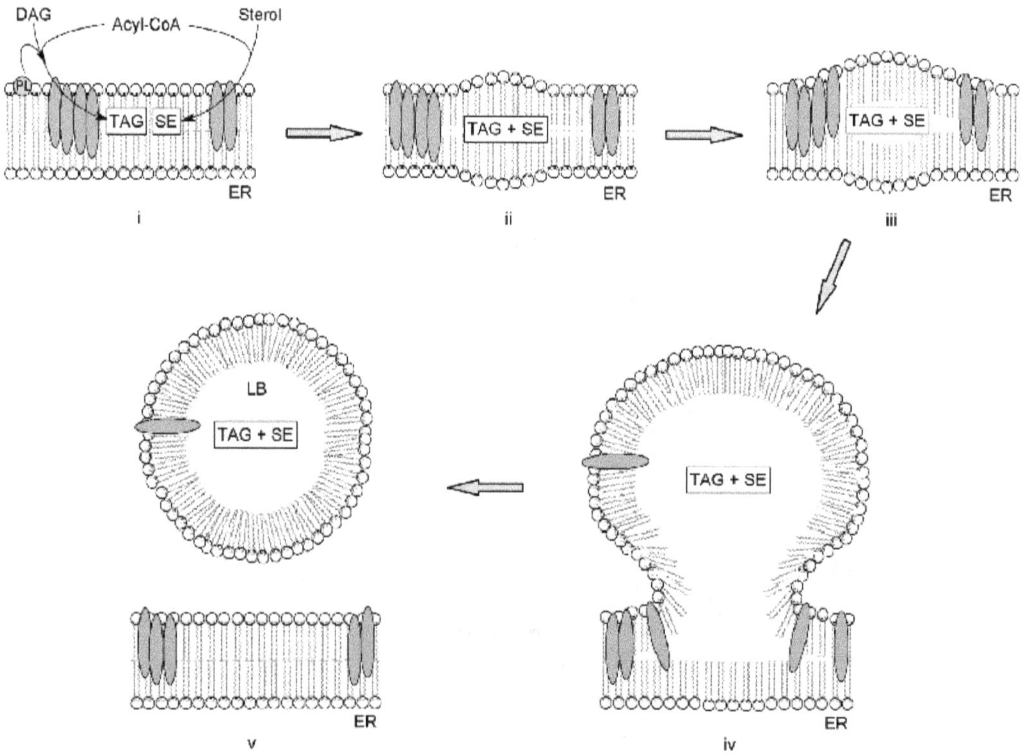

Fig. 5. Model of lipid bodies biogenesis from the membrane of the ER. TAG and SE accumulate between the two leaflets of the phospholipid bilayer (i to iii). The micro-droplet generated (iii, iv) evolve to lipid bodies (v) (Figure adapted from Czabany et al., 2007).

In most oleaginous yeasts, the neutral lipids of LB consist mostly of TAG (up to 90% or more) whereas a small fraction is represented by steryl esters. The presence of significant quantity of free fatty acids (FFA) within LP has been reported only for *Y. lipolytica*. In *S. cerevisiae*, which accumulates less than 15% lipids of its biomass, LB contain similar amounts of TAG and SE.

The core of LB, consisting of neutral lipids is surrounded by a phospholipid monolayer where several proteins are embedded. These proteins exert a key role in lipid metabolism, biosynthesis, and substrate trafficking. Upon requirement, storage lipids are mobilized from this compartment by triacylglycerol lipases and steryl ester hydrolases. The respective degradation products serve as energy sources and/or building blocks for membrane formation. In fact, FA hydrolyzed from TAG or SE are either channeled to the peroxisome, where β-oxidation takes place, or to phospholipid biosynthesis.

4. METABOLIC ENGINEERING OF OLEAGINOUS YEASTS

The availability of genome data and genetic tools, such as the possibility to integrate homologous or heterologous genes, opened up the possibility to use metabolic engineering to understand the molecular mechanisms involved in lipid accumulation or to increase the yield of stored lipids in *S. cerevisiae* and *Y. lipolytica*. Whereas *S. cerevisiae* has been used mostly as a model to investigate and understand the lipid metabolism, in *Y. lipolytica* several attempts have been done in order to address the carbon flux toward TAG production and accumulation. Similar approaches are precluded to other oleaginous fungi since they lack genome information and the necessary tools for gene manipulation and strain improvement.

In *Y. lipolytica*, the role of glycerol-3-phosphate (G3P) in triacylglycerol (TAG) biosynthesis and accumulation has been investigated (Beopoulos et al., 2008). In this yeast G3P is formed from glycerol by the glycerol kinase encoded by *GUT1*, or it is synthetized from dihydroxyacetone phosphate (DHAP) by the G3P dehydrogenase (*GPD1*). The antagonist reaction, which produces DHAP from G3P, is carried out in competition by a second isoform of the G3P dehydrogenase, encoded by *GUT2*. In order to force the conversion of DHAP into G3P, the gene *GPD1* was over-expressed and the gene *GUT2* was deleted.

A diverse strategy to increase lipid accumulation was based on the disruption of the β- oxidative metabolism, through the deletion of the 6 *POX* genes (*POX1 to POX6*) that encode the peroxisomal acyl-coenzyme oxidases (Mlickowa et al., 2004a; Mlickowa et al., 2004b; Beopoulos et al., 2008). As a whole, the best results in terms of percentage of lipids per dry biomass, were reached coupling the increased level of G3P with the disactivation of the β- oxidation pathway (65%) (Dulermo et al., 2011).

Metabolic engineering strategies have been recently exploited to expand the range of substrates used by oleaginous fungi, also through functional expression of heterologous genes. Recently, it has been found that inulin is a good material for bio-productions (Chi et al., 2009). In order to make the oleaginous yeast *Y. lipolytica* able to accumulate lipids on inulin containing materials, the *Kluyveromyces marxianus* exo-inulinase gene (*INU1*) was heterologously expressed on a high copy plasmid (Zhao et al., 2010). The inulinase was efficiently secreted by *Y. lipolytica*, and inulin was hydrolyzed, assimilated and converted into TAG.

5. CULTIVATION CONDITION OF OLEAGINOUS YEASTS

Lipid accumulation by oleaginous yeasts depends mostly on nutrient limitation conditions when excess carbon is present in the medium. Nutrient limitation prevents cells from being generated, while the carbon excess is converted into storage TAG. Published studies reports that phosphorus, magnesium, zinc, or iron limitation lead to lipid accumulation in model oleaginous yeasts (Hall & Ratledge, 1977; Beopoulos et al., 2009; Wu et al., 2010). However, nitrogen limitation is the most efficient form of nutrient limitation for lipogenesis induction, leading to the highest values of substrate/lipid conversion yield and lipid content within biomass (Hall & Ratledge, 1977; Wynn et al., 2001). Thus, nitrogen limitation is commonly used to induce lipogenesis in oleaginous fungi and the utilization of cultural media with appropriate C/N ratio is crucial to maximize lipid production.

Several studies focused on determining the optimal composition of cultural media for oleaginous fungi with the aim to optimize the performance of lipid-producing bioprocesses. The effect of the C/N ratio on lipid metabolism has been investigated for a number of oleaginous yeasts and molds, such as *Y. lipolytica* and many oleaginous species of *Rhodotorula*, *Candida*, *Apiotrichum/Cryptococcus*, *Mortierella* (Hall & Ratledge, 1977; Papanikolau et al., 2003; Granger et al., 1992; Wu et al., 2010; Park et al., 1990; Jang et al., 2005; Amaretti et al., 2010), and has been mathematically modeled for some of these organisms (Ykema et al., 1986; Granger et al., 1993; Economou et al., 2011). *Y. lipolytica* is the oleaginous microorganism for which information about the metabolic response to different C/N ratios is most abundant (Beopoulos et al., 2009a), particularly due to the availability of molecular tools for genetic engineering of this organism. Therefore, *Y. lipolytica* is regarded as a model organism for microbial oil production and the main traits of its metabolism can be used to give a general description of the metabolic response to different C/N ratios in the majority of oleaginous yeasts. With the increase of the C/N ratio, different metabolic behaviors were observed in *Y. lipolytica*: i) growth with mobilization of storage lipids, ii) growth of fat-free biomass, iii) growth with accumulation of lipids, and iv) growth with lipid accumulation and production of organic acids (Fig. 6).

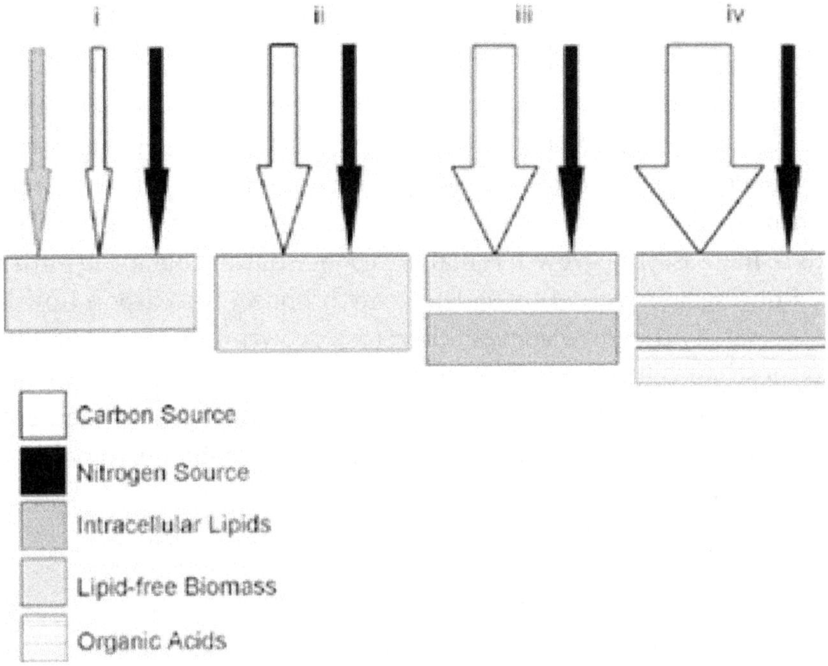

Fig. 6. Metabolic activity of oleaginous fungi (e.g. *Y. lipolitica*) as a function of carbon flow rate for a fixed nitrogen flow rate. Arrows indicate the consumption of nitrogen and carbon sources by the cells; squares indicate production rate. The dimension of arrows and squares is proportional to flow. (adapted from Beopoulos et al., 2009a)

If the medium is carbon limited or when the extracellular carbon supply gets exhausted, previously stored intracellular lipid can be mobilized and utilized by oleaginous microorganisms to sustain cells generation and production of lipid-free biomass (Park et al., 1990) (Fig. 6, i). If the medium is balanced and/or furnishes just the right amount of carbon flow to satisfy the growth need, balanced growth occurs without any accumulation of storage lipids (Fig. 6, ii). In conditions of carbon excess, a part of the carbon flow, which is proportional to nitrogen availability (Granger et al., 1993), is directed toward cells generation, whereas the carbon exceeding growth needs is channeled to the production of storage lipids (Fig. 6, iii). In some oleaginous fungi, the presence of a large carbon excess leads to the production of great amounts of organic acids, such as pyruvic acid and diverse TCA-cycle intermediates, at the expenses of lipid accumulation (Fig. 6, iv). In these latter conditions, *Y. lipolytica* produces citric acid (Levinson et al., 2007) but other oleaginous yeasts have never been reported to behave this way.

6. BATCH, FED-BATCH AND FERMENTATION PROCESSES

Batch, fed batch, and continuous modes of culture have been developed to culture oleaginous microorganisms. Lipid production in batch cultures is carried out in a cultural medium with a high initial C/N ratio, the carbon source being present in an adequate excess with respect to the nitrogen source. In fact, in this condition, the flow of carbon utilization is limited only by the substrate uptake system of the cell, while the changes in nitrogen concentration determine the passage from a phase of balanced growth to a phase of lipid accumulation, causing the process to proceed through two phases. As nitrogen is consumed from the culture the C/N ratio tends to increase, but growth remains exponential and balanced until nitrogen is not the limiting substrate. During the growth phase, the carbon flow is mostly channeled to satisfy the growth need, therefore growth is balanced and lipid- free biomass is mostly produced (Fig. 6 ii). As nitrogen concentration becomes limiting, the growth rate and the carbon flow toward biomass generation decrease, while lipid production is triggered, resulting in a shift of microbial metabolism into the lipogenic phase (Fig. 6 iii, Fig. 7).

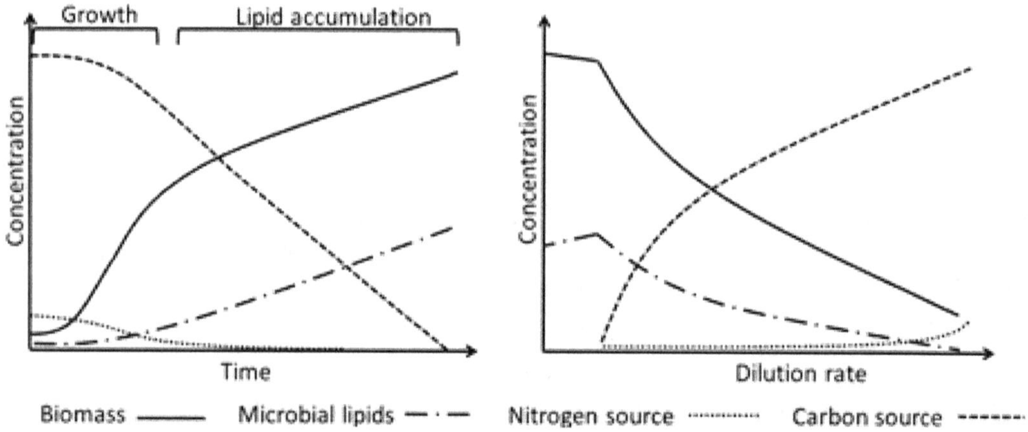

Fig. 7. Modeling and prediction of the timecourse of a batch fermentation (left) and the steady-state values of a continuous process (right) for microbial production of lipids. Axis are in arbitrary scales.

In batch cultures the initial C/N ratio of the cultural medium has a pivotal role in determining the bioprocess performance. In fact, both the rate and the yield of lipid production depend by the C/N ratio, which affects the duration of the exponential phase and the amount of biomass produced during growth. With a fixed carbon concentration, higher amounts of lipid-free biomass produced during the growth phase correspond to higher lipid production rates during the lipogenic phase, but to lower amounts of lipid content within cells and lipid/substrate conversion yields. Therefore, the initial C/N ratio needs to be optimized to maximize lipid productivity in batch cultures. The optimal C/N value is always high (e.g. in the range between 80 and 350 mol/mol) and strongly depends on the microorganism, the medium composition, the carbon source (e.g. glucose, glycerol, etc.), and the nitrogen source (e.g. diverse organic or inorganic sources). The minimal C/N ratio suitable for lipid accumulation can be estimated as $(Y_{X/S} \cdot q)^{-1}$, where $Y_{X/S}$ is the biomass/carbon source yield coefficient under conditions of carbon limitations (C-mol/C-mol) and q is the nitrogen/carbon content of biomass (N-mol/C-mol) (Ykema et al., 1986). However, it should be considered that extremely high C/N ratios may cause the production of unwanted byproducts, such as organic acids (Fig. 6 iv), or may lead to severe nitrogen deficiency, causing a rapid decrease in cells viability.

Unlike batch processes, in fed-batch mode, nutrients are fed into the bioreactor in a controlled manner, with the purpose to monitor and control the specific growth rate and the flows of nitrogen and carbon utilization. Through the judicious management of the feeding rate and composition, it is possible to control the C/N ratio within the culture and maintain the oleaginous microorganism in the optimal metabolic status, as appropriate, first for the growth phase, and later for the lipogenic phase. The lipogenic phase is the most extensive, corresponding to lipid production under nitrogen limitation, with constant C/N ratio, preventing loss of viability and acids production (Beopoulos et al., 2009a).

In continuous cultures, at the steady state, the assimilation of C and N sources and the microbial growth occur at constant rates, which ultimately depend by the dilution rate (D). The concentration of the substrates within the bioreactor is steady and depends by the dilution rate as well, the actual C/N ratio of the culture remaining constant unlike in batch cultures. Likewise in batch cultures, in continuous cultures the C/N ratio of the fresh medium needs to be higher than $(Y_{X/S} \cdot q)^{-1}$ to obtain some lipid accumulation (Ykema et al., 1986). However, at the steady-state with this medium, the C/N ratio within the culture is higher than in the fresh medium, due to nitrogen consumption. The extent of substrates utilization, and also the biomass and lipid concentrations are the highest at low D values and decrease with the increase of D (Fig. 7). While low D values promote lipid production and a more complete substrate utilization, on the contrary, the volumetric productivity of continuous processes is positively affected by the increase of the dilution rate (Ykema et al., 1986; Meeuwse et al., 2011b). Therefore, both the C/N content of the medium and the dilution rate need to be thoroughly tuned to maximize lipid productivity of continuous processes.

7. SUBSTRATES AND RAW MATERIAL

The demand for the inexpensive production of biofuels has intensified due to increasing concerns of climate change, depletion of petroleum-based fuels, and environmental problems. In a market economy, corporations aim to maximize profit, seeking the most competitive feedstock. To produce single-cell oils for biodiesel production, the carbon source has necessarily to be cheap and available in large quantities. Therefore, while the first investigations on oleaginous fungi most commonly employed glucose as carbon source, nowadays the production of single-cell oils is predominantly addressed to transformation of raw materials, by-products and surplus.

Glucose is the carbon source most commonly employed for growth of oleaginous fungi and lipid production (Boulton & Ratledge, 1984; Hansson & Dostalek, 1986; Hassan et al., 1993; Heredia & Ratledge, 1988; Jacob, 1991; Jacob, 1992; Johnson et al., 1992; Li et al., 2007; Pan et al., 1986; Ratledge, 2004; Rau et al., 2005; Saxena et al., 2008; Zhao et al., 2008). High glucose concentrations enhance the carbon flow that is directed toward TAG production, thus improving lipid production in several yeasts. However, growth of some yeasts (e.g. *R. toruloides)* is inhibited by high concentration of glucose, (Li et al., 2007). Furthermore, in batch cultures, initial glucose concentration also affects the fatty acids composition of the lipids (Amaretti et al., 2010).

Carbon sources other than glucose, such as xylose (Chistopher et al., 1983; Heredia & Ratledge, 1988;), lactose (Christopher et al., 1983; Daniel et al., 1999;), arabinose, mannose (Hansson & Dostalek, 1986), mannitol (Hansson & Dostalek, 1986), ethanol (Chistopher et al., 1983; Eroshin & Krylova, 1983), have been also investigated in the 80s and 90s for the production of microbial lipids.

Albeit glucose is a very good carbon source for lipid production with oleaginous fungi, molasses, which carbohydrate fraction is mainly composed of sucrose, glucose, and fructose, do not represent a promising raw material for lipid production, since they are characterized by a high nitrogen content which delays the unbalanced growth, where number of cells can not augment anymore and lipids are accumulated (Johnson et al., 1995).

Carbons sources obtained from ligno-cellulosic biomasses represent one of the most important potential to produce biodiesel. In fact, several waste biomasses containing forest residues, agricultural residues, food wastes, municipal wastes, and animal wastes can be utilized for the production of lignocellulosic based microbial lipids. Microbial oil production from sulphuric acid treated rice straw hydrolysate (SARSH) by the yeast *Trichosporon fermentans* pointed out the difficulty to perform the process of lipid accumulation in presence of the inhibitory compounds released during hydrolysis, such as acetic acid, furfural, 5-hydroxymethylfurfural, and water soluble lignin (Huang et al., 2009). Selected strains were able to grow on xylose and glucose (Zhu et al., 2008), but the crude hydrolyzate did not result an optimal substrate for a high yield process of lipid production. Cellulose and hemicellulose are generally hardly hydrolyzed and assimilated by yeasts, while they can be degraded and used as carbon source by filamentous fungi. A screening of endophytic fungi from the oleaginous plants was the selection of strains belonging to the genera *Microsphaeropsis, Phomopsis, Cephalosporium, Sclerocystis* and *Nigrospora* that simultaneously accumulated lipids (21.3 to 35.0% of dry weight) and produced cellulase (Peng & Chen, 2007). Albeit these strains could be exploited as microbial oil producers by utilising straw as substrate, they have never been claimed again as a SCO producers on lingo-cellulosic biomass. Attempts to carry out lipid production in Solid State Fermentation (SSF) on wheat straw have been performed exploiting a cellulolytic strain of *Aspergillus oryzae* (Lin et al., 2010). This strain is able to use cellulose as substrate and accumulate lipids in a low cost fermentation system on this abundant cellulosic by-product.

Other complex matrices have been used, such as solids from wheat bran fermentation (Jacob 1991), sewage sludge (Angerbauer et al., 2008), wastewaters of animal fat treatment (Papanikolaou et al., 2002), whey derivatives (Ykema et al., 1989; Vamvakaki et al., 2010), olive oil mill wastewaters (Yousuf et al., 2010), and tomato waste hydrolysate (Fakas et al., 2008).

Nowadays, lipid production with oleaginous yeasts is focused on selection and development of yeasts as converters of glycerol into lipid for biodiesel production, since it is the major side-product of the biodiesel production process. The biotransformation of glycerol into TAG is therefore regarded as a promising way to decrease the cost of biodiesel process through simultaneous reutilization of its major byproduct. In general, for every 100 kg of biodiesel produced, approximately 10 kg of crude glycerol are created. Crude glycerol is a mixture of glycerol (65–85%, w/w), methanol, and soap, and contains macro elements such as calcium, potassium, magnesium, sulfur and sodium. In order to minimize unknown variables introduced through the use of crude glycerol, several studies to determine whether or not glycerol could be used as substrate or co-substrate for growth have been conducted using purified glycerol.

A deep characterization of lipid accumulation on glycerol has been carried out with *Yarrowia lipolytica*, that is able to metabolize several important industrial and agro-industrial by-products such as raw glycerol, producing large amounts of SCO and organic acids (Papanikolaou et al., 2003; Papanikolaou & Aggelis, 2002; Rymowicz et al., 2010; Rywinska et al., 2009). Biochemistry of lipid production on glycerol has been investigated in this organism: glycerol passes into the microbial cell by facilitated diffusion and the conversion is carried out via phosphorylation pathway, with direct phosphorylation to G3P and subsequent dehydrogenation. Recently, *Y. lipolytica* has been subjected to targeted and purposeful alteration of G3P shuttle pathway to better utilize glycerol for lipid production. In the genetically manipulated strains, lipid accumulation resulted from a complex interrelation between different processes in diverse cell compartments, such as lipid synthesis in the cytosol, location and storage in ER and LB, mobilization and degradation processes (Dulermo & Nicaud, 2011).

Pure glycerol supported growth and lipid accumulation of *Rhodotorula glutinis* and *Candida freyschussii* (Easterling et al., 2009; Amaretti et al., 2011), being used as sole carbon and energy source or in addition to xylose or glucose. The diverse composition of the medium affected not only the lipid/biomass yield, but also the TAG composition, in terms of ration of saturated, monounsaturated, and polyunsaturated fatty acids (Easterling et al., 2009).

Attempts to convert crude glycerol into lipids have been successfully performed exploiting the oleaginous yeast *Cryptococcus curvatus* (Liang et al., 2010). Different processes have been developed with very efficient yields and productivities. In a 12 days two-stage fed-batch where raw glycerol was fed, the biomass density and the lipid content reached 32.9 g/l and 52%, respectively. Methanol of crude glycerol did not pose a significant inhibitory effect even though it was existent in the bioreactor. Lipid accumulated by *C. curvatus* on glycerol presented high amount of monounsaturated fatty acid, turning out as excellent substrate for transformation into biodiesel.

8. CONCISE SUMMARY

Oleaginous fungi, especially yeasts, exhibit remarkable efficiency in the formation of intracellular triacylglycerols (TAG), and its utilization by the biofuel sector is anticipated in the future. Nevertheless, the expenses associated with microbial lipids remain excessively expensive to compete with plant oils in biodiesel production. Inexpensive carbon sources must be utilized for the culture of these microorganisms, and the efficiency of the bioprocess requires enhancement regarding both yield and productivity. The investigation of natural biodiversity is a promising approach to discover new oleaginous organisms that utilize and accumulate fat from agro-industrial leftovers, specifically lingo-cellulosic biomass and crude glycerol from the biodiesel sector. Additional methodologies integrating genomic, transcriptomic, metabolomic, and lipidomic studies will certainly yield more comprehensive insights into lipid synthesis by oleaginous fungus. A metabolic engineering method has great potential; yet, it remains unattainable for the most oleaginous species, as their genomes have not been sequenced and genetic tools are not yet accessible.

REFERENCES

Adachi, D., Hama, S., Numata, T., Nakashima, K., Ogino, C., Fukuda, H., & Kondo, A., (2011) Development of an *Aspergillus oryzae* whole-cell biocatalyst coexpressing triglyceride and partial glyceride lipases for biodiesel production. *Bioresource Technology*. Vol. 102, No. 12, pp. 6723-6729, ISSN 0960-8524.

Ageitos, J. M., Vallejo, J. A., Veiga-Crespo, P., & Villa TG. (2011) Oily yeasts as oleaginous cell factories. *Applied Microbiology and Biotechnology*. Vol. 90, No. 4, pp. 1219-1227, ISSN 0175-7598.

Alvarez, H. M. & Steinbuchel, A. (2002) Triacylglycerols in prokaryotic microorganisms. *Applied Microbiology and Biotechnology*. Vol. 60, No. 4, pp. 367-376, ISSN 0175-7598.

Amaretti, A., Raimondi, S., Sala, M., Roncaglia, L., De Lucia, M., Leonardi, A., & Rossi, M. (2010) Single cell oils of the cold-adapted oleaginous yeast *Rhodotorula glacialis* DBVPG 4785. *Microbial Cell Factories*. Vol. 23, No. 9, pp. 73, ISSN 1475-2859.

Amaretti, A., Raimondi, S., Leonardi, A., & Rossi, M. (2011) Lipid production from glycerol by *Candida Freyschusii. Proceedings of FEMS 2011 14th Congress of European Microbiologists*, pp. 126, Geneva, Switzerland, June 26-30, 2011

Angerbauer C, Siebenhofer M, Mittelbach M, Guebitz GM (2008) Conversion of sewage sludge into lipids by *Lipomyces starkeyi* for biodiesel production. *Bioresource Technology*. Vol. 99, No. 8, pp. 3051–3056, ISSN 0960-8524.

Azócar, L., Ciudad, G., Heipieper, H. J., & Navia, R. (2010) Biotechnological processes for biodiesel production using alternative oils. *Applied Microbiology and Biotechnology*. Vol. 88, No. 3, pp. 621-636, ISSN 0175-7598.

Bankar, A. V., Kumar, A. R., & Zinjarde, S. S. (2009) Environmental and industrial applications of Yarrowia lipolytica. *Applied Microbiology and Biotechnology*. Vol. 84, No. 5, pp. 847-865, ISSN 0175-7598.

Beopoulos, A., Mrozova, Z., Thevenieau, F., Le Dall, M. T., Hapala, I., Papanikolaou, S., Chardot, T., & Nicaud, J. M. (2008) Control of lipid accumulation in the yeast *Yarrowia lipolytica*. *Applied and Environmental Microbiology*. Vol. 74, No. 24, pp. 7779- 7789, ISSN 0099-2240.

Beopoulos, A., Cescut, J., Haddouche, R., Uribelarrea, J. L., Molina-Jouve, C., & Nicaud, J. M. (2009a) *Yarrowia lipolytica* as a model for bio-oil production. *Progress in Lipid Research*. Vol. 48, No. 6, pp. 375-387, ISSN 0163-7827.

Beopoulos, A., Chardot, T., & Nicaud, J.M. (2009b) *Yarrowia lipolytica*: a model and a tool to understand the mechanisms implicated in lipid accumulation. *Biochimie*. Vol. 91, No. 6, pp. 692-696, ISSN 0300-9084.

Boulton, C. A. & Ratledge, C. (1984) *Cryptococcus terricolus*, an oleaginous yeast re-appraised. *Applied Microbiology and Biotechnology*. Vol. 20, No. 1, pp. 72-76, ISSN 0175-7598.

Butinar, L., Spencer-Martins, I., & Gunde-Cimerman, N. (2007) Yeasts in high Arctic glaciers: the discovery of a new habitat for eukaryotic microorganisms. *Antonie Van Leeuwenhoek*. Vol. 91, No. 3, pp. 277-289, ISSN 1572-9699.

Calderone, R. A. (2002) Candida and Candidiasis. Edited by R. A. Calderone, *American Society of Microbiology Press.* ISSN 1058-4838, Washington, DC, USA.

Chi, Z. M.,Chi, Z., Zhang, T., Liu, G. L.,& Yue,L. X. (2009) Inulinase expressing microorganisms and applications of inulinases. *Applied Microbiology and Biotechnology*. Vol. 82, No. 2, pp. 211-220, ISSN 0175-7598.

Christopher, T., Scragg, A. H., & Ratledge, C. (1983) A comparative study of citrate efflux from mitochondria of oleaginous and non oleaginous yeasts. *European Journal of Biochemistry* Vol. 130, No. 1, pp. 195–204, ISSN 0014-2956

Czabany, T., Athenstaedt, K., & Daum, G. (2007) Synthesis, storage and degradation of neutral lipids in yeast. *Biochimica et Biophysica Acta*. Vol. 1771, pp. 299–309. ISSN 0006-3002

Dahlqvist, A., Stahl, U., Lenman, M., Banas, A., Lee, M., Sandager, L., Ronne, H., & Stymne, S. (2009) Phospholipid:diacylglycerol acyltransferase: an enzyme that catalyzes the acyl-CoA-independent formation of triacylglycerol in yeast and plants. *Proceeding of the National Accademy of Science USA*. Vol. 6, No. 12, pp. 6487-6492, ISSN 0027- 8424

Daniel, H. J., Otto, R. T., Binder, M., Reuss, M., & Syldatk, C. (1999) Production of sophorolipids from whey: development of a two-stage process with *Cryptococcus curvatus* ATCC 20509 and *Candida bombicola* ATCC 22214 using deproteinized whey concentrates as substrate. *Applied Microbiology and Biotechnology*. Vol. 51, No. 1, pp. 40–45, ISSN 0175-7598

de Miguel, T., Calo, P., Díaz, A., & Villa, T. G. (1997) The genus Rhodosporidium: a potential source of beta-carotene. *Microbiologia*. Vol. 13, No. 1, pp. 67-70, ISSN 0213-4101

Dulermo, T. & Nicaud, J. M. (2011) Involvement of the G3P shuttle and β-oxidation pathway in the control of TAG synthesis and lipid accumulation in *Yarrowia lipolytica. Metabolic engineering*. Vol. 40, No. 4, pp. 483-488, ISSN 1096-7176

Easterling, E. R., French, W. T., Hernandez, R., & Licha, M. (2009) The effect of glycerol as a sole and secondary substrate on the growth and fatty acid composition of Rhodotorula glutinis. *Bioresource Technology*. Vol. 100, No. 1, pp. 356-361. ISSN 0960-8524

Economou, C.N., Aggelis, G., Pavlou, S., & Vayenas, D.V. (2011) Modeling of single-cell oil production under nitrogen-limited and substrate inhibition conditions. *Biotechnology and Bioengineering*. Vol. 108, No. 5, pp. 1049-1055, ISSN 0006-3592.

Fakas, S., Certik, M., Papanikolaou, S., Aggelis, G., Komaitis, M., & Galiotou-Panayotou, M. (2008) Gamma-linolenic acid production by *Cunninghamella echinulata* growing on complex organic nitrogen sources. *Bioresource Technology*. Vol. 99, No. 13, pp. 5986- 5990, ISSN 0960-8524

Findley, K., Rodriguez-Carres, M., Metin, B., Kroiss, J., Fonseca, A., Vilgalys, R., & Heitman, J. (2009) Phylogeny and phenotypic characterization of pathogenic Cryptococcus species and closely related saprobic taxa in the Tremellales. *Eukaryotic Cell.* Vol. 8, No. 3, pp. 353-361, ISSN 1535-9786

Gujjari, P., Suh, S. O., Coumes, K., Zhou, J. J. (2011) Characterization of oleaginous yeasts revealed two novel species: Trichosporon cacaoliposimilis sp. nov. and Trichosporon oleaginosus sp. nov. *Mycologia.* in press, ISSN 1557-2536.

Granger, L.M., Perlot, P., Goma, G., & Pareilleux, A. (1992) Kinetics of growth and fatty acid production of Rhodotorula glutinis. *Applied Microbiology and Biotechnology*. Vol. 37, No. 1, pp. 13-17, ISSN 0175-7598.

Granger, L.M., Perlot, P., Goma, G., & Pareilleux, A. (1993) Efficiency of fatty acid synthesis by oleaginous yeasts: Prediction of yield and fatty acid cell content from consumed C/N ratio by a simple method. *Journal of biochemical and microbiological technology and engineering*. Vol. 42, No. 10, pp. 1151-1156, ISSN 0006-3592.

Gutiérrez, A., López-García, S., & Garre, V. (2011) High reliability transformation of the basal fungus Mucor circinelloides by electroporation. *Journal of Microbiological Methods*. Vol. 84, No. 3, pp. 442-446, ISSN 0167-7012

Hall, M. J., & Ratledge, C. (1977) Lipid accumulation in an oleaginous yeast (Candida 107) growing on glucose under various conditions in a one- and two-stage continuous culture. *Applied and Environmental Microbiology*. Vol. 33, No. 3, pp. 577-584, ISSN 0099-2240.

Hansson, L. & Dostalek, M. (1986) Influence of cultivation conditions on lipid production by Cryptococcus albidus. *Applied Microbiology and Biotechnology*. Vol. 24, No. 1, pp. 12- 18, ISSN 0175-7598.

Hassan, M., Blanc, P. J., Granger, L. M., Pareilleux, A., & Goma, G. (1993) Lipid production by an unsaturated fatty acid auxotroph of the oleaginous yeast *Apiotrichum curvatum* grown in single-stated continuous culture. *Applied Microbiology and Biotechnology*. Vol. 40, No. 4, pp. 483-488, ISSN 0175-7598.

Heredia, L. & Ratledge, C. (1988) Simultaneous utilization of glucose and xylose by *Candida curvata* D in continuous culture. *Biotechnology Letters*. Vol. 10, No. 1, pp. 25-30, ISSN 0141-5492.

Holdsworth, J. E. & Ratledge, C. (1991) Triacylglycerol synthesis in the oleaginous yeast *Candida curvata* D. *Lipids*. Vol. 26, No. 2, pp. 111-118, ISSN 0024-4201.

Hu, C., Zhao, X., Zhao, J., Wu, S., & Zhao, Z. K. (2009) Effects of biomass hydrolysis by- products on oleaginous yeast *Rhodosporidium toruloides*. *Bioresource Technology*. Vol. 100, No. 20, pp. 4843-4847, ISSN 0960-8524.

Huang, C., Zong, M. H., Wu, H., & Liu, Q. P. (2009). Microbial oil production from rice straw hydrolysate by *Trichosporon fermentans*. *Bioresource Technology*. Vol. 100, No. 19, pp. 4535-4538, ISSN 0960-8524

Jacob Z. (1991) Enrichment of wheat bran by *Rhodotorula gracilis* through solid-state fermentation. *Folia Microbiologica*. Vol. 36, No. 1, pp. 86–91, ISSN 1874-9356.

Jacob, Z. (1992) Linear growth and lipid synthesis in the oleaginous yeast *Rhodotorula gracilis*. *Folia Microbiologica*. Vol. 37, No. 2, pp. 117–121, ISSN 1874-9356.

Jang, H.D., Lin, Y.Y., & Yang, S.S. (2005) Effect of culture media and conditions on polyunsaturated fatty acids production by *Mortierella alpina*. *Bioresource Technology*. Vol. 96, No. 15, pp. 1633-1644, ISSN 0960-8524.

Johnson, V. W., Singh, M., Saini, V. S., Adhikari, D. K., SIsta, V. R., & Yadav, N. K. (1992) Effect of pH on lipid accumulation by an oleaginous yeast: *Rhodotorula glutinis* IIP-30. *World Journal of Microbiology and Biotechnology* Vol. 8, pp. 382–384, ISSN 0959- 3993.

Johnson, V. W., Sigh, M., Saini V. S., Adhikari, D. K., Sista, V., & Yadav, N. K. (1995) Utilization of molasses for the production of fat by an oleaginous yeast, *Rhodotorula glutinis* IIP-30. *Journal of Industrial Microbiology and Biotechnology* Vol. 14, No. 1 pp. 1–4, ISSN 1476-5535

Kohlwein, S. D. (2010) Triacylglycerol homeostasis: insights from yeast. *Journal of Biological Chemistry*. Vol. 285, No. 21, pp. 15663-15667, ISSN 0021-9258.

Levinson, W.E., Kurtzman, C.P., & Min, T. (2007) Characterization of *Yarrowia lipolytica* and related species for citric acid production from glycerol. *Enzyme and Microbial Technology*. Vol. 41, No. 3, pp. 292-295, ISSN 0141-0229.

Liang, Y., Cui, Y., Trushenski, J., & Blackburn, J. W. (2010) Converting crude glycerol derived from yellow grease to lipids through yeast fermentation. *Bioresource Technology*. Vol. 101, No. 19, pp. 7581-7586, ISSN 0960-8524

Li, C. H., Cervantes, M., Springer, D. J., Boekhout, T., Ruiz-Vazquez, R. M., Torres-Martinez, S. R., Heitman, J., & Lee, S. C. (2011) Sporangiospore Size Dimorphism Is Linked to Virulence of Mucor circinelloides. *PLoS Pathogen*. Vol. 7, No. 6, e1002086, ISSN 1553-7366

Li, Q., Du, W., & Liu D. (2008) Perspectives of microbial oils for biodiesel production. *Applied Microbiology and Biotechnology*. Vol. 80, No. 5, pp. 749-756, ISSN 0175-7598.

Li, Y., Zhao, Z., & Bai, F. (2007) High-density cultivation of oleaginous yeast *Rhodosporidium toruloides* Y4 in fed-bach culture. *Enzyme and Microbial Technology*. Vol. 41, No. 3, pp. 312-317, ISSN 0141-0229.

Lin, H., Cheng, W., Ding, H. T., Chen, X. Zhou, Q. F., Zhao, Y. H. (2010) Direct microbial conversion of wheat straw into lipid by a cellulolytic fungus of *Aspergillus oryzae* A- 4 in solid-state fermentation. *Bioresource technology*. Vol. 101, No. 19, pp. 7556-7562, ISSN 0960-8524

Meesters, P. A. E. P., Huijberts, G. N. M., & Eggink, G. (1996) High-cell-density cultivation of the lipid accumulating yeast *Cryptococcus curvatus* using glycerol as a carbon source. *Applied Microbiology and Biotechnology*. Vol. 45, No. 5, pp. 749-756, ISSN 0175-7598.

Meeuwse, P., Tramper, J., & Rinzema, A. (2011a) Modeling lipid accumulation in oleaginous fungi in chemostat cultures: I. Development and validation of a chemostat model for *Umbelopsis isabellina*. *Bioprocess and Biosystems Engineering*. Epub ahead of print, ISSN 1615-7591.

Meeuwse, P., Tramper, J., & Rinzema A. (2011b) Modeling lipid accumulation in oleaginous fungi in chemostat cultures. II: Validation of the chemostat model using yeast culture data from literature. *Bioprocess and biosystems Engineering*. Epub ahead of print, ISSN 1615-7591.

Meng, X., , Yang, J., Xu, X., Zhang, L., Nie, Q., & Xian, M. (2009) Biodiesel production from oleaginous microorganisms. *Renewable Energy*. Vol. 34, No. 1, pp. 1-5, ISSN 0960-1481.

Mlickova, K., Luo, Y., d´Andrea, S., Pec, P., Chardot, T., & Nicaud, J.-M. (2004a). Acyl-CoA oxidase, a key step for lipid accumulation in the yeast *Yarrowia lipolytica*. *Journal of molecular catalysis. B, Enzymatic*. Vol. 28, No. 2-3, pp. 81-85 ISSN 1381-1177

Mlíckova, K., Roux, E., Athenstaedt, K., d'Andrea, S., Daum, G., Chardot, T., & Nicaud, J-M. (2004b). Lipid accumulation, lipid body formation, and acyl coenzyme A oxidases of the yeast *Yarrowia lipolytica*. *Applied and Environmental Microbiology*. Vol. 70, No. 7, pp. 3918-3924 ISSN 0099-2240

Pan, J. G., Kwak, N. Y., & Rhee, J. S. (1986) High density cell culture of *Rhodotorula glutinis* using oxygen-enriched air. *Biotechnology Letters*. Vol. 8, No. 10, pp. 715–718, ISSN 0141-5492

Papanikolaou, S., & Aggelis, G. (2002) Lipid production by *Yarrowia lipolytica* growing on industrial glycerol in a single-stage continuous culture. *Bioresource technology*. Vol 82, No. 1, pp. 43–49, ISSN 0960-8524

Papanikolaou, S., Chevalot, I., Komaitis, M., Marc, I., & Aggelis, G. (2002) Single cell oil production by *Yarrowia lipolytica* growing on an industrial derivative of animal fat in batch cultures. *Applied Microbiology and Biotechnology* Vol. 58, No. 3, pp. 308-312, ISSN 0175-7598

Papanikolaou, S., Muniglia, L., Chevalot, I., Aggelis, G., & Marc, I. (2003) Accumulation of a cocoa-butter-like lipid by *Yarrowia lipolytica* cultivated on agro-industrial residues. *Current Microbiology*. Vol. 46, No. 2, pp.124-130, ISSN 0343-8651.

Park, W.S., Murphy, P.A., & Glatz, B.A. (1990) Lipid metabolism and cell composition of the oleaginous yeast *Apiotrichum curvatum* grown at different carbon to nitrogen ratios. *Canadian Journal of Microbiology*. Vol. 36, No. 5, pp. 318-326, ISSN 0008-4166.

Peng, X. W. & Chen, H. Z. (2007) Microbial oil accumulation and cellulase secretion of the endophytic fungi from oleaginous plants. *Annals of Microbiology*. 57 Sakaiu,239-242, ISSN 1590-4261

Rajakumari, S., Grillitsch, K., & Daum, G. (2008) Synthesis and turnover of non-polar lipids in yeast. *Progress in Lipid Research*. Vol. 47, No. 3, pp. 157-171, ISSN 0163-7827

Ratledge, C. (2002) Regulation of lipid accumulation in oleaginous micro-organisms. *Biochemical Society Transactions*. Vol. 30, No. 6, pp. 1047-1050, ISSN 0300-5127

Ratledge, C., (2004) Fatty acid biosynthesis in microorganisms being used for single cell oil production. *Biochimie*. Vol. 86, No. 11, pp. 807–815, ISSN 0300-9084.

Ratledge, C. & Wynn, J. P. (2002) The biochemistry and molecular biology of lipid accumulation in oleaginous microorganisms. *Advances in Applied Microbiology*. Vol. 51, No. 1, pp. 1–44, ISSN 0065-2164.

Rymowicz, W., Fatykhova, A. R., Kamzolova, S. V., Rywińska, A., & Morgunov, I.G. (2010) Citric acid production from glycerol-containing waste of biodiesel industry by *Yarrowia lipolytica* in batch, repeated batch, and cell recycle regimes. *Applied Microbiology and Biotechnology*. Vol 87, No. 3, pp. 971-979. ISSN 0175-7598

Rywinska, A., Rymowicz, W., & Marcinkiewicz, M. (2010) Valorization of raw glycerol for citric acid production by *Yarrowia lipolytica* yeast. Electronic Journal of Biotechnology. Vol. 13, No. 4, ISSN 0717-3458

Rau, U., Nguyen, L. A., Roeper, H., Koch, H., & Lang, S. (2005) Fed-batch bioreactor production of mannosylerythritol lipids secreted by *Pseudozyma aphidis*. *Applied Microbiology and Biotechnology*. Vol. 80, No. :607–613, ISSN 0175-7598

Rossi, M., Buzzini, P., Cordisco, L., Amaretti, A., Sala, M., Raimondi, S., Ponzoni, C., Pagnoni, U. M., & Matteuzzi, D. (2009) Growth, lipid accumulation, and fatty acid composition in obligate psychrophilic, facultative psychrophilic, and mesophilic yeasts. *FEMS microbiology ecology*. Vol. 69, No. 3, pp. 363-372, ISSN 0168-6496

Sakuradani, E., Ando, A., Ogawa, J., & Shimizu, S. (2009) Improved production of various polyunsaturated fatty acids through filamentous fungus *Mortierella alpina* breeding. *Applied Microbiology and Biotechnology*. Vol. 84, No. 1, pp. 1-10. ISSN 0175-7598

Satyanarayana, T., & Kunze, G. (2010) Gatersleben, Germany (Eds.) Yeast Biotechnology: Diversity and Applications 746 pages ISBN: 1402082916

Saxena, V., Sharma, C. D., Bhagat, S. D., Saini, V. S., & Adhikari, D. K. (2008) Lipid and fatty acid biosynthesis by *Rhodotorula minuta*. *Journal of the American Oil Chemists Society*. Vol. 75, No. 4, pp. 501–505, ISSN 0003-021X.

Smit, M. S. (2004) Fungal epoxide hydrolases: new landmarks in sequence-activity space. *Trends in biotechnology*. Vol. 22, No. 3, pp. 123-129, ISSN 0167-7799

Tehlivets, O., Scheuringer, K., & Kohlwein, S. D. (2007) Fatty acid synthesis and elongation in yeast. *Biochimica et Biophysica Acta*. Vol. 1171, No. 3, pp. 255-270, ISSN 0006-3002

van der Walt, J. P. (1992) The Lipomycetaceae, a model family for phylogenetic studies. *Antonie Van Leeuwenhoek*. Vol. 62, No. 4, pp. 247-250, ISSN 0003-6072.

Vamvakaki, A. N., Kandarakis, I., Kaminarides, S., Komaitis, M. & Papanikolaou, S. (2010) Cheese whey as a renewable substrate for microbial lipid and biomass production by Zygomycetes *Engineering in Life Sciences* Vol. 10, No. 4, pp. 348–360, ISSN 1618- 0240

Vicente, G., Bautista, L. F., Rodríguez, R., Gutiérrez, F. J., Sádaba, I., Ruiz-Vázquez, R. M., Torres-Martínez, S., & Garre, V. (2009) Biodiesel production from biomass of an oleaginous fungus. *Biochemical Engineering Journal*. Vol. 48, No. 1, pp. 22–27, ISSN 1369-703X

Wang, S. L., Sun, J. S., Han, B. Z., & Wu, X. Z. (2007) Optimization of beta-carotene production by *Rhodotorula glutinis* using high hydrostatic pressure and response surface methodology. *Journal of Food Science*. Vol. 72, No. 8, pp. M325-329, 1911, ISSN 0022-1147

Wu, S., Hu, C., Jin, G., Zhao, X., & Zhao, Z.K. (2010) Phosphate-limitation mediated lipid production by *Rhodosporidium toruloides*. *Bioresource Technology*. Vol. 101, No. 15, pp. 6124-6129, ISSN 0960-8524

Wynn, J.P., Hamid, A. A., Li, Y, & Ratledge, C. (2001) Biochemical events leading to the diversion of carbon into storage lipids in the oleaginous fungi *Mucor circinelloides* and *Mortierella alpina*. *Microbiology*. Vol. 147, No. 10, pp. 2857-2864, ISSN 1350-0872.

Ykema, A., Verbree, E.C., Verseveld, H.W., & Smit, H. (1986) Mathematical modelling of lipid production by oleaginous yeasts in continuous cultures. *Antonie van Leeuwenhoek.* Vol. 52, No. 6, pp. 491-506, ISSN:1572-9699

Ykema, A., Kater, M. M. & Smit H. (1989) Lipid production in whey permeate by an unsaturated fatty acid mutant of the oleaginous yeast *Apiotrichum curvatum. Biotechnology Letters.* Vol.11, No. 7, pp. 477-482, ISSN 0141-5492

Yousuf, A., Sannino, F., Addorisio, V., Pirozzi, D. (2010) Microbial conversion of olive oil mill wastewaters into lipids suitable for biodiesel production. *Journal of agricultural and food chemistry.* Vol. 11 No. 58(15), pp. 8630-8635, ISSN 0021-8561

Yu, X., Zheng, Y., Dorgan, K. M., Chen, S. (2011) Oil production by oleaginous yeasts using the hydrolysate from pretreatment of wheat straw with dilute sulfuric acid. *Bioresource Technology.* Vol. 102, No. 10, pp. 6134-6140, ISSN 0960-8524

Zhang, Y., Adams, I. P., & Ratledge, C. (2007) Malic enzyme: the controlling activity for lipid production? Overexpression of malic enzyme in *Mucor circinelloides* leads to a 2.5- fold increase in lipid accumulation. *Microbiology.* Vol. 153, No. 7, pp. 2013-2025, ISSN 1350-0872

Zhang, J., Fang, X., Zhu, X. L., Li, Y., Xu, H. P., Zhao, B. F., Chen, L. & Zhang, X. D. (2011) Microbial lipid production by the oleaginous yeast *Cryptococcus curvatus* O3 grown in fed-batch culture. *Biomass and Bioenergy.* Vol. 35, No. 5, pp. 1906-1911, ISSN 0961- 9534

Zhao, C. H., Cuim W., Lium X. Y., Chi, Z. M., & Madzak, C. (2010) Expression of inulinase gene in the oleaginous yeast *Yarrowia lipolytica* and single cell oil production from inulin-containing materials. *Metabolic engineering.* Vol. 12, No. 6, pp. 510-517, ISSN 1096-7176

Zhao, X., Kong, X., Hua, Y., Feng, B., Zhao, Z. K. (2008) Medium optimization for lipid production through co-fermentation of glucose and xylose by the oleaginous yeast *Lipomyces starkeyi. European Journal of Lipid Science and Technology.* Vol. 110, No. 5, pp. 405–412, ISSN 1438-9312

Zhu, L. Y., Zong, M. H., & Wu, H, (2008) Efficient lipid production with *Trichosporon fermentans* and its use for biodiesel preparation. *Bioresource Technology.* Vol. 99, No. 16, pp. 7881-7885, ISSN 0960-8524

Zinoviev, S., Müller-Langer, F., Das, P., Bertero, N., Fornasiero, P., Kaltschmitt, M., Centi, G., & Miertus, S. (2010) Next-generation biofuels: Survey of emerging technologies and sustainability issues. *Chemistry & sustainability, energy & materials.* Vol. 3, No. 10, pp. 1106-1133, ISSN 1864-564X

5 Chapter

MICROBIAL BIODIESEL PRODUCTION - OIL FEEDSTOCKS DERIVED FROM MICROBIAL CELL CULTIVATIONS

1. INTRODUCTION: AN OVERVIEW

The price of crude oil has risen above $100 per barrel, resulting in significant adverse effects on both the global and national economies. In 2005, the United States generated 8.3 million barrels per day while consuming 20.8 million barrels per day, with the deficit being imported from foreign nations.For these fossil fuels, the U.S. used about 138 billion gallons of gasoline in 2006, accounting for about 44 percent of the world's gasoline consumption (EarthTrends, 2008). The annual U.S. usage of jet fuel was 21 billion gallons in 2006 (Energy Information Administration, 2008). The U.S. annual consumption of diesel fuel in 2006 was about 50 billion gallons (Energy Information Administration, 2008). Massive consumption of fossil fuels has already caused serious concern over global warming caused by greenhouse gases.

Biofuel offers an alternative to fossil fuels. It provides several benefits, such as alleviation from foreign oil dependence, carbon neutral process without greenhouse emission, and profits to local farmers. Bioethanol production from starch and lignocellulosic materials is a kind of an alternative to fossil fuels. It can be blended with gasoline in varying quantities up to pure ethanol (E100). The first generation of ethanol biofuel has been massively commercialized and dominated by the U.S. and Brazil. Fuel ethanol in the U.S. is primarily produced from corn, while Brazilian ethanol is produced mainly from sugarcane. These raw materials are in direct competition with human diet or the land to produce food, which triggers the controversy of food versus fuel. The second generation of ethanol is intended to be derived from lignocellulosic biomass, sourced from agricultural residues or other woody and herbaceous biomass from marginal lands. Extensive scientific study has been conducted over the past decade on this path to reduce total process costs, with a steady emphasis on commercialization.

2. BIODIESEL AND CURRENT FEEDSTOCKS

2.1 Biodiesel production

Another approach for alternative biofuels is biodiesel. The most common type of biodiesel is the methyl esters of fatty acid (FAME), obtained by transesterification of lipid with methanol or ethanol. It can be used in pure form (B100) or may be blended with fossil diesel at any rate. The commonly used biodiesel is B99 because 1% of fossil fuel is applied to inhibit mold growth, which shortens its shelf life. Biodiesel is around 5-8% less efficient than conventional fossil diesel; yet, biodiesel has potential as a total or partial (in the cold regions) replacement to the fossil diesel, compatible to the current diesel engine. There are numerous environmental benefits to replace fossil diesel to biodiesel since combustion of biodiesel emits far less pollutants comparing to fossil diesel (except the NO_x emission) and the entire process is close to carbon neutral considering that the plant oil used to produce biodiesel is synthesized in the agriculture from CO_2 in the air. The production and utilization of biodiesel are significant in many aspects, for instance, increasing oilseed crop market, providing domestic job opportunities to the rural community, and decreasing the dependence on imported oil; therefore biodiesel has been commercialized around the globe. Another type of biodiesel is the hydrocarbon generated with direct decarboxylation of fatty acid or lipid. Although it shows superior features than current FAME biodiesel in many aspects, the chemical decarboxylation process to produce diesel still needs further development compared to the transesterification process for biodiesel production.

2.2 Agricultural feedstocks: Vegetable oil, animal fat and recycled grease

The biodiesel industry suffered due to limited raw materials such as soybean oil or vegetable oil. Although there are numerous potential renewable and carbon neutral feedstocks for the production of biodiesel, none seems capable of displacing fossil diesel. Table 1 shows a comparison of the oil yields of some sources of biodiesel, the land area for their cultivation, and the percentage of existing United States (U.S.) cropping area to meet half of the transport fuel needs. Using soybean as feedstock for the production of biodiesel requires 326% of the U.S. cropping area to meet the 50% of all U.S. transport fuel needs. Based on this rough estimation, none of the terrestrial crops is able to completely substitute crude oil.

Table 1. Comparison of some sources of biodiesel (Chisti 2007)

Crop	Oil yield [Liters per hectare]	Land area needed [Mha][a]	Percent of existing US cropping area[a]
Corn	172	1540	846
Soybean	446	594	326
Canola	1190	223	122
Jatropha	1892	140	77
Coconut	2689	99	54
Oil palm	5950	45	24
Microalgaea	136,900	2	1.1
Microalgaec	58,700	4.5	2.5

[a] For meeting 50% of all transport fuel needs of the U.S.

[b] 70% oil (by wt) in biomass.

[c] 30% oil (by wt) in biomass.

Various raw materials are proposed to produce biodiesel, including waste cooking oil and various oil-accumulating plants, which could help solve the shortage for raw materials but are not sufficiently available. New approaches need to be developed to use an alternative biomass as the substrate for biofuel production, for example, biomass leftover from agricultural harvest or the biomass produced in non-traditional agricultural land as the source of sugars for biofuel and bioproducts. This potential has been gaining much attention recently as it is been defined as "second generation of bioenergy", compared to current technologies that primarily are limited to food materials. Oil accumulation with microalgae and other similar technologies started to show some potential to produce biofuel products with massive scales (Benemann, 1996; Chisti, 2007; Jarvis et al., 1994).

3. MICROORGANISMS FOR MICROBIAL LIPID PRODUCTION

3.1 Microalgae

Algae oil is being seriously considered because of the large oil yields it shows compared with other oilseeds. The Office of Fuels Development, a division of the Department of Energy, funded a program from 1978 through 1996 under the National Renewable Energy Laboratory known as the "Aquatic Species Program". The focus of this program was to investigate high-oil algaes that could be grown specifically for the purpose of widescale biodiesel production. NREL's research showed that one quad (7.5 billion gallons) of biodiesel could be produced from 200,000 hectares of desert land (200,000 hectares is equivalent to 780 square miles, roughly 500,000 acres). It would be preferable to spread the algae production around the country, to lessen the cost and energy used in transporting the feedstocks. Algae farms could also be constructed to use waste streams (either human waste or animal waste from animal farms) as a food source, which would spread algae production around the country. Nutrients can also be extracted from the algae for the production of a fertilizer high in nitrogen and phosphorous. By using waste streams (agricultural, farm animal waste, and human sewage) as the nutrient source, these farms essentially also provide a means of recycling nutrients from fertilizer to food to waste and back to fertilizer. Microalgae are generally autotrophic eukaryotic cells, and contain one or more types of chlorophyll plus additional pigments known as carotenoids and biloproteins (also called phycobilins). Carotenoids are yellow, orange, or red water-insoluble linear hydrocarbons; biloproteins are blue or red water-soluble pigment-protein complexes. The color of the different groups of algae depends on the ratio of these pigments. Almost all the microalgae cells contain chlorophyll, but the green color can be masked by the carotenoids, giving them a brown or red color. Microalgae cells are mostly unicellular, but some species are colonial or filamentous. They can grow autotrophically and/or heterotrophically, with a wide range of tolerance to different temperature, salinity, pH and nutrient availabilities. Cyanobacteria, although sometimes also referred as algae, specifically as blue-green algae, are by definition prokaryotic bacteria, instead of microalgae. The eukaryotes microalgae we are referring to include green algae, diatoms, yellow–green algae, golden algae, red algae, brown algae, dinoflagellates and others; and only limited species have the capability to accumulate high content of lipids in their cell biomass. Microbial species that can accumulate over 20% of lipids in their cell biomass are considered oleaginous species.

One of the characteristics of algae that separate them from other oilseeds is the quantity of lipids and fatty acids algae have as membrane components; some algae have been found to contain over 80% of lipids, which is of great interest for a sustainable feedstock for biodiesel production. Different microalgae species are listed in Table 2 as examples of lipid accumulation research. For example, *Chlorella vulgaris* is a commercially important green microalgae because it has the potential to serve as a food and energy source due to its high photosynthetic efficiency, which can, in theory, reach 8%. It can grow with both autotrophic and heterotrophic modes, and its mixotrophic growth rate is the sum of its autotrophic growth rate and heterotrophic growth rate, separately. *Chlorella protothecoides* is another single-cell green microalgae, which has high potential for the energy and food production. Heterotrophic growth of *C. protothecoides* supplied with acetate, glucose, or other organic compounds as carbon source, results in high biomass and high content of lipid in cells.

Table 2. Some examples of microalgae cultivation for oil accumulation

Microalgae	Cultures			Substrates	Growth Rate	Lipid Content
	AC	MC	HC			
Chlorella protothecoides	X		X	glucose, acetate / CO_2	3.74g/L - 144h	55.2%
Chlorella vulgaris	X	X	X	glucose, acetate, lactate / CO_2	0.098/h	
Crypthecodinium cohnii			X	glucose/ CO_2	40g/L - 60-90h	15-30%
Scenedesmus obliquus	X		X	glucose/ CO_2	double in 14h after adaptation	14-22%
Chlamydomonas reinhardtii	X	x	X	acetate/ CO_2	exponential during the first 20 Hr	21%
Micractinium pusillum		x	X	CO_2	0.94g/L - 24h	
Euglena Gracilis			X	CO2		14-20%
Schizochytrium sp				glycerol/ CO_2	130 -140h (28°C)	55%
Spirulina platensis	X		X	glucose/ CO_2	0.008 /h	
Botryococcus Braunii			X	CO_2	low growth rate	20 - 86%
Dunaliella salina	X	x		CO_2		~70%

AC: autotrophic cultures; MC: mixotrophic cultures; HC: heterotrophic cultures

Lipid accumulation occurs within the microalgae cells and it varies from strain and growth conditions. There are many nutritional and environmental factors controlling the cell growth and lipid contents, such as organic and inorganic carbon sources, nitrogen source, and other essential macro- and micro-nutrients like magnesium and copper, temperature, pH level, salinity, agitation speed (dissolved oxygen). Many microalgae species, for example, *C. protothecoides*, accumulated a higher content of lipids in cells and achieved higher growth rate when the culture was under heterotrophic mode. Like yeast and fungi, heterotrophic algae can accumulate biomass and lipids using organic carbon as its source instead of carbon dioxide and sunlight. Compared with autotrophic algae, the heterotrophic growth process has the advantages of no light limitation, a high degree of process control, higher productivity, and low costs for biomass harvesting (Barclay, Meager et al. 1994). Table 2 shows the oil production of autotrophically and heterotrophically cultured microalgae. Most heterotrophically cultured algae have greater than ten times the biomass concentration, while the lipid productivity is also significantly higher than the theoretical data for autotrophic cultivation. For certain algae strains, it was suggested the heterotrophically cultured cells exhibited better capability for biomass and lipid production. Miao and Wu (2006) reported that the oil content of heterotrophically cultured *C. protothecoides* was approximately four times greater than that in the corresponding autotrophic culture. Liu et al. (2010) demonstrated that the heterotrophically cultured cells of *Chlorella zofingiensis* showed 411% and 900% increases in dry cell weight and lipid yield, respectively, compared to autotrophically cultured cells. Moreover, biodiesel produced from heterotrophically cultured algae oils had similar properties to diesel fuel in terms of density, viscosity, heating value, and H/C ratio (Xu, Miao et al. 2006). In addition to lipid production, high value byproducts can be obtained from heterotrophically cultured microalgae, including polyunsaturated fatty acids and carotenoids (Chen and Chen 2006).

3.2 Yeast and fungi

Besides microalgae, many yeast and fungi species (e.g., *Mucor circillenous* or *Mortierella isabellina*) also can accumulate a high content of lipids (Xia 2011; Heredia-Arroyo, Wei et al. 2011). Many oleaginous yeasts were studied for lipid accumulation on different substrates, such as industrial glycerol (Meesters, Huijberts et al. 1996; Papanikolaou and Aggelis 2002), sewage sludge (Angerbauer, Siebenhofer et al. 2008), whey permeate (Ykema, Verbree et al. 1988; Akhtar, Gray et al. 1998), sugar cane molasses (Alvarez, Rodriguez et al. 1992), and rice straw hydrolysate (Huang, Zong et al. 2009). The use of non-starch biomass is critical so that lignocelluloses can be used for organic carbon supply without concern of using food crops for fuel sources. Recent studies detailed conversion of hemicellulose hydrolysate into lipids by oleaginous yeast strains and their tolerance degrees to lignocellulose degradation compounds (Chen, Li et al. 2009; Hu, Zhao et al. 2009; Huang, Zong et al. 2009). However, these strains were unable to efficiently produce lipids in the presence of inhibitors in the hydrolysate, necessitating detoxification treatment prior to fermentation, which increases the cost of the process. Thus, using strains capable of growing in the non-detoxified hydrolysate is necessary for viable microbial lipid production in an industrial context. In addition, previous reports indicate that temperature is an key factor in regulating the fatty acid composition in fungi (Kendrick and Ratledge 1992; Weinstein, Montiel et al. 2000).

Similarly, some oleaginous filamentous fungi can also produce lipids by utilizing glycerol, acetic acid, soluble starch, wheat straw, and wheat bran. Dey et al screening two endophytic oleaginous fungi *Colletotrichum sp.* and *Alternaria sp.* with lipid content 30% and 58% respectively (Dey, Banerjee et al. 2011). Fifteen eukaryotic microorganism were tested for waste glycerol assimilation to produce lipid. Fungi accumulated lipid inside their mycelia (lipid content ranging between

18.1 and 42.6%) (Chatzifragkou, Makri et al. 2011). Such capabilities provide potential to utilize sugars in the pretreated lignocellulosic materials hydrolysate. Moreover, because the fatty acid profile of microbial oils is quite similar to that of conventional vegetable oils, oleaginous filamentous fungi are suggested as a favorable feedstock for a sustainable biodiesel industry (Peng and Chen 2008; Zhao, Hu et al. 2010).

4. FEEDSTOCKS FOR MICROBIAL LIPID PRODUCTION

4.1 Light and carbon dioxide

Microalgae cells can generally utilize sunlight, carbon dioxide and nutrients from waste water for their cell growth (Brennan and Owende 2010). Lipid accumulation with microalgae cultivation is relatively efficient due to its high production efficiency and less demand of agricultural land. Most microalgal ponds have a solar energy conversion efficiency of 1–4% under normal operating conditions and higher efficiencies can be achieved with closed photo-bioreactor systems. There is a considerable margin for improvement, which is being targeted through accelerated breeding programs and genetic modification. Autotrophic microalgae cells also absorb CO_2 as their carbon source to support their cell growth, which makes the microalgae an attractive option for the biological CO_2 fixation. Atmospheric CO_2 accumulation, derived mainly from fossil fuel combustion, is proved as the leading cause of global warming. Current mitigation methods such as physicochemical adsorption, injection into deep oceans and geological formations are not economically feasible due to the high cost of implementing these methods and possible CO_2 leakage. After CO_2 is fixed via microalgae assimilation, the cell biomass can be utilized to generate methane gas, biochar, or the oil can be extracted to generate biodiesel. Microalgae can fix not only atmospheric CO_2, but also the CO_2 from industrial exhaust gases and from the carbonate salt, which are chemically fixed. Atmospheric CO_2 levels (0.0387%) are not sufficient to support the high microalgae growth rates and productivities needed for commercial biofuel production, adding CO_2 into autotrophic microalgae culture is an effective method to accelerate the microalgae growth rate. Utilization of CO_2, for example, flue gas from electrical plant, by means of microalgae alleviates the impact of CO_2 on the environment (greenhouse effect) and renders algal biomass production less expensive. It was also reported that the CO_2 content was reduced from 44–48% to 2.5–11% when the microalgae cultivation was integrated with anaerobic digestion to remove impurities of biogas produced from anaerobic digestion. Microalgae also assimilated other impurities such as ammonia and hydrogen sulfide; and the gas leaving the algae pond had 88–97% by volume of methane [23]. Although there are conflicting results from different references about the toxicity of ammonia, hydrogen sulfide and other impurities on the growth of microalgae cells, integration of microalgae cultivation together with this industrial processes can be more sustainable and economically feasible than the individual microalgae cultivation process to generate oil for biodiesel production.

4.2 Wastewater

Culturing microalgae with nutrients from wastewater, such as nitrogen and phosphate, can decrease the cost of the raw materials and also provide some environmental benefits. Agricultural effluent and municipal wastewater, even after treated with anaerobic digestion (AD), cannot be disposed directly because of their high nutrient level (Levine, Costanza- Robinson et al. 2011). In contrast, these wastewater can be considered as a cost-effective candidate of raw materials for biodiesel production (Siddiquee and Rohani 2011). Cultivation of microalgae, yeast, or fungi can be integrated with AD system to reduce the remaining COD, phosphate, and ammonia. Microalgae were also studied to absorb metal ion, waste pharmaceutical chemicals and dye into their cell biomass in order to remove the pollutants from wastewater once their cell biomass were stabilized and harvested. A typical example is the recently developed high-rate algae pond (HRAP) in the tertiary wastewater treatment facility (El Hamouri 2009). HRAP functions behind a two-step upflow anaerobic reactor (pre-treatment) and was followed by one maturation pond (MP) for polishing. The HRAP was revealed to have no activity for removing the COD from the wastewater; however, it removed 85% of total N and 63% of total P. Nitrogen removal was discovered due to the assimilation of microalgae for their growth, and denitrification did not play any role in removing the nitrogen in this process. Phosphorus removal in this process was attributed to chemical precipitation and biological assimilation (around 50% each). In removing ammonia, the HRAP is superior to the traditional bacterial nitrification-denitrification process, which requires the assimilation of extra-organic carbon as a carbon source. Phosphorus removal by microalgae is largely thought to be due to its uptake for normal growth, as an essential element required for making cellular constituents such as phospholipids, nucleotides, and nucleic acids. Under certain conditions microalgae can be triggered to uptake more phosphorus than is necessary for survival, in the form of polyphosphate (Powell, Shilton et al. 2011). Phosphorus removal by luxury uptake (amount of P uptake more than growth required) was confirmed to occur in the microalgae growing in the wastewater treatment facility; further research is needed in this prosperous field about the detailed mechanism and applications.

4.3 Lignocellulosic biomass

Several research studies revealed that organic carbon sources, if the algae have the capability to grow on heterotrophic mode, can significantly increase the cell growth rate and dramatically enhance the lipid content of the biomass. For example, *Chlorella protothecoides* can only accumulate 18-25% lipid in the autotrophic culture, while the lipid content can reach to 55.2% dry cell biomass if cultured with addition of organic carbons (Xiong, Li et al. 2008). However, the heterotrophic cultivation of microalgae mostly request pure mono- sugars, which are usually costly and limited. Alternative materials with relative abundance and zero or negative valued organic materials, such as lignocellulosic biomass from agriculture, are the only choice for this route. Agricultural feedstocks contribute a large part of renewable resource for biodiesel production, which are the target of biodiesel resource of cost reduction of biodiesel and non-human food resource discovery. Most of these substrates are locally available and thus are expected to support mainly small production facilities. Lignocellulosic materials are mainly composed of cellulose, hemicellulose, and lignin, which make up approximately 90% of the dry weight of most plant materials (Kumar, Barrett et al. 2009). Cellulose and hemicellulose can be converted to fermentable sugars for microbial lipid production. However, the direct enzymatic hydrolysis of cellulose and hemicellulose to sugars is impeded by the cell wall physico-chemical and structural composition. Thus, biomass pretreatment prior to enzymatic hydrolysis is essential to enhance enzymatic digestibility. Distinctly different from autotrophic microalgae cultures, this process is more similar to cellulosic ethanol production, where hydrolysis and fermentation are needed for conversion.

5. CONVERSION PROCESSES FOR MICROBIAL LIPID PRODUCTION

5.1 Autotrophic microalgae cultivation and lipid accumulation

Three cultivation processes were designed to culture microalgae and other oleaginous cells, including open-pond system, photobioreactor, and fermentation. The open-pond system is typically a closed loop with a pump to create microalgae flow in the channel. The channel is 0.2-0.5 meter deep, and the pump keeps the microalgae cell well mixed for continuous growth. This type of open-pond system has been used for several years because it is easy to operate and inexpensive to maintain. The open-pond system also can be upgraded to large-scale microalgae production. On the other hand, a high cell density of microalgae cannot be reached because of limited capability to assimilate the sun light and low carbon dioxide concentration of air. Increasing the CO_2 concentration by using flue gas instead of air can increase the microagale cell concentrations, but the final cell density is still limited to the mutual shading effects where light cannot penetrate through dense microalgae cell broth. Another problem is biological contamination during the long period of cultivation. The bacteria contamination or other non-oleaginous microalgae invasion can occur in stressed cultural conditions, where lipid accumulation usually is stimulated, such as nitrogen depletion or other nutrient imbalance. There is now extensive evidence that open-pond systems can operate for more than six months without significant contamination using a wide range of microalgae. Prolific strains of *Chlorella*, for example, are often dominant because they outgrow their competitors (and indeed can often be contaminants themselves in *Arthrospira* cultures or other microalgal strains). Extreme halophiles, such as *Dunaliella salina*, are also dominant in their optimal environments because they do not encounter much competition at high salinities. However, in the context of the wider microalgal industry, contamination issues are still of significant interest.

To enhance the productivity of microalgae, closed photobioreactor systems (tubular flat plate, Orcolumn are designed to increase the surface of microalgae broth exposed to sunlight. Closed photobioreactors are more costly than open-pond systems, but they have potential for higher productivity of cell biomass with less chance of contamination. The flat plate photobioreactors can receive greater sunlight for microalgae growth although there is potential for cell mixing. The microalgae cell density could reach up to 80g/L dry cell weight, significantly higher than the cell density of a pond system, which ranges within several g/L (Hu, 1998). Another design for the photoreactor is the tubular photoreactor, made with a diameter less than 0.1 m to maximize the sunlight harvest by microalgae. The tubular reactor can also expose the microalgae cells to sunlight from all the directions (Miron, Gomez et al. 1999; Ugwu, Ogbonna et al. 2002). There are a few reports about scale- up test of tubular photoreactor, such as the one in Hawaii with a size of $25M^3$ (Olaizola 2000), and 700 M^3 in Germany (Pulz 2001). However, the tubular photobioreactors cannot scale-up indefinitely because of oxygen accumulation, carbon dioxide limitation, and pH changes (Eriksen 2008). The third type of photoreactor is the column photoreactor, the most controllable type among three because it most closely resembles the traditional bioreactor. The column containing microalgae is vertical, and the air is bubbled from the bottom. Sunlight is provided horizontally (Eriksen 2008).

In addition to individual open-pond system and closed photobioreactor, the hybrid system of microalgae cultivation is currently under intense investigation because it combines the open-pond system and closed photobioreactor to increase the cell productivity and reduce the cost. The first stage is autotrophy to avoid biological contamination, and the second stage is heterotrophy, which provides the stress condition for lipid accumulation.

5.2 Oleaginous microbial fermentation and lipid accumulation

Besides the pond system and photobioreactor, heterotrophic cell cultivation, including heterotrophic microalgae, oleaginous yeast and fungi, is usually limited to industrial fermentation tanks. These cells can be cultured in a dark fermentor with optional sunlight, and can usually reach a very high cell density (up to 200g/L). Due to its excellent controllability of all operation parameters, most of the industrial microalgae cultivations for nutraceutical production (e.g., polyunsaturated fatty acid) have been switched to the heterotrophic fermentations where sugar is provided to produce high-valued products, such as Docosahexaenoic acid (DHA).

Table 3. Comparison of different algae cultivation systems (Alabi. 2009)

Factor	Raceway	Photobioreactor	Fermenter
Cell density in culture	Low	Medium	High
Limiting factor for growth	Light	Light	Oxygen
Culture volume necessary to harvest a unit weight of cells	High	Medium	Low
Surface area-to-volume ratio	High	Very high	Not applicable
Control over parameters	Low	Medium	Very high
Commercial availability	Readily available	Usually custom built	Readily available
Construction costs per unit volume produced	Medium	High	Low
Operating costs	Medium	High	Low
Technology base	Readily available	Under development	Readily available
Risk of contamination	High	Medium	Low
Evaporative water losses	High	High[33]	Low
Weather dependence	High	Medium	Low
Maintenance	Easy to maintain	Difficult to maintain	Requires specialized maintenance
Susceptibility to overheating	Low	High	N/A
Susceptibility to excessive O_2 levels	Low	High[34]	N/A
Ease of cleaning	Very easy	Difficult	Difficult (must be sterilized)
Ease of Scale-up	High	Variable[35]	High
Land requirement	High	Variable	Low
Applicability to different species	Low	High	Low

Table 3 compares three cultivation methods and shows that the process cost of fermentation can be high due to its requirement of raw materials and oxygen, and sterilization of culture media during the cell growth. It is readily available both in the lab and in the industry, but is only suitable to produce high-valued products, of which biofuel products are not. The key barriers to apply this technology to biofuel production is the cost and availability of raw materials. Considering the competition with human diet, sugars cannot serve as the raw material for biofuel production; and alternative materials such as lignocellulosic materials should be used for the heterotrophic oil production. If the oleaginous cells are capable of generating the hydrolytic enzymes for lignocelluloses degradation, it will be the big plus for biodiesel production via oleaginous fermentation the overall system. Otherwise, external hydrolytic enzymes have to be used to release the monosugar, followed by lipid accumulation via olgeaginous microorganisms. Separated hydrolysis and fermentation (SHF) is a common working model to have these two steps separated. Two bioreactors will be necessary because the hydrolytic degradation of lignocellulose is preferred at 50ºC, while the oleaginous microorganisms grow at much lower temperature (28^0C to 30^0C for most of the fungus). Simultaneous saccharification and fermentation is another working model currently under intense investigation, in which two steps are integrated into one.

Different fermentation processes are applied to obtain high productivity of lipids and high conversion ratio of substrate for the fermentation, such as batch cultivation, fed-batch cultivation, and continous cultivation. Fed-batch cultivatioin is a modified batch model that can reach high cell density and it has many applications in the fermentative lipid accumulation process. For example, *Rhorosporidum toruloides* reach much higher cell density with 48% lipid compared to its batch cultivation (Li, Zhao et al. 2007). The high productivity of fed-batch cultivation was conformed by *Phodotorula glutinis* (Xue, Miao et al. 2008), *C. curvatus* (Meesters, Huijberts et al. 1996), and *L. starkeyi* (Yamauchi, Mori et al. 1983). Continuous cultivation has advantages of easy maintenance and time-saving, although it is difficult to control the contamination. It has limited applicatioins in the fermentative lipid accumulation.

Besides the commonly used submerged cultivations, solid state fermentation, as a compact process for lipid production, showed many advantages, such as low requirements to the raw materails; low capital cost; low energy expenditure; less expensive downstream processing; less water usage and low water output; potential higher volumetric productivity; less fermentation space; easy operation and maintenance. The research for *Aspergillus oryzae* growing on rice bran and wheat bran through solid state fermentation resulted to the lipid content of cell biomass at about 10-11% (Da Silveira, Oliveira et al. 2010). The lipid yield reached

62.87 mg/gds in solid state fermentation on the 6th day after Plackett-Burman design (PBD) by *A. oryzae* A-4 (Lin, Cheng et al. 2010) . Currently, the solid state fermentation research is still in its infancy and many barriers are hindering this process from commercilization. The lipid yield is relatively low compared to submerge cultivation. Modern biotechnological approaches, such as heterogenous expression of hydrolytic enzymes and UV radiation, are available to enhance the hydrolytic enzymes production (Li, Yang et al. 2010; Awan, Tabbasam et al. 2011). Semi-solid state fermentation is used to avoid high sugar concentration on the surface of lignocellulose. An oleaginous fungus *M. isabellina* was cultured at semi-solid state fermentation with the results of 11g oil per 100g sweet sorghum (Economou, Makri et al. 2010) .

6. CELL HARVEST AND LIPID EXTRACTION

6.1 Cell harvest methods

The algae cell harvest from pond water and the subsequent water reuse have been one of the major obstacles for the algae-to-fuel approach. Microalgae cell harvest is technically challenging, especially considering the low cell densities (typically in the range of 0.3–5 g/L) of autotrophic microalgae due to limited light penetration, the small size of the oleaginous algal cells (typically in the range of 2–40 um), and their similar density to water (Li, Horsman et al. 2008). Oleaginous microalgae cells are usually suspended in the water and are hard to settle by natural gravity force due to their negative charges. The recovery of microalgae biomass generally requires one or more solid–liquid separation steps, and usually accounts for 20–30% of the total costs of production, according to one source (Uduman, Qi et al. 2010).

How to harvest microalgae cells is dependent on the characteristics of the microalgae, such as size and density(Olaizola 2003). All of the available harvest approaches, which include flocculation, flotation, centrifugal sedimentation, and filtration, have limitations for effective, cost-efficient production of biofuel (Shelef, Sukenik et al. 1984). For instance, flotation methods, based on the trapping of algae cells using dispersed micro-air bubbles, is very limited in its technical and economic viability. Most conventional and economical separation methods such as filtration and gravitational sedimentation are widely applied in wastewater treatment facilities to harvest relatively large (>70 μm) microalgae such as *Coelastrum* and *Spirulina*. However, they cannot be used to harvest algae species approaching bacterial dimensions (<30 μm) like *Scenedesmus*, *Dunaliella*, and *Chlorella* (Brennan and Owende 2010), to which most oleaginous microalgae species belong. Centrifugation is a method widely used to recover microalgae biomass, especially small- sized algae cells; however, its application is restricted to algae cultures for high-value metabolites due to intensive energy needs and high equipment maintenance requirements. While flocculation is used to harvest small-sized microalgae cells, it is a preparatory step to aggregate the microalgae cells and increase the particle size so that other harvesting methods such as filtration, centrifugation, or gravity sedimentation can be applied (Molina Grima, Belarbi et al. 2003). Several flocculants have been developed to facilitate the aggregation of microalgae cells, including multivalent metal salts like ferric chloride ($FeCl_3$), aluminium sulphate ($Al_2(SO_4)_3$), and ferric sulphate ($Fe_2(SO_4)_3$), and organic polymers such as Chitosan (Li, Horsman et al. 2008). Chemical flocculation can be reliably used to remove small algae cells from pond water by forming large-sized (1–5 mm) flocs (Sharma, Dhuldhoya et al. 2006). However, the chemical reactions are highly sensitive to pH and the high doses of flocculants required produce large amounts of sludge and may leave a residue in the treated effluent. In summary, most technologies including chemical and mechanical methods greatly increase operational costs for algal production and are only economically feasible for production of high-value products (Park, Craggs et al. 2011).

Besides traditional methods mentioned above, there are several new technology developments in this field. DOE-ARPA-E recently funded a research project for Algae Venture Systems (AVS) to develop a Harvesting, Dewatering, and Drying (AVS-HDD) technology by using the principles of liquid adhesion and capillary action to extract water from dilute microalgae solutions. Attached algal culture systems have been developed for growing microalgae on the surface of polystyrene foam (Wilkie and Mulbry 2002) (Johnson and Wen 2010) to simplify the cell harvest. New bioflocculants, which are more environmentally friendly, are also proposed to address the cost and environmental concerns for current flocculation method (Uduman, Qi et al. 2010). All these methods are innovative and will decrease the harvest cost to some extent if developed successfully, but heavy investments on equipment and chemical supplies are still needed.

Dr. Bo Hu's research group at University of Minnesota developed an innovative approach to enhance natural algae aggregation and to encourage simple gravity settling or filtration by co-culturing filamentous fungal cells at the end of the microalgae cultures. Instead of suspended culture, this approach uses pelletized or granulized culture where cells form pellets in culture medium. In submerged cultures, many filamentous microorganisms tend to aggregate and grow as pellets/granules. They are spherical or ellipsoidal masses of hyphae with variable internal structure, ranging from loosely packed hyphae, forming "fluffy" pellets, to tightly packed, compact, dense granules (Hu and Chen 2007; Hu and Chen 2008; Hu, Zhou et al. 2009; Chunjie Xia 2011). Besides merits from the cell immobilization, there are several other advantages, especially for the micro-oil production: a). easy to harvest cells, and b). easy to re-use pond water (Johnson and Wen 2010;

Xia 2011). As the first research group to introduce pelletized liquid fermentation (PLF) into biofuel production, this research group at University of Minnesota found key operational conditions that induce the fungal pelletization. They discovered that changing conditions during cell cultivation can force fungal cells to aggregate and form pellets. This method avoids traditional approaches that use $CaCO_3$ powder to induce the fungal pelletization (Liao, Liu et al. 2007; Liu, Liao et al. 2008), which are costly and cause solid waste disposal issues. Self aggregated pelletization/granulation dramatically improves mass transfer and cell cultivation performance and facilitates cell harvest and separation. A simple filtration can be used to separate the cell biomass from the fermentation broth. This approach brings tremendous advantages to decrease the harvest cost of biofuel production, especially when the raw materials only contain very diluted sugar, (which are the cases for many agricultural waste). This would appear to be the most promising option to achieve both a high-quality treated effluent in terms of total suspended solids and economically recovering algal biomass for biofuel use (Uduman, Qi et al. 2010). It will also be more environmentally sound than current procedures which may need chemical addition.

6.2 Lipid extraction methods

Oil extraction also contributes a large part of the cost in the process to generate microbial biodiesel. Several oil extraction technologies are currently available to process the microbial biomass in order to meet the requirement of being low cost, easy and safe to operate, and environmentally friendly.

6.2.1 Mechanical methods

Mechanical methods include pressing, bead milling, and homogenization. Pressing is a technology to harvest lipids out of cells by high pressure. Bead milling works in a container to destruct the cell wall by high speed small beads. Homogenization provides a sudden pressure change when cells go through an orifice. The mechanical technologies are often used in combination with solvent methods to separate the lipid from the cell biomass. The mechanical methods are energy intensive and better operated at the high cell density condition; in addition, pretreatments are necessary to obtain high recovery ratio (Greenwell, Laurens et al. 2010).

6.2.2 Solvent extraction methods

Solvent extraction is a commonly used method for soybean processing, and it is also used to extract lipids from microbial cells. Organic solvents should be insoluble in water, be easy to obtain, have a low boiling point, and be reusable. Current industrial solvents for microlipids accumulation include hexane, chloroform, acetone, benzene, and cyclohexane, can dissolve lipid without residual cell. The extraction process is significantly affected by operation condition, such as temperature and pressure. Accelerated solvent extraction (ASE) is named when the operation temperature is higher than that of solvent boiling point, which can be used for oil extraction from dry biomass (Cooney, Young et al. 2009). Mixture chloroform and methanol (Bligh and Hyer method) is the most common organic solvent to extract oil from biomass. This organic mixture can extract oil not only from dry biomass but also from wet biomass. However, the efficiency is different at certain condition (Zhu, Zhou et al. 2002). The efficiency of oil extraction was not working well at wet *Mortierella alpina* biomass. The process generated large amounts of wastewater and solvent often contaminated the final products. Simultaneous extraction and tranesterification is more efficient (15-20%) than the separate process (Belarbi, Molina et al. 2000); however, the important point of the simultaneous process is to balance the reaction time for the best components of product (Lewis, Nichols et al. 2000).

6.2.3 Supercritical fluid extraction

Supercritical fluid extraction takes advantage that some chemicals behave as both a liquid and a gas, and have increasing solvating power when they are raised above their critical temperature and pressure points. Carbon dioxide is the most commonly used supercritical fluid, sometimes modified by co-solvents such as ethanol or methanol. Critical temperature and critical pressure of carbon dioxide is at 31°C and 74 bar, respectively (Cooney, Young et al. 2009). Supercritical fluids produce highly purified extracts without using toxic solvent; and the process is fast and safe for thermally sensitive products. Supercritical CO_2 extraction efficiency is affected by four main factors: pressure, temperature, CO_2 flow rate, and extraction time. Ethanol (10 -15%), co-solvent, lead to similar results of Bligh and Hyer method at extracting oil from *Arthrospira maxima* and *Spirulina platensis* (Mendes, Reis et al. 2006; Sajilata, Singhal et al. 2008). The limitation of supercritical fluid extraction is high capital cost and high cost for maintainence.

6.2.4 Other methods

Besides the methods mentioned above, numerous technologies are being tested at different labs to harvest lipids from cells. Genetic engineering has been applied to improve the porosity of the cell membranes in order to increase the release of lipids directly from the cells (Greenwell, Laurens et al. 2010). Enzymes treatment and pulsed electric field technology are other effective methods to break the cell wall and membrane and enhance the mass transfer across the cell membrane for oil extraction (Shah, Sharma et al. 2004; Guderjan, Elez-Martinez et al. 2007). Microwave technology is a portential pretreatment method, which heats the cell components in order to increase the release of oil. Oil yield increased from 4.8% to 17.7% from microalgae *Crypthecodinium chnii* when microwave was applied (Cravotto, Boffa et al. 2008). Microwave technology is featured for its time-savings, but its disadvantages include the oxidative damage to products and its intense energy need. Sonnication is a timely and efficient method, free of toxic materials. Cavitation occurs when high voltage is applied into cell lipids. Vapour bubbles form with negative pressure and cause a violent collapse when compressed under positive pressure while growing; then the cell contents are released (Wei, Gao et al. 2008). The sonnication is, however, difficult to scale-up.

7. TECHNO-ECONOMIC ANALYSIS AND LIFE CYCLE ASSESSMENT

A complete techno-econoic analysis for the microbial biodiesel production is difficult, especially considering that most of the technologies are still in the early research stage. Initial investment into microalgal biofuels has mostly failed and several early start-up companies have closed. Different versions of economic analysis for microalgae biofuel production have been published recently, and Table 4 lists an analysis conducted by Seed Science Ltd, sporsored by the British Columbia Innovation Council in Canada.

Table 4 shows that although photobioreactor has a higher cell concentration and utilizes CO_2, its cost to produce lipid is the highest of all methods. Heterotrophic fermentation, however, appears to be the most economically feasible route to produce microbial biodiesel. Techno-economic analyses may vary from different research group, but their conclusions are similar. The biomass and oil generated from heterotrophic fermentation are more close to current fossile fuel cost. Heterotrophic fermentation relies less on local climate conditions and can be carried out in close fermentors, which may facilitate their commercialization.

Table 4. Cost comparison among different microalagae cultivation methods (Alabi. 2009)

	Raceway		Photobioreactor		Fermentor	
Initial invertment ($/L)	52		111		2	
Production cost						
Labor cost	$4.03	26.69%	$2.96	11.90%	$0.29	10.88%
Other production cost	$3.71	24.59%	$6.37	25.59%	$2.07	78.45%
Capital cost	$7.35	48.71%	$15.56	62.50%	$0.28	10.66%
Total cost	$15.09		$24.89		$2.63	
Credit from sale of algae cake*	$0.65		0.29		$0.05	
Net total cost	$14.44		$24.60		$2.58	
Lipid content	15%		25%		50%	
Cost per kg of algae	$2.66		$7.32		$1.54	

*Assumes that the algae cake is sold to an ethanol producer for its carbohydrate content

More effective, cost-efficient, and environmentally sound fermentation means to produce lipids are urgently needed, as well as adaptation of the fermentation cells to utilize lignocellulosic biomass. It is also widely indicated that currently microalgal biofuel systems are dependent on the production of coproducts (e.g., biochar, pigments, and nutriceuticals) for profitability. Considering the large scale of biofuel production, the market of the valuable byproducts will be the primary concern.

8. CONCISE SUMMARY

Despite the theoretical promise of microalgal biofuel systems to resolve the food vs fuel dilemma, no microbial biofuel system has yet attained economic viability. Microbial lipid productivity must substantially rise, and total costs must markedly reduce before this method can become economically viable.

REFERENCES

Akhtar, P., Gray, J.I.; et al. (1998). "Synthesis of lipids by certain yeast strains grown on whey permeate." *Journal of Food Lipids* 5(4): 283-297.

Alabi., A. O. (2009). "Microalgae technologies and processes for biofuels/bioenergy production in British Columbia."

Alvarez, R. M., B. Rodriguez, et al. (1992). "Lipid accumulation in Rhodotorula glutinis on sugar cane molasses in single-stage continuous culture." *World Journal of Microbiology & Biotechnology* 8(2): 214-215.

Angerbauer, C., M. Siebenhofer, et al. (2008). "Conversion of sewage sludge into lipids by Lipomyces starkeyi for biodiesel production." *Bioresource Technology* 99(8): 3051- 3056.

Awan, M. S., N. Tabbasam, et al. (2011). "Gamma radiation induced mutagenesis in Aspergillus niger to enhance its microbial fermentation activity for industrial enzyme production." *Molecular Biology Reports* 38(2): 1367-1374.

Barclay, W., K. Meager, et al. (1994). "Heterotrophic production of long chain omega-3 fatty acids utilizing algae and algae-like microorganisms." *J. Appl. Phycol.* 6(2): 123-129.

Belarbi, E. H., E. Molina, et al. (2000). "A process for high yield and scaleable recovery of high purity eicosapentaenoic acid esters from microalgae and fish oil." *Enzyme and Microbial Technology* 26(7): 516-529.

Brennan, L. and P. Owende (2010). "Biofuels from microalgae--A review of technologies for production, processing, and extractions of biofuels and co-products." *Renewable and Sustainable Energy Reviews* 14(2): 557-577.

Chatzifragkou, A., A. Makri, et al. (2011). "Biotechnological conversions of biodiesel derived waste glycerol by yeast and fungal species." *Energy* 36(2): 1097-1108.

Chen, G.-Q. and F. Chen (2006). "Growing Phototrophic Cells without Light." *Biotechnology Letters* 28(9): 607-616.

Chen, X., Z. Li, et al. (2009). "Screening of oleaginous yeast strains tolerant to lignocellulose degradation compounds." *Applied Biochemistry and Biotechnology* 159(3): 591-604.

Chisti, Y. (2007). "Biodiesel from microalgae." *Biotechnol Adv* 25(3): 294-306.

Cooney, M., G. Young, et al. (2009). "Extraction of Bio-oils from Microalgae." *Separation and Purification Reviews* 38(4): 291-325.

Cravotto, G., L. Boffa, et al. (2008). "Improved extraction of vegetable oils under high- intensity ultrasound and/or microwaves." *Ultrasonics Sonochemistry* 15(5): 898-902.

Da Silveira, C. M., M. D. Oliveira, et al. (2010). "Lipid content and fatty acid profile of defatted rice bran and wheat bran submitted to solid state fermentation by *Aspergillus oryzae*." *Boletim Do Centro De Pesquisa De Processamento De Alimentos* 28(1): 133-140.

Dey, P., J. Banerjee, et al. (2011). "Comparative lipid profiling of two endophytic fungal isolates - Colletotrichum sp. and Alternaria sp. having potential utilities as biodiesel feedstock." *Bioresource technology* 102(10): 5815-5823.

Economou, C. N., A. Makri, et al. (2010). "Semi-solid state fermentation of sweet sorghum for the biotechnological production of single cell oil." *Bioresource technology* 101(4): 1385-1388.

El Hamouri, B. (2009). "Rethinking natural, extensive systems for tertiary treatment purposes: The high-rate algae pond as an example." *Desalination and Water Treatment* 4(1-3): 128-134.

Eriksen, N. T. (2008). "The technology of microalgal culturing." *Biotechnology Letters* 30(9): 1525-1536.

Greenwell, H. C., L. M. L. Laurens, et al. (2010). "Placing microalgae on the biofuels priority list: a review of the technological challenges." *Journal of the Royal Society Interface* 7(46): 703-726.

Guderjan, M., P. Elez-Martinez, et al. (2007). "Application of pulsed electric fields at oil yield and content of functional food ingredients at the production of rapeseed oil." *Innovative Food Science & Emerging Technologies* 8(1): 55-62.

Heredia-Arroyo, T., W. Wei, et al. (2011). "Mixotrophic cultivation of Chlorella vulgaris and its potential application for the oil accumulation from non-sugar materials." *Biomass & Bioenergy* 35(5): 2245-2253.

Hu, B. and S. L. Chen (2007). "Pretreatment of methanogenic granules for immobilized hydrogen fermentation." *International Journal of Hydrogen Energy* 32(15): 3266-3273.

Hu, B. and S. L. Chen (2008). "Biological hydrogen production using chloroform-treated methanogenic granules." *Applied Biochemistry and Biotechnology* 148(1-3): 83-95.

Hu, B., X. Zhou, et al. (2009). "Changes in Microbial Community Composition Following Treatment of Methanogenic Granules with Chloroform." *Environmental Progress & Sustainable Energy* 28(1): 60-71.

Hu, C., X. Zhao, et al. (2009). "Effects of biomass hydrolysis by-products on oleaginous yeast Rhodosporidium toruloides." *Bioresource Technology* 100(20): 4843-4847.

Hu, Q., Kurano, N., Iwasaki, I., et al. (1998). Ultrahigh cell density culture of a marine green alga, Chlorococcum littorale in a flat plate photobioreactor. *Applied Microbiology Biotechnology* 49(6):655-662.

Huang, C., M.-h. Zong, et al. (2009). "Microbial oil production from rice straw hydrolysate by Trichosporon fermentans." *Bioresource Technology* 100(19): 4535-4538.

Johnson, M. B. and Z. Y. Wen (2010). "Development of an attached microalgal growth system for biofuel production." *Appl. Microbiol. Biotechnol.* 85(3): 525-534.

Kendrick, A. and C. Ratledge (1992). "Lipid formation in the oleaginous mould Entomophthora exitalis grown in continuous culture: effects of growth rate, temperature and dissolved oxygen tension on polyunsaturated fatty acids." *Applied Microbiology and Biotechnology* 37(1): 18-22.

Kumar, P., D. M. Barrett, et al. (2009). "Methods for Pretreatment of Lignocellulosic Biomass for Efficient Hydrolysis and Biofuel Production." *Industrial & Engineering Chemistry Research* 48(8): 3713-3729.

Levine, R. B., M. S. Costanza-Robinson, et al. (2011). "Neochloris oleoabundans grown on anaerobically digested dairy manure for concomitant nutrient removal and biodiesel feedstock production." *Biomass & Bioenergy* 35(1): 40-49.

Lewis, T., P. D. Nichols, et al. (2000). "Evaluation of extraction methods for recovery of fatty acids from lipid-producing microheterotrophs." *Journal of Microbiological Methods* 43(2): 107-116.

Li, X. H., H. J. Yang, et al. (2010). "Enhanced cellulase production of the Trichoderma viride mutated by microwave and ultraviolet." *Microbiological Research* 165(3): 190-198.

Li, Y. H., Z. B. Zhao, et al. (2007). "High-density cultivation of oleaginous yeast Rhodosporidium toruloides Y4 in fed-batch culture." *Enzyme and Microbial Technology* 41(3): 312-317.

Li, Y., M. Horsman, et al. (2008). "Biofuels from microalgae." *Biotechnol Prog* 24(4): 815-820.

Liao, W., Y. Liu, et al. (2007). "Studying pellet formation of a filamentous fungus Rhizopus oryzae to enhance organic acid production." *Applied Biochemistry and Biotechnology* 137: 689-701.

Lin, H., W. Cheng, et al. (2010). "Direct microbial conversion of wheat straw into lipid by a cellulolytic fungus of Aspergillus oryzae A-4 in solid-state fermentation." *Bioresource technology* 101(19): 7556-7562.

Liu, J., J. Huang, et al. (2010). "Differential lipid and fatty acid profiles of photoautotrophic and heterotrophic Chlorella zofingiensis: Assessment of algal oils for biodiesel production." *Bioresour. Technol.* 102(1): 106-110.

Liu, Y., W. Liao, et al. (2008). "Co-production of lactic acid and chitin using a pelletized filamentous fungus Rhizopus oryzae cultured on cull potatoes and glucose." *Journal of Applied Microbiology* 105(5): 1521-1528.

Meesters, P. A. E. P., G. N. M. Huijberts, et al. (1996). "High cell density cultivation of the lipid accumulation yeast Cryptococcus curvatus using glycerol as a carbon source." *Appl. Microbiol. Biotechnol.* 45(5): 575-579.

Meesters, P. A. E. P., G. N. M. Huijberts, et al. (1996). "High cell density cultivation of the lipid accumulation yeast Cryptococcus curvatus using glycerol as a carbon source." *Applied Microbiology and Biotechnology* 45(5): 575-579.

Mendes, R. L., A. D. Reis, et al. (2006). "Supercritical CO2 extraction of gamma-linolenic acid and other lipids from Arthrospira (Spirulina)maxima: Comparison with organic solvent extraction." *Food Chemistry* 99(1): 57-63.

Miao, X. and Q. Wu (2006). "Biodiesel production from heterotrophic microalgal oil." *Bioresour. Technol.* 97(6): 841-846.

Miron, A. S., A. C. Gomez, et al. (1999). "Comparative evaluation of compact photobioreactors for large-scale monoculture of microalgae." *Journal of Biotechnology* 70(1-3): 249-270.

Molina Grima, E., E. H. Belarbi, et al. (2003). "Recovery of microalgal biomass and metabolites: process options and economics." *Biotechnology Advances* 20(7-8): 491- 515.

Olaizola, M. (2000). "Commercial production of astaxanthin from Haematococcus pluvialis using 25,000-liter outdoor photobioreactors." *Journal of Applied Phycology* 12(3-5): 499-506.

Olaizola, M. (2003). "Commercial development of microalgal biotechnology: from the test tube to the marketplace." *Biomolecular Engineering* 20(4-6): 459-466.

Papanikolaou, S. and G. Aggelis (2002). "Lipid production by Yarrowia lipolytica growing on industrial glycerol in a single-stage continuous culture." *Bioresource Technology* 82(1): 43-49.

Park, J. B. K., R. J. Craggs, et al. (2011). "Wastewater treatment high rate algal ponds for biofuel production." *Bioresource technology* 102(1): 35-42.

Peng, X. and H. Chen (2008). "Single cell oil production in solid-state fermentation by Microsphaeropsis sp. from steam-exploded wheat straw mixed with wheat bran." *Bioresource Technology* 99(9): 3885-3889.

Powell, N., A. Shilton, et al. (2011). "Luxury uptake of phosphorus by microalgae in full- scale waste stabilisation ponds." *Water Science and Technology* 63(4): 704-709.

Pulz, O. (2001). "Photobioreactors: production systems for phototrophic microorganisms." *Appl. Microbiol. Biotechnol.* 57(3): 287-293.

Sajilata, M. G., R. S. Singhal, et al. (2008). "Supercritical CO2 extraction of gamma-linolenic acid (GLA) from Spirulina platensis ARM 740 using response surface methodology." *Journal of Food Engineering* 84(2): 321-326.

Samson, R. and A. Leduy (1985). "Multistage continuous cultivation of blue-green-alga spirulina-maxima in the flat tank photobioreactors with recycle." *Canadian Journal of Chemical Engineering* 63(1): 105-112.

Shah, S., A. Sharma, et al. (2004). "Extraction of oil from Jatropha curcas L. seed kernels by enzyme assisted three phase partitioning." *Industrial Crops and Products* 20(3): 275- 279.

Sharma, B. R., N. C. Dhuldhoya, et al. (2006). "Flocculants—an Ecofriendly Approach." *Journal of Polymers and the Environment* 14(2): 195-202.

Shelef, G., A. Sukenik, et al. (1984). Microalgae harvesting and processing: a literature review: Medium: ED; Size: Pages: 70.

Siddiquee, M. N. and S. Rohani (2011). "Lipid extraction and biodiesel production from municipal sewage sludges: A review." *Renewable & Sustainable Energy Reviews* 15(2): 1067-1072.

Uduman, N., Y. Qi, et al. (2010). "Dewatering of microalgal cultures: A major bottleneck to algae-based fuels." *Journal of Renewable and Sustainable Energy* 2(1).

Ugwu, C. U., J. C. Ogbonna, et al. (2002). "Improvement of mass transfer characteristics and productivities of inclined tubular photobioreactors by installation of internal static mixers." *Appl. Microbiol. Biotechnol.* 58(5): 600-607.

Wei, F., G. Z. Gao, et al. (2008). "Quantitative determination of oil content in small quantity of oilseed rape by ultrasound-assisted extraction combined with gas chromatography." *Ultrasonics Sonochemistry* 15(6): 938-942.

Weinstein, R. N., P. O. Montiel, et al. (2000). "Influence of growth temperature on lipid and soluble carbohydrate synthesis by fungi isolated from fellfield soil in the maritime Antarctic." *Mycologia* 92(2): 222-229.

Wilkie, A. C. and W. W. Mulbry (2002). "Recovery of dairy manure nutrients by benthic freshwater algae." *Bioresource Technology* 84(1): 81-91.

Xia, C. J., J. G. Zhang, et al. (2011). "A new cultivation method for microbial oil production: cell pelletization and lipid accumulation by Mucor circinelloides." *Biotechnology for Biofuels* 4.

Xiong, W., X. F. Li, et al. (2008). "High-density fermentation of microalga Chlorella protothecoides in bioreactor for microbio-diesel production." *Applied Microbiology and Biotechnology* 78(1): 29-36.

Xu, H., X. Miao, et al. (2006). "High quality biodiesel production from a microalga Chlorella protothecoides by heterotrophic growth in fermenters." *J. Biotechnol.* 126(4): 499-507.

Xue, F. Y., J. X. Miao, et al. (2008). "Studies on lipid production by Rhodotorula glutinis fermentation using monosodium glutamate wastewater as culture medium." *Bioresource technology* 99(13): 5923-5927.

Yamauchi, H., H. Mori, et al. (1983). "Mass-production of lipids by lipomyces-starkeyi in microcomputer-aided fed-batch culture." *Journal of Fermentation Technology* 61(3): 275-280.

Ykema, A., E. C. Verbree, et al. (1988). "Optimization of lipid production in the oleaginous yeast Apiotrichum curvatum in whey permeate." *Applied Microbiology and Biotechnology* 29(2-3): 211-218.

Zhao, X., C. Hu, et al. (2010). "Lipid production by Rhodosporidium toruloides Y4 using different substrate feeding strategies." *Journal of Industrial Microbiology and Biotechnology*: 1-6.

Zhu, M., P. P. Zhou, et al. (2002). "Extraction of lipids from Mortierella alpina and enrichment of arachidonic acid from the fungal lipids." *Bioresource technology* 84(1): 93-95.

6 Chapter

PRODUCTION OF ALGAL BIOMASS AND BIODIESEL

1. INTRODUCTION: AN OVERVIEW

Biodiesel has gained increased appeal recently due to its ecological advantages and its production from renewable resources. The primary obstacle to the commercialization of biodiesel is its cost. Utilization of spent cooking oil and algae as raw materials, along with the adaptation of a continuous transesterification process and the recovery of high-quality glycerol from biodiesel by-products, are major strategies to reduce biodiesel costs. There are four principal methods for producing biodiesel: direct usage and blending, microemulsions, thermal cracking (pyrolysis), and transesterification. The predominant technique employed is the transesterification of vegetable oils and animal fats. The transesterification process is influenced by the molar ratio of glycerides to alcohol, catalysts, reaction temperature, reaction duration, and the presence of free fatty acids and water concentration in oils or fats. This chapter will examine the superior potential of algae in biodiesel production relative to alternative sources.

2. ALGAE AS BIOLOGICAL MATERIAL

Microalgae are prokaryotic or eukaryotic photosynthetic microorganisms that can grow rapidly and live in harsh conditions due to their unicellular or simple multicellular structure. Examples of prokaryotic microorganisms are Cyanobacteria (Cyanophyceae) and eukaryotic microalgae are for example green algae (Chlorophyta) and diatoms (Bacillariophyta) [Richmond, 2004]. A more in depth description of microalgae is presented by Richmond [Richmond, 2004]. Microalgae are present in all existing earth ecosystems, not just aquatic but also terrestrial, representing a big variety of species living in a wide range of environmental conditions. It is estimated that more than 50,000 species exist, but only a limited number, of around 30,000, have been studied and analyzed [Richmond, 2004]. Algae are aquatic plants that lack the leaves, stem, roots, vascular systems, and sexual organs of the higher plants. They range in size from microscopic phytoplankton to gain kelp 200 feet long. They live in temperatures ranging from hot spring to arctic snows, and they come in various colors mostly green, brown and red. There are about 25,000 species of algae compared to 250,000 species of land plants.

Algae make up in quantity what they lack in diversity for the biomass of algae is immensely greater than that of terrestrial plants (Lowenstein, 1986). Phytoplankton comprises organisms such as diatome, dinoflagellates and macrophytes include: green, red and brown algae. As photosynthetic organisms, these groups play a key role in productivity of ocean and constitute the basis of marine food chain. On the other hand, the use of macroalgae as a potential source of high value chemicals and in therapeutic purpose has a long history.

Recently, macroalgae have been used as a noval food with potential nutritional benefits and in industry and medicine for various purposes.

Furthermore, macroalgae have shown to provide a rich source of natural bioactive compounds with antiviral, antifungal, antibacterial, antioxidant, anti-inflammatory, hypercholesterolemia, and hypolipidemic and antineoplasteic properties. Thus, there is a growing interest in the area of research on the positive effect of macroalgae on human health and other benefits. In Egypt, the macroalgae self grown on the craggy surface near to the seashore of the Mediterranean and Red Seas. Macroalgae have not used as healthy food, while in Japan and China the macroalgae are tradionally used in folk medicine and as a healthy food in addition to, biofuel production (Lee-Saung *et al.*, 2003). The present study was conducted to evaluate the potentialities of micro and macroalgae species for biodiesel production and study the effect of biotic and a biotic stress on biodiesel percentage and the difference between biodiesel production from vegetable sources and algae.

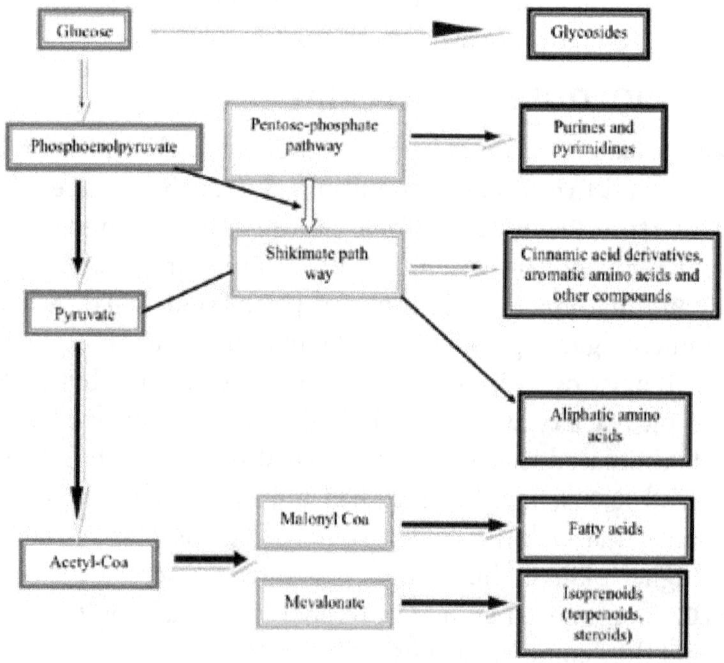

Fig. 1. Secondary and primary metabolites produced from algal cell

Algae were promising organisms for providing both novel biologically active substances and essential compounds for human nutrition (Mayer and Hamann, 2004). Therefore, an increasing supply for algal extracts, fractions or pure compounds for the economical sector was needed (Dos Santos *et al.*, 2005). In this regard, both secondary and primary metabolisms were studied as a prelude to future rational economic exploitation as show in Fig. 1.

3. DIESEL PRODUCTION PROBLEMS

The transportation and energy sectors are the major anthropogenic sources, responsible in European Union (EU) for more than 20% and 60% of greenhouse gas (GHG) emissions, respectively [European Environmental Agency, 2004]. Agriculture is the third largest anthropogenic source, representing about 9% of GHG emissions, where the most important gases are nitrous oxide (N_2O) and methane (CH_4) [European Environmental Agency, 2007]. It is expected that with the development of new growing economies, such as India and China, the global consumption of energy will raise and lead to more environmental damage [International Energy Agency, 2007].

GHG contributes not only to global warming (GW) but also to other impacts on the environment and human life. Oceans absorb approximately one-third of the CO_2 emitted each year by human activities and as its levels increase in the atmosphere, the amount dissolved in oceans will also increase turning the water pH gradually to more acidic. This pH decrease may cause the quick loss of coral reefs and of marine ecosystem biodiversity with huge implications in ocean life and consequently in earth life [Ormerod *et al.*, 2002].

As GW is a problem affecting different aspects of human life and the global environment, not only a single but a host of solutions is needed to address it. One side of the problem concerns the reduction of crude oil reserves and difficulties in their extraction and processing, leading to an increase of its cost [Laherrere, 2005]. This situation is particularly acute in the transportation sector, where currently there are no relevant alternatives to fossil fuels. To find clean and renewable energy sources ranks as one of the most challenging problems facing mankind in the medium to long term. The associated issues are intimately connected with economic development and prosperity, quality of life, global stability, and require from all stakeholders tough decisions and long term strategies. For example, many countries and regions around the world established targets for CO_2 reduction in order to meet the sustainability goals agreed under the Kyoto Protocol. Presently many options are being studied and implemented in practice, with different degrees of success, and in different phases of study and implementation. Examples include solar energy, either thermal or photovoltaic, hydroelectric, geothermal, wind, biofuels, and carbon sequestration, among others [Dewulf *et al.*, 2006]. Each one has its own advantages and problems and, depending on the area of application.

4. BIODIESEL INSTEAD OF DIESEL

One important goal is to take measures for transportation emissions reduction, such as the gradual replacement of fossil fuels by renewable energy sources, where biofuels are seen as real contributors to reach those goals, particularly in the short term. Biofuels production is expected to offer new opportunities to diversify income and fuel supply sources, to promote employment in rural areas, to develop long term replacement of fossil fuels, and to reduce GHG emissions, boosting the decarbonisation of transportation fuels and increasing the security of energy supply. The most common biofuels are biodiesel and bio-ethanol, which can replace diesel and gasoline, respectively, in today cars with little or none modifications of vehicle engines. They are mainly produced from biomass or renewable energy sources and contribute to lower combustion emissions than fossil fuels per equivalent power output. They can be produced using existing technologies and be distributed through the available distribution system. For this reason biofuels are currently pursued as a fuel alternative that can be easily applied until other options harder to implement, such as hydrogen, are available.

Although biofuels are still more expensive than fossil fuels their production is increasing in countries around the world. Encouraged by policy measures and biofuels targets for transport, its global production is estimated to be over 35 billion liters [COM, 2006]. The main alternative to diesel fuel in EU is biodiesel, representing 82% of total biofuels production and is still growing in Europe, Brazil, and United States, based on political and economic objectives. Biodiesel is produced from vegetable oils (edible or non-edible) or animal fats. Since vegetable oils may also be used for human consumption, it can lead to an increase in price of food-grade oils, causing the cost of biodiesel to increase and preventing its usage, even if it has advantages comparing with diesel fuel.

The potential market for biodiesel far surpasses the availability of plant oils not designated for other markets. For example, to fulfill a 10% target in EU from domestic production, the actual feedstocks supply is not enough to meet the current demand and the land requirements for biofuels production, would be more than the potential available arable land for bio-energy crops [Scarlat *et al.*, 2008]. The extensive plantation and pressure for land use change and increase of cultivated fields may lead to land competition and biodiversity loss, due to the cutting of existing forests and the utilization of ecological importance areas [Renewable Fuel Agency, 200]. Biodiesel may also be disadvantageous when replacing crops used for human consumption or if its feedstocks are cultivated in forests and other critical habitats with associated biological diversity. The negative impacts of global warming, now accepted as a serious problem by many people, have clearly been observed for past decade and seem to intensify every year. The release of the carbon oxides and related inorganic oxides are more than the amount that could be absorbed by the natural sinks in the world since 88% of the world energy demand is provided by carbon based non-renewable fuels (Baruch, 2008). It is vital to develop solutions to prevent and/or reduce the emission of greenhouse gases, such as carbon dioxide, to the atmosphere. Carbon dioxide neutral fuels like biodiesel could replace fossil fuels.

Biodiesel, an alternative diesel fuel, is made from renewable biological sources such as vegetable oils and animal fats. It is biodegradable and nontoxic, has low emission profiles and so is environmentally beneficial (Krawczyk, 1996). One hundred years ago, Rudolf Diesel tested vegetable oil as fuel for his engine (Shay, 1993). With the advent of cheap petroleum, appropriate crude oil fractions were refined to serve as fuel and diesel fuels and diesel engines evolved together. In the 1930s and 1940s vegetable oils were used as diesel fuels from time to time, but usually only in emergency situations. Recently, because of increases in crude oil prices, limited resources of fossil oil and environmental concerns there has been a renewed focus on vegetable oils and animal fats to make biodiesel fuels. Continued and increasing use of petroleum will intensify local air pollution and magnify the global warming problems caused by CO2 (Shay, 1993). In a particular case, such as the emission of pollutants in the closed environments of underground mines, biodiesel fuel has the potential to reduce the level of pollutants and the level of potential or probable carcinogens (Krawczyk, 1996). Edible vegetable oils such as canola, soybean, and corn have been used for biodiesel production and found to be a diesel substitute [Lang *et al.*, 2002]. However, a major obstacle in the commercialization of biodiesel production from edible vegetable oil is its high production cost, which is due to the higher cost of edible oil. Waste cooking oil, which is much less expensive than edible vegetable oil, is a promising alternative to edible vegetable oil [Canakci *et al.*, 2003]. Waste cooking oil and fats set forth significant disposal problems in many parts of the world. This environmental problem could be solved by proper utilization and management of waste cooking oil as a fuel. Many developed countries have set policies that penalize the disposal of waste cooking oil the waste drainage [Kulkarni *et al.*, 2006]. The Energy Information Administration in the United States estimated that around 100 million gallons of waste cooking oil is produced per day in USA, where the average per capita waste cooking oil was reported to be 9 pounds [Radich *et al.*, 2006]. The estimated amount of waste cooking oil collected in Europe is about 700,000– 100,000 tons/year [Supple *et al.*, 2002]

Biodiesel is made from biomass oils, mostly from vegetable oils. Biodiesel appears to be an attractive energy resource for several reasons. First, biodiesel is a renewable resource of energy that could be sustainably supplied. It is understood that the petroleum reserves are to be depleted in less than 50 years at the present rate of consumption [Sheehan *et al.*, 1998]. Second, biodiesel appears to have several favorable environmental properties resulting in no net increased release of carbon dioxide and very low sulfur content [Antolin *et al.*, 2002]. The release of sulfur content and carbon monoxide would be cut down by 30% and 10%, respectively, by using biodiesel as energy source. Using biodiesel as energy source, the gas generated during combustion could be reduced, and the decrease in carbon monoxide is owing to the relatively high oxygen content in biodiesel. Moreover, biodiesel contains no aromatic compounds and other chemical substances which are harmful to the environment. Recent investigation has indicated that the use of biodiesel can decrease 90% of air toxicity and 95% of cancers compared to common diesel source. Third, biodiesel appears to have significant economic potential because as a non-renewable fuel that fossil fuel prices will increase inescapability further in the future. Finally, biodiesel is better than diesel fuel in terms of flash point and biodegradability [Ma *et al.*, 1999].

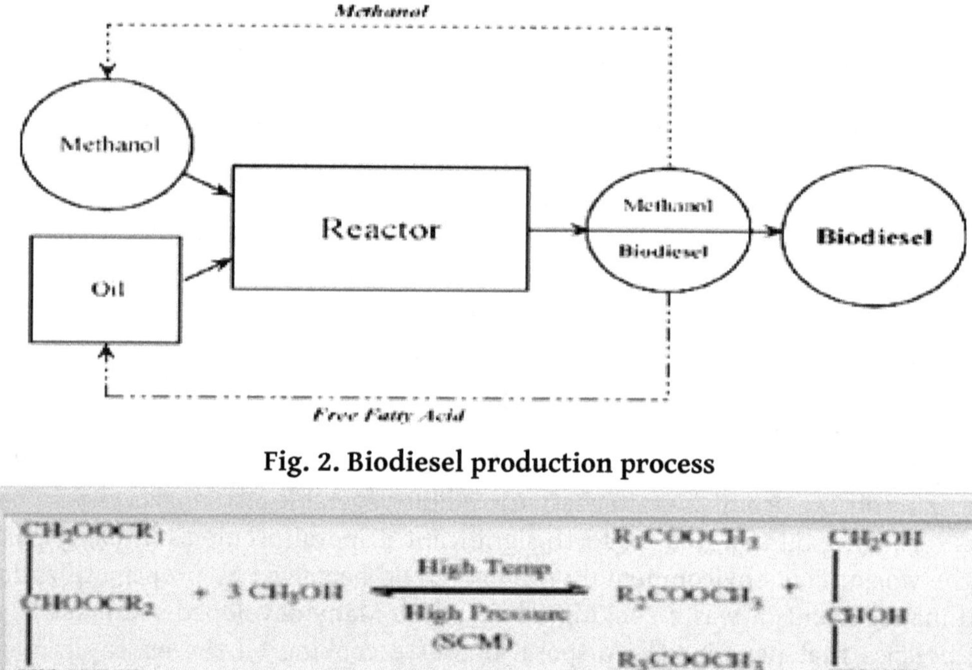

Fig. 2. Biodiesel production process

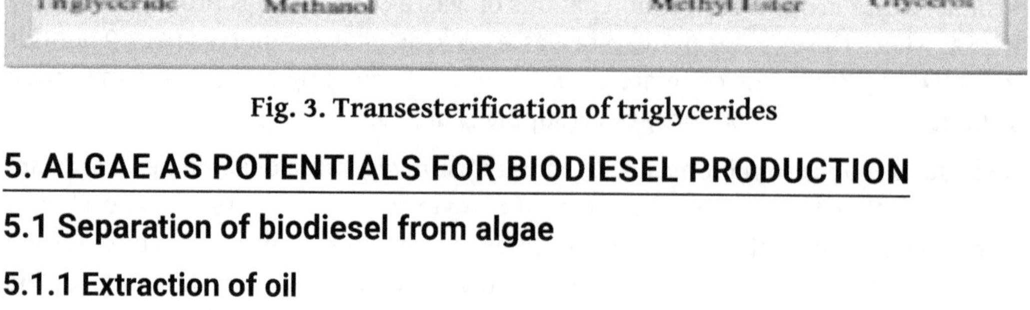

Fig. 3. Transesterification of triglycerides

5. ALGAE AS POTENTIALS FOR BIODIESEL PRODUCTION

5.1 Separation of biodiesel from algae

5.1.1 Extraction of oil

Extraction of oil was carried out using two extraction solvent systems to compare the oil content in each case and select the most suitable solvent system for the highest biodiesel yield (Afify *et al.*, 2010).

5.1.1.1 *Chloroform /methanol (2:1, v/v) method*

A known weight of each ground dried algal species (10 g dry weight) was mixed separately with the extraction solvent mixture; chloroform/methanol (100 ml, 2:1, v/v) for 20 min. using shaker, followed by the addition of mixture of chloroform/ water (50 ml, 1:1, v/v) for 10 min. filter and the algal residue was extracted three times by 100 ml chloroform followed by filtration (Fig.1) according to Bligh and Dayer (1959).

5.1.1.2 *Hexane/ether (1:1, v/v) method*

A known weight of each ground dried algal species (10 g dry weight) was mixed with the extraction solvent mixture, hexane/ether (100 ml, 1:1, v/v), kept to settle for 24 hrs, followed by filtration (Fig. 1) according to Hossain and Salleh (2008).

5.1.2 Transesterification and biodiesel production

The extracted oil was evaporated under vaccum to release the solvent mixture solutions using rotary evaporator at 40- 45 °C. Then, the oil produced from each algal species was mixed with a mixture of catalyst (0.25g NaOH) and 24 ml methanol, a process called transesterification (Fig. 2, 3,4, 5 and Table 2), with stirring properly for 20 min. The Mixture was kept for 3hrs in electric shaker at 3000 rpm. (National Biodiesel Board, 2002). After shaking the solution was kept for 16 hrs to settle the biodiesel and the sediment layers clearly. The biodiesel layer was separated from sedimentation by flask separator carefully. Quantity of sediments (glycerin, pigments, etc) was measured. Biodiesel (Fig. 6) was washed by 5% water many times until it becomes clear then Biodiesel was dried by using dryer and finally kept under the running fan for 12 h. the produced biodiesel was measured (using measuring cylinder), pH was recorded and stored for analysis.

6. BIODIESEL FROM ALGAE

Sustainable production of renewable energy is being hotly debated globally since it is increasingly understood that first generation biofuels, primarily produced from food crops and mostly oil seeds are limited in their ability to achieve targets for biofuel production, climate changemitigation and economic growth. These concerns have increased the interest in developing second generation biofuels produced from non-food feedstocks such as microalgae, which potentially offer greatest opportunities in the longer term. This paper reviews the current status of microalgae use for biodiesel production, including their cultivation, harvesting, and processing. The microalgae species most used for biodiesel production are presented and their main advantages described in comparison with other available biodiesel feedstocks. The various aspects associated with the design of microalgae production units are described, giving an overview of the current state of development of algae cultivation systems (photo-bioreactors and open ponds). Other potential applications and products from microalgae are also presented such as for biological sequestration of CO_2, wastewater treatment, in human health, as food additive, and for aquaculture (Mata *et al.*, 2010).

Biodiesel seem to be a viable choice but its most significant drawback is the cost of crop oils, such as canola oil, that accounts for 80% of total operating cost, used to produce biodiesel (Demirbas, 2007). Besides, the availability of the oil crop for the biodiesel production is limited (Chisti, 2008). Therefore, it is necessary to find new feedstock suitable for biodiesel production, which does not drain on the edible vegetable oil supply. One alternative to oil crops is the algae because they contain lipids suitable for esterification/ transesterification. Among many types of algae, microalgae seem to be promising (Table 1) because:

1. They have high growth rates; e.g., doubling in 24 h (Rittmann, 2008).
2. Their lipid content could be adjusted through changing growth medium composition (Naik *et al.*, 2006).
3. They could be harvested more than once in a year (Schenk *et al.*, 2008).
4. Salty or waste water could be used (Schenk *et al.*, 2008).
5. Atmospheric carbon dioxide is the carbon source for growth of microalgae (Schenk *et al.*, 2008).
6. Biodiesel from algal lipid is non-toxic and highly biodegradable (Schenk *et al.*, 2008).
7. Microalgae produce 15–300 times more oil for biodiesel production than traditional crops on an area basis (Chisti, 2007).

Table 1. Biochemical composition of algae expressed on a dry matter basis (Becker, 1994)

Strain	Protein	Carbohydrates	Lipid	Nucleic acid
Scenedesmus obliquus	50-56	10-17	12-14	3-6
Scenedesmus quadricauda	47	-	1.9	-
Scenedesmus dimorphus	8-18	21-52	16-40	-
Chlamydomonas rheinhardii	48	17	21	-
Chlorella vulgaris	51-58	12-17	14-22	4-5
Chlorella pyrenoidosa	57	26	2	-
Spirogyra sp.	6-20	33-64	11-21	-
Dunaliella bioculata	49	4	8	-
Dunaliella salina	57	32	6	-
Euglena gracilis	39-61	14-18	14-20	-
Prymnesium parvum	28-45	25-33	22-38	1-2
Tetraselmis maculata	52	15	3	-
Porphyridium cruentum	28-39	40-57	9-14	-
Spirulina platensis	46-63	8-14	4-9	2-5
Spirulina maxima	60-71	13-16	6-7	3-4.5
Synechoccus sp.	63	15	11	5
Anabaena cylindrica	43-56	25-30	4-7	-

Note: Algal-oil is very high in unsaturated fatty acids. Some UFA's found in different algal-species include: Arachidonic acid (AA), Eicospentaenoic acid (EPA), Docasahexaenoic acid (DHA), Gamma-linolenic acid (GLA) Linoleic acid (LA)

Algae are made up of eukaryotic cells. These are cellswith nuclei and organelles. All algae have plastids, the bodies with chlorophyll that carry out photosynthesis. But the various strains of algae have different combinations of chlorophyll molecules. Some have only Chlorophyll A, some A and B, while other strains, A and C [Benemann *et al.*, 1978]. Algae biomass contains three main components: proteins, carbohydrates, and natural oil. The chemical compositions of various microalgae are shown in Table 1. While the percentages vary with the type of algae, there are algae types that are comprised of up to 40% of their overall mass by fatty acids [Becker, 1994]. It is this fatty acid (oil) that can be extracted and converted into biodiesel.

Table. 2. Types of transesterification catalysts

Type of transesterification	Advantage	Disadvantage
Chemical catalysis	1-reaction condition can be well controlled	1-reaction temperature is relative high and the process is complex
	2-Large scale production	2-The later disposal process is complex
	3-The cost of the production process is cheap	3-The process need much energy
	4-The methanol produced in the process can be recycled	4-Need a installation for methanol recycle
	5-high conversion of the production	5-the waste water pollute the environment
Enzymatic catalyst	1-Moderate reaction condition	1-Limitation of enzyme in the conversion of short chain fatty acids
	2-The small amount of methanol required in the reaction	2-Chemicals arise in the process of production are poisons to enzyme
	3-Have no pollution to natural environment	
Supercritical fluid techniques	1-Easy to be controlled	1-High temperature and high pressure in the reaction condition leads to high coast for production and waste energy
	2-It is safe and fast	
	3-friendly to environment	

Fig. 4. Biodiesel from algae

Fig. 5 shows a schematic representation of the algal biodiesel value chain stages, starting with the selection of microalgae species depending on local specific conditions and the design and implementation of cultivation system for microalgae growth. Then, it follows the biomass harvesting, processing and oil extraction to supply the biodiesel production unit.

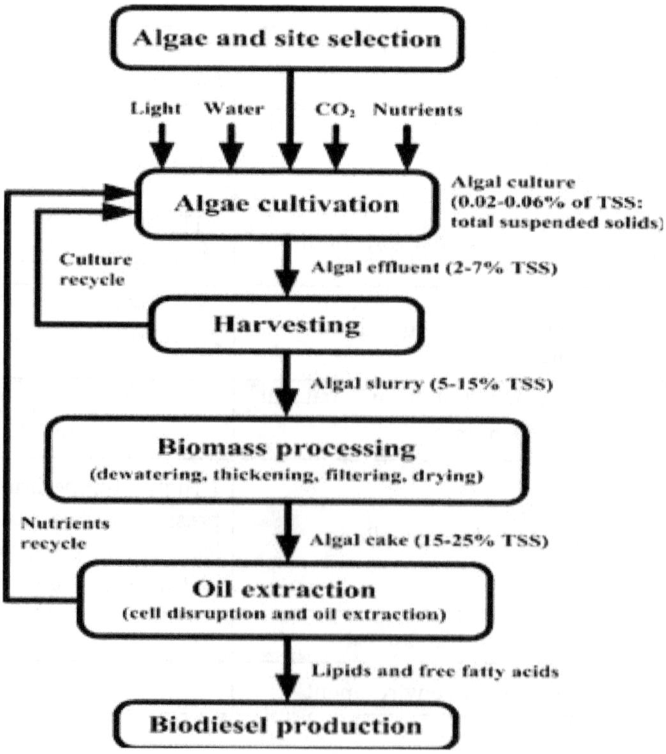

Fig. 5. Microalgae biodiesel value chain stages.

Algae's potential as a feedstock is dramatically growing in the biofuel market. Microalgae (to distinguish it from such macroalgae species as seaweed) have many desirable attributes as energy producers [Choe *et al.*, 2002]:

- Algae is the most promising non-food source of biofuels,
- Algae has a simple cellular structure,
- a lipid-rich composition (40–80% in dry weight),
- a rapid reproduction rate,
- Algae can grow in salt water and harsh conditions,
- Algae thrive on carbon dioxide from gas- and coal-fired power Plants,
- Algae biofuel contains no sulfur, is non-toxic and highly biodegradable.
- The utilization of microalgae for biofuels production can also serve other purposes. Some possibilities currently being considered are listed below.
- Removal of CO_2 from industrial flue gases by algae bio-fixation [Wang *et al.*, 2008], reducing the GHG emissions of a company or process while producing biodiesel. Wastewater treatment by removal of $NH^{+}4$, $NO^{-}3$, $PO^{-3}4$, making algae to grow using these water contaminants as nutrients [Wang *et al.*, 2008].
- After oil extraction the resulting algae biomass can be processed into ethanol, methane, livestock feed, used as organic fertilizer due to its high N:P ratio, or simply burned for energy cogeneration (electricity and heat) [Wang *et al.*, 2008];
- Combined with their ability to grow under harsher conditions, and their reduced needs for nutrients, they can be grown in areas unsuitable for agricultural purposes independently of the seasonal weather changes, thus not competing for arable land use, and can use wastewaters as the culture medium, not requiring the use of freshwater.
- Depending on the microalgae species other compounds may also be extracted, with valuable applications in different industrial sectors, including a large range of fine chemicals and bulk products, such as fats, polyunsaturated fatty acids, oil, natural dyes, sugars, pigments, antioxidants, high-value bioactive compounds, and other fine chemicals and biomass [Raja *et al.*, 2008].
- Because of this variety of high-value biological derivatives, with many possible commercial applications, microalgae can potentially revolutionize a large number of biotechnology areas including biofuels, cosmetics, pharmaceuticals, nutrition and food additives, aquaculture, and pollution prevention [Raja *et al.*, 2008].

7. ENVIRONMENTAL ADVANTAGES OF ALGAL BIOFUELS

In order to be a viable alternative energy source, a biofuel should provide a net energy gain, have environmental benefits, be economically competitive and be producible in large quantities without reducing food supplies [Hill, 2006]. In the subsections below we illustrate how the use of microalgae as feedstocks for biodiesel production can provide significant environmental benefits by reducing the land, pollutant and water footprints of biofuel production.

7.1 Advantages of biodiesel from algae oil (Table 3)

- Rapid growth rates
- Grows practically anywhere
- A high per-acre yield (7–31 times greater than the next best crop – palm oil)-
- No need to use crops such as palms to produce oil
- A certain species of algae can be harvested daily
- Algae biofuel contains no sulfur
- Algae biofuel is non-toxic
- Algae biofuel is highly bio-degradable
- Algae oil extracts can be used as livestock feed and even processed into ethanol
- High levels of polyunsaturates in algae biodiesel is suitable for cold weather climates
- Can reduce carbon emissions based on where it's grown

7.2 Disadvantages of biodiesel from algae oil

- Produces unstable biodiesel with many polyunsaturates
- Biodiesel performs poorly compared to it's mainstream alternative
- Relatively new technology

Table 3. Advantage and disadvantage of algae as biodiesel source compared with bacteria, yeast and waste oils

Type of organism	Advantage	Disadvantage
Microalgal oil	1-Fatty acid profile similar to vegetable oil	1-Most algal lipid have lower fuel value than diesel fuel
	2-Under certain condition it may be as high as 85% of the dry weight	2-The cost of cultivation is higher compared to common crop oil currently
	3-Short-time growth cycle	
	4-Composition is relative single in microalgae	

Type of organism	Advantage	Disadvantage
Bacteria oil	1-Fast growth rate	1-Most of bacteria can not yield lipids but complicated lipoid
Oleaginous yeast and mildews	1-Resource are abundant in the nature	1-Filteration and cultivation of yeasts and mildews with high-content are required
	2-High oil content in some species	2-Process of oils extracted is complex and new technology
	3-Short time growth cycle	3-The cost of cultivation is also higher compared to common crops currently
	4-Strong capability of growth in different cultivation on conditions	
Waste oils	1-The waste oil is cheap compared to crop oils	1-Conataing a lot of saturated fatty acids which is hard to converted to biodiesel by catalyst

8. COMPARISON BETWEEN BIODIESEL PRODUCTION FROM ALGAE AND VEGETABLES

Quantifying the land use changes associated with intensive biofuel feedstock production relies upon many assumptions [Chisti,. 2007], but it is clear that the accelerated cultivation of terrestrial plant biomass for biofuels will have an exceptionally large land footprint (Table 4). For example, the United States has the fourth largest absolute biodiesel potential of the 119 countries studied by Johnston and Holloway [Johnston, M. and Holloway, 2007]. However, recent work has suggested that the projected year 2016 demand for corn ethanol alone would require 43% of all U.S. land used for corn production in 2004 [Chisti,. 2007]. A related study concluded that the annual corn production needed to satisfy one half of all U.S. transportation fuel needs would require an area equivalent to more than eight times the U.S. land area that is presently used for crop production [Chisti,. 2007]. Other land-based crops would require less cropland, based on their oil content: oil palm (24% of current cropland area), coconut (54%), jatropha (77%), canola (122%) and soybean (326%) [Chisti,. 2007]. Moreover, recent work indicates that the ability of countries to grow terrestrial crops explicitly for the production of biofuels such as ethanol and biodiesel is significantly overestimated [Johnston, M. and Holloway, 2007], contributing to concerns that these biofuels are not feasible options for providing a significant fraction of global fuel demand.

Table 4. Comparison of estimated biodiesel production efficiencies from vascular plants and microalgae

Biodiesel feedstock	Area needed to meet global oil demand (10^6 hectans)	Area required as a percent of total global land	Area required as a percent of total arable global land
Cotton	15000	101	757
Soybean	10900	73	552
Mastard seed	8500	57	430
Sunflower	5100	34	258
Rapeseed/Canola	4100	27	207
Jatropha	2600	17	130[b]
Oil palm	820	5.5	41
Microalgae (10 g/m^3/day, 30%TAG)	410	2.7	21[c]
Microalgae (50 g/m^3/day, 50%TAG)	49	0.3	25[c]

[b]Jatropha is mainly grown on marginal land

[c]Assuring that microalgal ponds and bioreactors are located on non-arable land

9. THE PHYSICAL AND CHEMICAL PROPERTIES OF BIODIESEL PRODUCED FROM ALGAL CELL

Analysis of the produced biodiesel from the promising alga *Dictyochloropsis splendida* (Table 5). showed that the unsaturated fatty acids percentage was increased in alga cultivated in nitrogen free media (0.0g/l N) two times more than normal conditions (13.67, 4.81% respectively). However, the composition of fatty acids was different in these algae depending on its growth condition as showed in table 3. These results were in agreements with those reported by Wood (1974) relative to *Chlorophycean* species. Furthermore Ramos *et al.* (2009) reported that monounsaturated, polyunsaturated and saturated methyl esters were built in order to predict the critical parameters of European standard for any biodiesel, composition. The extent of unsaturation of microalgae oil and its content of fatty acids with more than four double bonds can be reduced easily by partial catalytic hydrogenation of the oil (Jang *et al.*, 2005, Dijkstra, 2006). Concerning the fatty acids contents of the produced biodiesel from microalgae, Chisti (2007) reported in his review that, microalgal oils differ from vegetable oils in being quite rich in polyunsaturated fatty acids with four or more double bands (Belarbi *et al.*, 2000) as eicosapentanoic acid (C20:5n-3) and docosahexaenoic acid (C22:6n-3) which occurred commonly in algal oils. The author added that, fatty acids and fatty acid methyl esters with four and more double bands are susceptible to oxidation during storage and this reduces their acceptability for use in biodiesel especially for vehicle use (European standard EN 14214 limits to 12%) while no such limitation exists for biodiesel intended for use as healing oil. In addition to the content of unsaturated fatty acids in the biodiesel also its iodine value (represented total unsaturation) must be taken in consideration (not exceeded 120 g iodine/100g biodiesel according to the European standard.

Table 5. Analysis of fatty acids of the obtained biodiesel from the promising green microalgae *Dictyochloropsis* sp

Fatty acids	*RT	Fatty acids percentage	
		Algae cultivated under normal conditions	Algae cultivated under free nitrogen media
C10:0 (Capric acid)	1.223	0.0	1.26
C14:0 (Myristic acid)	2.437	13.04	13.88
C16:0 (Palmitic acid)	2.860	81.14	69.59
C17:0 (Margeric acid)	3.240	1.01	1.21
C18:0 (stearic acid)	4.335	0.0	0.38
C18:1 (Oleic acid)	4.667	0.26	1.11
C18:2 (Linoleic acid)	5.333	4.39	12.14
C18:3 (linolenic acid)	6.948	0.15	0.42
Total saturated fatty acids		95.19	86.33
Total unsaturated fatty acids		4.81	13.67
TU/TS		0.05	0.16

*Retention time; TU/TS: total unsaturated/ total saturated fatty acids ratio.

10. ENHANCEMENT THE BIODIESEL PRODUCTION FROM ALGAE

Lipid productivity, the mass of lipid that can be produced per day, is dependent upon plant biomass production as well as the lipid content of this biomass. Algal biodiesel production will therefore be limited not only by the standing crop of microalgae, but also by its lipid content, which can vary from <1% to >50% dry weight [Shifrin, N.S. and Chisholm, 1980]. Given that a strong and predictable response of microalgal biomass to phosphorus enrichment has consistently been exhibited by freshwater ecosystems worldwide (Box 2), it can be expected that the volumetric lipid content (in mg L-1) of water contained in algal bioreactors should also in general increase with an increase in the total phosphorus content of the system, as has been reported for lakes by Berglund *et al.* [Berglund, 2001]. However, both the quantity and the quality of lipids produced will vary with the identity of the algal species that are present in the water, as well as with site-specific growth conditions. This variability probably reflects modifications in the properties of cellular membranes, and alterations in the relative rates of production and utilization of storage lipids [Roessler, 1990]. In the presence of moderate temperatures and sufficient light, many dozens of studies during the past several decades have revealed that algal lipid content is particularly sensitive to conditions of nutrient limitation . For example, silicon-starved diatoms can contain almost 90% more lipids than silicon-sufficient cells [Shifrin, N.S. and Chisholm, 1980]. However, silicon will be a growth-limiting nutrient only for the limited subset of microalgal species that have an absolute requirement of this element for their cellular growth. A stronger stimulation of lipid production occurs in response to conditions of nitrogen limitation, which potentially can occur in all known microalgae. Nitrogen-starved cells can contain as much as four times the lipid content of Nsufficient cells [Shifrin, N.S. and Chisholm, 1980], and maximizing the lipid production of pond bioreactors should therefore depend on their operators' ability to reliably and consistently induce N-limitation in the resident algal cells. Resource-ratio theory and the principles of ecological stoichiometry, provide additional new insights into the control of algal biomass and lipid production in pond bioreactors. the nutrient limitation status of microalgae can be directly controlled by regulating the ratio of nitrogen and phosphorus (N:P) supplied in the incoming nutrient feed: nitrogen limitation occurs at N:P supply ratios that lie below the optimal N:P ratio for microalgal growth, whereas phosphorus limitation occurs at ratios that exceed this ratio. A transition between N- and P-limitation of phytoplankton growth typically occurs in the range of N:P supply ratios between ca. 20:1 to ca. 50:1 by moles . Such shifts between N- and P-limitation have extremely important implications for algal biofuel production because diverse species of microalgae grown under nitrogen-limited conditions (i.e. low N:P supply ratios) can exhibit as much as three times the lipid content of cells grown under conditions of phosphorus limitation (high N:P supply ratios) . Both the total phosphorus concentration as well as the total nitrogen concentration in the nutrient feeds to pond bioreactors should therefore impact algal biodiesel production, because the N:P ratio of incoming nutrients will strongly influence algal biomass production as well as the cellular lipid content. Given the inverse relationship observed between N:P and cellular lipids , and the positive, hyperbolic relationship observed between N:P and microalgal biomass , we conclude that optimal lipid yields (in terms of mass of lipid produced per unit bioreactor volume per day) should occur at intermediate values of the N:P supply ratio. From the strong apparent interactions between the effects of nitrogen and carbon dioxide availability on microalgal lipids, we also conclude that the effects of N:P supply ratios on volumetric lipid production might be even greater if the bioreactors are simultaneously provided with supplemental CO2 (cf. Figure 2).

Table 6. Comparison between lipid percentage (%) produced by eight algal species using two different extraction system.

Algal species	Chloroform/methanol (2:1, v/v)	Hexane/ether (1:1, v/v)
Jania rubens	4.4±0.12	2.8±0.04
Galaxaura oblongata	2.5±0.09	2.4±0.01
Gelidium latifolium	3.0±0.0	3.1±0.02
Asporagopsis taxiformis	4.1±0.08	3.4±0.05
Ulva lactuca	4.2±0.1	3.5±0.1
Colpomenia sinuosa	3.5±0.05	2.3±0.03
Dictyochloropsis splendida	12.5±0.23	2.4±0.14
Spirulina platensis	9.2±0.25	3.0±0.10
LSD	0.3261	0.3261

Each value is presented as mean of triplet treatments, LSD: Least different significantly at P ≤ 0.05 according to Duncan's multiple range test.

Table 7. Total lipid, biodiesel, sediment percentage and biodiesel color of eight algal species

Algal sp.	Lipid %	Biodiesel%	Sediment %	Biodiesel color
Jania rubens	4.4±0.12	0.25±0.01	4.2^{a} ± 0.05	Light brown
Galaxaura oblongata	2.5±0.09	2.06±0.02	0.08±0.0	Light green
Gelidium latifolium	3.0±0.0	1.3±0.0	1.6±0.01	yellow
Asporagopsis taxiformis	4.1±0.08	3.64^{c} ± 0.10	0.40±0.01	Dark green
Ulva lactuca	4.2±0.1	3.8±0.12	0.44±0.0	Light green
Colpomenia sinuosa	3.5±0.05	3.1±0.05	0.31±0.05	yellow
Dictyochloropsis splendida	12.5±0.23	8.75±0.24	3.75±0.08	colorless
Spirulina platensis	9.2±0.25	7.5±0.30	1.66±0.06	Light green
LSD	0.3261	0.3314	0.1786	

Each value is presented as mean of triplet treatments, LSD: Least different significantly at P ≤ 0.05 according to Duncan's multiple range test.

Eight algal species (4 *Rhodo*, 1 *chloro* and 1 *phaeophycean* macroalgae, 1 *cyanobacterium* and 1 green microalga) were used for the production of biodiesel using two extraction solvent systems (Hexane/ether (1:1, v/v)) and (Chloroform/ methanol (2:1, v/v)) Table 6. Biochemical evaluations of algal species were carried out by estimating biomass, lipid, biodiesel and sediment (glycerin and pigments) percentages. Hexane/ ether (1:1, v/v) extraction solvent system resulted in low lipid recoveries (2.3-3.5% dry weight) while; chloroform/methanol (2: 1, v/v) extraction solvent system was proved to be more efficient for lipid and biodiesel extraction (2.5 – 12.5% dry weight) depending on algae species (Table 7). The green microalga *Dictyochloropsis splendida* extract produced the highest lipid and biodiesel yield (12.5

and 8.75% respectively) followed by the cyanobacterium *Spirulina maxima* (9.2 and 7.5 % respectively). On the other hand, the macroalga (red, brown and green) produced the lowest biodieselyield. The fatty acids of *Dictyochloropsis splendida* Geitler biodiesel were determined using gas liquid chromatography. Lipids, biodiesel and glycerol production of *Dictyochloropsis splendida* Geitler (the promising alga) were markedly enhanced by either increasing salt concentration or by nitrogen deficiency (Table 8) with maximum production of (26.8, 18.9 and 7.9 % respectively) at nitrogen starvation condition. (Afify *et al.*, 2010)

Natural biotic communities in outdoor bioreactors require the external provision of potentially growth-limiting resources (e.g. light, carbon dioxide and the essential mineral nutrients N and P). These resources act as “bottom-up” regulators of the potential microalgal biomass that can be produced. Once harvested, the cellular lipids in this microalgal biomass can be extracted and processed to create biodiesel fuels. The lipid content of microalgal biomass is not constant, however, and can be influenced by many factors, including nitrogen:phosphorus supply ratios, light, CO_2 and the hydraulic residence time of the bioreactor. Moreover, natural assemblages of microalgae are taxonomically diverse: some species are small and can easily be consumed by herbivorous zooplankton. Undesirable grazing losses of edible microalgae (and their cellular lipids) to large-bodied zooplankton can be reduced by adding zooplanktivorous fish, which can greatly restrict large-bodied zooplankton growth via sizeselective predation (“top-down” regulation).

Table 8. Total lipid, biodiesel, sediment percentage and biodiesel color of Dictyochloropsis sp cultivated under stress

Sample culture conditions	Lipid content (%)	Biodiesel content (%)	Glycerol+ pigments content (%)	Biodiesel color
Control (2.5 g/l NaCl and 25g/l $NaNO_3$)	12.50±0.36	8.75±0.25	3.75±0.12	Colorless
NaCl stress				
5 g/l	14.50±1.2	8.90±0.62	5.60±0.18	Colorless
7.5 g/l	17.00±0.53	11.94±0.98	5.06±0.22	Light green
10 g/l	17.50±0.36	11.38±0.80	5.11±0.24	Light green
Nitrogen stress				
12.5 g/l	15.40±2.10	8.90±0.36	6.50±0.30	Yellowish green
6.25g/l	16.20±1.8	10.01±1.0	6.19±0.12	Light Yellow
0.0g/l	26.80± 2.12	18.90±1.2	7.9±0.50	Yellow
LSD	0.3643	0.1681	0.1431	

Each value is presented as mean of triplet treatments, LSD: Least different significantly at $P \leq 0.05$ according to Duncan's multiple range test.

11.WASTEWATERNITROGENANDPHOSPHOROUSASMICROALGAE

NUTRIENTS

There is a unique opportunity to both treat wastewater and provide nutrients to algae using nutrient-rich effluent streams. By cultivating microalgae, which consume polluting nutrients in municipal wastewater, and abstracting and processing this resource, then the goals of sustainable fuel production and wastewater treatment can be combined (Andersen, 2005). Treated wastewater is rich in nitrogen and phosphorus, which if left to flow into waterways, can spawn unwanted algae blooms and result in eutrophication (Sebnem Aslan, 2006). These nutrients can instead be utilized by algae, which provide the co-benefit of producing biofuels and removing nitrogen and phosphorus as well as organic carbon (Mostafa and Ali, 2009). Wastewater treatment using algae has many advantages. It offers the feasibility to recycle these nutrients into algae biomass as a fertilizer and thus can offset treatment cost. Oxygen rich effluent is released into water bodies after wastewater treatment using algae (Becker, 2004).

Cyanobacteria strains (Anabaena flos aquae, Anabaena oryzae, Nostoc humifusum, Nostoc muscorum, Oscillatoria sp., Spirulina platensis, Phormedium fragile and Wollea saccata) and the green alga strain Chlorella vulgaris were obtained from the Microbiology Department, Soils, Water and Environment Res. Inst. (SWERI), Agric. Res., Center (ARC). Cyanobacteria strains were maintained in BG11 medium (Rippka *et al.*, 1979) except Spirulina platensis which was cultivated in Zarrouk medium (Zarrouk, 1966). While, Bold medium (Nichols and Bold, 1965) was used for the green alga Chlorella vulgaris. Cultures were incubated in a growth chamber under continuous shaking (150 rpm) and illumination (2000 lux) at 25 ± 1°C for 30 days. Shalaby *et al.* (2011). The effluent of the secondary treated sewage wastewater from Zenien Waste Water Treatment Plant (ZWWTP), Giza Governorate, Egypt was used after filtered using glass microfiber filter to remove large particles and indigenous bacteria for the experiment and the chemical and physical parameters were analysis as reported by APHA (1998) Table (2). The supplementation of $NaNO_3$, K_2HPO_4 and $FeSO_4.7H_2O$ in amounts equal to those of the standard BG11, Bold and Zarrouk were used as basal media. The algal strains were grown in 500 ml Erlenmeyer flasks containing 200 ml of 100% effluent supplemented with basal nutrients and 100% effluent without basal nutrients with/without sterilization and the synthetic media (BG11, Bold and Zarrouk) were used as control. Two per cent algal inoculums were added to each flask. The experiment was conducted in triplicates and cultures were incubated at 25 ºC ±1ºC, under continuous shaking (150 rpm) and illumination (2000 lux) for 15 days. This work aimed to evaluate the laboratory cultivation of nine algal strains belonging to Nostocales and Chlorellales in secondary treated municipal domestic wastewater for biomass and biodiesel production as shown in Table (9 and 10)

Table 9. Total lipids, biodiesel, glycerine+pigments percentage and color, pH of biodiesel from different microalgae species cultivated in different waste water

Algal species	Total lipids	Biodiesel content	Glycerin + pigments	Color	pH
Nostoc muscorum	16.80±3.62	12.52±1.74	4.28±1.74	Brown	7.4±0.33
Anabaena flous aquae	5.50±0.58	4.00±0.41	1.50±0.41	Red	6.9±0.95
Chlorella vulgaris	12.50±1.20	8.8±0.16	3.70±0.16	Green	8.1±1.0
Oscillatoria sp	8.00±0.58	4.30±0.32	3.70±0.32	Yellow	7.5±0.85
Spirulina platensis	10.0±0.11	7.80±0.17	2.20±0.17	Light green	8.0±0.32
Anabaena oryzae	7.40±0.90	4.50±0.10	2.90±0.10	Orange	7.3±0.96
Wollea sp	6.30±1.31	3.90±0.60	2.40±0.60	Yellow	7.8±0.35
Nostoc humifusum	14.80±2.40	10.20±1.30	4.6±1.30	Yellowish brown	7.5±0.50
Phormedium sp	12.20±1.66	10.10±1.50	2.10±1.50	Dark brown	7.1±0.0
LSD	0.159	0.151	0.151		1.659

Each value is presented as mean of triplet treatments, LSD: Least different significantly at $P \leq 0.05$ according to Duncan's multiple range tests.

T1: waste water without treatment; T2: waste water after sterilization; T3: waste water+ nutrients with sterilization T4: waste water+ nutrients without sterilization

Table 10. Total lipids, biodiesel, glycerine+pigments percentage and color, pH of biodiesel from different microalgae species cultivated in different waste water

Algal species	Optimal waste water treatment	Total lipids	Biodiesel	Glycerin + pigments
Nostoc muscorum	T3	12.50±2.65	7.40±0.74	5.10±0.74
Anabaena flous aquae	T3	7.40±0.95	5.00±0.61	2.40±0.61
Chlorella vulgaris	T3	13.20±1.87	8.50±1.74	4.70±1.74
Oscillatoria sp	T2	6.80±0.65	3.80±0.32	3.00±0.32
Spirulina platensis	T3	7.30±0.44	5.00±0.51	2.30±0.51
Anabaena oryzae	T4	8.00±0.16	4.70±0.12	3.30±0.12
Wollea sp	T3	7.20±1.32	4.00±0.22	3.23±0.22
Nostoc humifusum	T1	15.50±1.65	11.80±1.52	3.70±1.52
Phormedium sp	T2	11.60±0.88	8.40±0.65	3.20±0.65
LSD		0.159	0.159	0.152

Each value is presented as mean of triplet treatments, LSD: Least different significantly at $P \leq 0.05$ according to Duncan's multiple range tests.

T1: waste water without treatment; T2: waste water after sterilization; T3: waste water+ nutrients with sterilization T4: waste water+ nutrients without sterilization

12. ECONOMIC IMPORTANCE

Compared to biofuels from agricultural crops, the amount of land required would be minimal. Trials in ideal conditions show that fast-growing micro-algae can yield 1800–2000 gallons/(acre - year) of oil—compare this with 50 gallons for soyabeans, 130 gallons for rapeseed and _650 gallons for palm oil. It can grow on fresh or brackish water on marginal land so that it does not compete with areas for agricultural cultivation. As Sean Milmo points out in his article in Oils and Fats International [Milmo, 2008]; oil from algae on 20–40 M acres of marginal land would replace the entire US supply of imported oil, leaving 450 M acres of fertile soil in the country entirely for food production. Biomass can also be harvested from marine algae blooms and algae can even be cultivated in sewage and water treatment plants. However, most estimates of algal fuel productivity estimate that with current production technologies algal diesel can be manufactured for, at best, $4.54 per gallon using high density photobioreactors. In order to compete economically with petroleum diesel costs – and not accounting for any potential subsidy scheme, which is a likely possibility – requires the reduction of these costs to near $1.81 per gallon relative to 2006 fuel prices. These cost reduction figures take into account the fact that materials input and refining of fuels (in this case the algae vegetable oil) account for roughly 71% of total at pump fuel cost [Chisti, 2007]. Algal biodiesel becomes even more plausible given the potential for GHG regulation in the near future. Since for every ton of algal biomass produced, approximately 1.83 tons of carbon dioxide is fixed while petroleum diesel carries a massive negat balance, the competitiveness of algae diesel increases as GHG externalities are taken into account. Given certain research objectives these cost reductions are achievable in the near future. The National Renewable Energy Laboratory (NREL) outlines many such research objects including: increasing photosynthetic efficiency of algae species for high lipid production, control of mechanisms of algae biofocculation, understanding the effects of non-steady-state operating conditions, and methods of species selection and control [Sheehan *et al.*, 1998].

13. THE ISSUES ASSOCIATED WITH ALGAE

The majority of issues concerning marine microalgae cultures stem from predation by diverse protozoan species, including zooflagellates, ciliates, and rhizopods. Another issue is the proliferation of undesirable or toxic organisms, such as blue-green algae or dinoflagellates (red tides), which can lead to elevated toxicity for consumers and perhaps for humans. Instances include the extensive proliferation of green chlorococcalean algae, such as Synechocystis in freshwater, and the growth of Phaeodactylum in seawater, which is detrimental to bivalve molluscs. [De Pauw *et al.*, 1984].

14. ADDITIONAL APPLICATIONS OF ALGAE

In Western countries, algae are mostly utilized as a source for extracting alginates (from brown algae) and agar and carrageenans (from red algae). Furthermore, algae possess a plethora of bioactive chemicals, including phenolic compounds, alkaloids, plant acids, terpenoids, and glycosides, which may exhibit antioxidant, antibacterial, antiviral, and anticarcinogenic activities. (Plaza, *et al.*, 2008).

REFERENCES

Afify, AMM.; Shanab, SM.; Shalaby, EA. (2010). Enhancement of biodiesel production from different species of algae, grasas y aceites, 61 (4), octubre-diciembre, 416- 422.

Antolin, G.; Tinaut, F. V.; Briceno, Y. (2002). Optimisation of biodiesel production by sunflower oil transesterification. Bioresour Technol ., 83:111–4.

Baruch, J.J. (2008). Combating global warming while enhancing the future. Technol. Soc. 30, 111–121.

Becker, E.W. in: J. Baddiley, *et al.* (Eds.), Microalgae: Biotechnology and Microbiology, Cambridge Univ. Press, Cambridge, NY, 1994, p. 178.

Benemann, J.R.; Koopman, B.L.; Weissman, J.C.; Eisenberg, D.M.; Oswald, W.J. (1978). An Integrated System for the Conversion of Solar Energy with Sewage-grown Microalgae, Report, Contract D(0-3)-34, U.S. Dept. of Energy, SAN-003-4-2.

Berglund, O. *et al.* (2001) The effect of lake trophy on lipid content and PCB concentrations in planktonic food webs. Ecology 82, 1078–1088.

Bligh EG, Dayer WJ. 1959. A rapid method for total lipid extraction and purification. Can. J. Biochem and Physiol., 37: 911-7.

Canakci, M.; Van Gerpen, J. (2003). A pilot plant to produce biodiesel from high free fatty acid feedstocks. Trans ASAE, 46: 945–54.

Chisti, Y. (2007). Biodiesel from microalgae. Biotechnol. Adv. 25, 294–306.

Chisti, Y. (2008). Biodiesel from microalgae beats bioethanol. Cell Press 26, 126–131. Choe, S.H.; Jung, I.H.(2002). J. Ind. Eng. Chem. 8 (4): 297.

COM (2006) 34 final. An EU strategy for biofuels. Commission of the European Communities, Brussels, 8.2.2006.

Demirbas, A. (2007). Importance of biodiesel as transportation fuel. Energy Policy 35, 4661- 4670.

De Pauw, N.; Morales, J.; Persoone, G. (1984). Mass culture of microalgae in aquaculture systems: progress and constraints. Hydrobiologia,116/117: 121–34.

Dewulf J, Van Langenhove H. Renewables-based technology: sustainability assessment. John Wiley & Sons, Ltd; 2006.

Dos Santos, M. D.; Guaratini, T.; Lopes, J. L. C.; Colepicolo, P. and Lopes, N. P. (2005). Plant cell and microalgae culture. In: Modern Biotechnology in Medicinal Chemistry and Industry. Research Signpost, Kerala, India.

European Environmental Agency (EEA) Report N85. Copenhagen, Denmark; 2007. European Environmental Agency (EEA). Greenhouse gas emission trends and projections in

Europe 2004: progress by the EU and its Member States towards achieving their Kyoto Protocol targets. Report N85. Copenhagen, Denmark; 2004.

Johnston, M. and Holloway, T. (2007) A global comparison of national biodiesel production potentials. Environ. Sci. Technol., 41: 7967–7973.

Hill, J. *et al.* (2006) Environmental, economic, and energetic costs and benefits of biodiesel and ethanol biofuels. Proc. Natl. Acad. Sci. U. S. A.

Hossain ABM, Salleh A. (2008). Biodiesel fuel production from algae as renewable energy. Am. J. Biochem. and Biotechn., 4(3): 250-254.

International Energy Agency (IEA). World Energy Outlook 2007. China and India Insights, Paris, France; 2007.

Kulkarni, M. G.; Dalai, A. K. (2006). Waste cooking oil – an economical source for biodiesel: a review. Ind Eng Chem Res., 45: 2901–13.

Krawczyk, T. (1996). Biodiesel: Alternative fuel makes inroads but hurdles remain. INFORM, 7: 801-829.

Laherrere J. Forecasting production from discovery. In: ASPO; 2005.

Lang, X.; Dalai, A. K.; Bakhashi, N. N.; Reaney, M. J. (2002). Preparation and characterization of biodiesels from various bio-oils. Biores Technol., 80: 53– 62.

Lee-Saung, H.; Lee-Yeon, S.; Jung-Sang, H.; Kang-Sam, S. and Shin-Kuk, H. (2003). Antioxidant activities of fucosterol from the marine algae *Pelvetia siliquosa*. Archives of Pharmacal Research, 26: 719-722.

Lowenstein. J. (1986). The secret life of seaweeds. Oceans, 19: 72-75.

Ma, F.; Hanna, M. A. (1999). Bidiesel production: a review. Bioresour Technol., 70: 1–15.

Mata, T. M.; Martins, A. A.; Caetano, N. S. (2010). Microalgae for biodiesel production and other applications: A review. Renewable and Sustainable Energy Reviews 14, 217– 232.

Mayer, A. M. S. and Hamann, M.T. (2004). Marine pharmacology in 2000: marine compounds with antibacterial, anticoagulant, antifungal, anti-inflammatory, antimalarial, antiplatelet, antituberculosis, and antiviral activities; affecting the cardiovascular, immune, and nervous system and other miscellaneous mechanisms of action. Mar. Biotechnol. 6: 37–52.

Milmo, S. (2008). Oil Fat Int. 24 (2): 22.

National Biodiesel Board. 2002. USA. Available in www.biodiesel.org/.

Naik, S.N., Meher, L.C., Sagar, D.V. (2006). Technical aspects of biodiesel production by transesterification – a review. Renew. Sust. Energy Rev. 10, 248–268.

Ormerod WG, Freund P, Smith A, Davison J. Ocean storage of CO2. IEA greenhouse gas R&D programme. UK: International Energy Agency; 2002.

Radich, A. (2006). Biodiesel performance, costs, and use. US Energy Information Administration. <http://www.eia.doe.gov/oiaf/analysispaper/biodiesel/index.html>.

Raja, R.; Hemaiswarya, S.; Kumar, N.A.; Sridhar, S,; Rengasamy, R. A. (2008). perspective on the biotechnological potential of microalgae. Critical Reviews in Microbiology, 34(2):77–88.

Renewable Fuel Agency (RFA). The Gallagher review of the indirect effects of biofuels production; 2008.

Richmond A. Handbook of microalgal culture: biotechnology and applied phycology. Blackwell Science Ltd; 2004.

Rittmann, B.E. (2008). Opportunities for renewable bioenergy using microorganisms. Biotechnol. Bioeng. 100, 203–212.

Roessler, P.G. (1990) Environmental control of glycerolipid metabolism in microalgae: commercial implications and future research directions. J. Phycol. 26, 393–399.

Scarlat N, Dallemand JF, Pinilla FG. Impact on agricultural land resources of biofuels production and use in the European Union. In: Bioenergy: challenges and opportunities. International conference and exhibition on bioenergy; 2008.

Schenk, P.M., Thomas-Hall, S.R., Stephens, E., Marx, U.C., Mussgnug, J.H., Posten, C., Kruse, O., Hankamer, B. (2008). Second generation biofuels: high-efficiency microalgae for biodiesel production. Bioenergy Res. 1, 20–43.

Shay, E.G. (1993). Diesel fuel from vegetable oils: status and opportunities. Biomass and Bioenergy, 4: 227-242.

Sheehan, J.; Cambreco, J.; Graboski, M.; Shapouri, H. (1998). An overview of biodiesel and petroleum diesel life cycles. US Department of agriculture and Energy Report, p. 1–35.

Shifrin, N.S. and Chisholm, S.W. (1980) Phytoplankton lipids: environmental influences on production and possible commercial applications. In Algae Biomass (Shelef, G. and Soeder, C.J., eds), pp. 627–645, Elsevier

Wang, B.; Li, Y.; Wu, N.; Lan, CQ. (2008). CO_2 bio-mitigation using microalgae. Applied Microbiology and Biotechnology, 79(5):707–18.

7

Chapter

MICROALGAE AS FEEDSTOCKS FOR BIODIESEL SYNTHESIS

1. INTRODUCTION: AN OVERVIEW

Fossil fuels, including oil, coal, and gas, are integral to the contemporary global energy industry. According to the World Energy Outlook 2007, fossil fuels will continue to be the predominant energy sources, projected to satisfy around 84% of energy demand by 2030. Nonetheless, fossil fuels are non-renewable and will ultimately be depleted.. It has been recently estimated that the global oil, coal and gas last only approximately for 35, 100 and 37 years respectively, based on a modified Klass model (Shafiee & Topal, 2009). In order to sustain a stable energy supply in the future, it is necessary to develop other sources of energy, e.g., renewable energy. Renewable energy is derived from natural processes that are replenished constantly, including hydropower, wind power, solar energy, geothermal energy, biodiesel, etc. An estimated $150 billion was invested in renewable energy worldwide in 2009, around 2.5 times of the 2006 investment (Figure 1).

It is well known that transport is almost totally dependent on petroleum-based fuels, which will be depleted within 40 years. An alternative fuel to petrodiesel must be technically feasible, easily available, economically competitive, and environmentally acceptable (Demirbas, 2008). Biodiesel is such a candidate fuel for powering the transport vehicles. Biodiesel refers to a biomass-based diesel fuel consisting of long-chain alkyl (methyl, propyl or ethyl) esters. In addition to being comparable to petrodiesel in most technical aspects, biodiesel has the following distinct advantages over petrodiesel (Knothe, 2005a):

1. derived from renewable domestic resources, thus reducing dependence on and preserving petroleum;
2. biodegradable and reduced exhaust emissions, being environment-friendly;
3. higher flash point, being safer for handling and storage; and
4. excellent lubricity.

Like petrodiesel, biodiesel operates in compression ignition engines. Biodiesel is miscible with petrodiesel in all ratios. Currently, the blends of biodiesel and petrodiesel instead of net biodiesel have been widely used in many countries and no engine modification is required (Singhania et al., 2008). These blends of biodiesel with petrodiesel are usually denoted by acronyms, for example B20 which indicates a blend of 20% biodiesel with petrodiesel (Knothe, 2005a).

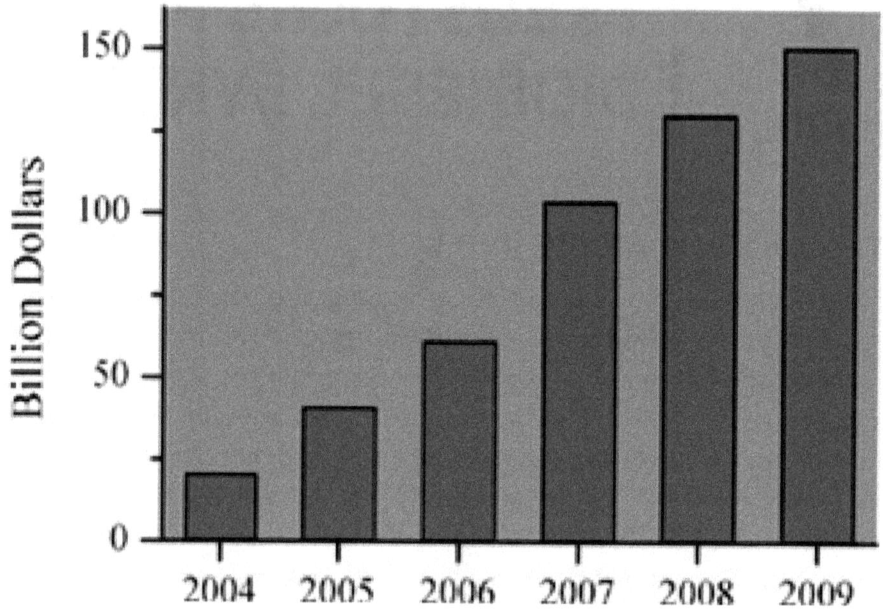

Fig. 1. Global investment in renewable energy, 2004-2009. Adapted from REN21 (2010)

The global markets for biodiesel are entering a period of rapid and transitional growth. In the year 2007, there were only 20 nations producing biodiesel for the needs of over 200 nations; by the year 2010, more than 200 nations become biodiesel producing nations and suppliers (Thurmond, 2008). Global biodiesel production has massively increased to 16.6 billion liters per year over the last nine years (Figure 2). Much of the growth is happening in just three countries: the United States, Brazil and Germany, which together account for over half of biodiesel (Checkbiotech, 2009). The International Energy Agency's report suggests that world production of biodiesel could top 25 million tons per year by 2012 if the recent trends continue.

Biodiesel can be produced from a variety of feedstocks, including plant oils, animal fats and waste oils as well as microalgae (Demiras, 2008). Each feedstock has its advantages and disadvantages in terms of oil content, fatty acid composition, biomass yield and geographic distribution. Depending on the origin and quality of feedstocks, changes may be required for the production process of biodiesel.

The use of plant oils as biodiesel feedstocks has been long recognized and well documented in numerous studies (Abdullah et al., 2009; de Oliveira et al., 2005; Graef et al., 2009; Hawash et al., 2009; Hill et al., 2006; Jain & Sharma, 2010; Nakpong & Wootthikanokkhan, 2010; Patil & Deng, 2009; Rashid & Anwar, 2008; Sahoo &

Das, 2009; Saka & Kusdiana, 2001). These feedstocks include the oils from soybean, rapeseed, palm, canola, peanut, cottonseed, sunflower and safflower. Based on the geographic distribution, soybean is the primary source for biodiesel in USA, palm oil is used as a significant biodiesel feedstock in Malaysia and Indonesia, and rapeseed is the most common base oil used in Europe for biodiesel production (Demiras, 2008). The vast majority of these plants are also used for food and feed production, which means that possible food versus fuel conflicts are present. Thus, the use of these plant oils as feedstocks for biodiesel seems insignificant for the developing countries which are importers of edible oils (Meher et al., 2008). In addition to these edible oils, various non-edible, tree-borne oils from jatropha, karanja, jojoba and neem are the potential biodiesel feedstocks (Jain & Sharma, 2009; Meher et al., 2008; Sahoo & Das, 2009). Jatropha and karanja are two oilseed plants that are not widely exploited due to the presence of toxic components in the oils. In India, they are popularly used as biodiesel feedstocks.

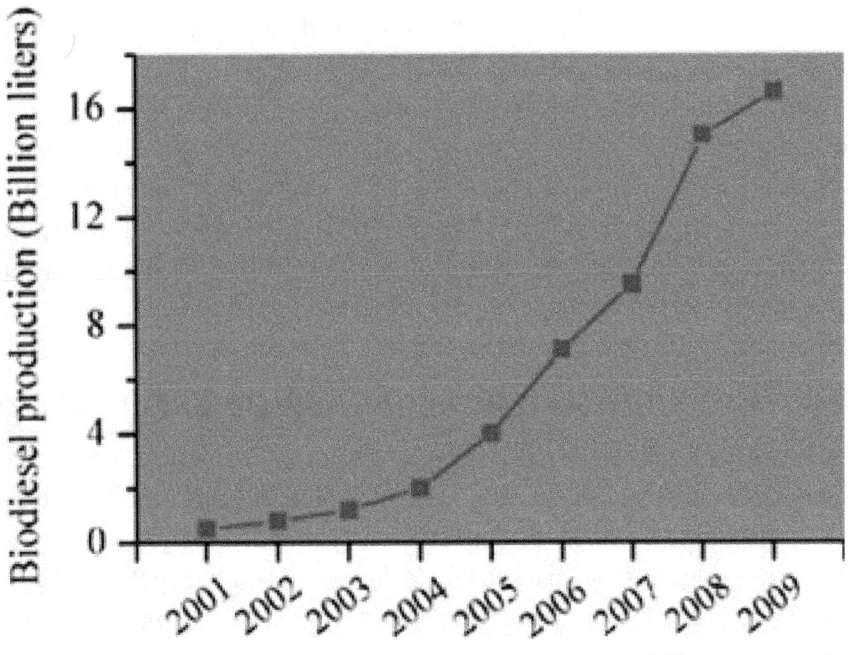

Fig. 2. Global biodiesel production, 2001-2009. Adapted from REN21 (2010)

In addition to the plant oils, animal fats and waste oils are the potential sources for commercial biodiesel production (Thompson et al., 2010). Among these feedstocks, tallow, lard, yellow grease and waste cooking oils have received most interest (Banerjee et al., 2009; Canakci, 2007; da Cunha et al., 2009; Dias et al., 2009; Diaz-Felix et al., 2009; Oner & Altun, 2009; Phan & Phan, 2008). However, animal fats and waste oils usually contain large amounts of free fatty acids, which can be as high as 41.8% (Canakci, 2007). Free fatty acids cannot be directly converted to biodiesel in alkali-catalyzed transesterificatoin but react with alkali to form soaps that inhibit the separation of biodiesel from glycerin and wash water fraction (Huang et al., 2010). A two-step process was developed for these high fatty acid feedstocks: acid-catalyzed pretreatment and alkali-catalyzed transesterificaton. Because animal fats and waste oils have relatively high level of saturation (Canakci, 2007), the biodiesel from these sources exhibits poor cold flow properties.

Microalgae represent a wide variety of aquatic photosynthetic organisms with the potential of producing high biomass and accumulating high level of oil. The production of biodiesel from microalgal oil has long been recognized and been evaluated in response to the United States Department of Energy for research in alternativerenewableenergy(Sheehanetal.,1998).Currently,thecommercialization of algae-derived biodiesel is still in its infancy stage. Using microalgae as biodiesel feedstocks has received unprecedentedly increasing interest, including but not restricted to microalgal strain selection and genetic engineering, mass cultivation for biomass production, lipid extraction and analysis, transesterification technologies, fuel properties and engine tests (Abou-Shanab et al., 2011; Brennan & Owende, 2010; Demirbas, 2009; Greenwell et al., 2010; Miao & Wu, 2006; Pruvost et al., 2011; Radakovits et al., 2010; Rodolfi et al., 2009; Ross et al., 2008; Sydney et al., 2011). Considering their unique characteristics, microalgae have been considered as the most promising feedstock of biodiesel that has the potential to displace fossil diesel (Chisti, 2007). This analysis primarily examines the possibilities of utilizing microalgae as biodiesel feedstocks, the biodiesel production process, and the feasibility of employing genetic engineering to enhance microalgal productivity.

2.POTENTIALOFUSINGMICROALGAEASBIODIESELFEEDSTOCKS

Microalgae represent a large and diverse group of prokaryotic or eukaryotic photosynthetic microorganisms that are in unicellular or multicellular form. Examples of prokaryotic microorganisms are cyanobacteria (commonly referred to as blue-green algae) that are closely related to Gram-negative bacteria and eukaryotic ones are for example green microalgae and diatoms (Graham et al., 2009). Microalgae can be found in a wide range of environmental conditions, including water, land, and even unusual environments such as snow and desert soils (Lee, 2008). It is estimated that there are more than 50,000 species around the world, among which only about 30,000 have been studied and analyzed (Mata et al., 2010). Extensive collections of microalgae have been established by researchers in different countries, including the Freshwater Microalgae Collection of University of Coimbra (Portugal), the Collection of the Goettingen University (Germany), the Provasoli-Guillard National Center for Culture of Marine Phytoplankton (CCMP, USA), the University of Texas Algal Culture Collection (USA), the CSIRO collection of Living Microalgae (CCLM, Australia), the National Institute for Environmental Studies Collection (NIES, Japan), the American Type Culture Collection (ATCC, USA), and the Freshwater Algae Culture Collection of Institute of Hydrobiology (China). Together more than 10,000 microalgal strains are available to be selected for use in a broad range of applications, for example, as biodiesel feedstocks.

The use of microalgae for biodiesel production has long been recognized and its potential has been widely reported by many research studies recently (Abou-Shanab et al., 2011; Afify et al., 2010; Ahmad et al., 2011; Cheng et al., 2009; Damiani et al., 2010; Gouveia et al., 2009; Liu et al., 2010; Rodolfi et al., 2009; Yoo et al., 2010). Microalgae reproduce themselves autotrophically using CO_2 from air and light through photosynthesis. Compared with higher plants, microalgae exhibit higher photosynthetic efficiency and grow much faster, finishing an entire growth cycle within a few days (Christi, 2007). Typical growth rates are presented in Figure 3 as the doubling time for each microalgal species. A low doubling time corresponds to a high specific growth rate. Microalgae double themselves with an average time of 26 h, and some can even reproduce within 8 h. Moreover, they can be adapted to grow in a broad range of environmental conditions, suggesting the possibility of finding species best suited to local environments which is not suitable for cultivating oil plants (e.g. palm, soybean and rapeseed).

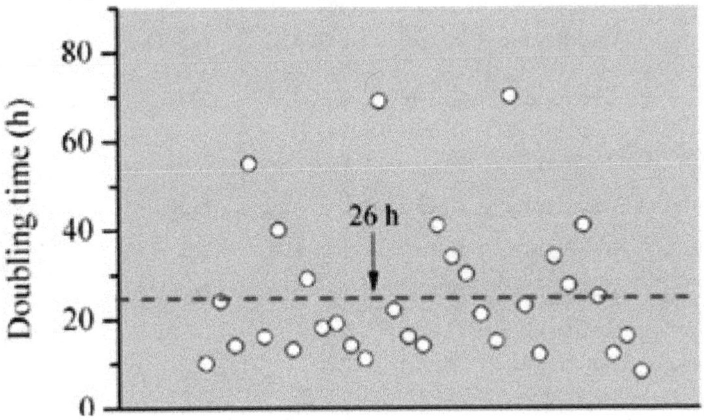

Fig. 3. Doubling time for some microalgal species. The dash line indicates the average value. T_d=ln(2)/μ, T_d, doubling time, μ, specific growth rate.

In addition to growth rate, lipid content is another important factor to assess the potential of microalgae for biodiesel production. Over the past few decades, thousands of algae and cyanobacterial species have been screened for high lipid production, and numerous oleaginous species have been isolated and characterized. The lipid contents of these oleaginous algae are species- and/or strains-dependent, vary greatly, and may reach as high as 68% of dry weight, as shown in Table 1.

Generally, microalgae synthesize a low content of lipids under nutrient replete conditions (Figure 4), with membrane lipids (e.g., phospholipids and glycolidips) being the main components; whereas under stress conditions such as nitrogen deficiency, a great increase in total lipids was observed (Figure 4) with neutral lipids in particular triacylglycerols (TAGs) being the dominant components (Hu, 2004). TAGs are considered to be superior to phospholipids or glycolipids for biodiesel feedstocks because of their higher percentage of fatty acids and lack of phosphate (Pruvost et al., 2009). Unlike higher plants in which individual classes of lipids may be synthesized and localized in a specific cell, tissue or organ, algae produce these different lipids in a single cell (Hu et al., 2008b). The synthesized TAGs are deposited in lipid bodies located in cytoplasm of algal cells (Damiani et al., 2010; Rabbani et al., 1998).

Table 1. Lipid content and productivity of various microalgal species.

Algal species	Culture conditions	Lipid content (%)	biomass productivity (g/L/day)	Lipid productivity (mg/L/day)	References
Chlorophyta					
Botryococcus braunii	Phototrophic	9.5-13.5	0.02-0.04	2.6-4.5	Chinnasamy et al., 2010
Botryococcus braunii	Phototrophic	17.85	0.346		Órpez et al., 2009
Botryococcus braunii	Phototrophic	24	0.077	21	Yoo et al., 2010
Botryococcus sp.	Phototrophic	15.8-35.9	0.14-0.22	21.3-46.9	Yeesang and Cheirsilp, 2011
Chlamydomonas reinhardtii	Mixotrophic	12.2-46	0.21-0.36	29-95	Li et al., 2010a
Chlorella ellipsoidea	Phototrophic	32	0.07	22.4	Abou-Shanab et al., 2011
Chlorella ellipsoidea	Phototrophic	15-43		11.4	Yang et al., 2011
Chlorella protothecoides	Heterotrophic	48.1-63.8	1.02-1.73	3432-6293	De la Hoz Siegler et al., 2011
Chlorella protothecoides	Heterotrophic	49	1.2	586.8	Gao et al., 2010
Chlorella saccharophila	Phototrophic	12.9-18.1	0.02	2.7-4.2	Chinnasamy et al., 2010
Chlorella sorokiniana	Phototrophic	19.3	0.23	44.7	Rodolfi et al., 2009
Chlorella sp.	Phototrophic	33.9	0.528	178.8	Chiu et al., 2008
Chlorella sp.	Phototrophic	22.4-66.1	0.08-0.34	51-124	Hsieh and Wu, 2009
Chlorella sp.	Phototrophic	34.1 [a]	0.053	22	Matsumoto et al., 2010
Chlorella sp.	Phototrophic	18.7	0.23	42.1	Rodolfi et al., 2009
Chlorella vulgaris	Phototrophic	20-42	0.21-0.35	44-147	Feng et al., 2011
Chlorella vulgaris	Phototrophic, Mixotrophic, heterotrophic	21-38	0.01-0.26	4-54	Liang et al., 2009
Chlorella vulgaris	Phototrophic	19.2	0.17	32.6	Rodolfi et al., 2009
Chlorella vulgaris	Phototrophic	26-52		11.6-13.2	Widjaja et al., 2009
Chlorella vulgaris	Phototrophic	35	0.117	41	Yeh et al., 2010
Chlorella zofingiensis	Heterotrophic	52	0.72	374.4	Liu et al., 2010

Algal species	Culture conditions	Lipid content (%)	biomass productivity (g/L/day)	Lipid productivity (mg/L/day)	References
Chlorella zofingiensis	Phototrophic	25.8	0.136	35.1	Liu et al., 2011
Chlorococcum sp.	Phototrophic	19.3	0.28	53.7	Rodolfi et al., 2009
Choricystis minor	Phototrophic	21-59.3	0.35	82	Sobczuk and Chisti, 2010
Dunaliella tertiolecta	Phototrophic	12.2-15.2	0.03-0.04	4.0-4.6	Chinnasamy et al., 2010
Dunaliella tertiolecta	Phototrophic	16.7	0.12	20	Gouveia and Oliveira, 2009
Haematococcus pluvialis	Phototrophic	15.6-34.9			Damiani et al., 2010
Micractinium pusillum	Phototrophic	24	0.108	25.7	Abou-Shanab et al., 2011
Neochloris oleabundans	Phototrophic	19-56	0.03-0.15	10.7-38.8	Gouveia et al., 2009
Neochloris oleabundans	Phototrophic	7-40.3	0.31-0.63	38-133	Li et al., 2008
Ourococcus multisporus	Phototrophic	52	0.045	23.3	Abou-Shanab et al., 2011
Parietochloris incisa	Phototrophic	18-34 [a]	0.23-0.47	46-160	Solovchenko et al., 2008
Pseudochlorococcum sp.	Phototrophic	24.6-52.1	0.234-0.76	53-350	Li et al., 2011a
Scenedesmus obliquus	Phototrophic	21-58	0.08-0.09	19-43.3	Abou-Shanab et al., 2011
Scenedesmus obliquus	Phototrophic	17.7	0.09	15.9	Gouveia and Oliveira, 2009
Scenedesmus obliquus	Phototrophic	12-38.9	0.20-0.29	35.1-78.7	Ho et al., 2010
Scenedesmus obliquus	Phototrophic, Mixotrophic	12.6-58.3	0.51	270	Mandal and Mallick, 2009
Scenedesmus rubescens like	Phototrophic	11.3-27 [a]	0.44-0.54	108-133	Lin and Lin, 2011
Scenedesmus quadricauda	Phototrophic	18.4	0.19	35.1	Rodolfi et al., 2009
Scenedesmus sp.	Phototrophic	22-53	0.08	20.3	Xin et al., 2010
Scenedesmus sp.	Phototrophic	18	0.203	39	Yoo et al., 2010
Scenedesmus sp.	Phototrophic	21.1	0.26	53.9	Rodolfi et al., 2009
Tetraselmis chui	Phototrophic	17.3-23.5	1-2.6	235-450	Araujo et al., 2011
Tetraselmis sp.	Phototrophic	8.7-33	0.21	22.86	Huerlimann et al., 2010
Tetraselmis suecica	Phototrophic	8.5-12.9	0.28-0.32	27-36.4	Rodolfi et al., 2009
Tetraselmis tetrathele	Phototrophic	29.2-30.3	3.1-4.4	905-1333	Araujo et al., 2011
Bacillariophyceae					
Chaetoceros calcitrans	Phototrophic	39.8	0.04	17.6	Rodolfi et al., 2009
Chaetoceros gracilis	Phototrophic	15.5-60.3	3.4-3.7	530-2210	Araujo et al., 2011
Chaetoceros muelleri	Phototrophic	11.7-25.3	1.2-2.7	1404-6831	Araujo et al., 2011
Chaetoceros muelleri	Phototrophic	33.6	0.07	21.8	Rodolfi et al., 2009
Cylindrotheca closterium	Phototrophic	17-30			Pruvost et al., 2011
Navicula sp.	Phototrophic	47.6 [a]	0.055	26.4	Matsumoto et al., 2010
Nitzschia cf. pusilla	Phototrophic	48	0.065	31.4	Abou-Shanab et al., 2011
Nitzschia laevis	Heterotrophic	12.8	2.02	258.6	Chen et al., 2008

Algal species	Culture conditions	Lipid content (%)	biomass productivity (g/L/day)	Lipid productivity (mg/L/day)	References
Nitzschia sp.	Phototrophic	32	0.013		Moazami et al., 2011
Phaeodactylum tricornutum	Phototrophic	18.7	0.24	44.8	Rodolfi et al., 2009
Skeletonema costatum	Phototrophic	21.1	0.08	17.4	Rodolfi et al., 2009
Skeletonema sp.	Phototrophic	31.8	0.09	27.3	Rodolfi et al., 2009
Thalassiosira pseudonana	Phototrophic	20.6	0.08	17.4	Rodolfi et al., 2009
Eustigmatophyceae					
Ellipsoidion sp.	Phototrophic	27.4	0.17	47.3	Rodolfi et al., 2009
Monodus subterraneus	Phototrophic	12.9-15 [a]	0.34-0.49	47.5-67.5	Khozin-Goldberg and Cohen, 2006
Monodus subterraneus	Phototrophic	16.1	0.19	30.4	Rodolfi et al., 2009
Nannochloropsis oculata	Phototrophic	22.8-23	2.4-3.4	547.2-782	Araujo et al., 2011
Nannochloropsis oculata	Phototrophic	26.2-30.7	0.37-0.50	84-151	Chiu et al., 2009
Nannochloropsis oculata	Phototrophic	7.9-15.9	0.06-0.13	9.1-16.4	Converti et al., 2009
Nannochloropsis sp.	Phototrophic	52	0.0465		Moazami et al., 2011
Nannochloropsis sp.	Phototrophic	23.1-37.8	0.06	20	Huerlimann et al., 2010
Nannochloropsis sp.	Phototrophic	28.7	0.09	25.8	Gouveia and Oliveira, 2009
Nannochloropsis sp.	Phototrophic	21.6-35.7	0.17-0.21	37.6-61	Rodolfi et al., 2009
Others					
Aphanothece microscopica	Heterotrophic	7.1-15.3	0.26-0.44	30-50	Queiroz et al., 2011
Crypthecodinium Cohnii	Heterotrophic	19.9	2.24	444.9	Couto et al., 2010
Isochrysis galbana	Phototrophic	24.6	0.057	14.02	Lin et al., 2007
Isochrysis sp.	Phototrophic	23.5-34.1	0.09	20.95	Huerlimann et al., 2010
Isochrysis sp.	Phototrophic	22.4-27.4	0.14-0.17	37.8	Rodolfi et al., 2009
Pavlova lutheri	Phototrophic	35.5	0.14	50.2	Rodolfi et al., 2009
Pavlova salina	Phototrophic	30.9	0.16	49.4	Rodolfi et al., 2009
Pavlova viridis	Phototrophic	24.8-32			Li et al., 2005
Pleurochrysis carterae	Phototrophic	9.7-12	0.03-0.04	2.7-4.4	Chinnasamy et al., 2010
Porphyridium cruentum	Phototrophic	9.5	0.37	34.8	Rodolfi et al., 2009
Rhodomonas sp.	Phototrophic	9.5-20.5	0.06	6.19	Huerlimann et al., 2010
Schizochytrium limacinum	Heterotrophic	50.3 [a]	3.48	1750	Ethier et al., 2011
Schizochytrium mangrovei	Heterotrophic	68 [a]	2.44	1659	Fan et al., 2007
Spirulina maxima	Phototrophic	4.1	0.21	8.6	Gouveia and Oliveira, 2009
Thalassiosira weissflogii	Phototrophic	6.3-13.2	0.5-1.5	31.5-198	Araujo et al., 2011

[a] Total fatty acid content

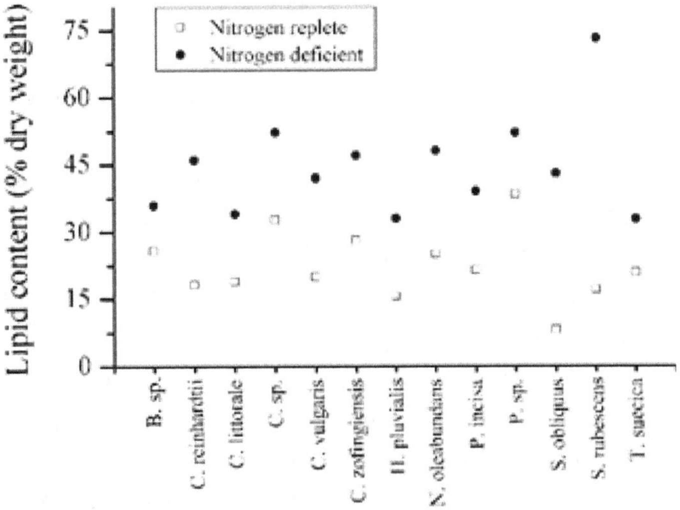

Fig. 4. Lipid content under nitrogen replete (open squares) and nitrogen deficient (filled circles) conditions for *Chlorophyta*. *B.* sp., *Botryococcus* sp. (Yeesang and Cheirsilp, 2011); *C. reinhardtii*, *Chlamydomonas reinhardtii* (Li et al., 2010); *C. littorale*, *Chlorocuccum littorale* (Ota et al., 2009); *C.* sp., *Chlorella* sp. (Hsieh and Wu, 2009); *C. vulgaris*, *Chlorella vulgaris* (Feng et al., 2011); *C. zofingiensis*, *Chlorella zofingiensis* (Liu et al., 2010); *H. pluvialis*, *Haematococcus pluvialis* (Damiani et al 2010); *N. oleabundans*, *Neochloris oleabundans* (Gouveia et al., 2009); *P. incisa*, *Parietochloris incisa* (Solovchenko et al., 2010); *P.* sp., *Pseudochlorococcum* sp. (Li et al., 2011); *S. obliquus*, *Scenedesmus obliquus* (Mandal and Mallick, 2009); *S. rubescens*, *Scenedesmus rubescens* (Mandal and Mallick, 2009); *T. suecica*, *Tetraselmis suecica* (Rodolfi et al., 2009).

The important properties of biodiesel such as cetane number, viscosity, cold flow, oxidative stability, are largely determined by the composition and structure of fatty acid esters which in turn are determined by the characteristics of fatty acids of biodiesel feedstocks, for exmaple carbon chain length and unsaturation degree (Knothe, 2005b). Fatty acids are either in saturated or unsaturated form, and the unsaturated fatty acids may vary in the number and position of double bones on the acyl chain. Based on the number of double bones, unsaturated fatty acids are clarified into monounsaturated fatty acids (MUFAs) and polyunsaturated fatty acids (PUFAs). The fatty acid profile of a great many algal species has been investigated and is shown in Table 2. The synthesized fatty acids in algae are commonly in medium length, ranging from 16 to 18 carbons, despite the great variation in fatty acid composition. Specifically, the major fatty acids are C16:0, C18:1 and C18:2 or C18:3 in green algae, C16:0 and C16:1 in diatoms and C16:0, C16:1, C18:1 and C18:2 in cyanobacteria. It is worthy to note that these data are obtained from algal species under specific conditions and vary greatly when algal cells are exposed to different environmental or nutritional conditions such as temperature, pH, light intensity, or nitrogen concentration (Guedes et al 2010; James et al., 2011; Sobczuk & Chisti,

2010; Tatsuzawa et al., 1996). Generally, saturated fatty esters possess high cetane number and superior oxidative stability; whereas unsaturated, especially polyunsaturated, fatty esters have improved low-temperature properties (Knothe, 2008). In this regard, it is suggested that the modification of fatty esters, for example the enhanced proportion of oleic acid (C18:1) ester, can provide a compromise solution between oxidative stability and low-temperature properties and therefore promote the quality of biodiesel (Knothe, 2009). Thus, microalgae with high oleic acid are suitable for biodiesel production.

Currently the commercial production of biodiesel is mainly from plant oils and animal fats. However, the plant oil derived biodiesel cannot realistically meet the demand of transport fuels because large arable lands are required for cultivation of oil plants, as demonstrated in Table 3. Based on the oil yield of different plants, the cropping area needed is calculated and expressed as a percentage of the total U.S. cropping area. If soybean, the popular oil crop in United States is used for biodiesel production to meet the existing transport fuel need, 5.2 times of U.S. cropland will need to be employed. Even the high-yielding oil plant palm is planted as the biodiesel feedstock, more than 50% of current U.S. arable lands have to be occupied. The requirement of huge arable lands and the resulted conflicts between food and oil make the biodiesel from plant oils unrealistic to completely replace the petroleum derived diesel in the foreseeable future. It is another case, however, if microalgae are used to produce biodiesel. As compared with the conventional oil plants, microalgae possess significant advantages in biomass production and oil yield and therefore the biodiesel productivity. In terms of land use, microalgae need much less than oil plants, thus eliminating the competition with food for arable lands (Table 3).

Table 2. Fatty acid composition of various algal species (% of total fatty acids)

Fatty acids Algal species	C12:0	C14:0	C15:0	C16:0	C16:1	C16:2	C16:3	C16:4	C17:0	C18:0	C18:1	C18:2	C18:3	C18:4	C20:0	C20:4	C20:5	C22:5	C22:6	efs
Chlorophyta																				
Botryococcus braunii				29.5	3.4					1	44.9	21.2								Yoo et al., 2010
Botryococcus sp.		3.95	1.56	30.04	0.94				1.54	12.02	37.68	5.01	7.35		0.63					Yeesang and Cheirsilp, 2011
Chlamydomonas reinhardtii				30.7	3	1.8	1.6	2.7		3.2	27.2	18.3	11		0.5					James et al., 2011
Chlorella ellipsoidea		2		26							4	40	23			5				Abou-Shanab et al 2011
Chlorella protothecoides				14.3	1				0.32	2.7	71.6	9.7								Cheng et al 2009
Chlorella pyrenoidosa		0.7		17.3	0.8	7	9.3			1.2	3.3	18.5	41.8							D'Oca et al 2011

Chlorella sorokiniana				25.4	3.1	10.7	4.1			1.4	12.4	34.4	7.1							Chen and Johns, 1991
Chlorella sp.	3.78		5.24	16.1	10.88	9.79			4.74	4.35	8.45	14.36	18.79							Li et al., 2011b
Chlorella vulgaris				24	2.1					1.3	24.8	47.8								Yoo et al., 2010
Chlorella zofingiensis				22.62	1.97	7.38	1.94	0.22		2.09	35.68	18.46	7.75	0.49						Liu et al., 2010
Chlorocuccum littorale				20.9	5.6			14.4			29.7	7.2	22.2							Ota et al., 2009
Choricystis minor				36					0.4	12.3	31.2	9.9	3.8		1.9					Sobczuk and Chisti, 2010
Dictyochlorop-sis splendida		13.88		69.59					1.21	0.38	1.11	12.14	0.42							Afify et al 2010
Dunaliella tertiolecta				26.4		2.3	1.27			0.6	16.8	13.1	39.6							Chen et al 2011
Haematococcus pluvialis	0.21	1.25		22.5	0.64				0.19	3.15	19.36	26.9	17.04		0.2	0.89	0.57			Damiani et al 2010
Micractinium pusillum				33	1						31	17	18							Abou-Shanab et al 2011
Neochloris oleabundans				23.3	0.6	1.6	2.4		0.2	4.5	43	17.8	5.8							Levine et al., 2011
Neochloris sp.		5.22		29.4					5.2	6.6	17.5	23.6	12.6							Moazami et al., 2011
Ourococcus multisporus		2		19	1					5	26	11	36							Abou-Shanab et al 2011
Parietochloris incise				9.1	0.7	0.6				2.1	15.1	9.3	1.6	1.2		58.9				Khoz-in-Goldberg et al., 2002
Scenedesmus obliquus		1.48		21.8	5.95	3.96	0.68	0.43		0.45	17.93	21.74	3.76	0.21						Gouveia and Olivei-ra, 2009
Scenedesmus sp.				36.3	4					2.7	25.9	31.1								Yoo et al., 2010
Tetraselmis sp.		0.6		27.8						0.9	28.2	9.3	23.9	3.7		0.9	3.4			Huerli-mann et al., 2010
Bacillarioph-yceae																				
Chaetoceros sp.		23.6		9.2	36.5	6.9	2.6		2		3		1.4	0.6		4.1	8		1	Renaud et al., 2002
Cyclotella cryptica		1.4		15.2	10.7						3.9	1.2	3.5				9.7		1.7	Pahl et al., 2010
Navicula sp.				45	52.7					0.6	1.1	0.6								Matsumoto et al., 2010
Nitzschia cf. pusilla		6		31	57	0.27							6							Abou-Shanab et al 2011
Nitzschia laevis		16.9		28.5	23.9					0.7	5.1	3.4	4.1			5	11.7			Chen et al 2008
Nitzschia sp.		9	3.5	37.4					4.6	5.3	16.9	11.6								Moazami et al., 2011
Cyanobacteria																				
Nostoc commune				23.5	22.5						5.6	21.1	14.1							Pushparaj et al., 2008

Nostoc flagelliforme		0.65		21.27	14.91					6.2	22.59	15.03	19.35							Liu et al., 2005
Spirulina				49.2	5.9					1.7	2.9	22.7	17.5							Chaiklahan et al 2008
Spirulina maxima		0.34		40.16	9.19		0.42	0.16		1.18	5.43	17.89	18.32	0.08	0.06					Gouveia and Oliveira, 2009
Synechocystis PCC6803				52	3				1		3	9	29	3						Wada and Murata, 1990
Eustigmatoph-yceae																				
Monodus subterraneus		3.3		19.8	34.3					9.7	9	0.8	0.7			2.8	15.5			Khoz-in-Goldberg and Cohen, 2006
Nannochlorop-sis oculata				62						11	5	8	15							Converti et al 2009
Nannochloropsis sp.				23.4						7.14	45.4	11.7	12.2							Moazami et al., 2011

In addition to biodiesel, microalgae can serve as sources of other renewable fuels such as biogas, bioethanol, bio-oil and syngas (Chisti, 2008; Demirbas, 2010; Mussgnug et al., 2010). Moreover, microalgal biomass contains significant amounts of proteins, carbohydrates and other high-value compounds that can be potentially used as feeds, foods and pharmaceuticals (Chisti, 2007). Thus, integrating the production of such co-products with biofuels will provide new insight into improving the production economics of microalgal biodiesel. Microalgae can also be used for sequestration of carbon dioxide from industrial flue gases and wastewater treatment by removal of nutrients (Chinnasamy et al 2010; Fulke et al., 2010; Levine et al., 2011; Yang et al., 2011). Coupled with these environment-beneficial approaches, the production potential of microalgae derived biodiesel is desirable.

Table 3. Comparison of microalgae with other biodiesel feedstocks.

Feedstocks	Oil content (% dry weight)	Oil yeild (L/ha year)	Land area needed (M ha)a	Percentage of existing US cropping areaa
Corn	44	172	3480	1912
Hemp	33	363	1650	906
Soybean	18	636	940	516
Jatropha	28	741	807	443
Camelina	42	915	650	357
Canola	41	974	610	335

Feedstocks	Oil content (% dry weight)	Oil yeild (L/ha year)	Land area needed (M ha)a	Percentage of existing US cropping areaa
Sunflower	40	1070	560	307
Castor	48	1307	450	247
Palm oil	36	5366	110	60.4
Microalgae (low oil content)	30	58,700	10.2	5.6
Microalgae (medium oil content)	50	97,800	6.1	3.4
Microalgae (high oil content)	70	136,900	4.4	2.4

[a]For meeting all transport fuel needs of the United States. Adapted from Chisti, 2007 and Mata et al., 2010.

3. BIODIESEL PRODUCTION FROM MICROALGAE

The biodiesel production from microalgal oil shares the same processes and technologies as those used for other feedstocks derived oils. However, microalgae are microorganisms living essentially in liquid environments and thus have particular cultivation, harvesting, and downstream processing techniques for efficient biodiesel production. The microalgal biodiesel production pipeline is schematically presented in Figure 5, including strain selection, mass culture, biomass harvesting and processing, oil extraction and transesterification.

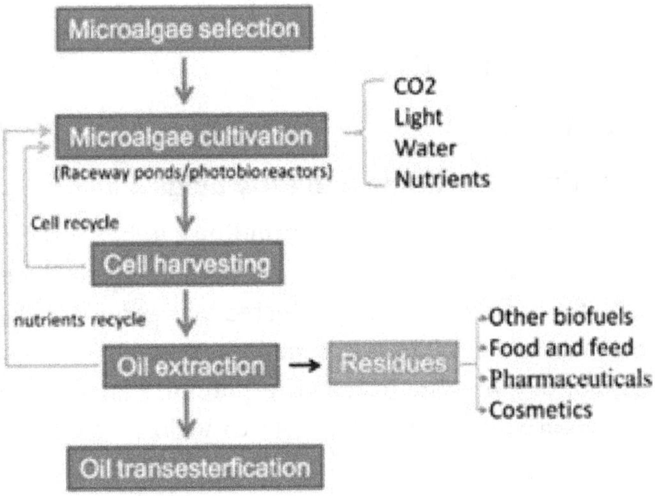

Fig. 5. Microalgal biodiesel production pipeline

3.1 Microalgae selection

There are more than 50,000 microalgal species around the world. Selection of an ideal species is of fundamental importance to the success of algal biodiesel production. Theoretically, an ideal species should own the following desirable characteristics: rapid growth rate, high oil content, wide tolerance of environmental conditions, CO2 tolerance and uptake, large cell size, easy of disruption, etc. However, it is unlikely for a single species to excel in all above mentioned characteristics. Thus, prioritization is required. Commonly, fast-growing strains with high oil content are placed on the priority list for biodiesel production. Fast growth makes sure the high biomass productivity and reduces the contamination risk owing to out-competition of slower growers. High oil content helps increase the process yield coefficient and reduce the cost of downstream extraction and purification. The selected species should be suitable for mass cultivation under local geographic and climatic conditions, for example, the inland prefers freshwater algae while the coastal place desires marine algal species. Ease of harvesting is an often-overlooked criterion and should be taken into account. Algal biomass harvest requires significant capital and accounts for up to 30% of total biomass production cost (Molina Grima et al., 2003). Therefore, it is desirable to choose algal species with properties that simplify harvesting, including large cell size, high specific gravity and autofloculation potential (Griffiths & Harrison, 2009). These properties can greatly influence the process economics for biodiesel production from algae. An additional algal characteristic is the suitability of lipids for biodiesel production; for example, neutral lipids in particular TAG are superior to polar lipids (phospholipids and glycolipids) for biodiesel and C18:1 h as advantages over other fatty acids for improving biodiesel quality (Knothe, 2009).

3.2 Microalgae cultivation

3.2.1 Factors affecting algal lipids and fatty acids

Microalgae require several things to grow, including a light source, carbon dioxide, water, and inorganic salts. The lipid content and fatty acid composition are species/strain- specific and can be greatly affected by a variety of medium nutrients and environmental factors. Carbon is the main component of algal biomass and accounts for *ca* 50% of dry weight. CO_2 is the common carbon source for algal growth. But some algal species are also able to utilize organic carbon sources, for example sugars and glycerol (Easterling et al., 2009; Liu et al., 2010). Sugars particularly glucose are preferred and can be used to boost production of both algal biomass and lipids (Liu et al., 2010). Nitrogen is an important nutrient affecting lipid metabolism in algae. The influence of nitrogen concentration on lipid and fatty acid production has been investigated in numerous algal species.

Nitrate was suggested to be superior to other nitrogen sources such as urea and ammonium for algal lipid production (Li et al., 2008). Generally, low concentration of nitrogen in the medium favors the accumulation of lipids particularly TAGs and total fatty acids. But in some cases, nitrogen starvation caused decreased synthesis of lipids and fatty acids (Saha et al., 2003). Nitrogen concentration also affects algal fatty acid composition. For example, in cyanobacteria, increased levels of C16:0 and C18:1 and decreased C18:2 levels were observed in response to nitrogen deprivation (Piorreck & Pohl, 1984). In the marine alga *Pavlova viridis*, nitrogen depletion resulted in an increase in saturated, monounsaturated fatty acids and C22:6 (n-3) contents (Li et al., 2005). Nitrogen starvation brought about a strong increase in the proportion of C20:4 (n-6) in the green algal *Parietochloris incisa* (Solovchenko et al., 2008). Similar to nitrogen, silicon is a key nutrient that affects lipid metabolism of diatoms, and can promote the accumulation of neutral lipids as well as of saturated and monounsaturated fatty acids when depleted from culture medium (Roessler, 1988). Other types of nutrient deficiency include phosphorus and sulfur limitations are also able to enhance lipid accumulation in algae (Khozin-Goldberg & Cohen, 2006; Li et al., 2010b; Sato et al., 2000). These types of nutrient deficiency, however, do not always lead to elevated overall lipid production, because they at the same time exert negative effect on algal growth and contribute to the reduced algal biomass production that compromises the enhanced lipid yield resulting from increased lipid content. Therefore, the manipulation of these nutrients needs to be optimized to induce lipid accumulation while maintaining algal growth for maximal production of lipids. Iron is a micro-nutrient required in a tiny amount for ensuring algal growth. Within a certain range of concentrations, high concentrations of iron benefit algal growth as well as cellular lipid accumulation and thus the overall lipid yield in the green alga *Chlorella vulgaris* (Liu et al., 2008).

Among the environmental factors, light is an important one that has a marked effect on the lipid production and fatty acid composition in algae (Brown et al., 1996; Damiani et al., 2010; Khotimchenko & Yakovleva, 2005; Napolitano, 1994; Sukenik et al., 1989; Zhekisheva et al., 2002, 2005). Generally, low light intensity favors the formation of polar lipids such as the membrane lipids associated with the chloroplast; whereas high light intensity benefits the accumulation of neutral storage lipids in particular TAGs. In *H. pluvialis*, for example, high light intensity resulted in a great increase of both neutral and polar lipids, but the increase extent of neutral lipids was much greater than that of polar lipids, leading to the dominant proportion of neutral lipids in the total lipids (Zhekisheva et al., 2002, 2005). Although the effect of light intensity on fatty acid composition differs among the algal species and/or strains, there is a general trend that the increase of light intensity contributes to the enhanced proportions of saturated and monounsaturated fatty acids and the concurrently the reduced proportion of polyunsaturated fatty acids (Damiani et al., 2010; Sukenik et al., 1989; Zhekisheva et al., 2002, 2005). Temperature is another important environmental factor that affects profiles of algal lipids and fatty acids. In response to temperature shift, algae commonly alter the physical properties and thermal responses of membrane lipids to maintain fluidity and function of membranes (Somerville, 1995). In general, increased temperature causes increased fatty acid saturation and at the same time decreased fatty acid unsaturation. For example, C14:0, C16:0, C18:0 and C18:2 increased and C18:3 (n-3), C18:4, C20:5 and C22:6 decreased in *Rhodomonas* sp., and C16:0 increased and C18:4 decreased in *Cryptomonas* sp. when temperature increased (Renaud et al., 2002). As for the effect of temperature on cellular lipid content, it differs in a species-dependent manner. In response to increased temperature, algae may show an increase (Boussiba et al., 1987), no significant change or even a decrease (Renaud et al., 2002) in lipid contents. Other environmental factors such as salinity, pH and dissolved O_2 are also important and able to affect algal lipid metabolism.

In addition to the nutritional and environmental factors, growth phase and aging of the culture affect algal lipids and fatty acids. Commonly, algae accumulate more lipids at stationary phase than at logarithmic phase (Bigogno et al., 2002; Mansour et al., 2003). Associated with the growth phase transition from logarithmic to stationary phase, increased proportions of C16:0 and C18:1 and decreased proportions of PUFAs are often observed. Besides, it is suggested that algal lipids and fatty acids can be greatly affected by cultivation modes. Algae growing under heterotrophic mode usually produce more lipids in particular TAG and higher proportion of C18:1 than under photoautotrophic mode (Liu et al., 2011).

3.2.2 Raceway ponds and photobioreactors

Currently, the commonly used culture systems for large-scale production of algal biomass are open ponds and enclosed photobioreactors. An open pond culture system usually consists of a series of raceways-type of ponds placed outdoors. In this system, the shallow pond is usually about one foot deep; algae are cultured under conditions identical to their natural environment. The pond is designed in a raceway configuration, in which a paddle wheel provides circulation and mixing of the algal cells and nutrients (Chisti, 2007). The raceways are typically made from poured concrete, or they are simply dug into the earth and lined with a plastic liner to prevent the ground from soaking up the liquid. Compared with photobioreactors, open ponds cost less to build and operate, and are more durable with a large production capacity. However, the open pond system has its intrinsic disadvantages including rapid water loss due to evaporation, contamination with unwanted algal species as well as organisms that feed on algae, and low biomass productivity. In addition, optimal culture conditions are difficult to maintain in open ponds and recovering the biomass from such a dilute culture is expensive.

Unlike open ponds, enclosed photobioreactors are flexible systems that can be employed to overcome the problems of evaporation, contamination and low biomass productivity encountered in open ponds (Mata et al., 2010). These systems are made of transparent materials with a large surface area-to-volume ratio, and generally placed outdoors using natural light for illumination. The tubular photobioreactor is the most widely used one, which consists of an array of straight transparent tubes aligned with the sun's rays (Chisti, 2007). The tubes are generally no more than 10 cm in diameter to maximize sunlight penetration. The medium broth is circulated through a pump to the tubes, where it is exposed to light for photosynthesis, and then back to a reservoir. In some photobioreactors, the tubes are coiled to form what is known as a helical tubular photobioreactor. Artificial illumination can be used for photobioreactor. But it adds to the production cost and thus is used for the production of high value products instead of biodiesel feedstock. The algal biomass is prevented from settling by maintaining a highly turbulent flow within the reactor using either a mechanical pump or an airlift pump (Chisti, 2007). The result of photosynthesis will generate oxygen. The oxygen levels will accumulate in the closed photobioreactor and inhibit the growth of algae. Therefore, the culture must periodically be returned to a degassing zone, an area where the algal broth is bubbled with air to remove the excess oxygen. In addition, carbon dioxide must be fed into the system to provide carbon source and maintain culture pH for algal growth. Photobioreactors require cooling during daylight hours and temperature regulation in night hours. This may be done through heat exchangers located either in the tubes themselves or in the degassing column.

Table 4. Comparison of open ponds and photobioreactors for microalgae cultivation (Mata et al., 2010)

Culture systems	Open ponds	Enclosed bioreactors
Contamination control	Difficult	Easy
Contamination risk	High	Reduced
Sterility	None	Achievable
Process control	Difficult	Easy
Species control	Difficult	Easy
Mixing	Very poor	Uniform
Operation regime	Batch or semi-continuous	Batch or semi-continuous
Area/volume ration	Low	High
Algal cell density	Low	High
Investment	Low	Hight
Operation cost	Low	High
Light utilization efficiency	Poor	High

Culture systems	Open ponds	Enclosed bioreactors
Temperature control	difficult	More uniform temperature
Productivity	Low	High
Hydrodynamic stress on algae	Very low	Low-high
Evaporation of growth medium	High	Low
Gas transfer control	Low	High
O2 inhibition	< bioreactors	Great problem
Scale-up	Difficult	Difficult

Table 4 shows the comparison between open ponds and photobioreactors for microalgae cultivation.

Photobioreactors have obvious advantages over open ponds: offer better control, prevent contamination and evaporation, reduce carbon dioxide losses and allow to achieve higher biomass productivities. However, enclosed photobioreactors cost high to build and operate and the scale-up is difficult, limiting the number of large-scale commercial systems operating globally to high-value production runs (Greenwell et al., 2010). In this context, a hybrid photobioreactor-open pond system is proposed: using photobioreactors to produce contaminant-free inoculants for large open ponds.

3.3 Biomass harvesting and concentration

Algal harvesting is the concentration of diluted algal suspension into a thick algal paste, with the aim of obtaining slurry with at least 2–7% algal suspension on dry matter basis. Biomass harvest is a very challenging process and may contribute to 20-30% of the total biomass production cost (Molina Grima et al., 2003). The most common harvesting methods include sedimentation, filtration, centrifugation, sometimes with a pre-step of flocculation or flocculation-flotation. Flocculation is employed to aggregate the microalgal cells into larger clumps to enhance the harvest efficiency by gravity sedimentation, filtration, or centrifugation (Molina Grima et al., 2003). The selection of a harvesting process for a particular strain depends on size and properties of the algal strain. The selected harvest method must be able to handle a large volume of algal culture broth.

Filtration is the most commonly used method for harvesting algal biomass. The process can range from micro-strainers to pressure filtration and ultra-filtration systems. Vacuum filtration is feasible for harvesting large microalgae such as *Coelastrum proboscideum* and *Spirulina platensis* but unsuitable for recovering small size algal cells such as *Scenedesmus, Dunaliella*, or *Chlorella* (Molina Grima et al., 2003). Membrane-based microfiltration and ultrafiltration have also been used for harvesting algal cells for some specific application purposes, but overall, they are more expensive. Centrifugation is an accelerated sedimentation process for algae harvesting. Generally, centrifugation has high capital and operation costs, but its efficiency is much higher than natural sedimentation. Because of its high cost, centrifugation as an algae harvesting method is usually considered only feasible for high value products rather than biofuels.

3.4 Biomass processing for oil extraction

After harvesting, chemicals in the biomass may be subject to degradation induced by the process itself and also by internal enzyme in the algal cells. For example, lipase contained in the cells can rapidly hydrolyze cellular lipids into free fatty acids that are not suitable for biodiesel production. Therefore, the harvested biomass need be processed rapidly. Drying is a major step to keep the quality of the oil. In addition, the solvent-based oil extraction can be difficult when wet biomass is used. Various drying methods such as sun drying, spray drying, freeze drying, and drum drying can be used for drying algal biomass (Mata et al., 2010). Due to the high water content of algal biomass, sun-drying is not a very effective method for algal powder production. Spray drying and freeze drying are rapid and effective, but also expensive and not economically feasible for biofuel production. Because of the high energy required, drying is considered as one of the main economical bottlenecks in the entire process.

There are several approaches for extracting oil from the dry algal biomass, including solvent extraction, osmotic shock, ultrasonic extraction and supercritical CO_2 extraction. Oil extraction from dried biomass can be performed in two steps, mechanical crushing followed by solvent extraction in which hexane is the main solvent used. For example, after the oil extraction using an expeller, the leftover pulp can be mixed with cyclohexane to extract the remaining oil. The oil dissolves in the cyclohexane and the pulp is filtered out from the solution. These two stages are able to extract more than 95% of the total oil present in the algae. Oil extraction from algal cells can also be facilitated by osmotic shock or ultrasonic treatment to break the cells. Osmotic shock is a sudden reduction in osmotic pressure causing cells to rupture and release cellular components including oil. The algae lacking the cell wall are suitable for this process. In the ultrasonic treatment, the collapsing cavitation bubbles near to the cell walls cause cell walls to break and release the oil into the solvent. Supercritical CO_2 is another way for efficient extraction of algal oil, but the high energy demand is a limitation for commercialization of this technology (Herrero et al., 2010).

3.5 Oil transesterification

Algal oil contained in algal cells can be converted into biodiesel through transesterification. Transesterification is a chemical conversion process involving reacting triglycerides of vegetable oils or animal fats catalytically with a short-chain alcohol (typically methanol or ethanol) to form fatty acid esters and glycerol

(Figure 6). This reaction occurs stepwise with the first conversion of triglycerides to diglycerides and then to monoglycerides and finally to glycerol. The complete transesterification of 1 mol of triglycerides requires 3 mol of alcohol, producing 1 mol of glycerol and 3 mol of fatty esters. Considering that the reaction is reversible, large excess of alcohol is used in industrial processes to ensure the direction of fatty acid esters. Methanol is the preferred alcohol for industrial use because of its low cost, although other alcohols like ethanol, propanol and butanol are also commonly used.

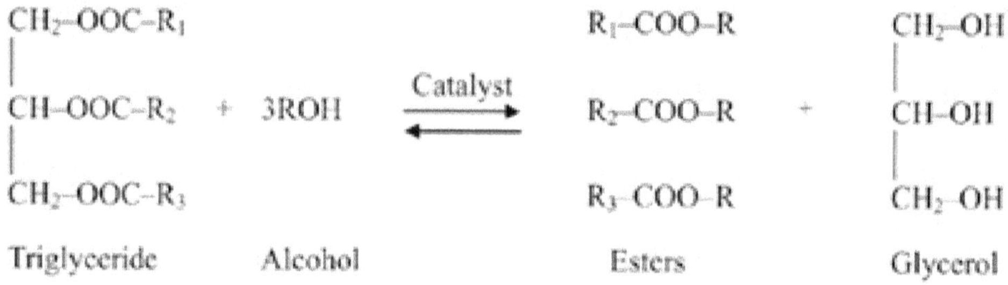

Fig. 6. Transesterification of oil to biodiesel. R_{1-3} indicates hydrocarbon groups.

In addition to heat, a catalyst is needed to facilitate the transesterification. The transesterification of triglycerides can be catalyzed by acids, alkalis or enzymes. Acid transesterification is considered suitable for the conversion of feedstocks with high free fatty acids but its reaction rate is low (Gerpen, 2005). In contrast, alkali-catalyzed transesterification has a much higher reaction rate, approximately 4000 times faster than the acid-catalyzed one (Fukuda et al., 2001). In this context, alkalis (sodium hydroxide and potassium hydroxide) are preferred as catalysts for industrial production of biodiesel. The use of lipases as transesterification catalysts has also attracted much attention as it produces high purity product and enables easy separation from the byproduct glycerol (Ranganathan et al., 2008). However, the cost of enzyme is still relatively high and remains a barrier for its industrial implementation. In addition, it has been proposed that biodiesel can be prepared from oil via transesterification with supercritical methanol (Demirbas, 2002).

4. GENETIC ENGINEERING OF MICROALGAE

4.1 Microalgal lipid biosynthesis

Although lipid metabolism, in particular the biosynthesis of fatty acids and TAG, is poorly understood in algae, it is generally recognized that the basic pathways for fatty acid and TAG biosynthesis are similar to those demonstrated in higher plants.

Algae synthesize the *de novo* fatty acids in the chloroplast using a single set of enzymes. A simplified schedule for saturated fatty acid biosynthesis is shown in Figure 7. Acetyl-CoA is the basic building block of the acyl chain and serves as a substrate for acetyl CoA carboxylation and as well as a substrate for the initial condensation reaction. The formation of malonyl CoA from acetyl CoA is generally regarded as the first reaction of fatty acid biosynthesis, which is catalyzed by acetyl CoA carboxylase (ACCase). The malonyl group of malonyl CoA is transferred to a protein co-factor, acyl carrier protein (ACP), resulting in the formation of malonyl ACP that enters into a series of condensation reactions with acyl ACP (or acetyl CoA) acceptors. The first condensation reaction is catalyzed by 3-ketoayl ACP synthase III (KAS III), forming a four-carbon product. KAS I and KAS II catalyze the subsequent condensations. After each condensation, the 3-ketoacyl-ACP product is reduced, dehydrated, and reduced again, by 3-ketoacyl-ACP reductase, 3-hydroxyacyl-ACP dehydratase, and enoyl-ACP reductase, respectively, to form a saturated fatty acid. To produce an unsaturated fatty acid, a double bond is introduced onto the acyl chain by the soluble enzyme stearoyl ACP desaturase (SAD). Unlike plants, some algae produce long- chain acyl ACPs (C_{20}-C_{22}) that derive from the further elongation and/or desaturation of C_{18}. The fatty acid elongation is terminated when the acyl group is released from ACP by an acyl-ACP thioesterase that hydrolyzes the acyl ACP and produces free fatty acids or by acyltransferases that transfer the fatty acid from ACP to glycerol-3-phosphate or monoacylglycerol-3-phosphate. These released fatty acids serve as precursors for the synthesis of cellular membranes and neutral storage lipids like TAG.

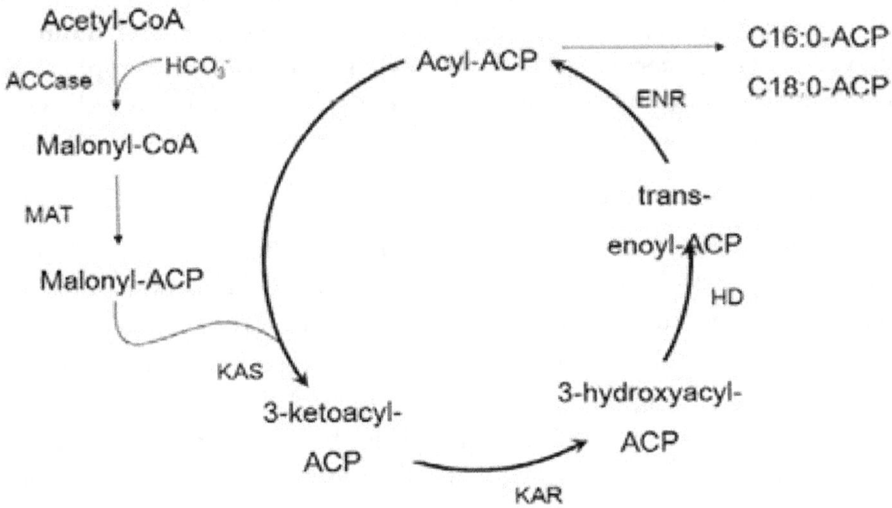

Figure 7. Acetyl-CoA is the basic building block of the acyl chain and serves as a substrate for acetyl CoA acetyl-CoA carboxylase; ACP, acyl carrier protein; CoA, coenzyme A; ENR, enoyl-ACP reductase; HD, 3-hydroxyacyl-ACP dehydratase; KAR, 3-ketoacyl-ACP reductase; KAS, 3-ketoacyl-ACP synthase; MAT, malonyl-CoA:ACP transacylase.

It has been proposed that the biosynthesis of TAG occurs in cytosol via the direct glycerol pathway (Figure 8). Generally, acyl-CoAs sequentially react with the hydroxyl groups in glycerol-3-phosphate to form phosphatidic acid via lysophosphatidic acid. These two reactions are catalyzed by glycerol-3-phospate acyl transferase and lysophosphatidic acid acyl transferase respectively. Dephosphorylation of phosphatidic acid results in the release of DAG which accepts a third acyl from CoA to form TAG. This final step is catalyzed by diacylglycerol acyltransferase, an enzymatic reaction that is unique to TAG synthesis. In addition, an alternative pathway that is independent of acyl-CoA may also be present in algae for TAG biosynthesis (Dahlqvist et al., 2000). This pathway employs phospholipids as acyl donors and diacylglycerols as the acceptors and might be activated when algal cells are exposed to stress conditions, under which algae usually undergo rapid degradation of the photosynthetic membranes and concurrent accumulation of cytosolic TAG-enriched lipid bodies (Hu et al., 2008b).

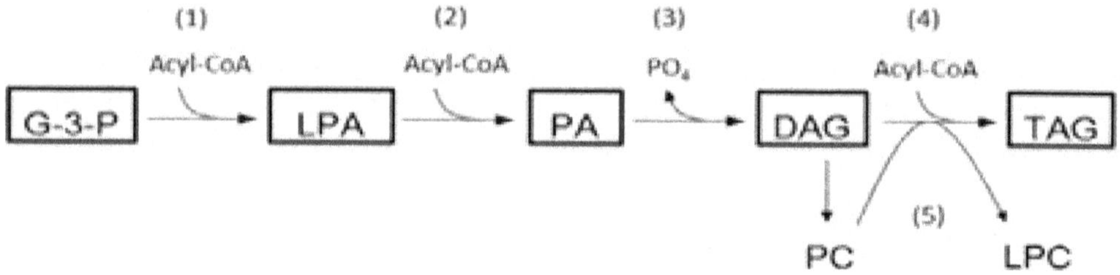

Fig. 8. Simplified illustration of the TAG biosynthesis in algae. DAG, diacylglycerol; LPA, lysophosphatidic acid; LPC, lysophosphatidylcholine; G-3-P, glycerol-3-phosphate; PA, phosphatidic acid; PC, phosphatidylcholine; TAG, triacylglycerol. (1) glycerol-3-phosphate acyl transferase, (2) lysophosphatidic acid acyl transferase, (3) phosphatidic acid phosphatase, (4) diacylglycerol acyl transferase, and (5) phospholipid:diacylglycerol acyltransferase.

4.2 Genetic engineering of microalgal lipids

Genetic engineering is a feasible and complimentary approach to increase algal productivity and improve the economics of algal biodiesel production. This has long been recognized but it seems that so far little progress has been made. The lack of full or near-full genome sequences and robust transformation systems makes genetic engineering of algae lag much behind that of bacteria, fungi and higher eukaryotes. Although certain algal species have been reported for efficient transformation, it proves to be difficult to produce stable transformants of algae. Currently, sophisticated genetic engineering whereby several genes are concurrently down-regulated or overexpressed is only really applicable to the green alga *Chlamydomonas reinhardtii.* This situation, however, is likely to change because of the growing scientific and commercial interest in other algal species that are of great potential for industrial applications.

Understanding the algal lipid biosynthesis is of great help to engineer algal lipid production. Although lipid metabolism in algae is not as fully understood as that in higher plants, they have similar lipid biosynthetic pathway as mentioned above.

Theoretically, overexpression of the genes involved in fatty acid synthesis is able to increase lipid accumulation, in that fatty acids required as precursors for lipid biosynthesis are produced in excess. However, overexpressoin of the native ACCase, the rate-limiting enzyme catalyzing the first committed step of fatty acid biosynthesis in many organisms, could not increase the lipid production in diatom (Dunahay et al., 1995). It is possible that under high flux conditions through ACCase, the condensing enzymes or other factors may begin to limit fatty acid synthesis rate. Therefore, more complete control may come from certain transcription factors that can increase expression of the entire pathway. Another feasible approach of increasing cellular lipid contents is to inhibit metabolic pathways that lead to other carbon storage compounds, such as starch. Starch synthesis shares common carbon precursors with lipid synthesis in algae. Blocking starch synthesis is able to redirect carbon flux to lipid biosynthetic pathway, resulting in overproduction of fatty acids and thus total lipids (Li et al., 2010a). Neutral lipids in particular TAG surpass other lipids for biodiesel production, attracting the interest of enhancing cellular TAG contents through genetic engineering. Overexpression of genes involved in TAG assembly, e.g., glycerol-3-phosphate acyltransferase, lysophosphatidic acid acyltransferase, or diacylglycerol acyltransferase, all significantly increase TAG production in plants. Such strategies may also be applicable to algae for enhancing TAG levels. Commonly, algae produce larger amounts of lipids under unfavorable conditions than logarithmic growing condition. Enhancing lipid biosynthesis through genetic engineering, therefore, is likely to reduce algal proliferation and biomass production. In this context, the use of inducible promoters could overcome the problem because the transgenic expression can only be activated when a high cell density is achieved. The important properties of biodiesel such as cetane number, viscosity, cold flow, oxidative stability, are largely determined by the composition and structure of fatty acid esters which in turn are determined by the characteristics of fatty acids of biodiesel feedstocks, for example carbon chain length and unsaturation degree (Knothe, 2005b). Thus, the genetic modification of algal fatty acid composition is of also great interest. Generally, saturated fatty esters possess high cetane number and superior oxidative stability; whereas unsaturated, especially polyunsaturated fatty esters have improved low-temperature properties (Knothe, 2008). In this regard, it is suggested that the modification of fatty esters, for example the enhanced proportion of oleic acid (C18:1) ester, can provide a compromise solution between oxidative stability and low-temperature properties and therefore promote the quality of biodiesel (Knothe, 2008, 2009). Oleic acid is converted to linoleic acid (C18:2) in a single desaturation step, catalyzed by a Δ12 desaturase enzyme encoded by the *FAD2* gene. Inactivation of this desaturation step can greatly increase the proportion of oleic acid in soybean and may represent a possible strategy for elevated accumulation of oleic acid in algae.

Genetic engineering can also be used potentially to improve tolerance of algae to stress factors such as temperature, salinity and pH. These improved attributes will allow for the cost reduction in algal biomass production and be beneficial for growing selected algae under extreme conditions that limit the proliferation of invasive species. Photoinhibition is another technical challenge to be addressed by genetic engineering. When the light intensities exceed the value for maximum photosynthetic efficiency, algae show photoinhibition, a common phenomenon for phototrophy under which the growth rate slows down. Engineered algae with a higher threshold of light inhibition will significantly improve the economics of biodiesel production.

The engineering of algae for biodiesel production remains in its nascent stages. Substantial progress has been attained solely in the genetic modification of certain model algae. Many of these advancements are anticipated to be applicable to industrially significant algal species in the future, enabling the utilization of modified algae as cellular factories for commercial biodiesel production. Nonetheless, other problems persist and must be resolved before algal biodiesel may become economically viable..

5. CONCISE SUMMARY

Microalgae possess the capability to generate lucrative biodiesel that may ultimately supplant petroleum-based fuels. Nevertheless, algal-biodiesel production remains prohibitively expensive for commercialization, as no algal strains currently exist that encompass all the requisite benefits for attaining substantial oil yields through the cost-effective open pond cultivation system. Contemporary research is constrained to the identification of optimal microalgal species, the enhancement of mass cultivation, biomass harvesting, and oil extraction methodologies, all of which contribute to elevated biodiesel production costs from microalgae. Future cost-reduction initiatives for algal-biofuel production should concentrate on the technological advancement of oil-rich algae by improving algal biology (regarding biomass yield and oil content) and optimizing culture systems, in conjunction with sophisticated genetic engineering techniques and waste utilization. Besides oils, microalgae are rich in proteins, carbohydrates, and various nutrients or bioactive substances that may be utilized as feed, food, and medications. Integrating the production of co-products with biodiesel is an attractive method for reducing the cost of algae biofuel production.

6. REFERENCES

Abdullah, A.Z.; Salamatinia, B.; Mootabadi, H.& Bhatia, S. (2009). Current status and policies on biodiesel indus try in Malaysia as the world's leading producer of palm oil. *Energy Policy*, 37, 5440-5448.

Abou-Shanab, R.A.I.; Hwang, J.-H.; Cho, Y.; Min, B.& Jeon, B.-H. (2011). Characterization of microalgal species isolated from fresh water bodies as a potential source for biodiesel production. *Applied Energy*, In Press, DOI: 10.1016/j.apenergy.2011.01.060.

Afify, A.; Shalaby, E.A.& Shanab, S.M.M. (2010). Enhancement of biodiesel production from different species of algae. *Grasas Y Aceites*, 61, 416-422.

Ahmad, A.L.; Yasin, N.H.M.; Derek, C.J.C.& Lim, J.K. (2011). Microalgae as a sustainable energy source for biodiesel production: A review. *Renewable and Sustainable Energy Reviews*, 15, 584-593.

Araujo, G.S.; Matos, L.J.B.L.; Gonçalves, L.R.B.; Fernandes, F.A.N.& Farias, W.R.L. (2011). Bioprospecting for oil producing microalgal strains: Evaluation of oil and biomass production for ten microalgal strains. *Bioresource Technology*, In Press, DOI: 10.1016/j.biortech.2011.01.089.

Banerjee, A.& Chakraborty, R. (2009). Parametric sensitivity in transesterification of waste cooking oil for biodiesel production--A review. *Resources, Conservation and Recycling*, 53, 490-497.

Bigogno, C.; Khozin-Goldberg, I.; Boussiba, S.; Vonshak, A.& Cohen, Z. (2002). Lipid and fatty acid composition of the green oleaginous alga *Parietochloris incisa*, the richest plant source of arachidonic acid. *Phytochemistry*, 60, 497-503.

Boussiba, S.; Vonshak, A.; Cohen, Z.; Avissar, Y.& Richmond, A. (1987). Lipid and biomass production by the halotolerant microalga *Nannochloropsis salina*. *Biomass*, 12, 37-47.

Brennan, L.& Owende, P. (2010). Biofuels from microalgae--A review of technologies for production, processing, and extractions of biofuels and co-products. *Renewable and Sustainable Energy Reviews*, 14, 557-577.

Brown, M.R.; Dunstan, G.A.; Norwood, S.J.& Miller, K.A. (1996). Effects of harvest stage and light on the biochemical composition of the diatom *Thalassiosira pseudonana*. *Journal of Phycology*, 32, 64-73.

Canakci, M. (2007). The potential of restaurant waste lipids as biodiesel feedstocks. *Bioresource Technology*, 98, 183-190.

Chaiklahan, R.; Chirasuwan, N.; Loha, V.& Bunnag, B. (2008). Lipid and fatty acids extraction from the cyanobacterium *Spirulina*. *Scienceasia*, 34, 299-305.

Checkbiotech (2009) Massive increase in global biofuel production. Available from: http://bioenergy.checkbiotech.org/news/massive_increase_global_biofuel_produ ction.

Chen, C.-Y.; Yeh, K.-L.; Aisyah, R.; Lee, D.-J.& Chang, J.-S. (2011). Cultivation, photobioreactor design and harvesting of microalgae for biodiesel production: A critical review. *Bioresource Technology*, 102, 71-81.

Chen, F.& Johns, M. (1991). Effect of C/N ratio and aeration on the fatty acid composition of heterotrophic *Chlorella sorokiniana*. *Journal of Applied Phycology*, 3, 203-209.

Chen, G.-Q.; Jiang, Y.& Chen, F. (2008). Salt-induced alterations in lipid composition of diatom *Nitzshia laevis* (Bacillariophyceae). *Journal of Phycology*, 44, 1309-1314.

Cheng, Y.; Zhou, W.; Gao, C.; Lan, K.; Gao, Y.& Wu, Q. (2009). Biodiesel production from Jerusalem artichoke (*Helianthus Tuberosus L.*) tuber by heterotrophic microalgae *Chlorella protothecoides*. *Journal of Chemical Technology & Biotechnology*, 84, 777-781.

Chinnasamy, S.; Bhatnagar, A.; Hunt, R.W.& Das, K.C. (2010). Microalgae cultivation in a wastewater dominated by carpet mill effluents for biofuel applications. *Bioresource Technology*, 101, 3097-3105.

Chisti, Y. (2007). Biodiesel from microalgae. *Biotechnology Advances*, 25, 294-306.

Chisti, Y. (2008). Biodiesel from microalgae beats bioethanol. *Trends in Biotechnology*, 26, 126- 31.

Chiu, S.Y.; Kao, C.Y.; Chen, C.H.; Kuan, T.C.; Ong, S.C.& Lin, C.S. (2008). Reduction of CO2 by a high-density culture of *Chlorella sp* in a semicontinuous photobioreactor. *Bioresource Technology*, 99, 3389-3396.

Chiu, S.-Y.; Kao, C.-Y.; Tsai, M.-T.; Ong, S.-C.; Chen, C.-H.& Lin, C.-S. (2009). Lipid accumulation and CO2 utilization of *Nannochloropsis oculata* in response to CO2 aeration. *Bioresource Technology*, 100, 833-838.

Converti, A.; Casazza, A.A.; Ortiz, E.Y.; Perego, P.& Del Borghi, M. (2009). Effect of temperature and nitrogen concentration on the growth and lipid content of *Nannochloropsis oculata* and *Chlorella vulgaris* for biodiesel production. *Chemical Engineering and Processing: Process Intensification*, 48, 1146-1151.

Couto, R.M.; Simoes, P.C.; Reis, A.; Da Silva, T.L.; Martins, V.H.& Sanchez-Vicente, Y. (2010). Supercritical fluid extraction of lipids from the heterotrophic microalga *Crypthecodinium cohnii*. *Engineering in Life Sciences*, 10, 158-164.

da Cunha, M.E.; Krause, L.C.; Moraes, M.S.A.; Faccini, C.S.; Jacques, R.A.; Almeida, S.R.; Rodrigues, M.R.A.& Caramao, E.B. (2009). Beef tallow biodiesel produced in a pilot scale. *Fuel Process Technology*, 90, 570-575.

Dahlqvist, A.; Stahl, U.; Lenman, M.; Banas, A.; Lee, M.; Sandager, L.; Ronne, H.& Stymne, H. (2000). Phospholipid : diacylglycerol acyltransferase: An enzyme that catalyzes the acyl-CoA-independent formation of triacylglycerol in yeast and plants. *Proceedings of the National Academy of Sciences*, 97, 6487-6492.

Damiani, M.C.; Popovich, C.A.; Constenla, D.& Leonardi, P.I. (2010). Lipid analysis in *Haematococcus pluvialis* to assess its potential use as a biodiesel feedstock. *Bioresource Technology*, 101, 3801-3807.

De la Hoz Siegler, H.; Ben-Zvi, A.; Burrell, R.E.& McCaffrey, W.C. (2011). The dynamics of heterotrophic algal cultures. *Bioresource Technology*, In Press, DOI: 10.1016/j.biortech.2011.01.081.

de Oliveira, D.; Di Luccio, M.; Faccio, C.; Dalla Rosa, C.; Bender, J.; Lipke, N.; Amroginski, C.; Dariva, C.& de Oliveira, J. (2005). Optimization of alkaline transesterification of soybean oil and castor oil for biodiesel production. *Applied Biochemistry and Biotechnology*, 122, 553-560.

Demirbas, A. (2002). Biodiesel from vegetable oils via transesterification in supercritical methanol. *Energy Conversion and Management*, 43, 2349-2356.

Demirbas, A. (2008). *Biodiesel-a realistic fuel alternative for diesel engines*. Springer - Verlag, London.

Demirbas, A. (2009). Progress and recent trends in biodiesel fuels. *Energy Conversion and Management*, 50, 14-34.

Demirbas, A. (2010). Use of algae as biofuel sources. *Energy Conversion and Management*, 51, 2738-2749.

Dias, J.M.; Alvim-Ferraz, M.C.M.& Almeida, M.F. (2009). Production of biodiesel from acid waste lard. *Bioresource Technology*, 100, 6355-6361.

Diaz-Felix, W.; Riley, M.R.; Zimmt, W.& Kazz, M. (2009). Pretreatment of yellow grease for efficient production of fatty acid methyl esters. *Biomass and Bioenergy*, 33, 558-563.

D'Oca, M.G.M.; Viêgas, C.V.; Lemões, J.S.; Miyasaki, E.K.; Morón-Villarreyes, J.A.; Primel, E.G.& Abreu, P.C. (2011). Production of FAMEs from several microalgal lipidic extracts and direct transesterification of the *Chlorella pyrenoidosa*. *Biomass and Bioenergy*, In Press, DOI: 10.1016/j.biombioe.2010.12.047.

Dunahay, T.G.; Jarvis, E.E.& Roessler, P.G. (1995). Genetic transformation of the diatoms *Cyclotella cryptica* and *Navicula saprophila*. *Journal of Phycology*, 31, 1004-1012.

Easterling, E.R.; French, W.T.; Hernandez, R.& Licha, M. (2009). The effect of glycerol as a sole and secondary substrate on the growth and fatty acid composition of *Rhodotorula glutinis*. *Bioresource Technology*, 100, 356-361.

Ethier, S.; Woisard, K.; Vaughan, D.& Wen, Z. (2011). Continuous culture of the microalgae Schizochytrium limacinum on biodiesel-derived crude glycerol for producing docosahexaenoic acid. *Bioresource Technology*, 102, 88-93.

Fan, K.-W.; Jiang, Y.; Faan, Y.-W.& Chen, F. (2007). Lipid characterization of Mangrove thraustochytrid - *Schizochytrium mangrovei*. *Journal of Agricultural and Food Chemistry*, 55, 2906-2910.

Feng, Y.; Li, C.& Zhang, D. (2011). Lipid production of Chlorella vulgaris cultured in artificial wastewater medium. *Bioresource Technology*, 102, 101-105.

Fukuda, H.; Kondo, A.& Noda, H. (2001). Biodiesel fuel production by transesterification of oils. *Journal of Bioscience and Bioengineering*, 92, 405-416.

Fulke, A.B.; Mudliar, S.N.; Yadav, R.; Shekh, A.; Srinivasan, N.; Ramanan, R.; Krishnamurthi, K.; Devi, S.S.& Chakrabarti, T. (2010). Bio-mitigation of CO2, calcite formation and simultaneous biodiesel precursors production using *Chlorella* sp. *Bioresource Technology*, 101, 8473-8476.

Gao, C.; Zhai, Y.; Ding, Y.& Wu, Q. (2010). Application of sweet sorghum for biodiesel production by heterotrophic microalga *Chlorella protothecoides*. *Applied Energy*, 87, 756-761.

Gerpen, J.V. (2005). Biodiesel processing and production. *Fuel Process Technology*, 86, 1097- 1107.

Gouveia, L.; Marques, A.; da Silva, T.& Reis, A. (2009). *Neochloris oleabundans* UTEX #1185: a suitable renewable lipid source for biofuel production. *Journal of Industrial Microbiology and Biotechnology*, 36, 821-826.

Gouveia, L.& Oliveira, A. (2009). Microalgae as a raw material for biofuels production. *Journal of Industrial Microbiology and Biotechnology*, 36, 269-274.

Graef, G.; LaVallee, B.J.; Tenopir, P.; Tat, M.; Schweiger, B.; Kinney, A.J.; Gerpen, J.H.V.& Clemente, T.E. (2009). A high-oleic-acid and low-palmitic-acid soybean: agronomic performance and evaluation as a feedstock for biodiesel. *Plant Biotechnology Journal*, 7, 411-421.

Graham, L.E.; Wilcox, L.W.& Graham, J. (2009). *Algae. 2nd ed.* Benjamin Cummings, San Francisco, CA.

Greenwell, H.C.; Laurens, L.M.L.; Shields, R.J.; Lovitt, R.W.& Flynn, K.J. (2010). Placing microalgae on the biofuels priority list: a review of the technological challenges. *Journal of The Royal Society Interface*, 7, 703-726.

Griffiths, M.& Harrison, S. (2009). Lipid productivity as a key characteristic for choosing algal species for biodiesel production. *Journal of Applied Phycology*, 21, 493-507.

Guedes, A.; Meireles, L.; Amaro, H.& Malcata, F. (2010). Changes in lipid class and fatty acid composition of cultures of *Pavlova lutheri*, in response to light intensity. *Journal of the American Oil Chemists' Society*, 87, 791-801.

Hawash, S.; Kamal, N.; Zaher, F.; Kenawi, O.& Diwani, G.E. (2009). Biodiesel fuel from Jatropha oil via non-catalytic supercritical methanol transesterification. *Fuel*, 88, 579-582.

Herrero, M.; Mendiola, J.A.; Cifuentes, A.& Ibáñez, E. (2010). Supercritical fluid extraction: Recent advances and applications. *Journal of Chromatography A*, 1217, 2495-2511.

Hill, J.; Nelson, E.; Tilman, D.; Polasky, S.& Tiffany, D. (2006). Environmental, economic, and energetic costs and benefits of biodiesel and ethanol biofuels. *Proceedings of the National Academy of Sciences*, 103, 11206-11210.

Ho, S.-H.; Chen, W.-M.& Chang, J.-S. (2010). *Scenedesmus obliquus* CNW-N as a potential candidate for CO2 mitigation and biodiesel production. *Bioresource Technology*, 101, 8725-8730.

Hsieh, C.-H.& Wu, W.-T. (2009). Cultivation of microalgae for oil production with a cultivation strategy of urea limitation. *Bioresource Technology*, 100, 3921-3926.

Hu, C.; Li, M.; Li, J.; Zhu, Q.& Liu, Z. (2008a). Variation of lipid and fatty acid compositions of the marine microalga *Pavlova viridis* (Prymnesiophyceae) under laboratory and outdoor culture conditions. *World Journal of Microbiology and Biotechnology*, 24, 1209- 1214.

Hu, Q. (2004). Environmental effects on cell composition. in: *Handbook of microalgal culture*, (Ed.) Richmond, A., Blackwell. Oxford, pp. 83-93.

Hu, Q.; Sommerfeld, M.; Jarvis, E.; Ghirardi, M.; Posewitz, M.; Seibert, M.& Darzins, A. (2008b). Microalgal triacylglycerols as feedstocks for biofuel production: perspectives and advances. *Plant Journal*, 54, 621-639.

Huang, G.; Chen, F.; Wei, D.; Zhang, X.& Chen, G. (2010). Biodiesel production by microalgal biotechnology. *Applied Energy*, 87, 38-46.

Huerlimann, R.; de Nys, R.& Heimann, K. (2010). Growth, lipid content, productivity, and fatty acid composition of tropical microalgae for scale-Up production. *Biotechnology and Bioengineering*, 107, 245-257.

Jain, S.& Sharma, M.P. (2010). Prospects of biodiesel from Jatropha in India: A review. *Renewable and Sustainable Energy Reviews*, 14, 763-771.

James, G.O.; Hocart, C.H.; Hillier, W.; Chen, H.; Kordbacheh, F.; Price, G.D.& Djordjevic, M.A. (2011). Fatty acid profiling of Chlamydomonas reinhardtii under nitrogen deprivation. *Bioresource Technology*, 102, 3343-3351.

Kawachi, M.; Inouye, I.; Honda, D.; O'Kelly, C.J.; Bailey, J.C.; Bidigare, R.R.& Andersen, R.A. (2002). The Pinguiophyceae classis nova, a new class of photosynthetic stramenopiles whose members produce large amounts of omega-3 fatty acids. *Phycological Research*, 50, 31-47.

Khozin-Goldberg, I.; Bigogno, C.; Shrestha, P.& Cohen, Z. (2002). Nitrogen starvation induces the accumulation of arachidonic acid in the freshwater green alga *Parietochloris incisa* (trebuxiophyceae). *Journal of Phycology*, 38, 991-994.

Khozin-Goldberg, I.& Cohen, Z. (2006). The effect of phosphate starvation on the lipid and fatty acid composition of the fresh water eustigmatophyte *Monodus subterraneus*. *Phytochemistry*, 67, 696-701.

Knothe, G. (2005b). Dependence of biodiesel fuel properties on the structure of fatty acid alkyl esters. *Fuel Process Technology*, 86, 1059-1070.

Knothe, G. (2005a). Introduction: what is biodiesel. in: *The biodiesel handbook*, (Eds.) Knothe, G.;Gerpen, J.V.&Krahl, J., AOCS Press. Champaign, pp. 1-3.

Knothe, G. (2008). "Designer" biodiesel: Optimizing fatty ester composition to improve fuel properties. *Energy & Fuel*, 22, 1358-1364.

Knothe, G. (2009). Improving biodiesel fuel properties by modifying fatty ester composition. *Energy and Environmental Science*, 2, 759-766.

Lee, R.E. (2008). *Phycology. Forth ed.* Cambridge University Press, Cambridge.

Levine, R.B.; Costanza-Robinson, M.S.& Spatafora, G.A. (2011). *Neochloris oleoabundans* grown on anaerobically digested dairy manure for concomitant nutrient removal and biodiesel feedstock production. *Biomass and Bioenergy*, In Press, DOI: 10.1016/j.biombioe.2010.08.035.

Li, M.; Gong, R.; Rao, X.; Liu, Z.& Wang, X. (2005). Effects of nitrate concentration on growth and fatty acid composition of the marine microalga *Pavlova viridis* (Prymnesiophyceae). *Annals of Microbiology*, 55, 51-55.

Li, X.; Hu, H.-Y.; Gan, K.& Sun, Y.-X. (2010a). Effects of different nitrogen and phosphorus concentrations on the growth, nutrient uptake, and lipid accumulation of a freshwater microalga *Scenedesmus* sp. *Bioresource Technology*, 101, 5494-5500.

Li, Y.; Chen, Y.-F.; Chen, P.; Min, M.; Zhou, W.; Martinez, B.; Zhu, J.& Ruan, R. (2011b). Characterization of a microalga Chlorella sp. well adapted to highly concentrated municipal wastewater for nutrient removal and biodiesel production. *Bioresource Technology*, In Press, DOI: 10.1016/j.biortech.2011.01.091.

Li, Y.; Han, D.; Hu, G.; Sommerfeld, M.& Hu, Q. (2010b). Inhibition of starch synthesis results in overproduction of lipids in *Chlamydomonas reinhardtii*. *Biotechnology and Bioengineering*, 107, 258-268.

Li, Y.; Han, D.; Sommerfeld, M.& Hu, Q. (2011a). Photosynthetic carbon partitioning and lipid production in the oleaginous microalga *Pseudochlorococcum* sp. (Chlorophyceae) under nitrogen-limited conditions. *Bioresource Technology*, 102, 123-129.

Li, Y.; Horsman, M.; Wang, B.; Wu, N.& Lan, C. (2008). Effects of nitrogen sources on cell growth and lipid accumulation of green alga *Neochloris oleoabundans*. *Appl Microbiol Biotechnol*, 81, 629-636.

Liang, Y.; Sarkany, N.& Cui, Y. (2009). Biomass and lipid productivities of *Chlorella vulgaris* under autotrophic, heterotrophic and mixotrophic growth conditions. *Biotechnology Letters*, 31, 1043-1049.

Lin, Q.& Lin, J. (2011). Effects of nitrogen source and concentration on biomass and oil production of a *Scenedesmus rubescens* like microalga. *Bioresource Technology*, 102, 1615-1621.

Lin, Y.-H.; Chang, F.-L.; Tsao, C.-Y.& Leu, J.-Y. (2007). Influence of growth phase and nutrient source on fatty acid composition of *Isochrysis galbana* CCMP 1324 in a batch photoreactor. *Biochemical Engineering Journal*, 37, 166-176.

Liu, J.; Huang, J.; Fan, K.W.; Jiang, Y.; Zhong, Y.; Sun, Z.& Chen, F. (2010). Production potential of *Chlorella zofingienesis* as a feedstock for biodiesel. *Bioresource Technology*, 101, 8658-8663.

Liu, J.; Huang, J.; Sun, Z.; Zhong, Y.; Jiang, Y.& Chen, F. (2011). Differential lipid and fatty acid profiles of photoautotrophic and heterotrophic Chlorella zofingiensis: Assessment of algal oils for biodiesel production. *Bioresource Technology*, 102, 106- 110.

Liu, X.-J.; Jiang, Y.& Chen, F. (2005). Fatty acid profile of the edible filamentous cyanobacterium *Nostoc flagelliforme* at different temperatures and developmental stages in liquid suspension culture. *Process Biochemistry*, 40, 371-377.

Liu, Z.-Y.; Wang, G.-C.& Zhou, B.-C. (2008). Effect of iron on growth and lipid accumulation in *Chlorella vulgaris*. *Bioresour Technol*, 99, 4717-4722.

Mandal, S.& Mallick, N. (2009). Microalga *Scenedesmus obliquus* as a potential source for biodiesel production. *Applied Microbiology and Biotechnology*, 84, 281-291.

Mansour, M.P.; Volkman, J.K.& Blackburn, S.I. (2003). The effect of growth phase on the lipid class, fatty acid and sterol composition in the marine dinoflagellate, *Gymnodinium* sp. in batch culture. *Phytochemistry*, 63, 145-153.

Mata, T.M.; Martins, A.A.& Caetano, N.S. (2010). Microalgae for biodiesel production and other applications: A review. *Renewable and Sustainable Energy Reviews*, 14, 217-232.

Matsumoto, M.; Sugiyama, H.; Maeda, Y.; Sato, R.; Tanaka, T.& Matsunaga, T. (2010). Marine diatom, *Navicula* sp. strain JPCC DA0580 and marine green alga, *Chlorella* sp. strain NKG400014 as potential sources for biodiesel production. *Applied Biochemistry and Biotechnology*, 161, 483-490.

Meher, L.C.; Naik, S.N.; Naik, M.K.& Dalai, A.K. (2008). Biodiesel production using karanja (*Pongamia pinnata*) and jatropha (*Jatropha curcas*) seed oil. in: *Handbook of plant-based biofuels*, (Ed.) Pandey, A., CRC Press. Boca Raton, FL pp. 255-266.

Miao, X.& Wu, Q. (2006). Biodiesel production from heterotrophic microalgal oil. *Bioresource Technology*, 97, 841-846.

Moazami, N.; Ranjbar, R.; Ashori, A.; Tangestani, M.& Nejad, A.S. (2011). Biomass and lipid productivities of marine microalgae isolated from the Persian Gulf and the Qeshm Island. *Biomass and Bioenergy*, In Press, DOI: 10.1016/j.biombioe.2011.01.039.

Molina Grima, E.; Belarbi, E.H.; Acién Fernández, F.G.; Robles Medina, A.& Chisti, Y. (2003). Recovery of microalgal biomass and metabolites: process options and economics. *Biotechnology Advances*, 20, 491-515.

Mussgnug, J.H.; Klassen, V.; Schlüter, A.& Kruse, O. (2010). Microalgae as substrates for fermentative biogas production in a combined biorefinery concept. *Journal of Biotechnology*, 150, 51-56.

Nakpong, P.& Wootthikanokkhan, S. (2010). High free fatty acid coconut oil as a potential feedstock for biodiesel production in Thailand. *Renewable Energy*, 35, 1682-1687.

Napolitano, G.E. (1994). The relationship of lipids with light and chlorophyll measurements in freshwater algae and periphyton. *Journal of Phycology*, 30, 943-950.

Oh, S.H.; Han, J.G.; Kim, Y.; Ha, J.H.; Kim, S.S.; Jeong, M.H.; Jeong, H.S.; Kim, N.Y.; Cho, J.S.; Yoon, W.B.; Lee, S.Y.; Kang, D.H.& Lee, H.Y. (2009). Lipid production in *Porphyridium cruentum* grown under different culture conditions. *Journal of Bioscience and Bioengineering*, 108, 429-434.

Oner, C.& Altun, S. (2009). Biodiesel production from inedible animal tallow and an experimental investigation of its use as alternative fuel in a direct injection diesel engine. *Applied Energy*, 86, 2114-2120.

Órpez, R.; Martínez, M.E.; Hodaifa, G.; El Yousfi, F.; Jbari, N.& Sánchez, S. (2009). Growth of the microalga *Botryococcus braunii* in secondarily treated sewage. *Desalination*, 246, 625-630.

Ota, M.; Kato, Y.; Watanabe, H.; Watanabe, M.; Sato, Y.; Smith Jr, R.L.& Inomata, H. (2009). Fatty acid production from a highly CO_2 tolerant alga, *Chlorocuccum littorale*, in the presence of inorganic carbon and nitrate. *Bioresource Technology*,

100, 5237-5242.

Pahl, S.L.; Lewis, D.M.; Chen, F.& King, K.D. (2010). Heterotrophic growth and nutritional aspects of the diatom *Cyclotella cryptica* (Bacillariophyceae): Effect of some environmental factors. *Journal of Bioscience and Bioengineering*, 109, 235-239.

Patil, P.D.& Deng, S. (2009). Optimization of biodiesel production from edible and non- edible vegetable oils. *Fuel*, 88, 1302-1306.

Phan, A.N.& Phan, T.M. (2008). Biodiesel production from waste cooking oils. *Fuel*, 87, 3490- 3496.

Piorreck, M.& Pohl, P. (1984). Preparatory experiments for the axenic mass-culture of microalgae. 2. Formation of biomass, total protein, chlorophylls, lipids and fatty- acids in green and blue green-algae during one growth-phase. *Phytochem*, 23, 217- 223.

Pruvost, J.; Van Vooren, G.; Cogne, G.& Legrand, J. (2009). Investigation of biomass and lipids production with *Neochloris oleoabundans* in photobioreactor. *Bioresource Technology*, 100, 5988-5995.

Pruvost, J.; Van Vooren, G.; Le Gouic, B.; Couzinet-Mossion, A.& Legrand, J. (2011). Systematic investigation of biomass and lipid productivity by microalgae in photobioreactors for biodiesel application. *Bioresource Technology*, 102, 150-158.

Pushparaj, B.; Buccioni, A.; Paperi, R.; Piccardi, R.; Ena, A.; Carlozzi, P.& Sili, C. (2008). Fatty acid composition of Antarctic cyanobacteria. *Phycologia*, 47, 430-434.

Queiroz, M.I.; Hornes, M.O.; da Silva-Manetti, A.G.& Jacob-Lopes, E. (2011). Single-cell oil production by cyanobacterium Aphanothece microscopica Nägeli cultivated heterotrophically in fish processing wastewater. *Applied Energy*, In Press, DOI: 10.1016/j.apenergy.2010.12.047.

Rabbani, S.; Beyer, P.; Lintig, J.v.; Hugueney, P.& Kleinig, H. (1998). Induced beta -carotene synthesis driven by triacylglycerol deposition in the unicellular alga *Dunaliella bardawil*. *Plant Physiology*, 116, 1239-1248.

Radakovits, R.; Jinkerson, R.E.; Darzins, A.& Posewitz, M.C. (2010). Genetic Engineering of Algae for Enhanced Biofuel Production. *Eukaryotic Cell*, 9, 486-501.

Ranganathan, S.V.; Narasimhan, S.L.& Muthukumar, K. (2008). An overview of enzymatic production of biodiesel. *Bioresource Technology*, 99, 3975-3981.

Rashid, U.& Anwar, F. (2008). Production of biodiesel through optimized alkaline-catalyzed transesterification of rapeseed oil. *Fuel*, 87, 265-273.

REN21. (2010) Renewables Global Status Report. Available from: http://www.ren21.net/Portals/97/documents/GSR/REN21_GSR_2010_full_revised%20Sept 2010.pdf.

Renaud, S.M.; Thinh, L.-V.; Lambrinidis, G.& Parry, D.L. (2002). Effect of temperature on growth, chemical composition and fatty acid composition of tropical Australian microalgae grown in batch cultures. *Aquaculture*, 211, 195-214.

Rodolfi, L.; Zittelli, G.C.; Bassi, N.; Padovani, G.; Biondi, N.; Bonini, G.& Tredici, M.R. (2009). Microalgae for oil: Strain selection, induction of lipid synthesis and outdoor mass cultivation in a low-cost photobioreactor. *Biotechnology and Bioengneering*, 102, 100-112.

Roessler, P.G. (1988). Changes in the activities of various lipid and carbohydrate biosynthetic enzymes in the diatom *Cyclotella cryptica* in response to silicon deficiency. *Archives of Biochemistry and Biophysics*, 267, 521-528.

Ross, A.B.; Jones, J.M.; Kubacki, M.L.& Bridgeman, T. (2008). Classification of macroalgae as fuel and its thermochemical behaviour. *Bioresource Technology*, 99, 6494-6504.

Saha, S.K.; Uma, L.& Subramanian, G. (2003). Nitrogen stress induced changes in the marine cyanobacterium *Oscillatoria willei* BDU 130511. *FEMS Microbiolgy Ecolology*, 45, 263- 72.

Sahoo, P.K.& Das, L.M. (2009). Process optimization for biodiesel production from Jatropha, Karanja and Polanga oils. *Fuel*, 88, 1588-1594.

Saka, S.& Kusdiana, D. (2001). Biodiesel fuel from rapeseed oil as prepared in supercritical methanol. *Fuel*, 80, 225-231.

Sato, N.; Hagio, M.; Wada, H.& Tsuzuki, M. (2000). Environmental effects on acidic lipids of thylakoid membranes. in: *Recent advances in biochemistry of plant lipids*, (Eds.) Harwood, J.L.&Quinn, P.J., Portland Press. London, pp. 912-914.

Shafiee, S.& Topal, E. (2009). When will fossil fuel reserves be diminished? *Energy Policy*, 37, 181-189.

Sheehan, J.; Dunahay, T.; Benemann, J.& Roessler, P. 1998. A look back at the U.S. Department of Energy's aquatic species programme - Biodiesel from algae, Report No. NREL/TP-580-24190; National Renewable Energy Laboratory: Golden, CO.

Singhania, R.R.; Parameswaran, B.& Pandey, A. (2008). Plant-based bioufuels: an introduction. in: *Handbook of plant-based biofuels*, (Ed.) Pandey, A., CRC Press. Boca Raton, FL, pp. 3-12.

Sobczuk, T.M.& Chisti, Y. (2010). Potential fuel oils from the microalga Choricystis minor. *Journal of Chemical Technology and Biotechnology*, 85, 100-108.

Solovchenko, A.; Khozin-Goldberg, I.; Didi-Cohen, S.; Cohen, Z.& Merzlyak, M. (2008). Effects of light intensity and nitrogen starvation on growth, total fatty acids and arachidonic acid in the green microalga *Parietochloris incisa*. *Journal of Applied Phycology*, 20, 245-251.

Somerville, C. (1995). Direct tests of the role of membrane lipid composition in low- temperature-induced photoinhibition and chilling sensitivity in plants and cyanobacteria. *Proceedings of the National Academy of Sciences*, 92, 6215-8.

Sukenik, A.; Carmeli, Y.& Berner, T. (1989). Regulation of fatty acid composition by irradiance level in the eustigmatophyte *Nannochloropsis* sp. *Journal of Phycology*, 25, 686-692.

8

Chapter

ECO-PHYSIOLOGICAL CONSTRAI TS AND TECHNOLOGICAL INNOVATIONS FOR BIODIESEL PRODUCTION FROM MICROALGAE

1. INTRODUCTION: AN OVERVIEW

The interplay of depleting fossil fuel reserves (peak oil) and escalating diesel prices poses a significant problem to most nations regarding national energy security and the assurance of sustainable energy supplies. The urgent challenges have catalyzed an expedited exploration of renewable energy to discover new and revolutionary liquid biofuels for the future. (IEA 2011). Any such liquid biofuels from renewable resources will need to have a lower environmental footprint than fossil fuel derived liquid biofuels in order to meet key sustainability criteria (Nuffield 2011). The most abundant available natural renewable resource on planet earth is solar energy. Photosynthetic organisms such as plants, algae (macro- and micro-algae) and some bacteria have been selected through evolution to convert solar energy to storable forms of energy. Such photosynthetic organisms can constitute a renewable resource which can effectively harvest and convert solar energy to a variety of energy-dense biofuels. In the case of microalgae, at least US$ 300 million has been committed to facilitate phycology research on bioprospecting microalgal diversity and evaluation of the feasibility of different microalgal species and strains for biofuels production (Sheehan et al., 1998). The use of oil crops such as palm, soy, and oilseed rape as feedstocks for biodiesel production has provided the basis for the first generation of biofuels. However, the cultivation of plants on arable land for biofuel production can compete with the use of the same land for food and animal feed production – the so called "food vs fuel" land-use competition dilemma. In addition, biofuel crops also have significant water and nutrient requirements which can adversely affect their sustainability when Life Cycle Analysis (LCA) is conducting across their value chain. For instance, microalgal production systems reliant on dwindling freshwater supplies will face sustainability problems if they are scaled up (Wigmosta et al., 2011). As a result microalgal systems based on saltwater or waste water are likely to be more sustainable. One approach being pursued for circumventing the 'food vs fuel' dilemma associated with first generation

Biofuel involves biological processing of lignocellulosic biomass as the basis for development of second generation biofuel systems. The development of commercial-scale efficient conversion technologies for exploitation of biological wastes as an obvious source for biofuel generation is a major focus of research and development efforts associated with this generation of biofuel. One of the third generation biofuels under development aims to exploit the photosynthetic capability of microalgae for the conversion of solar energy into energy-dense biomass. A major advantage of microalgae over the use of crop biomass for biodiesel production is the lower land area requirements for production of an equivalent amount of fuel. As understanding of microalgal genomes and biochemistry increases, opportunities are emerging for development of fourth generation biofuels where metabolic engineering of microbes leads to more effective domestication of microalgae for biofuel production. Currently, the commercial production of biofuels from microalgae is constrained by inadequate systems for biomass production, harvesting, extraction, and oil recovery that can economically unify all operational units from cultivation to biofuel product recovery. This chapter examines the constraints of specific operational units concerning initiatives aimed at developing fourth generation microalgal biofuels that are both commercially and environmentally viable.

2. MICROALGAE BIOMASS GENERATION

2.1 Open cultivation

Large scale microalgal biomass production can be achieved either through open pond cultivation under natural sunlight or under the controlled conditions of a photobioreactor. In the USA, the history of mass production of microalgae dates back at least to 1953 with the production of *Scenedesmus* species in Washington. Many systems for cultivating microalgae on a large scale have been suggested in many countries including the USA, Germany, Japan, Israel, the UK, the Czech Republic and others. Typically, microalgae are first grown in inorganic nutrients and then, in a second phase, are cultivated is done using waste water streams.

Commercial cultivation of microalgae can be done in a range of different ways including (a) open cultivation using natural sunlight, (b) closed cultivation using natural light and (c) closed cultivation using artificial light (in photobioreactors). Each of these systems has advantages and disadvantages, and the choice of system depends on the degree of parameter control needed to produce the desired product and on the value of the end- product (Apt and Behrens, 1999). The most commonly used artificial open pond systems consist of large shallow ponds, tanks, circular ponds and raceway ponds (Ugwu et al., 2008). The construction and operation costs of such open cultivation systems are considerably less but are challenging to operate on a year round basis due to seasonal climatic variations. While open pond culture is cheaper than culture in closed photobioreactors (Borowitzka 1999), it is currently limited to a relatively small number of microalgal species. Rectangular ponds with a paddle wheel (raceway ponds) are the most widely used for the production of *Spirulina* sp., *Dunaliella salina* and *Haematococcus* sp. and currently represent the most efficient design for the large scale culture of most species of microalgae. Individual ponds are tyically up to 1 ha in area, with an average depth of about 20- 30 cm (Andersen 2005). The need to provide adequate light to the algal cells and maintaining an adequate water depth for mixing of the microalgae are important considerations for determining the pond depth. The diurnal natural light cycle results in the exposure of microalgae to limiting, saturating and over-saturating light conditions. High irradiances throughout the year and moderate temperatures are optimal for outdoor microalgae cultivation. For example, the geographical location of southern Spain with an average of 10-12 hours of sunlight per day, and a mean solar irradiance ranging from 400 μmol photons $m^{-2}s^{-1}$ during winter to 1800 μmol photons $m^{-2}s^{-1}$ is considered highly suitable for outdoor cultivation of microalgae. The maximal areal productivity of microalgae in outdoor conditions ranges from 20 to 30 $gm^{-2}d^{-1}$ (Cuaresma et al., 2011). To date, light-to-biomass conversion efficiencies of 1-4 % have been achieved for microalgae grown in conventional open pond cultivation systems. Because the scaling-up of microalgal biomass production in open raceway ponds is relatively easy, such systems are primarily considered for commercial applications. However, differences in weather variables such as solar irradiance, rainfall, and temperature significantly affect prospects for open cultivation of microalgae at different geographical locations. Temperature influences the rate of various reactions of photosynthesis (Raven, 1988). Therefore, microalgae exhibit an optimal growth within a narrow temperature range and die above a certain threshold temperature (Béchet et al., 2011). In addition, temperature is an important factor that affect the rate of evaporation from shallow algal ponds. In addition to changing the physical environment of open ponds, rainfall can lead to microbial contamination that inhibits microalgal growth (Hase et al., 2000).

The paddle wheels installed in open ponds are used to circulate the water, while compressed air can be introduced into the bottom of a pond to agitate the water, bringing microalgae from the lower levels upwards. Raceway channels are typically built in concrete or compacted earth, and are often lined with white plastic. During daylight, the microalgal culture is fed continuously in front of the paddlewheel where the flow begins. The biomass is harvested behind the paddlewheel, on completion of the circulation loop. The paddlewheel operates continuously to prevent sedimentation and flocculation (Chisti, 2007). The largest raceway- based biomass production facility currently occupies an area of 440,000 m^2 (Spolaore et al., 2006). This facility is owned by Earthrise Nutritional (www.earthrise.com) and is used to produce cyanobacterial biomass for food. In India, Pary Nutraceuticals (part of the Chennai- based Murugappa group) has been focusing on microalgal research and development and are commercially producing *Spirulina* for nutraceuticals.

2.2 Closed cultivation systems for microalgae

Several different closed systems using natural sunlight have been described for microalgae (Richmond et al. 1993, Molina Grima et al., 1995, Spektorova et al., 1997). In such systems, microalgae are grown in transparent glass or plastic vessels, and the vessels are placed under natural illumination. A higher surface to volume ratio is provided, so microalgal cell densities are often higher than in open ponds. However, these systems are also subject to variations in light intensity and temperature that make cultivation reproducibility problematic. In addition, removal of oxygen from the culture and the provision of adequate temperature control (especially if energy is required for cooling) pose a major problem with such closed systems. Large scale indoor cultivation using highly-controlled photobioreactors or fermentors have also been used successfully for microalgal biomass production. The wide range of different types of closed photobioreactors (PBRs) include vertical-column, flat-plate and tubular PBRs (Ugwu et al., 2008). These provide the ability to control and optimize culture parameters, and as a result such photobioreactors are suitable for culturing many different species of microalgae. The basic features which must be considered when designing a photobioreactor are: source of light, churning rate of algae (to avoid biomass sedimentation and for uniform availability of nutrients and light), material for construction, CO_2 supply, and removal of O_2, pH and temperature control (Kaur et al., 2010). Whether, closed or open systems will be optimal for commercial cultivation of different species (or strains) of microalgae is difficult to determine. However, it is clear that photobioreactors will play a critical role to feed open ponds with a high-cell-density unialgal inoculum (Cheng et al., 2011).

3. THERMODYNAMIC EFFICIENCY OF PHOTOSYNTHESIS IN MICROALGAE

Photosynthesis is a chemical reaction governed by the laws of thermodynamics. Assuming a microalgal cell as 'boundary' and the process of photosynthesis as 'system', then according to the law of thermodynamics, the two kinds of work associated with this chemical reaction are electrical work and work of expansion. In a biological system such as microalgae, the production of ATP derived by the transfer of charges across the biofluidic membranes can be called electrical work. The growth or increase in size of cell and cellular components (including oil bodies) is the 'work of expansion'. In the very familiar photosynthetic reaction (Albarrán-Zavala & Angulo-Brown, 2007);

$$6CO_2 + 12\,H_2O \longrightarrow C_6H_{12}O_6 + 6H_2O + 6O_2$$

$$\Delta G^0 = 2880.31\ kJ/mol_{C6H12O6} \qquad \text{at } \lambda = 680\ nm$$

3.1 Photosynthetic conversion efficiency

In outdoor cultivation systems, the microalgal biomass productivity derived through photosynthesis depends on the solar energy input. The estimated yearly average solar energy density, including both direct beam radiation and diffuse scattered radiation is 10,038 MJ/m^{2}year. To account for non-sunny weather conditions, a more realistic theoretical maximum solar energy density is obtained after reducing this value by 10%, which corresponds to a value of 9034 MJ/m^{2}year. However, the actual value will exhibit temporal and spatial variation depending on the geographical location and will generally be lower (Cooney et al., 2011). The fraction of the solar energy spectrum (SEarth ~ 9034 MJ/m^{2}year) is further reduced by 45% to calculate the value of photosynthetically active radiation (PAR) that supports photosynthesis ($S_{EarthPAR}$ ~ 4065 MJ/m^{2}year). PAR is expressed in terms of photon flux as it reaches surface of microalgal cells in the form of photons, the energy of which varies inversely with the wavelength. The upper theoretical limit for the average PAR spectrum photon flux energy ($E_{MaxAvePAR}$) is 0.2253 MJ/mol that corresponds to λ531 nm (green) (Weyer et al., 2009). Hence, the available photon flux reaching the earth surface and which is available for photosynthesis is calculated by the following formula (Cooney et al., 2011):

$$PF_{PAR} = \frac{S_{EarthPAR} \sim 4065\ MJ/m^2/year}{E_{MaxAvePAR} \sim 0.2253\ MJ/mol\ photon} = 18,043\ moles\ photons/m^2year$$

3.2 Maximum theoretical photosynthetic efficiency

The most cited values for maximum photosynthetic efficiencies in microalgae are in the range of 17-23% (Gordon & Polle 2007, Zemke et al., 2010). Cooney and coworkers (2011) illustrate how to obtain the maximum theoretical value of photosynthetic efficiency from the available PF_{PAR} of 18,043 moles photons/m^2 year. The theoretical maximum is calculated by considering both photon transmission (η_{PTE}) and photon conversion efficiencies (η_{PUE}) as 100%. Thus, according to the following equation:

$$\text{Photons utilized} = PF_{PAR} \times \eta_{PTE} \times \eta_{PUE} = 18,043\ moles\ photons/m^2\ year \times 1 \times 1$$

In a microalgal cell, these photons power the photosynthetic production of carbohydrates (CH_2O) which have an average energy content of 0.4825MJ/mole (Weyer et al., 2009). On average 10 photons are required to derive one mole of CH_2O. Hence, the total energy consumed during the photosynthetic conversion reaction is obtained as follows (Cooney et al 2011).

$$E_{CARB} = \frac{(18,043\ moles\ photons/m^2year \times 1 \times 1).\ (0.4825MJ/mole)}{10} = 871\ MJ/m^2year$$

The estimated total photosynthetic efficiency during the conversion of PAR to microalgal biomass is calculated by dividing E_{CARB} (871 MJ/m²year) and $S_{EarthPAR}$ (4065 MJ/m²year), assuming that bioconversion of carbohydrates is 100% efficient. This gives a value of 21.4%, which is stated as the overall maximum theoretical photosynthetic efficiency relative to PAR. High lipid productivity depends on both the microalgal biomass areal productivity and the lipid content that can be generated from the microalgal strain. The lipid productivity is the most important factor influencing the cost of biodiesel production. High lipid content microalgal species and strains also favor the efficiency of biomass processing during oil extraction.

3.3 Technological innovations in illumination sources

3.3.1 Light Emitting Diodes (LEDs)

A light source with narrow spectral output that overlaps the photosynthetic absorption spectrum improves the energy conversion as the emission of light at unusable frequencies is eliminated. Light-emitting diodes (LEDs) are the only light source that currently meet this criterion. LEDs have the ability to produce high light levels with low radiant heat output and maintain useful light output for years. Thus, LEDs can have a very significantly longer life of 100,000 h as compared to 8000 h of fluorescent lights. These advantages make LEDs ideal for microalgal growth in controlled environments of growth chambers. The optimal wavelength conditions will vary from species to species of microalgae (Chen et al 2011). For example, the highest specific growth rate and biomass production from the photosynthetic cultivation of *Spirulina platensis* was obtained using red LED. The superimposed pattern of luminescence spectrum of blue LED (450-470 nm) and that of red LED (650-665 nm) corresponds to the light absorption spectrum of carotenoids and chlorophyll (Yeh & Chung, 2009). Therefore, the red LED favors microalgal growth but switching to illumination with blue LED improved the rate of astaxanthin production by *Haematococcus pluvialis* (Katsuda et al. 2004). Flashing light from blue LEDs is also a promising illumination method for *H. pluvialis* growth and astaxanthin production (Katsuda et al 2006). The use of flashing LED as sources of intermittent light in indoor algal culture can yield a major gain in energy economy comparing to fluorescent light sources (Matthijs et al., 1996). The research results by Nedbal et al. (1996) also suggest that algal growth rates in intermittent light can be higher than those in equivalent continuous light.

Red LEDs were found to reduce the average cell volume of *Chlorella vulgaris* without affecting the total biomass production (Lee and Palsson, 1994). However, under the exposure of fluorescent light, cells regained their normal size.

3.2.2 Optical fiber technology

The use of optical fibers as internal light sources could increase light efficiency whilst simultaneously reducing electricity consumption. For example, solar energy excited optical fiber requires only 1.0 kW-h of electricity (Chen et al., 2011). Spatial (i.e. orientation and dimensions of the photobioreactor) and temporal variations (i.e. due to weather conditions) greatly affect the availability of sunlight. Therefore, internal illumination by optical fiber is unstable. To circumvent this problem, Chen et al. (2011) have conceptualized a photobioreactor that combines optical fiber and multi-LED light sources with both solar panel and wind power generators. This has a potential to be developed into a commercially viable microalgae cultivation system with significantly reduced electricity consumption.

4. LIMITATIONS AND IMPROVEMENTS OF PHOTOSYNTHETIC BIOMASS PRODUCTION OF MICROALGAE

4.1 Low light intensity and distribution

There are several key parameters which determine the microalgal productivity in a photobioreactor. These are lighting, mixing, water, CO_2 pressure, O_2 removal, nutrient supply, temperature, and pH (Kunjapur and Eldrige, 2010). Under nutrient-sufficient and optimal temperature conditions, the maximal culture productivity of photoautotrophic microorganisms is solely limited by the light (Richmond, 2004). The penetration of visible spectrum of light in the microalgal cultures decreases as the cell density increases (Figure 1). The appropriate intensity, duration, and wavelength of light must be provided to enhance the microalgal growth in photobioreactors. Supra-optimal light conditions lead to photoinhibition and sub-optimal light becomes a growth limiting factor. In both conditions, microalgal productivity will be lowered. The photosynthetic conversion efficiency of microalgae will generally be lower than theoretically expected under optimal conditions due to insufficient capacity to utilize the incident radiation (Zhu et al., 2008). The distribution of solar radiation over a greater photosynthetic area can spatially dilute the light in the light saturation zone, thereby reducing the mutual shading of the cells in the culture resulting in higher growth rate and lower accessory pigments content. The distribution of solar radiation can be increased by maintaining the surface to volume ratio as high as possible. The temporal and spectral distribution of irradiation and photon flux density is the main physical parameter that determines the photosynthetic productivity of microalgae. The solar conversion efficiency of microalgal mass culture grown under full sunlight is limited because of two reasons: 1) the photon absorption rate of the chlorophyll antennae of upper layers of cells far exceeds the rate of their utilization hence there is a loss of excess photons as fluorescence and heat leading to photoinhibition; 2) the deprivation of functional photons in the deeper cell layers, which is strongly attenuated due to the filtering effect of upper cells (Naus & Melis, 1991, Neidhardt et al., 1998). Gordon and Polle (2007) argued that a microalgal biomass productivity of 100 g m^{-2} h^{-1} could be obtained solely by improving the flux tolerance rather than by raising intrinsic photosynthetic efficiency.

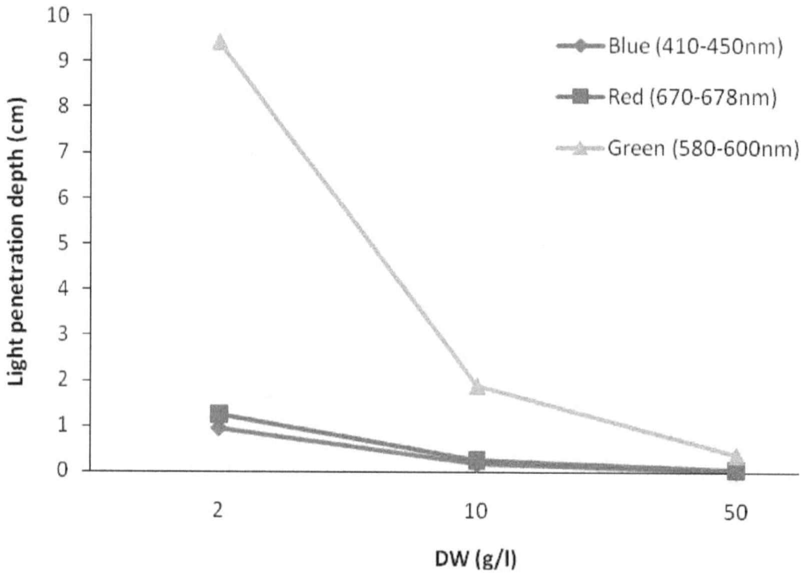

Fig. 1. Effect of biomass concentration on the penetration of incident light into cultures of *Nannochloropsis* sp. (from Richmond 2004).

4.2 Improvements of intrinsic photosynthetic efficiency

The photosynthetic efficiency of microalgae can potentially reach its theoretical maximum, which is calculated to be about 9-10% of total incident solar energy or 20-22% of PAR, being converted into biomass (Beilen, 2010). Such projected ultrahigh microalgal biomass yields of 100 g dry weight $m^{-2}h^{-1}$ can be realized in photobioreactors with sufficiently thin channels, ultradense cultures, and rapid light/dark cycles wherein optimal synchronization of photonic input with rate limiting dark reaction times is exploited (Gordon and Polle, 2007). However, this does not take into account the intrinsic conversion efficiency of photosynthesis, which is only likely to be improved upon through genetic engineering or synthetic biology. The integration of molecular and photobioreactor engineering is likely the only possible way of obtaining near-theoretical levels of algal biomass productivity while simultaneously augmenting lipid content. At the unicellular level, genetic modification of microalgal photophysiology could decrease light absorption, leading to enhanced availability of functional irradiance at the population level. The PSII and PSI in the light harnessing complex of green algae are associated with large numbers of chlorophyll a and chlorophyll b molecules, which are called antenna molecules. During photosynthetic biomass production in photobioreactors, high photon flux densities saturate the antenna molecules of upper cell layers with excessive photons which do not participate in the photosynthetic biomass production. These excess photons dissipate their energy as heat or fluorescence (photoinhibition) and reduce the overall solar to biomass conversion efficiency of the microalgal culture. Moreover, the lower layers do not receive appropriate amount of photons because of the

filtration of light by the cells of upper layer, which accounts for a further loss in the overall biomass productivities (Figure 2). Genetic modifications resulting in truncated chlorophyll antennae size could restrict the high photon absorption by the light harvesting complex. In this context, Polle et al. (2003) have cloned and functionally characterized the Chl antenna size regulatory *Tla1* gene in *Chlamydomonas reinhardtii.* The partially truncated chlorophyll antenna size of the *tla1* mutant prevents the over- absorption of irradiance by cells, thus avoiding wasteful heat losses (Polle et al., 2003). In *Dunaliella salina,* a highly truncated light-harvesting Chl antenna size resulted inaggravatedphotosynthetic productivity and greater oxygen production under mass microalgal culture (Melis et al., 1999). The *Stm3LR3* mutants of *C. reinhardtii* generated by RNAi technology demonstrated down-regulation of the entire LHC antenna system. The *Stm3LR3* mutant showed reduced fluorescence, increased photosynthetic quantum yield, increased resistance to photoinhibition and faster growth rate under high light levels (Mussgnug et al., 2007).

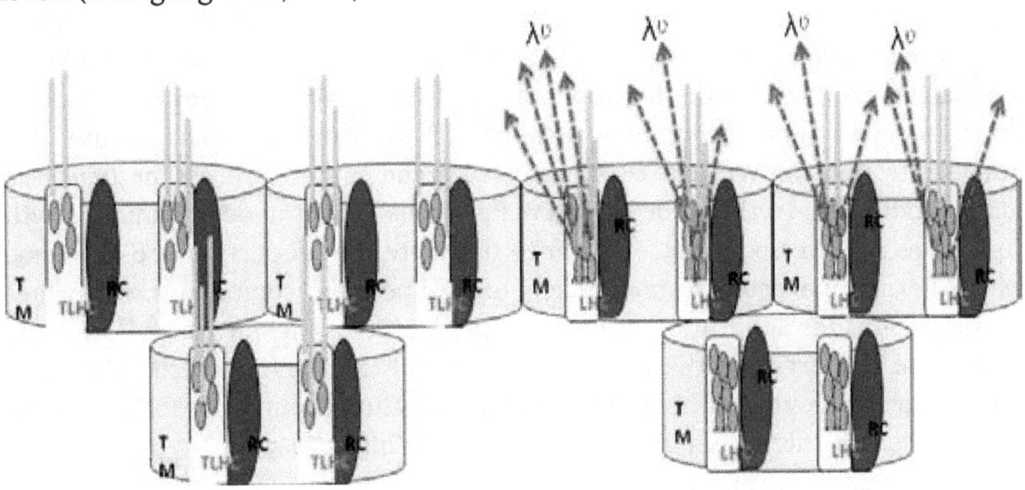

Fig. 2. A diagrammatic representation of wild-type and genetically truncated light harvesting complexes of microalgae. The incident light falling on the antenna molecules in the LHC are wasted as heat and fluorescence, while the lower layer cells are deprived of light. The modified TM has fewer antenna molecules in the

TLHC that allows the absorption of light by the cells in the deeper layers. TM (thylakoid membrane), RC (reaction center), LHC (light harvesting complex), TLHC (truncated light harvesting complex).

5. MICROALGAL BIOMASS HARVESTING

Conventional harvesting processes for microalgae include dewatering, extraction and purification of biomass. Bulk harvesting of microalgal biomass can be performed by centrifugation, flocculation, gravity sedimentation and/or filtration. Biomass harvesting is one of the most energy intensive processes, and can require high capital investments. Harvesting techniques tend to vary from species to species as various factors (namely density, size and the value of the microalgal end product) will typically inform the most appropriate method. Acoustic focusing, hybrid capacitive deionization or electrophoresis and use of novel materials for conventional membranes and flocculent systems are amongst the range of new innovative strategies that are currently under investigation (Cheng & Ogden, 2011).

6. BIOMASS CONVERSION TO BIODIESEL

6.1 Lipid extraction and biodiesel formation from microalgal biomass

Conventional methodologies for lipid extraction involve the use of toxic organic solvents such as chloroform, methanol and hexane. While the solvent extraction process is effective it is difficult to adopt on a large scale. Novel methods for lipid extraction involve technologies such as acoustics, sonication, the use of mesoporous nanomaterials, and amphiphilic solvents (Cheng & Ogden 2011). Super critical fluid extraction has been reported to be safer and faster than the conventional solvent extractions (Andrich et al., 2005). Another technique of "milking" microalgae manipulates the hydrophobicity of the solvent system, which allows the extraction of lipids from living algal cells. A flat panel two-phase bioreactor designed by Hejazi & Wijffels (2004) was used in the milking process for *Dunaliella salina* production. In this process, microalgal cells grown under optimal growth conditions are stressed by excess light to stimulate the production of β-carotene, which is then extracted from the cells using lipophilic compounds. Important considerations for application of this "milking" process includes: a) cell wall and membrane properties of the microalgal strain; b) location and accumulation of the product inside the cell; and c) biocompatibility and chemical properties of the solvent used for the "milking" process (Hejazi & Wijffels 2004).

Lipid conversion to biodiesel can easily be achieved by chemical trans-esterification, enzymatic conversion, and catalytic cracking. Chemically, biodiesel is comprised of monoalkyl esters of fatty acids that are derived from triacylglycerols. These triacylglycerides can be produced from crop or microalgae oils, animal fats, and waste cooking oils. Such biodiesel is miscible with petroleum diesel and thus suitable blends of biodiesel-diesel can be obtained. These are denoted as BXX, where XX is the percent of biodiesel in the blend. For example, B40 is 40% biodiesel in a diesel-biodiesel blend (Tat et al., 2007).

6.2 Fuel properties of microalgal biodiesel

There are several properties which determine the suitability of biodiesel as a biofuel including cetane number, kinematic viscosity, cold flow and oxidative stability (Ramos et al, 2009). These properties are greatly influenced by the fatty acid compositions of the feedstock oils (Figure 3, 4). Therefore, to determine the best composition of biodiesel, it is necessary to study the lipid profile of potential biomass feedstocks. The most common fatty acid methyl esters present in most biodiesel are palmitic acid (16:0), stearic acid (18:0), oleic acid (18:1), linoleic acid (18:2) and linolenic acid (18:3) (Knothe, 2008). Biodiesel obtained upon trans-esterification of these common fatty acids has many advantages over petroleum-derived diesel fuel. However, there are several performance problems with biodiesel, notably poor cold flow properties, lower cetane number and insufficient oxidative stability (Knothe, 2009). Ignition delay time and combustion quality of a diesel fuel is determined by the cetane number. An adequate cetane number is required for better engine performance and a high cetane number is also associated with biodiesel with good cold start properties and reduced NO_x exhaust emissions (Ramos et al., 2009; Knothe, 2009; Ladommatos et al., 1996).

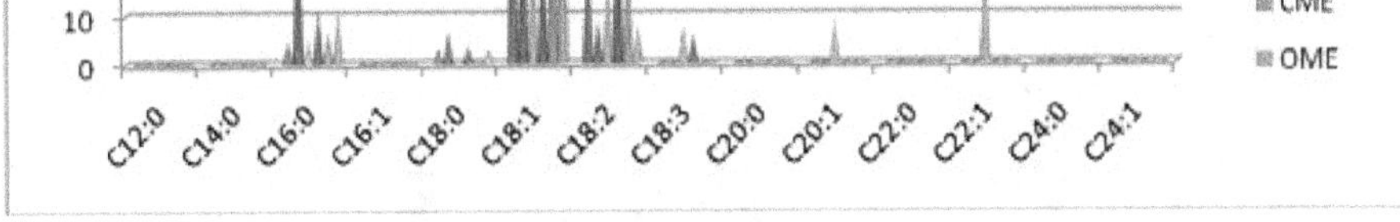

Fig. 3. Fatty acid composition of various vegetable oil crop feedstocks. (Data from Ramos et al 2009). HOSME (high oleic sunflower methyl ester), PME (palm methyl ester), RME (rape methyl ester), SME (soy methyl ester), CME (corn methyl ester), OME (olive methyl ester).

The cetane number of biodiesel can be determined using established standards such as ASTM D975 in the USA. Cetane is the common name for a long straight-chain hydrocarbon, hexadecane ($C_{16}H_{34}$), which is the high quality standard on the cetane scale with an assigned Cetane number of 100. In contrast, the low quality standard is a highly branched compound heptamethylnonane ($C_{16}H_{34}$) which exhibits poor ignition quality and has an assigned cetane number of 15 (Gopinath et al 2010). According to ASTM D975, conventional diesel fuel requires a minimum cetane number of 40 whereas a minimum of 47 has been prescribed for biodiesel (ASTM D6751) or (EN14214). For biodiesel, the cetane number depends on the microalgal feedstock biomass from which the oil is derived and the alcohol that is used during the trans-esterification process. The cetane number decreases with increasing unsaturation and increases with increasing chain length without branching of the CH_2 moities. However, the straight chain fatty acid methyl esters with carbon numbers of 6, 10, 12, 14, 16 and 18 show a non-linear increase with carbon number. Esters of saturated fatty acids (such as palmitic and stearic acids) give high cetane numbers. In summary, cetane number increases with chain length, decreases with unsaturation (or the number of double bonds) and decreases as double bonds and carbonyl group move toward the centre of the chain (Graboski & McCormik, 1998). Oils from different microalgal feedstocks will have different fatty acid compositions and such fatty acids are different with respect to the chain length, degree of unsaturation, or presence of other chemical moieties. While esters of long chain saturated fatty acid show higher cetane number, they can have a high cloud point that results in nozzle clogging (see below). On the other hand, esters of unsaturated fatty acid show low cetane number and are prone to oxidation. Among the non-algal biodiesel feedstocks, palm oil is reported to have the highest cetane number (61), whereas peanut, sunflower and corn oils typically have a cetane values of 53 (Ramos et al., 2009).

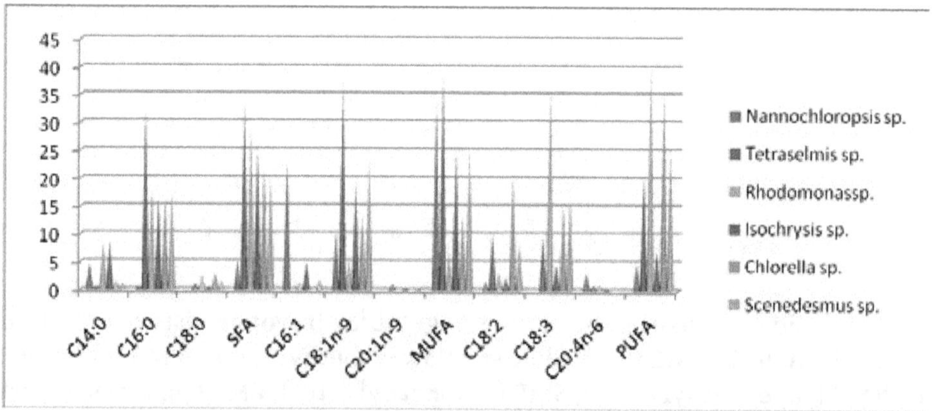

Fig. 4. Fatty acid composition of different microalgal species (Huerlimann et al., 2010). Data for *Chlorella* sp and *Scenedesmus* sp obtained from the author's investigations.

To improve the properties of biodiesel, it is necessary to enrich the biodiesel with fatty esters with desirable properties. Genetic engineering can provide important opportunities in this regard. For example, a soybean transgenic crop variety designated 335-13 has been genetically altered to increase concentration of oleic acid by more than 85%, with a corresponding reduction of palmitic acid content to less than 4% (Tat et al., 2007). Saturated and long chain fatty acids gave a high cetane number, which increases with increasing saturation and the chain length (Knothe et al., 2003; Bajpai & Tyagi, 2006; Knothe, 2009). Esters of highly unsaturated fatty acids such as linoleic (18:2) and linolenic (18:3) acids lower the cetane number (Knothe et al., 2003). According to Knothe et al. (2003) high cetane numbers were observed for esters of saturated fatty acids such as palmitic (C16:0) and stearic (C18:0) acids. Feedstock oils rich in these fatty acid compounds would have higher cetane values.

A higher content of polyunsaturated fatty esters in oil derived from the feedstock can reduce the quality of biodiesel upon storage due to oxidation in the presence of air, light, heat, peroxides, trace metals, or even the structural features of the fatty acids themselves. Thus, oxidation stability is another major issue affecting the use of biodiesel fuels (Ramos et al., 2009; Knothe, 2009). The number and position of double bonds in the chains of unsaturated fatty esters are both factors affecting the susceptible to autoxidation reaction (Bajpai & Tyagi, 2006; Frankel, 2005). Most biodiesel fuels contain significant amounts of alkyl esters of oleic, linoleic and linolenic acids, which influence the oxidation stability of the fuels. The relative rates of autoxidation for oleic acid methyl ester, linoleic acid methyl ester and linolenic acid methyl ester have been reported to be 1, 41 and 98 respectively (Knothe et al., 2005; Frankel, 2005; Ramos et al., 2009; Knothe, 2009).

The high viscosity of non-esterified vegetable oils leads to operational problems in diesel engines by increasing the level of engine deposits. Reduction of the high viscosity of vegetable oils is facilitated by the production of alkyl esters from the oil by transesterification. Since viscosity increases with decreasing temperature, handling of fuels in lower temperatures is facilitated by this lower viscosity as well (Knothe, 2009). This can be achieved by increasing the length and degree of saturation of the carbon chains (Knothe et al., 2005). Wax settling and plugging of filters and fuel lines are typical problems associated with biodiesel fuels at low temperatures. The maximum temperature at which the first solids appear for a particular fuel is known as the cloud point (CP) and such solids can lead to fuel filter plugging. The pour point (PP) is typically a few degrees below the cloud point and represents the temperature at which the fuel can no longer be poured (Dunn & Bagby, 1996). Key properties of biodiesel fuels at low-temperature are determined by the cold filter plugging point (CFPP) and low-temperature flow test (LTFT) (Dunn & Bagby, 1996; Knothe, 2009). The cloud point (CP) and CFPP are included in the biodiesel standards but as "soft" specifications (Knothe, 2009). For instance, The cloud point in ATSM D6751 requires a report, while the CFPP in UNE-EN 14214 varies with time of year and geographic location.

The properties of biodiesel at low-temperature are correlated with the properties of individual fatty acids, which mostly depend on the saturated ester content. In contrast, the effect of unsaturated fatty acids is considered negligible (Imahara et al., 2006; Ramos et al., 2009). Saturated fatty acids have significantly higher melting points than unsaturated fatty acids and in a mixture saturated fatty acids will crystallize at higher temperature. Therefore, biodiesel fuels derived from fats or oils with significant amounts of saturated fatty compounds display higher values of CP and CFPP (Knothe, 2003).

7. CONCISE SUMMARY

Biofuels can be categorized into oxygenated types (such as ethanol and biodiesel) and hydrocarbon types (including diesel, jet fuel, and gasoline). This classification delineates the various generations of biofuels: the 1st generation, which derives biofuels from natural vegetable oils and greases, and the 2nd generation, which utilizes lignocellulosic biomass and algal-derived fuels. Two biomass crops, *Jatropha* and *Camelina* bridged, the 1st and 2nd generations of biofuels. The next generations of biofuels will be based on the innovative technologies that improve the processing of biomass into various other types of biofuels and improving the existing feedstock species of biofuel using metabolic/ genetic engineering. For example, application of heat and pressure on algae/biomass/waste using innovative approaches like hydrothermal, catalytic and biological biomass conversions for the creation of cost-effective biofuels as a replacement for fossil fuels. Moreover, the next generation fuels are direct replacements for petroleum and are compatible with the existing infrastructure of the petrochemical industry. Genetic modification of microalgae improves their photosynthetic biomass conversion efficiency and hence can lead to higher biomass productivities, which is necessary for economic scalability. The improvements in the existing infrastructure for microalgae biomass production by photo engineering approaches will also play a key role towards the commercial application of next generation microalgae biofuels. The substitution of petroleum-based fuels with bio-based products is contingent upon several critical factors, including the selection of appropriate bio-based products, modifications to processes or enhancements for indirect substitutions, technological advancements to reduce costs of processing steps, scalability of biomass production and bioproduct distribution, and the availability of adequate and productive land.

REFERENCES

Albarrán-Zavala E & Angulo-Brown F. (2007) A simple thermodynamic analysis of photosynthesis. *Entropy*, 9, 152-168.

Andersen R. A. (2005) Algal Culturing Techniques. Elsevier Academic Press, London, U K.

Andrich G., Nesti U., Venturi F., Zinnai A. & Fiorentini R. (2005) Superctitical fluid extraction of bioactive lipids from microalga *Nannochloropsis* sp. *European J Lipid Science Tech.,* 107, 381-386.

Apt K. E. & Behrens P. W. (1999) Commercial developments in microalgal biotechnology. *J. Phycol.,* 35, 215-226.

Bajpai D. & Tyagi V.K. (2006) Biodiesel: source, production, composition, properties and its benefits. *J. Oleoginous Science* 55, 487–502.

Beilen J. B. V. (2010) Why microalgal biofuels won't save the internal combustion machine. *Biofuels, Bioprod. Bioref.*, 4, 41-52.

Béchet Q., Shilton A., Park J.B.K., Craggs R.J. & Guieysse B. (2011) Universal temperature model for shallow algal ponds provides improved accuracy. *Environmental Science Technology*, 45, 3702-3709.

Borowitzka M. A. (1999) Pharmaceuticals and agrochemicals from microalgae. In: Chemicals from microalgae. Ed. Cohen Z, Taylor & Francis. pp, 313-52.

Chen C.Y., Yeh K. Aisyah R., Lee D. & Chang J. (2011) Cultivation, photobioreactor design and harvesting of microalgae for biodiesel production: A critical review. *Bioresource Technology*, 102, 71-81.

Cheng Kuan-Chen & Ogden K. L. (2011) Algal biofuels: the research. *SBE Supplement: Algal Biofuels, AIChE*, 42-47, www.aiche.org/cep

Chisti Y. (2007) Biodiesel from microalgae. *Biotech. Advances.,* 25,294-306.

Cooney M. J., Young G. & Pate R. (2011) Bio-oil from photosynthetic microalgae: case study. *Bioresource Technology*, 102, 166-177.

Cuaresma M., Janssen M., Vilchez C., & Wijffels R. H. (2011) Horizontal or vertical photobioreactors? How to improve microalgae photosynthetic efficiency. *Bioresource Technology*, 102, 5219-5137.

Dunn R.O. & Bagby M.O. (1996) Low-temperature filterability properties of alternative diesel fuels from vegetable oils. Liquid fuel and industrial product from renewable resources. In: *Proc.Third Liquid Fuel Conference*, ASAE 95–103.

Frankel E.N.(2005) Lipid Oxidation. The Oily Press. U.K: PJ Barnes and Associates: Bridgwater.

Gopinath A., Puhan S. & Nagarajan G. (2010) Effect of biodiesel structural configuration on its ignition quality. *International J. Energy Environment*, 1 (2), 295-306.

Gordon J. F. & Polle J. E. W. (2007) Ultrahigh bioproductivity from algae. *Appl Microbiol Biotechnol.*, 76,969-975.

Graboski M.S. & McCormik R.L. (1998) Combustion of fat and vegetable oil derived fuels in diesel engines. *Prog. Energy Combus. Sci*, 24, 125-164.

Hase R, Oikawa H, Sasao C, Morita M & Watanabe Y (2000) Photosynthetic production of microalgal biomass in a raceway system under greenhouse conditions in Sendai city. *Journal of Bioscience and Bioengineering*,89 (2), 157-163.

Hejazi M. A. & Wijffels R. H. (2004) Milking of microalgae. *Trends Biotech.*, 22 (4), 189-194.

Huerlimann R, de Nys R, Heimann K (2010) Growth, lipid content, productivity, and fatty acid composition of tropical microalgae for scale-up production. *Biotechnol. Bioeng.* 107(2), 245–257.

IEA (2011) Technology Report: Biofuels for transport. International energy Agency, 9 rue de la Fédération, 75739 Pris Cedex 15, France. www.iea.org

Imahara H, Minami E, Saka S. (2006) Thermodynamic study on cloud point of biodiesel with its fatty acid composition. *Fuel*, 85, 1666–70.

Katsuda, T. Lababpour, A. Shimahara, K. & Katoh, S. (2004) Astaxanthin production by *Haematococcus pluvialis* under illumination with LEDs. *Enzyme Microb. Technol.*, 35, 81-86.

Katsuda, T. Shimahara, K. Yamagami , K. Ranjbar, R. & Katoh, S. (2006) Effect of flashing light from blue light emitting diodes on cell growth and astaxanthin production of *Haematococcus pluvialis*. *J. Biosci. Bioeng.*,102, 442-446.

Kaur S., Gogoi H.K., Srivastava R. B. & Kalita M.C. (2010) Algal Biodiesel: Procedures and Resources for laboratory study. In: *Algal Biotechnology: New Vistas*. Ed. M.K. Das, Daya Publishing House, Delhi. ISBN-13. 9788170356479, 978-8170356479

Knothe G., Matheaus,A.C. & Ryan III T.W. (2003) Cetane numbers of branched and straight chain fatty esters determined in an ignition quality tester. *Fuel* 82, 971–975.

Knothe G. 2009. Improving biodiesel fuel properties by modifying fatty ester composition. *Energy Environment Science* 2, 759-66.

Knothe G. (2008) "Designer" biodiesel: optimizing fatty ester composition to improve fuel properties. *Energy & Fuels*, 22,1358-1364.

Knothe, G., Gerpen, J.V. & Krahl, J. 2005. The Biodiesel Handbook. AOCS Press, Champaign, Illinois.

Kunjapur A. M. & Eldridge R. B. (2010) Photobioreactor design for commercial biofuel production from microalgae. *Ind. Eng. Chem. Res.*49, 3516-3526.

Ladommatos N., Parsi M. & Knowles A. (1996) The effect of fuel cetane improver on diesel pollutant emissions. *Fuel* 75, 8–14.

Lee, C.G. & Palsson, B. O. (1994) High-density algal photobioreactors using light-emitting diodes. *Biotechnol Bioeng.,* 44, 1161-1167.

Matthijs, H.C.P. Balke, H. Van Hes, U.M. Kroon, B.M.A Mur, L.R. & Binot, R.A. (1996) Application of light-emitting diodes in bioreactors: flashing light effects and energy economy in algal culture. *Biotechnol Bioeng,* 50, 98-107.

Melis Anastasios, N. J. & Benemann J. R. (1999). "*Dunaliella salina* (Chlorophyta) with small chlorophyll antenna sizes exhibit higher photosynthetic productivities and photon use efficiencies than normally pigmented cells.*Journal of Phycology,* 10, 515-525.

Molina Grima E., Sanchez Perez J. A., Garcia Camacho F., Fernandez Sevilla J. M., Acien Fernandez F. G. & Urda Cardona J. (1995) Biomass and icosapentanoic acid productivities from an outdoor batch culture of *Phaeodactylum tricornatum* UTEX 640 in an airlift tubular photobioreactor. *Appl. Microbiol. Biotechnol.,* 42,658-63.

Mussgnug J. H., S. T.-H., Jens Rupprecht, Foo A., Klassen V., McDowall A., Schenk P.M., Kruse O. & Hankamer B. (2007). Engineering photosynthetic light capture: impacts on improved solar energy to biomass conversion. *Plant Biotechnology Journal,* 5, 802- 814.

Naus J & Melis A. (1991). Changes of photosystem stoichiometry during cell growth in *Dunaliella salina* cultures.*Plant Cell Physiol.,* 32, 569-575.

Nedbal, L. Tichy, V. Xiong, F. & Grobbelaar J.U. (1996) Microscopic green algae and cyanobacteria in high-frequency intermittent light.*J Appl. Phys,* 8, 325-333.

NeidhardtJ., ZhangL. & Melis A. (1998) Photosystem II repair and chloroplast recovery from irradiance stress: relationship between chronic photoinhibition, light-harvesting chlorophyll antenna size and photosynthetic productivity in Dunaliella salina (green algae).*Photosynth.Res.,* 56, 175-184.

Nuffied Council on Bioethics (2011). Biofuels: ethical issues. ISBN. 978-1-904384-22-9, www.nuffieldbioethics.org

Polle Juergen E.W., S.-D. K. & Melis A. (2003). tla1, a DNA insertional transformant of the green alga Chlamydomonas reinhardtii with a truncated light-harvesting chlorophyll antenna size.*Planta,* 217, 49-59.

Ramos, M.J., Fernández, C. M., Casas, A., Rodríguez, L. & Pérez, A. 2009. Influence of fatty acid composition of raw materials on biodiesel properties. *Bioresource Technology* 100, 261–68.

Raven J.A. (1988) Temperature and algal growth. *New Phytol.,* 110 (4), 441-461.

Richmond A. (2004) Principles for attaining maximal microalgal productivity in photobioreactors: an overview. *Hydrobiologia,* 512, 33-37.

Richmond A., Boussiba S., Vonshak A. & Kopel R (1993) A new tubular reactor for mass production of microalgae outdoors. *J. Appl. Phycol.*, 5, 327-32.

Sheehan J., Dunahay T., Benemann J. & Roessler P. (1998) A look back at the US Department of Energy's Aquatic Species Program- Biodiesel from Algae. National Renewable Laboratory, Golden, CO, Report NREL/TP-580-24190.

Spektorova L., Creswell R. L. &Vaughan D. (1997) Closed tubular cultivators: an innovative system for commercial culture of microalgae. *World Aquaculture,* 28, 39-48.

Spolaore P., Joannis-Cassan C., Duran E. & Isambert A. (2006) Commercial applications of microalgae. J. *Bioscience and Bioengineering,* 101, 87-96.

Tat M.E., Wang P.S., Gerpen J.H.V. & Clemente T.E. (2007) Exhaust emissions from an engine fueled with biodiesel from high-oleic soybeans. *J Am Oil Chem Soc,* 84, 865- 869.

Ugwu C.U., Aoyagi H. & Uchiyama H. (2008) Photobioreactors for mass cultivation of algae. *Bioresource Technology*, 99, 4021-4028.

Weyer K., Daniel B., Darzins A. & Willson B. (2009) Theoretical maximum algal oil production. *Bioenergy Research,* 3, 204-213.

Wigmosta M. S., Coleman A. M., Skaggs R. J., Huesemann M. H. & Lane L. J. (2011) National microalgae biofuel production potential and resource demand. *Water Resource Research,* 47, W00H04.

Zhu Xin-Ghang., Stephen P. L. & Donald R. O. (2008) What is the maximum efficiency with which photosynthesis can convert solar energy into biomass? *Current Opinion in Biotechnology*, 19, 153-159.

Yeh N. & Chung Jen-Ping (2009) High-brightness LEDs- energy efficient lighting sources and their potential in indoor plant cultivation. *Renewable Sustainable Energy Reveiws*, 13, 2175-2180.

Zemke P. E., Wood B. D. & Dye D. J. (2010) Considerations for the maximum production rates of triacylglycerol from microalgae. *Biomass Bioenergy,* 34, 145-151.

9 Chapter

BENEFITS AND OBSTACLES OF MICROALGAE AS A SOURCE OF OIL FOR BIODIESEL

1. INTRODUCTION: AN OVERVIEW

Microalgal oil is presently regarded as a viable alternative feedstock for biodiesel. The current demand for oil in biofuel production significantly surpasses the availability, necessitating alternate biomass sources. Microalgae have several advantages over land-based crops in terms of oil production. Their uncomplicated unicellular composition and elevated photosynthetic efficiency enable a possibly greater oil yield per unit area compared to the most productive oilseed crops. Algae can be cultivated on marginal terrain with brackish or saline water, so avoiding competition for resources with traditional agriculture. They do not necessitate herbicides or pesticides, and their cultivation may be integrated with the absorption of CO2 from industrial waste streams and the elimination of surplus nutrients from wastewater. (Hodaifa et al., 2008; An et al., 2003). In addition to oil production, potentially valuable co-products such as pigments, antioxidants, nutraceuticals, fertilizer or feeds could be produced (Mata et al., 2010; Rodolfi et al., 2009).

Despite these advantages, algal fuel is not currently in widespread use, largely due to its high cost of production (Chisti, 2007; Miao & Wu, 2006). Despite strong interest from the commercial and scientific sectors, there are currently no industrial facilities producing biodiesel from algae (Lardon et al., 2009). One of the major economic and technological bottlenecks in the process is biomass and lipid production by the algae (Borowitzka, 1992; Sheehan et al. 1998; Tsukahara & Sawayama, 2005). Productive strains and optimized culture conditions able to produce cells with a simultaneously high growth rate and lipid content are required. The high cost and energy demand of harvesting unicellular algae also remains a major challenge. The small cell size (often < 10 μm in diameter) and dilute biomass produced requires innovative solutions to minimize the consumption of water and energy as well as processing costs (Rodolfi et al., 2009).

2. MICROALGAE

The term 'algae' is used to describe a huge variety of prokaryotic (strictly termed Cyanobacteria) and eukaryotic organisms with a range of morphologies and phylogenies. They represent a wide array of species, inhabiting environments from deserts to the Arctic Ocean, including both salt and fresh water. They vary in colour, shape and size, from picoplankton (0.2 to 2 μm) to giant kelp fronds up to 60 m in length (Barsanti & Gualtieri, 2006). Macroalgae (e.g. seaweeds) are generally large (can be seen without the aid of a microscope), multicellular and often show some form of cellular specialisation. Microalgae are usually less than 2 mm in diameter and unicellular or colonial. Microalgae have been investigated for a variety of commercial applications. Annual global microalgal production is currently estimated at about 10 000 metric tons, with the main algae cultivated being *Spirulina* (accounting for roughly half of the worldwide algal production), *Chlorella*, *Dunaliella* and *Haematococcus*.

Algae have been investigated as a source of energy in many different contexts, from direct combustion to the production of hydrogen gas. Anaerobic digestion can be applied for the generation of methane or biogas (Golueke et al., 1957). Algal species with high oil content are particularly attractive as a feedstock for biodiesel production. Research into algae for the mass-production of oil has focused on the microalgae due to their high lipid content compared to macroalgae. Most algal species considered for biodiesel production are either green algae (Chlorophyta) or diatoms (Bacillariophyta) (Sheehan et al., 1998). They are generally photosynthetic, but several species are able to grow heterotrophically or mixotrophically (Barsanti & Gualtieri, 2006).

Microalgae have higher growth rates than land-based plants. Due to their simple cellular structure and existence in an aqueous environment, the entire cell surface is available for light capture and mass transfer, leading to high rates of substrate uptake and photosynthetic efficiency (Miao & Wu, 2006; Sheehan et al., 1998). In contrast to land-based oil crops, where only the seeds are harvested, each algal cell contains lipid and hence the yield of product from biomass is much higher (Becker, 1994). Due to these differences, the oil yield per area of microalgal cultures potentially exceeds that of the best oilseed crops (Table 1).

Table 1. Average productivities of some common oil seed crops compared to algae

Oil source	Yield (L.m-2.yr-1)	Reference
Algae	4.7 to 14	Sheehan et al., 1998
Palm	0.54	Mata et al., 2010
Jatropha	0.19	Sazdanoff, 2006
Rapeseed	0.12	Sazdanoff, 2006
Sunflower	0.09	Sazdanoff, 2006
Soya	0.04	Sazdanoff, 2006

3. BIODIESEL FROM MICROALGAE

Microalgal lipids can be extracted to yield oil similar to that from land-based oilseed crops. The amount and composition of the oil varies between algal species. Algal oil can be converted to biodiesel through the same methods applied to vegetable oil. The idea of using microalgae as a source of transportation fuel is not new. Research in this field has been conducted since the 1950s (Oswald & Golueke, 1960). In the 1970s, several large, publicly funded research programs were set up in the USA, Australia and Japan (Regan & Gartside, 1983; Sheehan et al., 1998). The US Department of Energy invested more than US$ 25 million between 1978 and 1996 in the Aquatic Species Program to develop biodiesel production from algae (Sheehan et al., 1998). The main focus of the program was the production of biodiesel from high lipid-content algae grown in open ponds, utilizing waste CO_2 from coal fired power plants. Over 3000 species were collected and many of them screened for lipid content.

Early in the program, it was observed that environmental stress, particularly nutrient limitation (nitrogen for green algae and silicon for diatoms) led to an increase in accumulation of lipids. Promising species were investigated to determine the mechanism of this 'lipid trigger'. Researchers in the program were the first to isolate the enzyme Acetyl CoA Carboxylase from a diatom. This enzyme catalyzes the first committed step in the lipid synthesis pathway. Acetyl CoA Carboxylase was over-expressed successfully in algae; however, the anticipated increase in oil production was not demonstrated. The program close out report (Sheehan et al., 1998) concluded that, although algae used significantly less land and water than traditional crops, and sufficient resources did exist for algal fuel to completely replace conventional diesel, the high cost of microalgae production remained an obstacle. Even with the most optimistic lipid yields, production would only have become cost effective if petro-diesel had risen to twice its 1998 price.

The last decade has seen a renewal of interest in biofuels and microalgae as a feedstock source. An increase in oil prices, additional pressure to find alternatives to dwindling oil supplies and an urgent need to cut carbon emissions contributing to global warming has led to a renewed interest in algae as a source of energy, particularly lipid producing algae as a source of biodiesel.

4. MICROALGAL LIPIDS

The main components of algae cells are proteins, carbohydrates and lipids (Becker, 1994). Microalgae naturally produce lipids as part of the structure of the cell (e.g. in cell membranes and as signalling molecules), and as a storage compound, similar to fat stores in animals and humans (Tsukahara & Sawayama, 2005). The term lipid encompasses a variety of compounds with different chemical structures (e.g.

esters, waxes, cholesterol). The most common lipids are composed of a glycerol molecule bound to three fatty acids, known as triacylglycerol or TAG, or to two fatty acids with the third position taken up by a phosphate (phospholipids) or carbohydrate (glycolipids) group. Fatty acids consist of a long unbranched carbon chain. They are classified according to the number of carbon atoms in the chain and the number of double bonds, for example saturated (no double bonds), monounsaturated (one double bond) or polyunsaturated (more than one double bond). Microalgae commonly contain fatty acids ranging from C12 to C24, often with C16 and C18 unsaturates. Certain species contain significant amounts of polyunsaturated fatty acids.

Storage lipids, generally in the form of TAG, accumulate in lipid vesicles called oil bodies in the cytoplasm. Most fast-growing species have relatively low lipid content during normal growth, with these lipids mainly consisting of phospho- or glycolipids associated with cell membranes. Under certain conditions, generally triggered by stress or the cessation of growth, lipid content can increase to over 60% of cell dry weight (DW), mostly composed of TAG (Shifrin & Chisholm, 1981; Piorreck et al., 1984; Spoehr & Milner, 1949; De la Pena, 2007; Becker, 1994).

TAGs are the most suitable class of lipids for biodiesel production. Phospholipids are particularly undesirable as they increase consumption of catalyst and act as emulsifiers, impeding phase separation during transesterification (Mittelbach & Remschmidt, 2004; Van Gerpen, 2005). Phospholipids, and some sulphur-containing glycolipids, also increase the phosphorous and sulphur content of the fuel respectively, which must both be below 10 $mg.L^{-1}$ to meet the European biodiesel standard EN 14214. The type of fatty acids found in the oil can have a profound effect on the biodiesel quality. The fatty acid chain length and degree of saturation (determined by the number of double bonds) affects properties such as the viscosity, cold flow plug point, iodine number and cetane number of the fuel (Ramos et al., 2009). For biodiesel production, it is therefore important to maximize not only total lipid production, but also TAG content and appropriate fatty acid profile.

Lipid synthesis relies on carbon compounds generated from CO_2 by photosynthesis, as well as energy and reducing power (in the form of ATP and NAD(P)H respectively). The latter are produced during the light reactions of photosynthesis, while CO_2 uptake is mediated by the Calvin cycle during the dark reactions of photosynthesis. The output of the Calvin cycle is a three-carbon compound (glyceraldehyde 3-phosphate), which is converted through glycolysis into acetyl CoA. The conversion of acetyl CoA to malonyl CoA is the first committed step in lipid biosynthesis (Livne & Sukenik, 1992). Throughout metabolism there are a number of branch points at which metabolic intermediates are partitioned between the synthesis of lipids and other products such as carbohydrates and proteins (Lv et al., 2010). For example, acetyl CoA is a substrate for lipid synthesis as well as entry into the TCA cycle, which generates energy and biosynthetic precursors for proteins and nucleic acids. Both external and internal constraints, such as the availability of nutrients and the enzymatic reaction rates, limit the supply of metabolic intermediates. The production of storage lipids is particularly energy and resource intensive (Dennis et al., 1998; Roessler, 1990) and therefore usually occurs at conditions of reduced growth.

5. CULTIVATION OF MICROALGAE

The use of microalgae for energy generation requires large-scale, low-cost production. This demands cheap, scalable reactor design with efficient provision of the requirements for high algal productivity. Design considerations include optimum surface area to volume ratio for light provision, optimal mixing to keep cells in suspension and for distribution of nutrients, control over water balance and sterility, as well as maintenance of favorable temperature. A wide variety of reactor designs have been proposed, each with advantages and drawbacks.

5.1 Reactor systems

Microalgal production is a technology halfway between agriculture, which requires large areas for sunlight capture, and fermentation, which involves liquid culture of microorganisms (Becker, 1994). As light does not penetrate more than a few centimetres through a dense algal culture, scale-up is based on surface area rather than volume (Scott et al., 2010). Many different types of algal cultivation systems have been developed, but they can be divided into two main categories: open and closed.

Open systems consist of natural waters such as lakes, ponds and lagoons, or artificial ponds and containers that are open to the atmosphere. Most commercial production to date has taken place in open ponds as these systems are easy and cheap to construct (Pulz, 2001). The most common technical design is the raceway pond: an oblong, looped pond mixed by a paddlewheel, with water depths of 15 to 20 cm (Becker, 1994). Biomass concentrations of between 0.1 and 1 $g.L^{-1}$ and biomass productivities of between 50 and 100 $mg.L^{-1}.day^{-1}$ are possible (Chisti, 2007; Pulz, 2001). The main advantages of open systems are their low cost and ease of construction and operation. They also offer the potential for integration with wastewater treatment processes or aquaculture systems (Chen, 1996).

Disadvantages of open systems include contamination with unwanted species such as foreign algae, yeast, bacteria and predators, evaporation of water, diffusion of CO_2 to the atmosphere and low control over environmental conditions, particularly temperature and solar irradiation (Becker, 1994; Pulz, 2001). In addition, the relatively low cell densities achieved can lead to higher cost of cell recovery (Chen, 1996). Only a few microalgal species have been successfully mass cultivated in open ponds. These tend to be either fast-growers that naturally outcompete contaminating algae (e.g. *Chlorella* and *Scenedesmus*), or species that grow in a specialised environment such as high salt (e.g. *Dunaliella salina*) or high pH (*Spirulina platensis*), which limits growth of competitors and predators (Chen, 1996). Due to the lack of control over cultivation conditions resulting in low productivity, and the fact that many desirable species cannot be effectively maintained in open systems, attempts have been made to overcome some of these limitations through the use of enclosed reactor systems.

Closed systems, or photobioreactors, consist of containers, tubes or clear plastic bags of various sizes, lengths and orientations (Pulz, 2001). Commonly used designs include vertical flat-plate reactors and tubular reactors, either pumped mechanically or by airlift (Scott et al., 2010). Closed reactors offer a much higher degree of control over process parameters, leading to improved heat and mass transfer, and thus higher biomass yields. They can also offer a much higher surface area to volume ratio for light provision, better control of gas transfer, reduction of evaporation and easier installation in any open space (Chen, 1996). Additionally, the risk of contamination is reduced, CO_2 can be contained, production conditions can be reproduced and temperature can be controlled.

Productivity in closed systems can be much higher than open systems, with biomass concentrations of up to 8 $g.L^{-1}$ and productivities of between 800 and 1300 $mg.L^{-1}.day^{-1}$ (Pulz, 2001). However, they are generally much more costly to build and more energy demanding to operate than open systems (Table 2). Closed systems can also have problems with fouling and oxygen build-up. Large systems can be difficult to clean and sterilize and long sections of enclosed tubing may require oxygen purging. High oxygen concentrations cause the key enzyme Rubisco to bind oxygen instead of carbon dioxide, leading to photorespiration instead of photosynthesis (Dennis et al., 1998). Although closed bioreactors offer a much higher degree of control over process parameters and can have higher yields, it is uncertain whether the increased productivity can offset the higher cost and energy requirements. For a commodity product such as vegetable oil for biodiesel, low cost, high volume production is demanded, while quality is less critical (Pulz, 2001). In this case, the more favourable economics and energy requirements of open ponds may well outweigh the advantages of closed reactors.

A hybrid system combining the cost effectiveness of open ponds with the controlled environment of closed systems is appealing and has been tested in a few cases. Generally production is divided into an initial growth or inoculum production stage in closed reactors, followed by a stress or scaling up stage in open ponds (Huntley & Redalje, 2006).

Table 2. Comparison of open ponds and closed photobioreactors. Adapted from Pulz (2001).

Parameter	Open	Closed
Control over process parameters	Low	High
Contamination risk	High	Low
Water loss due to evaporation	High	Low
CO2 loss	High	Low
O2 build-up	Low	High
Area required	High	Low
Productivity	Low	High
Consistency and reproducibility	Low	High
Weather dependence	High	Low
Cost	Low	High
Energy required	Low	High

5.2 Cultivation parameters

Several factors need to be considered in the cultivation of algal biomass. These include the provision of light, carbon and nutrients such as nitrate, phosphate and trace metals, the mixing regime, maintenance of optimal temperature, removal of O_2 and control of pH and salinity (Becker, 1994; Grobbelaar, 2000; Mata et al., 2010). The optimal and tolerated ranges tend to be species specific, and may vary according to the desired product.

5.2.1 Temperature

Light and temperature are among the most difficult parameters to optimise in large-scale outdoor culture systems. Daily and annual fluctuations in temperature can lead to significant decreases in productivity. Optimal growth temperatures are generally between 20 and 30°C (Chisti, 2008). Many algal species can tolerate temperatures of up to 15°C lower than their optimum, with reduced growth rates, but a temperature of only a few degrees higher than optimal can lead to cell death (Mata et al., 2010). Closed systems in particular often suffer from overheating during hot days, when temperatures inside the reactor can reach in excess of 50°C. Heat exchangers or evaporative water-cooling systems may be employed to counteract this (Mata et al., 2010). Low seasonal and evening temperatures can also lead to significant losses in productivity.

5.2.2 Light and mixing

The efficient production of algal biomass relies on the optimal provision of light energy to all cells within the culture. Most algal growth systems become light limited at high cell densities. Due to absorption and shading by the cells, light only penetrates a few centimetres into a dense algal culture (Richmond, 2004). The average provision of light is linked to reactor depth or diameter, cell concentration and mixing. A larger surface area to volume ratio, usually achieved through areas of thin panelling or narrow tubing, results in higher light provision.

Photosynthetic efficiency is highest at low light intensities. At high light levels, although photosynthetic rate may be faster, there is less efficient use of absorbed light energy.

Above the saturation point, damage to photosynthetic machinery can occur in a process known as photoinhibition (Scott et al., 2010). In a dense culture exposed to direct sunlight, cells at the surface are likely to be photoinhibited, while those at the centre of the reactor are in the dark. Mixing is therefore important not only in preventing cell settling and improving mass transfer, but also exposing cells from within a dense culture to light at the surface.

The frequency of light-dark cycling has been reported to affect algal productivity (Grobbelaar, 1994; Grobbelaar, 2000). Algae are less likely to become photoinhibited when the light is supplied in short bursts because the photosystems have time to recover during the dark period (Nedbal et al., 1996). While high rates of mixing facilitate rapid circulation of cells between light and dark zones in the reactor, high liquid velocities can damage algal cells due to increased shear stress (Mata et al., 2010). High rates of mechanical mixing or gas sparging also have large energy requirements, jeopardizing the process energy balance and increasing costs (Richardson, 2011).

5.2.3 Gas exchange

In order to maintain a high photosynthetic rate, the influx of carbon and energy must be non-limiting. In photoautotrophic growth, energy is provided by light and carbon in the form of CO_2. In order to be taken up by cells, the CO_2 must dissolve in the water. The rate of dissolution is determined by the CO_2 concentration gradient as well as by the temperature, rate of gas sparging and surface area of contact between the liquid and gas (a function of agitation and bubble size). Reactor geometry, methods of gas introduction and reactor mixing can all influence the rate of CO_2 delivery (Bailey & Ollis, 1977). Certain strains of microalgae can tolerate up to 12% CO_2 (Pulz, 2001). The 0.03% CO_2 content of ambient air is suboptimal for photosynthesis (Pulz, 2001), hence for optimal microalgal growth, additional CO_2

must be provided. This is usually done by direct injection of a CO_2 enriched air stream. As the addition of CO_2 acidifies the medium, care must be taken not to adversely decrease the pH (Anderson, 2005). It is debatable whether direct gas injection is the optimal method of CO_2 delivery. Efficiencies of carbon uptake are very low at high CO_2 concentrations, as most CO_2 exits the top of the reactor. Novel strategies of CO_2 provision include microporous hollow fibre membranes and separate gas exchanger systems (Carvalho et al., 2006).

5.2.4 Salinity, nutrients and pH

The major nutrient requirements for microalgal growth are nitrogen and phosphorous, with certain diatoms, silicoflagellates and chrysophytes also requiring silicon (Anderson, 2005). Requirements of nutrients, pH and osmolarity are species dependent. Deviation from optimal levels may cause a decrease in biomass productivity, but can have other advantages, for example, high salinity may limit contamination. Sufficient supply of all essential nutrients is a prerequisite for efficient photosynthesis and growth, but limitation of key nutrients (e.g. nitrate, phosphate or silica) may cause accumulation of desired products such as lipid.

5.2.5 Nutritional mode

Most microalgae are photoautotrophs (utilizing sunlight as their source of energy and CO_2 as a carbon source). This is the most common growth mode employed in algal cultivation (Chen, 1996). However, several species (e.g. *Chlorella, Chlamydomonas, Phaeodactylum, Nitzschia, Tetraselmis* and *Crypthecodinium*) are also capable of heterotrophic growth (utilizing organic carbon such as glucose, acetate or glycerol as the sole source or carbon and energy) or mixotrophic growth (photoautotrophic growth supplemented by an organic carbon source).

The advantages of using an organic carbon substrate are that it decreases dependence on light provision, allowing growth in conventional fermenters in the dark. Optimal growth conditions can be maintained, allowing higher cell concentrations and hence increased volumetric productivities to be reached (Chen, 1996). Higher productivities of both biomass and lipid have been reported under heterotrophic growth compared to autotrophic (Ceron Garcia, 2000; Miao & Wu, 2006). Disadvantages of feeding an organic carbon source include the fact that there are a limited number of algal species that can utilize organic carbon sources, the risk of bacterial contamination is greatly increased and the carbon substrate adds an additional cost, along with the environmental burden of its production. The use of a substrate such as glucose, commonly sourced from crop plants, adds a trophic level to the process, thereby removing the simplicity of the concept of microalgae as cellular factories producing liquid fuel from pure sunlight and CO_2.

5.2.6 Cultivation strategy

The optimal cultivation strategy (e.g. batch, fed-batch or continuous cultivation mode) is determined by the kinetics of growth, product accumulation and substrate uptake (Shuler & Kargi, 2005). For production of a primary product such as protein or biomass for food or feed, optimisation of biomass productivity is the main objective. In this case, batch or continuous systems are generally used. For production of a secondary product such as carotenoids or storage lipids, the use of two or more production stages to enhance yield has been proposed (Ben-Amotz, 1995; Huntley & Redalje, 2006; Richmond, 2004). The first stage is designed to optimize growth, while the second stage provides conditions that retard growth and encourage product synthesis, usually by applying some form of stress, e.g. nutrient deprivation in the case of lipid accumulation. Another potential two-stage strategy that could enhance lipid productivity is an initial photosynthetic stage, followed by a second heterotrophic phase, where feeding with an organic carbon source such as glucose may boost lipid content.

6. HARVESTING AND PROCESSING

The economic recovery of microalgal biomass remains a major challenge. Microalgae for biofuel are a low value product suspended in large volumes of water. Harvesting contributes 20 to 40% of the total cost of biomass production (Gudin & Therpenier, 1986; Molina Grima et al., 2003).

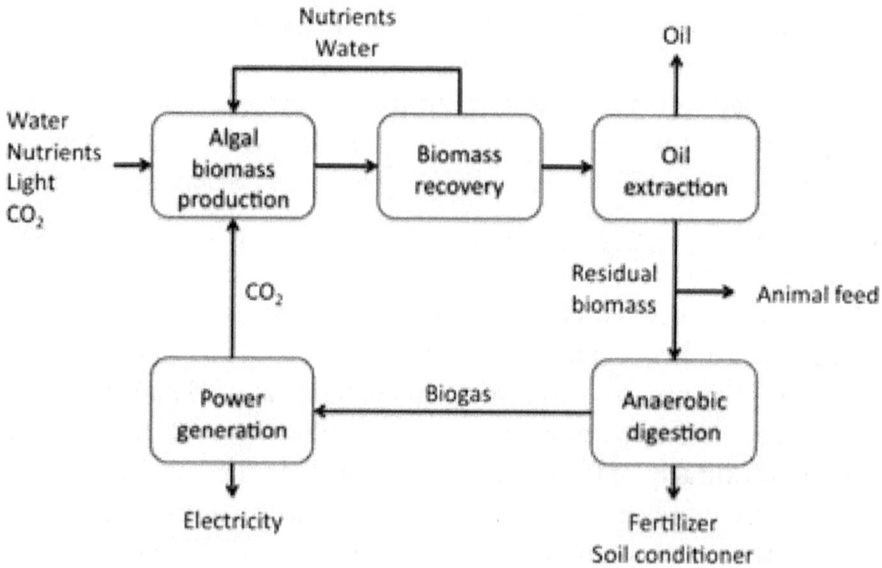

Fig. 1. Conceptual overview of microalgae process options (adapted from Chisti, 2008)

The difficulty in separation can be attributed to the small size of the cells (3 to 300 μm, Henderson et al., 2008), their neutral buoyancy and the fact that photoautotrophic microalgal cultures are relatively dilute, achieving concentrations in the order of 1 to 8 $g.L^{-1}$ (Pulz, 2001). Each algal species presents unique challenges due to the array of sizes, shapes, densities and cell surface properties encountered. A low-cost, energy efficient method with a high recovery efficiency and concentration is required, minimizing cell damage and allowing for water and nutrient recycle (Fig. 1).

6.1 Factors affecting separation

Several natural properties of microalgal cells affect the choice and efficiency of harvesting methods. Factors relevant to separation include density, surface charge, size, shape, hydrophobicity, salinity of the medium, adhesion and cohesion properties and settling or floating velocities. Table 3 highlights the variability in some of these parameters between species, indicating that species-specific solutions may be required. Algal cell characteristics can vary with culture age and growth conditions. For example, changes in biochemical composition such as lipid content could affect the buoyancy of algal cells. Surface charge, the chemical structure of the cell wall and the amount and composition of the extracellular organic matrix (EOM) can vary with growth phase and greatly influence the degree to which cells repel or stick to one another (Bernhardt & Clasen, 1994; Henderson et al., 2008). Danquah et al. (2009) found a strong correlation between growth phase and settling efficiency, with improved filtration, flocculation and sedimentation rates during the stationary phase.

Table 3. Characteristics relevant to harvesting of some microalgae (Henderson et al., 2008). Zeta potential is a measure of the degree of repulsion between adjacent particles due to surface charge.

	Density ($kg.m^{-3}$)	Zeta potential (mV)	Culturing pH	Morphology	Diameter, length (μm)
Microcystis	1200	-7.5 to -26	5.6 - 9.5	Globular sphere	3-7
Chlorella vulgaris	1070	-17.4	7	Single cell spherical	3.5
Cyclotella sp.	1140	-19.8 to -22.3	4 - 10	Chains of spheres	6.1
Syendra acus	1100	-30 to -40	7.6	Needles	4.5-6, 100-300

Morphological characteristics that influence harvesting include cell motility, size, shape, cell wall elongations such as spines and flagella, colony formation, and the presence of extracellular mucilage layers or capsules (Petrusevski et al., 1995; Jarvis et al., 2009). Larger particles allow for easier separation due to increased surface area and mass. Filamentous morphology or appendages also allow for easier filtration as cells cannot pass through filter pores. Density affects sedimentation and flotation. Most algae have a specific gravity close to that of water, rendering them with a neutral buoyancy. Some cyanobacteria can adjust their density through gas vacuoles (Anderson, 2005), rendering sedimentation difficult but enabling potential surface collection.

Cell surface charges influence the electrostatic interactions between cells and between cells and surfaces or bubbles. This directly affects the adhesion, adsorption, flotation and flocculation properties of algal cells. Most algae have a negatively charged cell surface, leading to electrostatic repulsion between cell walls. Addition of positively charged ions to the solution can help to neutralize the negative surface charge and aid cell flocculation. Changing the pH of the solution can also cause flocculation (Chen et al., 1998). Hydrophobicity is another, non-electric property affecting interaction of algal cells with each other and external surfaces. Most algae are naturally hydrophilic (Fattom & Shilo, 1984), but this can be altered by surfactants and pH (Jameson, 1999). Increasing the hydrophobicity of cells could cause them to adhere to bubbles, filters or other separation catalysts.

6.2 Harvesting methods

Harvesting requires one or more solid-liquid separation techniques (Molina Grima et al., 2003). In order to achieve the levels of concentration required, various chemical, biological and physical separation steps may be necessary. Common methods of cell harvesting include flocculation, filtration, sedimentation, centrifugation and flotation (Mata et al., 2010). The small cell size of microalgae makes them difficult to dewater. Flocculation is used to 'clump' the cells, grouping them together to form larger particle sizes. This is often suggested as a pretreatment step prior to filtration, sedimentation or flotation. Flocculation occurs when the repulsion between cells is reduced, allowing them to either aggregate directly onto each other or through an intermediate bridging surface. The extent of flocculation is dependent on pH, temperature, density, hydrophobicity, surface charge and culture age (Lee et al., 1998). Flocculation can be induced by addition of positively charged ions or polymers, e.g. minerals such as lime, calcium and salts, metal salts such as aluminium sulphate and ferric chloride, and naturally occurring flocculants such as starch derivatives and tannins. Drawbacks to the use of chemical flocculants are the high dosages required, the need for pH correction (Pushparaj et al., 1993) and the contamination of the biomass and media with the flocculant, meaning that media cannot be recycled without removal of the chemical. Autoflocculation can be induced through pH change (Csordas & Wang, 2004), nutrient limitation (Schenk et al., 2008), excretion of macromolecules (Benemann et al., 1980) or aggregation between microalgae and bacteria (Lee et al., 2009).

Conventional filtration is only effective for larger (> 70 μm) or filamentous species such as *Coelastrum* and *Spirulina* (Brennan & Owende, 2010, Lee et al., 2009). For smaller cells, micro- filtration, ultra-filtration and membrane-filtration can be used, though usually only for small volumes (Brennan & Owende, 2010, Petrusevski et al., 1995, Borowitzka, 1997). Fouling (accumulation of material on the surface of the membrane, slowing filtration) is a major problem. If filtration were to be considered for mass production, a high driving force for separation (high pressure or suction) would be required, which necessitates a high energy input. Microstraining (filtration by natural gravity using low speed rotating drum filters) is a promising method due to ease of operation and low energy consumption (Mohn, 1980). Another option is cross-flow membrane filtration (Zhang et al., 2010). Using a tangential, turbulent flow of liquid across the membrane prevents clogging of the filter with cells. The efficiency of the process is very dependent on cell morphology and the transmembrane pressure (Petrusevski et al., 1995). A more unconventional approach is magnetic filtration. Here addition of magnetic metals, either taken up by algal cells, or used to flocculate them, could allow capture using a magnetic field (Bitton et al., 1975).

Sedimentation is the process whereby solid particles suspended in a fluid are settled under the influence of gravity or some other force. In microalgae, it depends on coagulation or flocculation of cells to produce flocs with a large enough size (> 70 μm) or high enough density to induce settling (Vlaski et al., 1997). Sedimentation is typically used in wastewater treatment. It is suitable for large throughput volumes and has low operational costs. Flocculation, using a dense substance such as calcium carbonate, can greatly reduce settling time. Ultrasound (acoustic energy) can be used to induce aggregation and facilitate sedimentation (Bosma et al., 2003), however the energy requirement may be too high for large-scale use.

Centrifugation is essentially sedimentation under a rotational force rather than gravity. The efficiency of centrifugation depends on the size and density of the particles, the speed of the rotor, the time of centrifugation and the volume and density of the liquid. Almost all microalgae can be harvested by centrifugation. It is a highly efficient and reliable method, can separate a mixture of cells of different densities and does not require the addition of chemicals, but has a high energy consumption (Chisti, 2007). It is routinely used for recovery of high value products, or for small scale research operations, although large, flow-though centrifuges can be used to process large volumes. Many algae require speeds of up to 13 000 g which results in high shear forces (Harun et al., 2010; Knuckey et al., 2006) and can damage sensitive cells.

Flotation operates by passing bubbles through a solid-liquid mixture. The particles become attached to the bubble surface and are carried to the top of the liquid where they accumulate. The concentrated biomass can be skimmed off (Uduman et al., 2010). Flotation is considered to be faster and more efficient than sedimentation

(Henderson et al., 2008). It is associated with low space requirements and moderate cost. Addition of chemical coagulants or flotation agents is often required to overcome the natural repulsion between the negatively charged algal particles and air bubbles. The pH and ionic strength of the medium are important factors to optimize this recovery technique.

6.3 Processing

After harvesting, the major challenge is in releasing the lipids from their intracellular location in the most energy efficient and economical way possible. Algal lipids must be separated from the rest of the biomass (carbohydrates, proteins, nucleic acids, pigments) and water. Common harvesting methods generally produce a slurry or paste containing between 5 and 25% solids (Shelef et al., 1984). Removing the rest of the water is thought to be one of the most expensive steps with literature values ranging from 20 to 75% of the total processing cost (Uduman et al., 2010, Molina Grima et al., 2003). Shelef et al. (1984) highlight a number of possible techniques for drying biomass: flash drying, rotary driers, toroidal driers, spray drying, freeze-drying and sun drying. Because of the high water content, sun-drying is not an effective method and spray-drying is not economically feasible for low value products (Mata et al., 2010). The selection of drying technique is dependent on the scale of operation, the speed required and the downstream extraction process (Mohn, 1980).

Lipid extraction can be done in a number of ways. Solvent extraction techniques are popular, but the cost and toxicity of the solvent (e.g. hexane) is of concern and solvent recovery requires significant energy input. Other methods involve disruption of the cell wall, usually by enzymatic, chemical or physical means (e.g. homogenization, bead milling, sonication (Mata et al., 2010)), allowing the released oil to float to the top of the solution. Ultrasound and microwave assisted extraction methods have been investigated (Cravotto et al., 2008). Supercritical CO_2 extraction is an efficient process, but is too expensive and energy intensive for anything but lab-scale production. Direct transesterification (production of biodiesel directly from algal biomass) is also possible. Some of these techniques do not require dry biomass, but the larger the water content of the algal slurry, the greater the energy and solvent input required.

Once the algal oil is extracted, it can be treated as conventional vegetable oil in biodiesel production. Direct pyrolysis, liquefaction or gasification of algal biomass have also been suggested as means of producing fuel molecules. One of the concerns for biodiesel production through transesterification, shared with any biodiesel feedstock, is the quality of the biodiesel produced. Biodiesel must meet certain international regulations, for example, the ASTM international standards or the EN14214 in Europe. It has been calculated that the fatty acid profile of certain microalgal species will produce biodiesel that does not meet these specification, therefore blending or additives may be required (Stansell, 2011).

7. ECONOMIC AND ENVIRONMENTAL FEASIBILITY

In order to be economically feasible, microalgal biodiesel must be cost competitive with petroleum-based fuels. We have investigated the relationship between algal lipid productivity and cost in order to determine the range of productivities that need to be achieved for economic viability. Based on values from Chisti (2007), a model was set up to estimate cost per litre of algal oil as a function of algal biomass productivity and lipid content. Where the cost of producing a litre of algal biodiesel was below the price of a litre of fossil-fuel derived diesel, it was considered economically viable (i.e. no profit margin was introduced). The price of fossil-fuel derived diesel is partly dependent on the price of crude oil, which has varied widely in the last few years, hence several scenarios were evaluated.

Assumptions made in the execution of the model were:

1. Cost per kg algal biomass: US$ 0.6 for raceway ponds, and US$ 0.47 for photobioreactors (Chisti, 2007)
2. In order to be economically viable, the cost of algal oil per litre must be less than 6.9×10^{-3} times the cost of crude oil in US$ per barrel (Chisti, 2007)
3. Density of algal oil: 0.86 $g.cm^{-3}$ (Barsanti & Gualtieri, 2007)

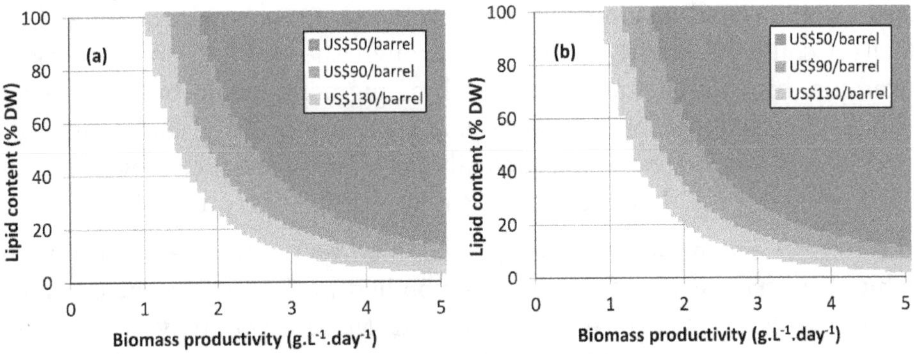

Fig. 2. Lipid contents and biomass productivities required for economic feasibility in (a) large-scale, outdoor raceway ponds and (b) large-scale, outdoor photobioreactors. Dark grey region: productivities economically feasible at US$ 50 per barrel crude oil (cost of algal oil per L lower than cost of regular diesel per L). Additional region for crude oil price US$ 90 per barrel = mid-grey and US$ 130 = light grey

The economic model was run for three prices of crude oil, based on fluctuations over the last few years. These scenarios of 'high' ($ 130), 'medium' ($ 90) and 'low' ($ 50) cost of crude oil per barrel gave the price limits for algal oil of 0.90, 0.62 and 0.35 US$ per L respectively. The results of the model are shown in Fig. 2a (raceway ponds) and 2b (closed photobioreactors).

Based on this model, the results for raceway ponds show that algal biodiesel will not be economically feasible, either in ponds or photobioreactors, at current costs below a biomass productivity of 1 $g.L^{-1}.day^{-1}$. Assuming a maximum realistically achievable lipid content of 50% DW, algal biodiesel becomes economically feasible at biomass productivities of 1.5 $g.L^{-1}.day^{-1}$ (US$ 130 per barrel crude oil), close to 2 $g.L^{-1}.day^{-1}$ (US$ 90), and 2.5 $g.L^{-1}.day^{-1}$ (US$ 50) in raceway ponds. At lower lipid contents, higher biomass productivity is required, e.g. at a lipid content of 25% DW, algal biodiesel only becomes cost effective at 2 $g.L^{-1}.day^{-1}$ for US$ 130 per barrel. The model for photobioreactors is based on a lower cost per kg algal biomass than raceway ponds, hence economic feasibility is reached at slightly lower biomass productivities and lipid contents, e.g. at a biomass productivity of 2 $g.L^{-1}.day^{-1}$, a lipid content of only 20% DW is required to be viable at US$ 130 per barrel crude oil.

Currently reported biomass productivities in outdoor raceway ponds average around 0.17 $g.L^{-1}.day^{-1}$, with a lipid content of 26% DW (Griffiths and Harrison, 2009), which is far from being economically feasible. Biomass productivities for closed photobioreactors (1.33 $g.L^{-1}.day^{-1}$) are closer to being within the economically viable range, if they can be maintained in the long term, concurrent with sufficiently high lipid content. As a reflection of this, there are currently no industrial facilities producing biodiesel from microalgae (Lardon, 2009). For cultivation to be economically viable, productivities must be increased, costs lowered, or additional income streams developed. The economics of algal biofuel production could be greatly improved through the production of co-products. For example, high value compounds such as pigments could be produced along with lipid. The residual biomass after lipid extraction could be sold as animal feed, fertilizer or soil conditioner, anaerobically digested to produce biogas, gasified or merely burned to provide some of the heat or electricity required in the process.

In addition to economic feasibility, algal biodiesel must be environmentally desirable. It is critical that the energy embodied in the fuel produced is greater than the energy input required to produce it. Net energy analysis and life cycle analysis (LCA) are tools used to quantify the environmental burdens at every stage of production, from growth of the biomass to combustion of the fuel. Lardon et al. (2009) conducted a life-cycle analysis of a hypothetical algal biodiesel production facility. Two different culture conditions: fertilizer feeding and nitrogen starvation, as well as two different extraction options: dry or wet, were investigated. The study confirmed the potential of microalgae as an energy source, but highlighted the necessity of decreasing energy and fertilizer consumption. Energy inputs, such as the energy required for mixing and pumping, the embodied energy in the materials used and the energy cost of harvesting and processing must be minimized. Recycling of material and energy from waste streams is also important wherever feasible (Scott et al., 2010). The use of nitrogen stress, as well as the optimization of wet extraction were indicated as desirable options. The anaerobic digestion of residual biomass was also suggested as a way of reducing external energy usage and recycling of nutrients.

We conducted a LCA on a hypothetical algal biodiesel process. Biomass production in three different reactor types (open ponds and two types of closed reactor: horizontal tubular and vertical tubular) was evaluated. In all cases, harvesting was modeled as an initial settling step followed by centrifugation. Hexane extraction was used to recover the oil, with the residual biomass sent for anaerobic digestion and the resulting energy from biogas production recycled to the process. The hexane was recovered and the oil converted to biodiesel using an enzymatic process. The basis chosen was production of 1000 kg of biodiesel from *Phaeodactylum tricornutum.* The net energy return (the energy embodied in the biodiesel produced divided by the energy input required) was positive (1.5) for the open pond, neutral (0.97) for the horizontal tubular reactor and negative (0.12) for the vertical tubular reactor. In this model, open ponds were the most energetically favorable reactor type, yielding 50% more energy than was put in. Horizontal tubular reactors required an energy input equivalent to the output, and vertical tubular reactors were the most unfavorable, requiring several times the energy input as that in the product, where system optimization was not conducted.

The overriding energy input in the process was found to be that required to run the reactor. Reactor energy was by far the most dominant determinant of the overall process energy requirement. This was largest in the vertical tubular reactor as these were continually mixed by gas sparging. Energy required for pumping between unit processes was also significant, particularly at lower biomass concentrations due to the larger volume of culture to be processed. The major energy inputs in downstream processing were that embodied in the lime used as a flocculation agent, and the energy required for solvent recovery. Lipid productivity and species choice had a significant impact on the energy balance.

8. OPTIMIZING LIPID PRODUCTIVITY

Increasing microalgal lipid productivity improves both the economics and energy balance of the process. The land area and size of culture vessels required, as well as the energy and water requirements for large-scale algal culture are strongly dependent on algal productivity. With a higher productivity, lower cultivation, mixing, pumping and harvesting volumes would be required to yield the same amount of product, resulting in lower cost and energy requirements. More concentrated cell suspensions could also make downstream processing more efficient. The genetic characteristics of an algal species determine the range of its productivity. The levels reached in practice within this range are determined by the culture conditions. The two main approaches to enhancing productivity are: 1. selection of highly productive algal species and 2. designing and maintaining optimal conditions for productivity.

The choice of algal strain is a key consideration. The diversity of algal species is much greater than that of land plants (Scott et al., 2010) allowing selection of species best suited to the local environment and goals of the project. Although there have been several screening programs, building on the work of the Aquatic

Species Program (Sheehan et al., 1998), the majority of strains remain untested, few species have been studied in depth and the data reported in the literature is often not comparable due to the different experimental procedures used. We conducted a broad literature review of the growth rates and lipid contents of 55 promising microalgal species under both nutrient replete and limited conditions. The original study (Griffiths & Harrison, 2009) has been extended here through the use of two key assumptions to convert data into common units of biomass and lipid productivity.

Lipid productivity is determined by both growth rate and lipid content. Lipid content (P) was typically reported as percentage dry weight (% DW). Data presented in pg lipid.cell^{-1} was discarded if no cell weight was available for conversion. Growth rates were reported as doubling time (T_d) or specific growth rate (μ). These were inter-converted according to Equation 1.

$$T_d = \frac{\ln 2}{\mu} \tag{1}$$

Standard units of g.L^{-1}.day^{-1} were chosen for biomass productivity. Specific growth rate (μ, in units of day^{-1}) can be converted to volumetric biomass productivity (Q_V, in g.L^{-1}.day^{-1}) where the biomass concentration (X, in g.L^{-1}) is known (Equation 2). Biomass productivity is often reported on the basis of surface area (Q_A), in units of g.m^{-2}.day^{-1}. This can be converted to Q_V using Equation 3 where the depth (D, in m) of the culture vessel can be calculated from the reactor geometry.

$$QV = \mu \times X \tag{2}$$

$$Q_V = \frac{Q_A}{D \times 1000} \tag{3}$$

Lipid productivity (Q_P) was infrequently reported in the literature, and was generally reported in g.L^{-1}.day^{-1} or mg.L^{-1}.day^{-1}. This parameter could be calculated from volumetric biomass productivity (Q_V, in g.L^{-1}.day^{-1}) and lipid content (P in % DW) where appropriate data were available (Equation 4).

$$Q_P = Q_V \times P \tag{4}$$

The calculation of lipid productivity for the majority of species necessitated two assumptions:

1. Conversion of areal productivities (in g.m-2.day-1) to volumetric productivities (g.L-1.day-1), using an average depth of 0.1 m, based on best fit of the data
2. Conversion of specific growth rate to biomass productivity using an average biomass concentration of 0.15 g.L-1, based on typical experimental results.

The average literature values for the 55 species are shown in Table 4. Among the species with the highest reported lipid productivity were *Neochloris oleoabundans*,

Navicula pelliculosa, *Amphora*, *Cylindrotheca* and *Chlorella sorokiniana* (Fig. 3). Other findings were that green algae (Chlorophyta) generally showed an increase in lipid content when nitrogen deficient, whereas

Table 4. Growth and lipid parameters of 55 species of microalgae, along with their taxonomy and media type (adapted from Griffiths and Harrison, 2009).

			Lipid content			Biomass productivity			Lipid productivity	
			N replete	N deficient	T_d	Q_a	Q_v	Ave Q_v	Calculated	Literature
Species	Taxa[a]	Media[b]	% dw	% dw	days	$g.m^{-2}.day^{-1}$	$g.L^{-1}.day^{-1}$	$g.L^{-1}.day^{-1}$	$mg.L^{-1}.day^{-1}$	$mg.L^{-1}.day^{-1}$
Amphiprora hyalina	B	M	22	28	0.41			0.30	67	
Amphora	B	M	51		0.83	40.0		0.23	117	160
Anabaena cylindrica	Cy	F	5	5	1.00			0.10	5	
Ankistrodesmus falcatus	C	F	24	32	0.33	31.6	0.46	0.36	85	
Chaetoceros calcitrans	O	M	40				0.04	0.04	16	18
Chaetoceros muelleri	O	M	19	27	0.46		0.07	0.26	50	22
Chlamydomonas applanata	C	F	18	33						
Chlamydomonas reinhardtii	C	F	21		0.26			0.40	83	
Chlorella emersonii	C	F	29	63	0.80		0.03	0.08	23	
Chlorella minutissima	C	M	31	57	1.60		0.03	0.05	15	
Chlorella protothecoides	C	F	13	23	1.68			0.07	8	
Chlorella pyrenoidosa	C	F	16	64	0.28			0.47	76	
Chlorella sorokiniana	C	F	18	18	0.35		0.55	0.62	110	45
Chlorella vulgaris	C	F	24	42	0.70	10.7	0.11	0.16	40	30
Crypthecodinium cohnii	D	M	25		0.38			0.28	70	
Cyclotella cryptica	O	M	18	34	0.56			0.20	36	
Cylindrotheca	B	M	27	27	0.30			0.43	114	
Dunaliella primolecta	Pr	S	23	14		9.1		0.09	21	
Dunaliella salina	Pr	S	19	10	0.44			0.27	53	
Dunaliella tertiolecta	Pr	S	15	18	0.48			0.22	35	
Euglena gracilis	Eg	F	20	35	0.60			0.18	37	
Hymenomonas carterae	H	M	20	14	1.71			0.06	12	
Isochrysis galbana	H	M	25	29	0.89	11.5	0.16	0.15	37	38
Monodopsis subterranea	E	F	25	13			0.19	0.19	48	30
Monoraphidium minutum	C	F	22	52	0.35			0.30	65	
Nannochloris	C	M/F	28	30	0.49	31.9	0.23	0.27	74	77
Nannochloropsis	E	M	31	41	1.20		0.27	0.24	72	52
Nannochloropsis salina	E	M	27	46		13.9		0.14	38	
Navicula acceptata	B	F	33	35	0.42			0.29	96	
Navicula pelliculosa	B	F	27	45	0.23			0.46	124	
Navicula saprophila	B	F	24	51	0.38			0.28	68	
Neochloris oleoabundans	C	F	36	42			0.46	0.46	164	136
Nitzschia communis	B	M			0.96			0.18		
Nitzschia dissipata	B	M	28	46	0.39			0.27	73	
Nitzschia frustulum	B	M	26							
Nitzschia palea	B	M	47	40						48
Oscillatoria	Cy	F	7	13	0.28			0.37	27	
Ourococcus	C	F	27	50	3.01			0.03	9	
Pavlova lutheri	H	M	36				0.21	0.21	75	50
Pavlova salina	H	M	31				0.16	0.16	49	49
Phaeodactylum tricornutum	B	M	21	26	1.02	20.0	0.34	0.18	38	45
Porphyridium purpureum	R	M	11		0.45		0.23	0.23	24	35
Prymnesium parvum	H	M	30		0.74			0.14	42	
Scenedesmus dimorphus	C	F	26		0.46			0.23	57	
Scenedesmus obliquus	C	F	21	42	2.74		0.12	0.10	22	
Scenedesmus quadricauda	C	F	18				0.19	0.19	35	35
Selenastrum gracile	C	F	21	28						
Skeletonema costatum	O	M	16	25	0.66		0.08	0.15	24	17
Spirulina maxima	Cy	S	7		1.34			0.16	11	
Spirulina platensis	Cy	S	13	10	0.60	25.0		0.23	29	
Synechococcus	Cy	M	11		0.36			0.29	32	75
Tetraselmis suecica	P	M	17	26	1.51	28.1	0.59	0.39	65	32
Thalassiosira pseudonana	O	M	16	26	0.49		0.08	0.26	43	17
Thalassiosira weissflogii	O	M	22	24	0.58			0.18	41	
Tribonema	O	M	12	16	1.82		0.51	0.28	33	
			Average	Average	Average	Average	Average	Average	Average	Average
Total			23	32	0.80	22.2	0.23	0.23	52	51
Freshwater			21	36	0.82	21.2	0.24	0.26	54	35
Marine			25	31	0.82	22.7	0.21	0.21	49	47
Chlorophyta			23	41	1.01	24.7	0.24	0.25	58	65
Other taxa			25	30	0.72	20.4	0.24	0.23	57	50
Cyanobacteria			8	9	0.72	25.0		0.23	21	75

[a] Key to taxa: C = Chlorophyta, Cy = Cyanobacteria, D = Dinophyta, E = Eustigmatophyta, Eg = Euglenozoa, H = Haptophyta, O = Ochrophyta, Pr = Prasinophyta, [b] Key to media: F = Freshwater, M = Marine, S = Saline

The average of literature values for lipid content under nitrogen (N) replete and deficient growth conditions, doubling time (T_d), and areal (Q_A) and volumetric (Q_V) biomass productivities are shown in columns 4 to 8. Average biomass productivity calculated from T_d, μ, Q_A and Q_V is shown in column 9, and calculated and literature lipid productivity in columns 10 and 11 respectively. Blanks represent no data available diatoms and other taxa were more variable in their response, although all those subjected to silicon deprivation showed an increase in lipid content. This increase in lipid content, however, does not necessarily translate into increased lipid productivity due to decreased growth rates under nutrient stress conditions. Response of biomass productivity to nutrient deprivation is variable between species and further investigation is necessary.

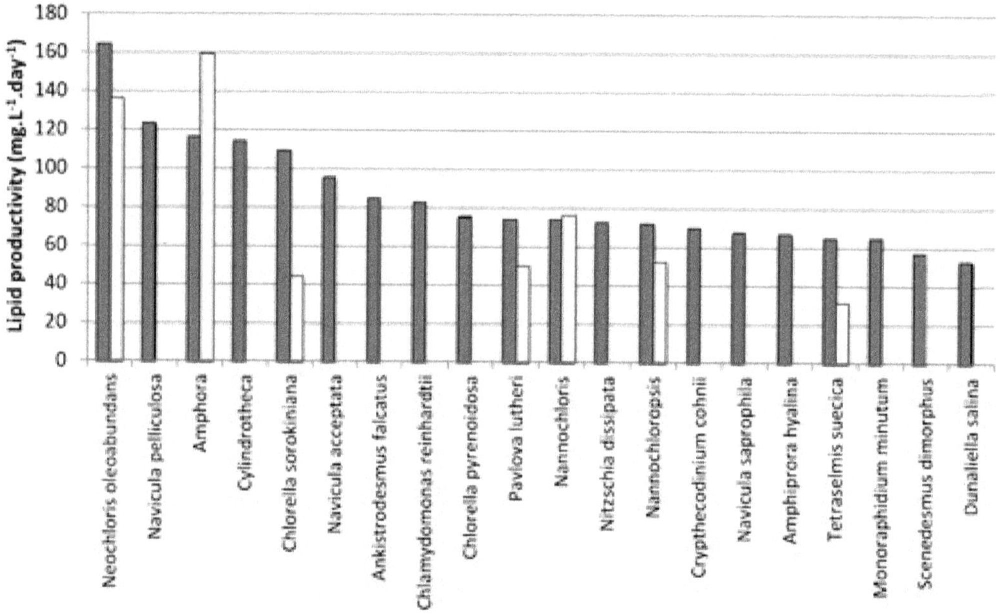

Fig. 3. Average calculated (grey bars) and literature (empty bars) biomass productivity for the 20 most productive species investigated (adapted from Griffiths & Harrison, 2009)

In Fig. 4, the impact of biomass productivity and lipid content on calculated lipid productivity is analyzed through correlation. A relationship is demonstrated between lipid productivity and biomass productivity. All species with a high biomass productivity (above 0.4 $g.L^{-1}.day^{-1}$), and all but one above 0.3 $g.L^{-1}.day^{-1}$, have a high lipid productivity, greater than 60 $mg.L^{-1}.day^{-1}$. However, there are a few species with high lipid productivity despite an average biomass productivity, indicating that lipid content is also a factor. Lipid content correlates poorly with lipid productivity, indicating that lipid content alone is not a good indicator of suitability for biodiesel production. There are several species with low lipid productivity despite an above- average lipid content (> 22%). The species with high lipid productivities (> 60 mg.L^{-1}.day^{-1}) range in lipid content from 16% DW to 51%. Further, species with high lipid content (> 30%) vary in lipid productivity between 15 and 164 mg.L^{-1}.day^{-1}.

Once the species has been chosen, the next critical factor is the optimisation of culture conditions. In addition to optimal temperature and pH, conditions that maximize autotrophic growth rate are optimal light, carbon and nutrient supply. Microalgal lipid accumulation is affected by a number of environmental factors (Guschina & Harwood 2006; Roessler 1990), and often enhanced by conditions that apply a 'stress' to the cells. Lipids appear to be synthesised in response to conditions when energy input (rate of photosynthesis) exceeds the capacity for energy use (cell growth and division) (Roessler 1990). Enhanced cell lipid content has been found under conditions of nutrient deprivation (Hsieh & Wu, 2009; Illman et al., 2000; Li et al., 2008; Shifrin & Chisholm, 1981; Takagi et al., 2000), high light intensity (Rodolfi et al., 2009), high temperature (Converti et al., 2009); high salt concentration (Takagi et al., 2000) and high iron concentration (Liu et al., 2008).

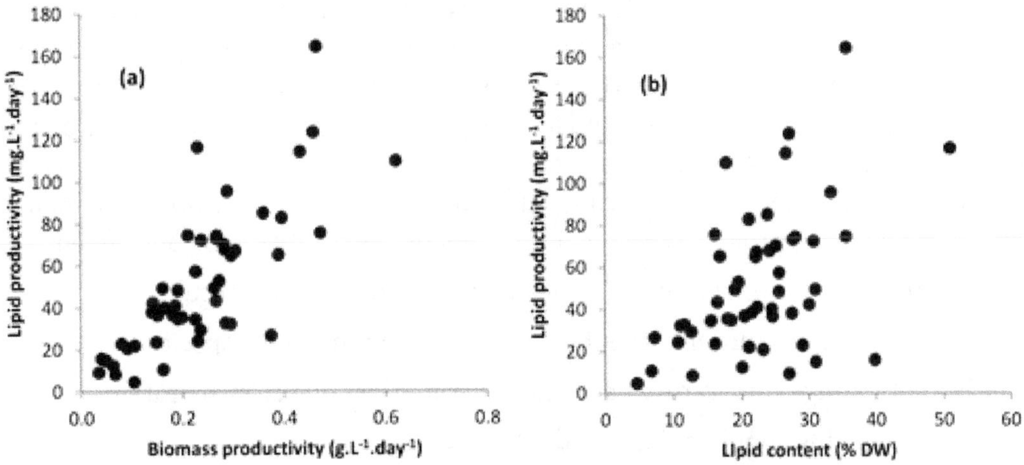

Fig. 4. Correlation of calculated lipid productivity with (a) biomass productivity and (b) lipid content under nutrient replete conditions

Nitrogen (N) deprivation is the most frequently reported method of enhancing lipid content, as it is cheap and easy to manipulate. N deficiency has a reliable and strong influence on lipid content in many species (Chelf, 1990; Rodolfi et al., 2009; Shifrin & Chisholm, 1981). Unfortunately, stress conditions that enhance lipid content, such as nitrogen deprivation, typically also decrease the growth rate, and thus the net effect on lipid productivity must be ascertained (Lardon et al., 2009). Maximum biomass productivity and lipid content in *Chlorella vulgaris* occur under different conditions of nitrogen availability, suggesting that a two-stage cultivation strategy may be advantageous. From studies we have conducted on *C. vulgaris*, it appears that an intermediate level of nitrogen limitation creates the optimum balance between biomass and lipid production. The optimum cultivation strategy tested was batch culture, using a low starting nitrate concentration (between 250 and 300 $mg.L^{-1}$ nitrate), ensuring that nitrogen in the medium was depleted towards the end of exponential growth. Other cultivation strategies (e.g. two-stage batch, fed-batch or continuous) were found not to improve upon the productivity achieved in N limited batch culture.

Although high lipid productivity is a key factor in species selection, other characteristics such as ease of cultivation, tolerance of a range of environmental conditions (particularly temperature and salinity), flue-gas contaminants and high O_2 concentrations, as well as resistance to contaminants and predators are likely to be equally as important.

9. CONCISE SUMMARY

Algal biodiesel remains a promising, sustainable, and carbon-neutral source of transportation fuel. The technical viability of algal biodiesel has been established (Miao & Wu, 2006; Xiong et al., 2008), although the economic aspects and energy requirements of production necessitate significant enhancement. The necessary changes appear attainable through the enhancement of productivity, the reduction of cost and energy demand for key processes and the application of the biorefinery concept (co-production of valuable products or processes). Current research is focussed on achieving this through a combination of biological and engineering approaches. The major challenges currently being addressed are:

- Increasing productivity in large-scale outdoor microalgal culture
- Minimizing contamination by predators and other algal species
- Mitigating temperature changes and water loss due to evaporation
- Optimizing supply of light and CO_2
- Developing cheap and efficient reactor designs
- Developing cost and energy-efficient methods of harvesting dilute suspensions of small microalgal cells
- Decreasing the overall energy and cost requirements, particularly for pumping, gas transfer, mixing, harvesting and dewatering
- Improving resource utilization and productivity through a biorefinery approach
- Producing valuable co-products
- Decreasing environmental footprint through recycling of water, energy and nutrients. These topics have captured the imagination of several researchers and some innovative solutions are being investigated.

> The overall goal of biofuel production is to optimise the conversion of sunlight energy to liquid fuel. In algal cultivation, techniques to improve light delivery include manipulating the reactor design, the use of optics to deliver light to the centre of the reactor, optimising fluid dynamics to expose all cells to frequent light flashes, increasing the efficiency of photosynthesis and carbon capture (e.g. enhancing the carbon concentrating mechanism), and using mixed-species cultures to utilise different intensities or wavelengths of light (Scott et al., 2010).

One of the major problems with light delivery is poor penetration of light into dense cultures due to mutual shading by the cells. Under high light conditions, microalgal cells absorb more light than they can use, shading those below them and dissipating the excess energy as fluorescence or heat. In nature, this confers individual cells an evolutionary advantage, however, in mass production systems it is undesirable as it decreases overall productivity. It would be advantageous to minimize the size of the chlorophyll antennae in cells at the surface, so as to permit greater light penetration to cells beneath (Melis, 2009). Reducing the size of the light harvesting complexes through genetic modification has been shown to improve productivity (Nakajima et al., 2001). The goal now is to engineer cells that change antennae size according to light intensity.

Although the TAG content of cells can be enhanced by manipulation of the nutrient supply, there is a tradeoff between growth and lipid production. For optimum productivity, cells that can maintain a simultaneously high growth rate and lipid content are required. Strategies to achieve this include screening for novel species, and genetic engineering of well characterised strains. The genes and proteins involved in regulation of lipid production pathways are currently being investigated through synthetic biology and the modelling of carbon flux through metabolism. Key enzymes and branch-points can then be manipulated to improve productivity. For example, carbohydrate and lipid production compete directly for carbon precursors. Shunting carbon away from starch synthesis by down regulation of the enzyme ADP-glucose pyrophosphorylase in *Chlamydomonas* has been shown to enhance TAG content 10-fold (Li et al., 2010).

The challenge of harvesting small algae cells from dilute suspensions has yet to be solved in a cheap, energy efficient manner. Ideally the addition of chemical agents that impede the recycling of the culture medium and nutrients should be avoided. A series of methods is likely to be used e.g. flocculation followed by sedimentation, or settling followed by centrifugation. Promising ideas for harvesting techniques include concentration using sound waves and triggering of autoflocculation on command. Another attractive idea is direct product excretion, where algae secrete fuel molecules into the medium as they are produced, allowing continuous production and harvesting without cell disruption. The Cyanobacterium *Synechocystis* has recently been successfully modified to excrete fatty acids (Liu, 2011).

The utilization of nutrients from waste sources (e.g., CO2 from flue gas and nitrate and phosphate from wastewater) may diminish prices and energy consumption while also aiding in environmental restoration. Possible co-products encompass fine chemicals such as astaxanthin, beta-carotene, omega-3 fatty acids, polyunsaturated fatty acids, nutraceuticals, therapeutic proteins, cosmetics, aquafeed, and animal feed (Mata et al., 2010). Algae may be engineered to produce alternative fuels such as ethanol, butanol, isopropanol, and hydrocarbons (Radakovits et al., 2010), or the downstream processing of algae might be adapted to convert the entire biomass into energy-rich fuels using thermal methods.

10. REFERENCES

An, J-Y, Sim, S-J, Lee JS & Kim BW. (2003). Hydrocarbon production from secondarily treated piggery wastewater by the green alga *Botryococcus braunii*. *Journal of Applied Phycology*, 15(2-3): 185-191

Anderson, RA. (2005). *Algal Culturing Techniques*. Elsevier, London

Bailey, J & Ollis, D. (1977). *Biochemical Engineering Fundamentals*. McGraw-Hill, New York Barsanti, L & Gualtieri, P. (2006). *Algae: Anatomy, Biochemisty and Biotechnology*. CRC Press, Taylor and Francis Group, Florida

Becker, EW. (1994). *Microalgae: Biotechnology and Microbiology*. Cambridge University Press, Cambridge

Ben-Amotz, A. (1995). New Mode of *Dunaliella* Biotechnology: Two-Phase Growth for B- Carotene Productio. *Journal of Applied Phycology*, 7(1):65-68

Benemann, J, Koopman, B, Weissman, J, Eisenberg, D & Gobell R. (1980). Development of Microalgae Wastewater Treatment and Harvesting Technologies in California. In *Algae Biomass: Production and Use*. Edited by Shelef, G & Soeder, C. pp. 457-495 . Elsevier, Amsterdam

Bernhardt, H & Clasen, J. (1994). Investigations into the Flocculation Mechanisms of Small Algal Cells. *Journal of Water Supply: Research and Technology - AQUA*, 43(5): 222-232

Bitton, G, Fox, JL & Strickland, HG. (1975). Removal of Algae from Florida Lakes by Magnetic Filtration. *Applied and Environmental Microbiology*, 30(6):905-908

Borowitzka, MA. (1992). Algal Biotechnology Products and Processes - Matching Science and Economics. *Journal of Applied Phycology*, 4:267-279

Borowitzka, MA. (1997). Microalgae for Aquaculture - Opportunities and Constraints. *Journal of Applied Phycology*, 9(5):393-401

Bosma, R, van Spronsen, WA, Tramper, J & Wijffels, RH. (2003). Ultrasound, a New Separation Technique to Harvest Microalgae. *Journal of Applied Phycology*, 15(2- 3):143-153

Brennan, L & Owende, P. (2010). Biofuels from microalgae: A Review of Technologies for Production, Processing, and Extractions of Biofuels and Co-Products. *Renewable and Sustainable Energy Reviews*, 14(2):557-577

Carvalho, AP, Meireles, LA & Malcata, FX. (2006). Microalgal Reactors: A Review of Enclosed System Designs and Performances. *Biotechnology Progress*, 22(6):1490-1506

Ceron Garcia, MC, Fernandez Sevilla, JM, Acien Fernandez, FG, Molina Grima, E & Garcia Camacho, F. (2000). Mixotrophic Growth of *Phaeodactylum tricornutum* on Glycerol: Growth Rate and Fatty Acid Profile. *Journal of Applied Phycology*, 12:239-248

Chelf, P. (1990). Environmental Control of Lipid and Biomass Production in Two Diatom Species. *Journal of Applied Phycology*, 2(2):121-129

Chen, F. (1996). High Cell Density Culture of Microalgae in Heterotrophic Growth. *Trends in Biotechnology*, 14(11):421-426

Chen, YM, Liu, JC & Ju, Y. (1998). Flotation Removal of Algae from Water. *Colloids and Surfaces B: Biointerfaces*, 12(1):49-55

Chisti, Y. (2007). Biodiesel from Microalgae. *Biotechnology Advances*, 26(3):294-306

Chisti, Y. (2008). Biodiesel from Microalgae Beats Bioethanol. *Trends in Biotechnology*, 25(3):126-31

Converti, A, Casazza, A, Ortiz, EY, Perego, P & Del Borghi, M. (2009). Effect of Temperature and Nitrogen Concentration on the Growth and Lipid Content of Nannochloropsis Oculata and *Chlorella vulgaris* for Biodiesel Production. *Chemical Engineering and Processing: Process Intensification*, 48(6):1146-1151

Csordas, A. (2004). An Integrated Photobioreactor and Foam Fractionation Unit for the Growth and Harvest of *Chaetoceros* spp. in Open Systems. *Aquaculture Engineering*, 30(1-2):15-30

Danquah, MK, Gladman, B, Moheimani, N & Forde, GM. (2009). Microalgal Growth Characteristics and Subsequent Influence on Dewatering Efficiency. *Chemical Engineering Journal*, 151(1-3):73-78

De la Pena, M. (2007). Cell Growth and Nutritive Value of the Tropical Benthic Diatom, *Amphora* sp., at Varying Levels of Nutrients and Light Intensity, and Different Culture Locations. *Journal of Applied Phycology*, 19(6):647-655

Dennis, DT, Turpin, DH, Lefebvre, DD & Layzell, DB. (1998). *Plant metabolism*. Longman, Singapore

Fattom, A & Shilo, M. (1984). Hydrophobicity as an Adhesion Mechanism of Benthic Cyanobacteria. *Applied and Environmental Microbiology,* 47(1):135-143

Golueke, CG, Oswald, WJ & Gotaas, HB. (1957). Anaerobic Digestion of Algae. *Applied Microbiology*, 5(1):47-55

Griffiths, MJ & Harrison, STL. (2009). Lipid Productivity as a Key Characteristic for Choosing Algal Species for Biodiesel Production. *Journal of Applied Phycology*, 21(5):493-507

Grobbelaar, JU. (1994). Turbulence in Mass Algal Cultures and the Role of light/ dark Fluctuations. *Journal of Applied Phycology,* 6(3):331-335

Grobbelaar, JU. (2000). Physiological and Technological Considerations for Optimising Mass Algal Cultures. *Journal of Applied Phycology*, 12:201-206

Gudin, C & Therpenier, C. (1986). Bioconversion of Solar Energy into Organic Chemicals by Microalgae. *Advanced Biotechnology Processes,* 6:73-110

Harun, R, Singh, M, Forde, GM & Danquah, MK. (2010). Bioprocess Engineering of Microalgae to Produce a Variety of Consumer Products. *Renewable and Sustainable Energy Reviews* 14(3):1037-1047

Henderson, R, Parsons, SA & Jefferson, B. (2008). The Impact of Algal Properties and Pre- Oxidation on Solid/Liquid Separation of Algae. *Water Research*, 42:1827-1845

Hodaifa, G, Martínez, M & Sánchez, S. (2008). Use of industrial wastewater from olive-oil extraction for biomass production of *Scenedesmus obliquus*. *Bioresource Technology*, 99(5):1111-1117

Hsieh, C & Wu, W. (2009). Cultivation of Microalgae for Oil Production with a Cultivation Strategy of Urea Limitation. *Bioresource Technology*, 100(17):3921-3926

Huntley, ME & Redalje, DG. (2006). CO_2 Mitigation and Renewable Oil from Photosynthetic Microbes: A New Appraisal. *Mitigation and Adaptation Strategies for Global Change,* 12(4):573-608

Illman, A, Scragg, AH & Shales, SW. (2000). Increase in *Chlorella* Strains Calorific Values when Grown in Low Nitrogen Medium. *Enzyme MicrobialTechnology*, 27(8):631-635

Lee, SJ, Kim, S, Kim, J, Kwon, G, Yoon, B & Oh, H. (1998). Effects of Harvesting Method and Growth Stage on the Flocculation of the Green Alga *Botryococcus braunii. Letters in Applied Microbiology*, 27(1):14-18

Jameson, GJ. (1999). Hydrophobicity and Floc Density in Induced-Air Flotation for Water Treatment. *Colloids and Surfaces A: Physicochemical and engineering aspects,* 151:269- 281

Jarvis, P, Buckingham, P, Holden, B & Jefferson, B. (2009). Low Energy Ballasted Flotation. *Water Research*, 43:3427-3434

Knuckey, R, Brown, M, Robert, R & Frampton, D. (2006). Production of Microalgal Concentrates by Flocculation and their Assessment as Aquaculture Feeds. *Aquaculture Engineering*, 35(3):300-313

Lardon, L, Heì lias, A, Sialve, B, Steyer, JP & Bernard, O. (2009). Life-Cycle Assessment of Biodiesel Production from Microalgae. *Environmental science & technology*, 43(17):6475-6481

Lee, AK, Lewis, DM & Ashman, PJ. (2008). Microbial Flocculation, a Potentially Low-Cost Harvesting Technique for Marine Microalgae for the Production of Biodiesel. *Journal of Applied Phycology*, 21(5):559-567

Li, Y, Han, D, Hu, G, Dauvillee, D, Sommerfeld, M, Ball, S & Hu, Q. (2010). Chlamydomonas Starchless Mutant Defective in ADP-Glucose Pyrophosphorylase Hyper-Accumulates Triacylglycerol. *Metabolic Engineering*, 12(4):387-91

Li, Y, Horsman, M, Wang, B, Wu, N & Lan, CQ. (2008). Effects of Nitrogen Sources on Cell Growth and Lipid Accumulation of Green Alga Neochloris Oleoabundans. *Applied microbiology and biotechnology*, 81(4):629-36

Liu, X, Sheng, J & Curtiss, R. (2011). Fatty Acid Production in Genetically Modified Cyanobacteria. *Proceedings of the National Academy of Sciences of the United States of America*, 108(17):6899-904

Liu, Z, Wang, G & Zhou, B. (2008). Effect of Iron on Growth and Lipid Accumulation in *Chlorella vulgaris*. *Bioresource Technology*, 99(11):4717-22

Livne, A & Sukenik, A. (1992). Lipid Synthesis and Abundance of Acetyl CoA Carboxylase in *Isochrysis galbana* (Prymnesiophyceae) Following Nitrogen Starvation. *Plant and Cell Physiology*, 33(8):1175-1181

Lv, J, Cheng, L, Xu, X, Zhang, L & Chen, H. (2010). Enhanced Lipid Production of *Chlorella vulgaris* by Adjustment of Cultivation Conditions. *Bioresource Technology*, 101(17):6797-804.

Mata, TM, Martins, AA & Caetano, NS. (2010). Microalgae for Biodiesel Production and Other Applications: A Review. *Renewable and Sustainable Energy Reviews*, 14(1):217-232

Melis, A. (2009). Solar Energy Conversion Efficiencies in Photosynthesis: Minimizing the Chlorophyll Antennae to Maximize Efficiency. *Plant Science*, 177(4):272-280

Miao, X & Wu, Q. (2006). Biodiesel Production from Heterotrophic Microalgal Oil. *Bioresource Technology*, 97(6):841-846

Mittelbach, M & Remschmidt, C. (2004). *Biodiesel - The Comprehensive Handbook.* Martin Mittelbach

Mohn, FH. (1980). Experiences and Strategies in the Recovery of Biomass from Mass Cultures of Microalgae. In *Algae biomass: Production and Use*. Edited by Shelef, G & Soeder, C. pp. 547-571. Elsevier, Amsterdam

Molina Grima, E, Belarbi, E, Acien Fernandez, FG, Robles Medina, A & Chisti, Y. (2003). Recovery of Microalgal Biomass and Metabolites: Process Options and Economics. *Biotechnology Advances,* 20(7-8):491-515

Nakajima, Y, Tsuzuki, M &Ueda, R. (2001). Improved Productivity by Reduction of the Content of Light-Harvesting Pigment in *Chlamydomonas perigranulata. Journal of Applied Phycology,* 13(2):95-101

Nedbal, L, Tichy,V, Xiong, F & Grobbelaar, JU. (1996). Microscopic Green Algae and Cyanobacteria in High-Frequency Intermittent Light. *Journal of Applied Phycology,* 8(4-5):325-333

Oswald, WJ & Golueke, C. (1960). Biological Transformation of Solar Energy. *Advanced Applied Microbiology*, 2:223-262

Petrusevski, B, Bolier, G, van Breemen, A & Alaerts, GJ. (1995). Tangential Flow Filtration: A Method to Concentrate Freshwater Algae. *Water Research,* 29:1419-1424

Piorreck, M, Baasch, K & Pohl, P. (1984). Biomass Production, Total Protein, Chlorophylls, Lipids and Fatty Acids of Freshwater Green and Blue-Green Algae Under Different Nitrogen Regimes. *Phytochemistry,* 23(2):207-216

Pulz, O. (2001). Photobioreactors: Production Systems for Phototrophic Microorganisms. *Applied Microbiology and Biotechnology*, 57(3):287-293

Pushparaj, B, Pelosi, E, Torzillo, G & Materassi, R. (1993). Microbial Biomass Recovery using a Synthetic Cationic Polymer. *Bioresource Technology*, 43(1):59-62

Radakovits, R, Jinkerson, RE, Darzins, A & Posewitz, MC. (2010). Genetic Engineering of Algae for Enhanced Biofuel Production. *Eukaryotic Cell,* 9(4):486-501

Ramos, MJ, Fernandez, CM, Casas, A, Rodriguez, L & Perez, A. (2009) Influence of Fatty Acid Composition of Raw Materials on Biodiesel Properties. *Bioresource Technology*, 100(1):261-268

Regan, DL & Gartside, G. (1983). *Liquid Fuels from Microalgae in Australia.* CSIRO, Melbourne Richardson, C. 2011. *Investigating the role of reactor design to maximise the environmental benefit of algal oil for biodiesel.* Master's thesis. University of Cape Town.

Richmond, A. (2004). *Microalgal culture: biotechnology and applied phycology*. Blackwell Science, Oxford

Rodolfi, L, Chini Zittelli, G, Bassi, N, Padovani, G, Biondi, N, Bonini, G & Tredici, MR. (2009). Microalgae for Oil: Strain Selection, Induction of Lipid Synthesis and

Outdoor Mass Cultivation in a Low-Cost Photobioreactor. *Biotechnology and Bioengineering*, 102(1):100-112

Roessler, P. (1990). Environmental Control of Glycerolipid Metabolism in Microalgae: Commercial Implications and Future Research Directions. *Journal of Phycology*, 26:393-399

Sazdanoff, N. (2006). *Modeling and simulation of the algae to biodiesel fuel cycle.* Honour's thesis, Ohio State University

Schenk, PM, Thomas-Hall, S, Stephens, E, Marx, UC, Mussgnug, JH, Posten, C, Kruse, O & Hankamer, B. (2008). Second Generation Biofuels: High-Efficiency Microalgae for Biodiesel Production. *BioEnergy Research* 1(1):20-43

Scott, SA, Davey, MP, Dennis, JS, Horst, I, Howe, CJ, Lea-Smith, D & Smith, AG. (2010). Biodiesel from Algae: Challenges and Prospects. *Current Opinions in Biotechnology*, 21:277-286

Sheehan, J, Dunahay, T, Benemann, J & Roessler, P. (1998). *A Look Back at the U.S. Department of Energy's Aquatic Species Program: Biodiesel from Algae.* Close-Out report. National Renewable Energy Lab, Department of Energy, Golden, Colorado, U.S.A. Report number NREL/TP-580-24190, dated July 1998

Shelef G, Sukenik A, Green M (1984) *Microalgae Harvesting and Processing: a Literature Review.* SERI report.

Shifrin, NS & Chisholm, SW. (1981). Phytoplankton Lipids: Interspecific Differences and Effects of Nitrate, Silicate and Light-Dark Cycles. *Journal of Phycology*, 374-384

Shuler, ML & Kargi, F. (2005). *Bioprocess Engineering: Basic Concepts.* Pearson Education, Singapore

Spoehr, HA & Milner, HW. (1949). The Chemical Composition of *Chlorella*, Effect of Environmental Conditions. *Plant Physiology*, 24:120-149

Stansell, G, Gray, VM & Sym, S. (2011). Microalgal Fatty Acid Composition: Implications for Biodiesel Quality. *Journal of Applied Phycology,* In press. DOI: 10.1007/s10811-011- 9696-x

Takagi, M, Watanabe, K, Yamaberi, K & Yoshida, T. (2000). Limited Feeding of Potassium Nitrate for Intracellular Lipid and Triglyceride Accumulation of *Nannochloris* sp. UTEX LB1999. *Applied Microbiology and Biotechnology*, 54(1):112-7

Tsukahara, K & Sawayama, S. (2005). Liquid Fuel Production using Microalgae. *Journal of the Japan Petroleum Institute,* 48(5):251-259

Uduman, N, Qi, Y, Danquah, MK, Forde, GM & Hoadley, A. (2010). Dewatering of Microalgal Cultures: A Major Bottleneck to Algae-Based Fuels. *Journal of Renewable and Sustainable Energy,* 2(1)

Van Gerpen, J. (2005). Biodiesel Processing and Production. *Fuel Processing Technology* 86:1097-1107

Vlaski, A, van Breemen, A & Alaers, G. (1997). The Role of Particle Size and Density in Dissolved Air Flotation and Sedimentation. *Water Science and Technology,* 36(4):177-189

Xiong, W, Li, X, Xiang, J & Wu, Q. (2008). High-Density Fermentation of Microalga *Chlorella protothecoides* in Bioreactor for Microbio-Diesel Production. *Applied Microbiology and Biotechnology*, 1:29-36

Zhang, X, Hu, Q, Sommerfeld, M, Puruhito, E & Chen, Y. (2010). Harvesting Algal Biomass for Biofuels using Ultrafiltration Membranes. *Bioresource Technology*, 101(14):5297-304

10 Chapter

AN INTEGRATED WASTE-FREE BIOMASS UTILIZATION SYSTEM FOR AN INCREASED PRODUCTIVITY OF BIOFUEL & BIOENERGY

1. INTRODUCTION: AN OVERVIEW

The escalation in the production and exploitation of biomass and other renewable energy sources constitutes a significant challenge for the energy sector. It, however, necessitates expectations for ecologically and economically viable manufacturing processes. We present a comprehensive system of established and novel technologies designed for the conversion of biomass into biofuels. This encompasses conventional and biobutanol-based biodiesels, biogas and electricity generation, as well as an agricultural production system utilizing biomass power plant waste for fertilizer. Three types of agricultural production systems are essential for the agricultural component of the integrated system, notably:

A – plants for combustion in biomass power plants (energy grass) B – plants for production of vegetable oils for biodiesel production

C – plants for conversion of sugar derivatives to price alcohols, mainly butanol as a diesel fuel source

Depending on the climate, the soil type, the agricultural experiences, and the type of the plants (A,B,C), the produced biomass materials can fulfill more than one requirement as it can be seen in Fig. 1. Depending on the constituents of the biomass (cellulose, starch, lignin, oil, proteins), the energy production can be performed via direct combustion or, after digestion in biogas systems, by using the biogas. The biomass power plants, biogas combustion plants/engines produce hot water, steam and electricity. In plants type B soybean, rape, sunflower or likes are pressed to obtain the oil, while the pressing cake can be used as optimal raw material for biogas plants due to its high protein content, while the stalk can also be used as solid fuel (after drying with the low heating value warm water) in biomass power plants. Typically, green biomass can be employed in biogas facilities, whereas dried biomass serves as solid fuel for combustion in power plants. The byproducts of sugar derivative-producing plants (such as sugar sorghum and corn) can provide a biomass power plant with their desiccated stalks. The comprehensive waste management in these energy-generating units and the recycling of other byproducts (potassium sulfate, calcium sulfate, or biomass power plant ash) within the integrated system as fertilizers for agriculture promote sustainable biomass and fuel production.

2. ENERGY ASPECTS OF BIOMASS UTILIZATION

The intensive production of biomass as raw material for fuel or bioenergy production would lead to fast exhausting of the soil and dramatic increasing the production costs without intensive fertilization. Except nitrogen, all of the nutrients (P, K) and microelements can be recycled by reprocessing the residues of the biomass work up or biomass utilizing energy producing technologies. The nitrogen fertilization, however, always requires fossil energy source, since the base material of the two most typical nitrogen fertilizer (ammonia and urea) is the natural gas. It is one of the main reasons of the opposite statements about energy intake and output balance of the biomass based fuel and energy productions. Otherwise, the conversion of the waste to fertilizers (to supply other elements like P, K, S and microelements) should also be the integrated part of the sustainable and economic biomass production system.

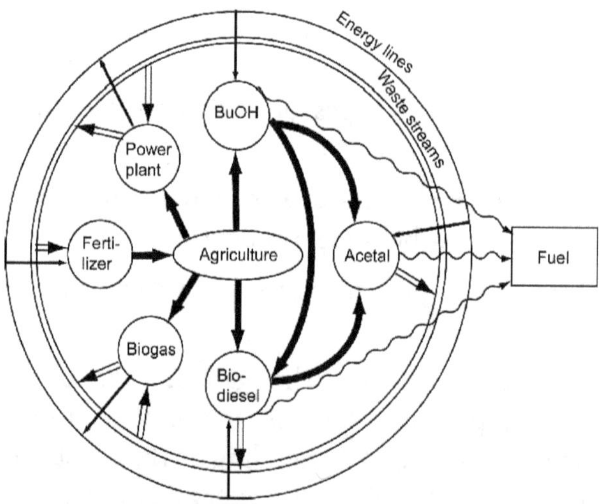

Fig. 1. Waste free biofuel production cycle

The review of the energy inputs required for the production of the raw materials like corn, switchgrass or sunflower, or the cost of the biofuel production form these agricultural materials (Pimental and Petzek, 2005) unambiguously show, that the climate (crop amount), type of the agricultural plant and the type of the processing technologies basically determines the feasibility of an energy positive biofuel production. The energy saving in the production of the various biofuel components from various sources is compared to the production costs related to petrochemical raw materials (Arlie, 1983) is given in Table 1. Selection of the biomass produced and the fuel type prepared from this biomass requires that environmental, economic and energetic viewpoints (Hill et al., 2006) should also be taken into consideration. The social-economic viewpoints play also key role in this decision. That is obvious, that there is not any type of biomass plant which could completely be turned into biofuel, only integrated systems can solve this problem, when more than one type of biomass plants are in synchronized operation. More than one energy producing system is used to utilize the various type of green biomass, and more than one type of biofuel are produced from the various parts of each type of biomass plants. At the same time the energy producing plants or the waste producing and nutrient reprocessing plant (fertilizer plant) can use the wastes of each technological step as raw material to ensure the recyclability.

Table 1. Energy gain of sugar-based biofuel components (ton of oil equivalent/ton)

Product	Petrochemical source	Biomass	Energy gain
MeOH	Natural gas	Wood	0.60
EtOH	Ethylene	Sugarcane	0.73
		Artichoke	0.88
		Corn	0.65
n-BuOH	Propylene	Sugarcane	0.10
		Artichoke	0.75
		Wheat straw	0.80
Acetone	Propylene	Sugarcane	0.80
		Artichoke	1.45
		Wheat straw	1.51

Since the energetically favorable components as BuOH, EtOH and acetone can be produced from biomass the use of these in biofuel including biodiesel production seems to be essential and unavoidable step. The EU biofuel standards now declare that rapeseed oil and ethanol as raw materials are permitted and standardized as biofuels within European Community. Due to the limits of the agricultural area and the productivity of rape in this climate and low energy content of ethanol, however, requires changing this statement and the use of other type of biofuels should be also permitted. Thus, in our integrated system we incorporate new type of blend materials, mainly butanol for replacing the ethanol, and new kind of blends are produced from the wastes of biodiesel and biobutanol production as well. In order to increase the amount of biodiesel produced from one unit of vegetable oil, butyl ester production is suggested instead of the methyl esters, and butoxylation or blending with pure butanol are also possible increments in increasing the efficiency of the biofuel production related to one hectare of the agricultural area.

3. NEW TRENDS IN BIODIESEL TECHNOLOGIES

The new developments in biodiesel production technologies are focused on simplification of the existing technologies by using co-solvents (Guan et al., 2009b), supercritical solvents (Geuens et al., 2008), new catalysts (including heterogeneous and biocatalysts) (Fu and Vasudevan, 2009; Soriano et al., 2009; Leung et al., 2010) and the efforts are directed on founding new blends and raw materials (including the waste glycerol (Guerrero-Perez, et. al., 2009)) to increase the amount of biofuel produced on a given agricultural area. Some of the improved methods are discussed below.

3.1 Trans-esterification with phase mixing materials

Base catalyzed trans-esterification of the vegetable oils with methanol is a slow process in the two-phase system due to the mass transfer limitation. The butanolysis is much faster because it takes place in homogeneous phase. Therefore, the possibilities to create a homogeneous mixture by using an appropriate co-solvent which homogenizes the vegetable oil and methanol has extensively been studied (Guan et al., 2009a; 2009b; Leung et al., 2010; Meng et al., 2009; Soiano et al., 2009). The best solvents are the ether-type compounds like THF, methyl tertiary butyl ether, Me_2O or diethers. Phase diagrams of different vegetable oils - ethereal solvents (1,4-dioxane, THF, Et_2O, diisopropyl ether and MTBE)-methanol have been studied (Boocok et al., 1996a) and THF was found to be the preferred solvent on the basis of the volume required by the miscibility and boiling point considerations. Initially the reactions are very fast , but they slow down drastically due to the polarity changes in the mixed phase containing esters and decreasing amount of alcohol [Boocock et al., 1998). The decreasing polarity of the intermediate mixture could be avoided only by using large excess of methanol (methanol/oil molar ratio 27) when the methyl ester production exceed 94.4 % in 7 min even at room temperature. The separation of the glycerol phase in the presence of THF was much more rapid as in the case of two-phase trans-esterification reaction mixtures (Boocock et al., 1996a). The use of cyclic THF or 1,4-dioxane provides a possibility to decrease the amount of the co-solvent and perform the trans-esterification at higher oil : co- solvent ratios. The THF and the dioxane are miscible with vegetable oils and methanol in any proportion and they have hydrogen bonding ability (Boocock et al., 1996b). In spite of that 5 % dioxane ensures almost complete reaction within 30 min at room temperature at a 6:1 methanol : oil molar ratio, the 1,4-dioxane ring opens up during the trans-esterification reaction at the presence of 1 % KOH catalyst, thus it cannot be recycled at all.

Dimethyl ether can be separated easily with depressurizing the reaction system and Me_2O as a polar compound ensures a sufficiently high trans-esterification reaction rate at room temperature. The effects of the reaction conditions on the reaction time indicated that under the usual conditions, such as 1 % of KOH and 2-fold molar excess of methanol (6:1 methanol : oil molar ratio), the reactions can be performed within 5-10 min which would provide a good technological base for the continuous production of biodiesels (Guan et al, 2009a and 2009b). The gaseous state of form of the Me_2O requires pressurized reactors and the risk of explosion is very high.

Recently a new technology has been developed (Kótai et al., 2008) for the phase-mixing of methanol (or other alcohols) and vegetable oils in trans-esterification reactions with alkoxyalkanol phase transfer agents. These are hemi-ethers of glycols. They are polar end- group, but they have oil-soluble alkyl chain as well. The best candidate is the butylglycol (2- butoxy-ethanol), which can act at even 1 % concentration, and at the 6:1 MeOH : rapeseed oil molar ratio with 1 % KOH a catalyst. The reaction is almost completed within 30 min even at room temperature. By using 5 % of butylglycol, the reaction time is 5-10 min. Since the butyl glycol acts also as an alcohol (not only as an ether), thus not only methyl esters but 2-butoxyethyl esters – $R\text{-}C(O)\text{-}O\text{-}C_2H_4\text{-}OC_4H_9$ - are also formed. These esters are formed in an amount of 2-3 %w/w and act as fuel components. In this way, the phase mixing agents built into the ester phase and they contribute to the mass of the biodiesel and do not need to recover it which simplifies the production technology. The catalyst solution prepared from KOH, butylglycol and methanol was used in our plant scale experiment performed in Hungary in 2009, when a continuous trans-esterification process with continuous separator was put into operation. The method could be combined with the ion exchange type removal and recycling of the neutralization agent ($KHSO_4$, chapter 3. 2), because the reaction takes place at room temperature and the amount of soaps formed and appeared in the ester phase was very small. The catalyst distribution between the ester and glycerol phase is around 2:98. The ester phase has been neutralized with $KHSO_4$, when the potassium content is decreased with the continuous operation mode below 50 ppm without any further washing. The further purification steps, washing with water and removing the residual MeOH in vacuum, are the same as in the classical biodiesel technologies, however, the amount of the dissolved MeOH, due to the lack of soaps and residual catalyst is much lower than in the usual technologies. The flow-sheet of the technology is shown in Fig 2.

In order to decrease the length of the tube reactor and the residence time of the mixture in the apparatus, a two-stage trans-esterification seems to be the most reliable, when after decreasing the rate of the reaction, after the first separator a further amount of methanol and catalyst are added, when the reaction rate is suddenly increases: it is attributed to the extra methanol ensuring a large excess for the residual triglyceride, thus the conversion reaches 98 % in 20-30 min reaction time. We have used 50 m^3 tanks for the esterification with intensive stirring.

Separators for removing the glycerol phase, and the same volume of the separator was used to separate the neutralization agent and the ester phase which was mixed in a tube after exit of the first separator and before entering into the second one.

3.2 Removal of the residual catalyst from the biodiesel (decontamination)

In spite of the efforts to produce solid phase non-soluble alkaline catalysts or highly active acidic (super-acidic) catalysts (Di Serio et al, 2008; Leung et al., 2010; Soriano et al., 2009), the homogeneous catalytic (KOH or NaOMe catalyzed) trans-esterification of vegetable oils have been the most commonly used method in the biodiesel industry (Huber et al., 2006). However, soap formation during the alkaline-catalyzed trans-esterification is the most problematic by-reaction. In case of low water and carboxylic acid containing vegetable oils the main source of the soap formation is the hydrolysis of the formed methyl esters during washing, which is a strongly pH dependent process. Thus the neutralization preceding the washing is an essential step to minimize the saponification by-reactions. The amount and type of the formed soap is strongly affected by the separation characteristics of the glycerol and the ester phase. The acid treatment is generally needed to start or quicken the phase separation process producing aqueous glycerol solution. It is well known that the distribution of the catalyst between the glycerol and the methyl ester phase is strongly depends on the temperature, type of the catalyst, excess of methanol and the composition of the two separated phases [Chiu et al., 2005; Di Felice et al., 2008; Zhou and Boocock, 2006). Table 2 shows the distribution of 1 % KOH and 0.5 % of H_2SO_4 distribution at different temperatures depending on the amount of methanol at biodiesel : glycerol molar ratio of 1:3.

Table 2. Distribution of the KOH catalyst between the ester and glycerol phase in the function of temperature and MeOH excess ($K= C_{glycerol} / C_{ester\ phase}$)

Catalyst	T, °C	MeOH, mol	K	Catalyst	T, °C	MeOH, mol	K
KOH	25	0	98	H_2SO_4	25	0	60
	25	3	95		25	3	53
	25	6	77		25	6	46
	75	0	47		75	0	31
	75	3	45		75	3	28
	75	6	35		75	6	24

It can be seen that increasing temperature increases the amount of the catalyst in the ester phase. The amount of dissolved methanol also increases the amount of dissolved catalyst. The K values of the distribution of methanol, however, 10.9 and 7.5 at 3 and 8.5 and 4.8 at 6 moles of methanol towards 3 mol of biodiesel and 1 moles of glycerol at 25 and 75 °C, respectively. In the presence of 1 % KOH, these values changed to 2.29, 1.64, 1.90 and 1.48, respectively.

Due to the room-temperature reaction, there is low catalyst and low methanol concentration in the ester phase, thus the amount of the required neutralization agent is also low. Considering the use of a continuous separator, the relative volumes of the separated phases should be adjusted between 10-1:1, so it is possible to use dilute (2 %) $KHSO_4$ solution. The neutralization acts as the first washing step, when the potassium content in the biodiesel phase was proved to be up to 8 ppm without further aqueous washing. Instead of mineral acids we used an acidic salt as hydrogen ion sources, namely, potassium hydrogen sulfate ($KHSO_4$). Potassium hydrogen sulfate has as strong acidic function as the pure sulfuric acid without the disadvantages of the sulfuric acid, e.g. it is a solid crystalline mass can be stored without risk and can be dissolved in water without extreme heat generation, and no acidic vapors as in the case of hydrochloric acid can be felt. Due to the low amount of soaps in the ester phase, the dissolved methanol content of the ester phase is lower than in case of the classical biodiesel technologies. Potassium hydrogen sulfate as a strong "acid" reacts with KOH catalyst and soaps easily and spontaneously as

$$KOR + KHSO_4 = K_2SO_4 + HOR \quad (1)$$

where R = alkylcarbonyl radical (potassium soaps) or H (KOH). The aqeous $KHSO_4$ solution decomposes soaps immediately without emulsion formation. The formed potassium sulfate is neutral, soluble in water, insoluble in the ester phase and has no phase transfer property at all. The concentration of the formed potassium sulfate is low (~2-3 %), since dilute $KHSO_4$ solutions is used for neutralization of the potassium compounds in the ester phase. The aqueous phase will contain potassium sulfate, glycerol, methanol, and other water soluble components. The dilute solution can easily be ion-exchanged with strong acidic cationic exchangers as Varion KSM resin. The glycerol and the methanol containing aqueous phase is a strong polar solution, thus regeneration and recycling of the $KHSO_4$ should be taken place. However, stopping and controlling the operation of the ion exchanger at the stage of the hydrogen sulfate formation is difficult. Therefore, the K_2SO_4 containing material stream is divided into two equal parts. One part of the K_2SO_4 is ion-exchanged in a common way with the formation of H_2SO_4 solution (the K-ions are bound by the resin phase). The dilute sulfuric acid is combined with the other stream of the K_2SO_4, when $KHSO_4$ forms which can be recycled without further treatment (Kótai et al., 2008a).

$$K_2SO_4 + H_2SO_4 = 2KHSO_4 \quad (2)$$

The exhausted ion exchanger can be regenerated with the 20 % sulfuric acid solution consumed also for the formal neutralization. The potassium sulfate solution obtained during the regeneration process of the ion exchanger resin can be used as a fertilizer component. The ion exchanger can be regenerated and used again several thousand times.

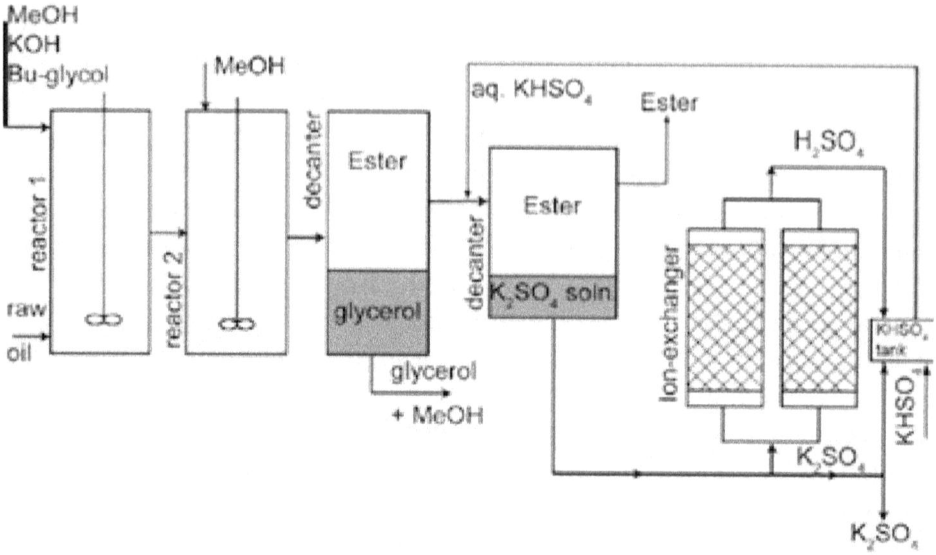

Fig. 2. Flow-sheet of biodiesel synthesis

However, the recycling of the $KHSO_4$ solutions is recommended only 2-20 times due to the accumulation of glycerol and methanol in the aqueous $KHSO_4$ solutions. The process flow is shown in Fig. 2.

The completely neutralized $KHSO_4$ solution turns into K_2SO_4, therefore it can be used as a fertilizer component together with the K_2SO_4 originated from the ion exchange regeneration cycle. Similarly, the potassium hydroxide content of the glycerol phase may be utilized in the same way after processing its glycerol and methanol content with H_2SO_4 catalyst into fuel. In this way, the starting potassium based catalyst (KOH) is used not only as a catalyst but as a fertilizer source in an advantageous sulfate form, which helps to avoid the chloride accumulation in the soil and supplies the sulfur deficiency of the soil, especially in the case of rape plant cultures of high sulfur demand (Scherer, 2001).

3.3 Trans-esterification of high acid containing vegetable oils

Production of biodiesels from non-food quality vegetable oils with high fatty acid content is an important area of the biodiesel developments. The common alkaline catalysts cannot be used economically, because they reacts with the fatty acids and form soaps causing high catalyst consumption and a decrease in the catalyst activity. The alkali soaps are the cause of problems in the phase separation and the decreasing in the yield of esters. The calcium soaps do not dissolve in water and not surface active agents, do not cause problems in the phase separation and using calcium methanolate as catalyst in the trans-esterification of the vegetable oils is a well-known process (Lindquist, 1991). Due to the insolubility of CaO and $Ca(OH)_2$ in methanol, the CaO and the $Ca(OH)_2$ are known as less active or inactive compounds for the vegetable oil trans-esterification (Di Serio, et al., 2008), but the activation of calcium oxide to act as a catalysts in this reaction leads to good results (Kótai et al., 2006, 2008b). The free acid content of the vegetable oils and CaO were immediately reacted with formation of calcium soaps, and the $Ca(OH)_2$ formed from the excess of CaO played role in the trans-esterification. The best results (quantitative conversion of triglycerides into methyl esters) was obtained by using calcium hydroxide (5.6 %) prepared in situ by slaking of CaO at a 3:1 CaO to H_2O mass ratio and at a 10:4 oil: methanol ratio for 1 h under reflux. Excess of CaO is essential in this reaction. Using half amount of catalyst and methanol, the conversion was 90 % only, and 9% triglyceride and 1 % diglyceride could be detected in the ester phase. In case of high fatty acid containing starting vegetable oils the calcium soaps formed is partially soluble in the methanol containing ester phase, thus after separation of the soap containing glycerol the ester phase was treated with a slight excess of concentrated sulfuric acid. The sulfuric acid has multiple roles in the system,

1. it forms calcium sulfate and releases the bound fatty acids
2. catalyzes the reaction of the liberated fatty acids and the methanol excess used in the trans-esterification
3. destroys the instable polyunsaturated compounds which could cause deposition in the engines
4. decreases the iodine number and the ash content in the ester phase

Since the optimal trans-esterification can be performed at 2-4 fold excess of methanol, the residual methanol content means a very high (30-60 fold) excess toward the liberated fatty acids taking part in the acid catalyzed trans-esterification (Khelevina 1968). The anhydrous $CaSO_4$ formed in the neutralization step promotes the esterification reaction of the liberated fatty acids due to binding of the water formed in the equilibrium esterification reactions:

$$R\text{-}COOH + MeOH = RCOOMe + H_2O \quad (3)$$

$$CaSO_4 + 0.5H_2O = CaSO_4.0.5H_2O \quad (4)$$

Although the calcium sulfate has di-hydrate, in this system only the hemihydrate has formed (Kótai et al., 2008b). The solid $CaSO_4.0.5H_2O$ can be used as sulfur fertilizer and improve the properties of unbound (sandy like) soils due to its binding properties with water (plaster of paris). During the removal of the polyunsaturated (Benjumea et al., 2011) labile compounds by sulfuric acid various compounds having acidic character form. The acid number of the ester phase is around 3.2 mg KOH/g oil. It is essential to remove these acids from the ester phase. It has been tried with filtering this ester phase through calcium or iron carbonate containing mineral granulates. This method, however, have not been effective, and in case of calcium carbonate some oil soluble products were also formed which increased the ash content of the ester. By using Varion ADA type resin (alkyl- ammonium type ion exchanger in OH form) the acidic constituent could easily be removed. The acid number was found to be 1.25 mg KOH/g oil, practically the same as was found for this oil after complete vacuum distillation (1.24 mg KOH/g oil) (Kótai et al., 2008b).

4. BUTANOL AS RAW MATERIAL IN BIODIESEL PRODUCTION

By using of butanol as blending material for gasoline and diesel fuels has many advantages towards fuel ethanol (Andersen et al., 2010; Bruno et al., 2009, Duck and Bruce, 1945, Workman et al., 1983), additionally, it can also be used as reactive component in biodiesel production. Butanol can substitute methanol as an alcohol for esterification (Wahlen et al., 2008; Nimcevic et al., 2000, Stoldt and Dave, 1998), it can be an acetal forming compounds for transforming of acetone or other ketones into acetals, or can easily be converted to butyraldehyde for the conversion of glycerol formed during trans-esterification of oils into 1,3-dioxane- or dioxolane type fuel additives (Silva et al., 2010)

4.1 Butanol as blending or reactive component in biodiesel production

The butanol or the butanol - acetone - ethanol mixture produced in the ABE fermentation have been tested already during the World War II as a blending components for gasoline powered engines (Duck and Bruce, 1945). The engines can be powered with gasoline containing butanol up to 40 % without any technical modification of the engine. The characteristics of pure butanol and ABE solvent mixtures as gasoline or diesel fuel additives have been tested in detail. Generally, the blended mixtures produce almost the same power and thermal efficiency as the gasoline (Schrock and Clark, 1983). The same time the blending has a positive effect via substantially decreasing the NO_x content of the exhaust gases.

Since the modern mobile agricultural equipment used in the production of biomass is mostly diesel powered, both ethanol and butanol have been tested for using them as blending components of diesel fuels. By the addition of butanol or ABE blends to diesel fuel the thermal efficiency could be increased, the exhaust gas temperatures are lowered and the soot formation is decreased. The operational parameters of the engine have been studied in detail (Workman et al., 1983). Butanol proved to be an alternative diesel fuel blend. By using butanol as a reactive component in vegetable oil butyl ester preparation or preparation of butanol based acetals, the residual butanol content does not has to be removed from the reaction mixture because it can act as blending component. A special case of the extraction of butanol from aqueous solutions during ABE fermentation is when the extractant is vegatable oil (Welsh and Williams, 1989). The butanol to be extracted can be reacted easily with the extractant. Since the vegetable oil alkyl esters are also extractants of the butanol (Crabbe et al., 2001, Ishizaki et al., 1999), this method can easily be integrated into a catalyst free supercritical trans-esterification technology, when the partially trans-esterified vegetable oil is recycled into the extraction process until its complete transformation into butylester. Since the butanol content extracted in the last step does not need to be separated from the butyl ester product, the method is advantageously integrated into the waste free biodiesel production system.

The use of butanol as a reactive component in biodiesel production has a lot of advantage. First of all, the convenient base catalyzed reaction takes place in homogeneous phase, thus the reaction is faster, the separation of the glycerol is better and the temperature of the reaction can be lowered. The excess of the butanol does not need to be removed from the ester phase. The viscosity of the butylester mixture prepared from soya oil was found to be 4.50 mm^2/s at 40 °C, the cetane index was 69. The cloud point of butyl-biodiesel is -3 °C. The flash point and pour point were found to be 44 and −13 °C, respectively (Wahlen et al., 2008). Since butanol reacts with free acids faster than methanol, the high acid containing vegetable oils (free fatty acid content of vegetable oils varies from 7 to 40 %), the waste cooked oils or other high free acid containing oils can also be used as raw materials. These high free fatty acid containing oils could not be trans-esterified economically with methanol and basic catalysts, and in the presence of acidic catalysts the reactions are very slow.

Solid phase heterogeneous catalysts have not widely been available for use them in industrial scale for these type of oils (Di Serio et al., 2008). The acid catalysts simultaneously catalyze the esterification of the free acids and the trans-esterification of the glycerides however, butyl esters are formed more easier than methyl esters (Wahlen et al., 2008). Methanol, ethanol, n-propanol and n-butanol have been reacted with oleic acid as a model for free fatty acids at 4:1 alcohol/acid molar ratio in the presence of 5 % sulfuric acid (as catalyst) at 80 °C for 16 min. The conversion was best (90%) for n-butanol, and the worst in the case of methanol (85%). The difference in trans-esterification activity of C_{1-4} alcohols in the presence of H_2SO_4 catalyst is more significant. The methanol can react with the soybean oil (10:1 methanol/bound fatty acid ratio) at 60 °C with 5 % sulfuric acid as catalyst only with 2 % conversion within 32 min. By using 12:1 alcohol/soybean oil ratio and 80 °C temperature, the methanol and ethanol gave 18 % conversion in 16 min, while the propanol and butanol showed 50 % conversion during the same time. The reaction of butanol with vegetable oils at a mixed feedstock containing oleic acid and soya oil with a ratio of 5:1-1:5 required minimum 2:1 butanol/fatty acid (free and bound) molar ratio at 110 °C in the presence of 5% H_2SO_4 catalyst . Using microwave heating at 6:1 butanol/soybean oil ratio in the presence of 3 % H_2SO_4, a 98 % conversion was achieved within 50 min. By using microwave heating the trans-esterification reaction of the vegetable oils with butanol can be performed without any catalyst under supercritical conditions (Geuens et al., 2008). Since the butanol boiling point is higher than methanol boiling point, the reactions takes place at higher temperatures without using extremely large pressures. The best results were achieved at 310 °C and 80 bar pressure in SiC coated tube reactor. The lack of the catalyst results very a small amount of glycerol without soap formation. The excess of butanol does not need to be separated from the ester phase or can be flashed out from the glycerol phase for recycling.

By using n-butanol instead of methanol and butoxylation of the unsaturated alkyl chain improve the ratio of the fossil energy used to produce a unit of renewable energy source. The highly unsaturated oils cause gum and deposit formation, but their epoxidation with peroxy-acetic acid and contacting the epoxides with n-butanol in the presence of 2 % sulfuric acid as catalyst at 80 C°, results 100% conversion of the epoxides. The selectivity is 87 %, and the 46 % conversion of the unsaturated alkyl chains does not cause an increase in the cloud point (Smith et al., 2009). As it can be seen, the butanol increases the amount of the biofuel produced from raw vegetable oil, by molar weight increasing referring to methyl esters (Table 3., Nimcevic et al., 2000), by incorporating butoxy groups into the unsaturated alkyl chains and by mixing the excess butanol with the formed fuel.

Table 3. Alcohol inputs in the production of biodiesel

Ester	Combustion value		Alcohol molar fraction in the ester molecule
	MJ/kg	MJ/kmol	
Methyl	39.83	14156	8.7
Ethyl	40.03	14787	12.2
Butyl	40.52	16103	18.4

In addition to these possibilities, in case of high acid containing raw vegetable oils, the acid content can also be transformed into butyl-type biofuel and does not need to be recovered as soaps. The abovementioned possibilities can be applied as parts of an integrated system together with other techniques to improve biofuel production, e.g. during the conversion of vegetable oils into methyl esters, the free fatty acids liberated from calcium soaps can be esterified with butanol instead of methanol as well (see Chapter 3.3).

4.2 Transformation of the wastes of butanol and biodiesel production into fuel

Glycerol is a very hygroscopic material and its combustion heat is low due to its high oxygen content. Neither its viscosity nor its hygroscopic nature or the miscibility properties indicate direct applicability as a fuel component. However, the glycerol is a reactive compound, thus the glycerol formed in the biodiesel synthesis can be transformed into lower oxygen containing compounds or to their mixture by various reactions (Guerro-Perez et al., 2009, Mota et al., 2009). In order to decrease the oxygen content of the products formed, the most reliable way is water elimination. It can be performed by reduction or by condensation reactions performed with reactants containing O= or HO-functions. The structure and reactivity ensure a series of water elimination (intra or intermolecular) reactions and formation of a variety of compounds.

The glycerol can act as a multifunctional primary and secondary alcohol and can easily be dimerized or polymerised into compounds with residual alcohol functions and alcoholic type reactivity. The glycerol can also be reacted with various other alcohol derivatives (with methanol residue from trans-esterification or with ethanol or butanol from ABE fermentation) into ethers. Transformation into cyclic acetals by using oxo-compounds e.g. acetone from ABE fermentation or acetone - acetaldehyde - butyraldehyde mixtures from the oxidation of not separated ABE products can also be performed. The formed acetals are cyclic dioxolane and dioxane type primary or secondary alcohols, or their stereoisomers (if R_1 and R_2 are not the same), respectively (Ferreira, et. al., 2010; Kótai and Angyal, 2011).

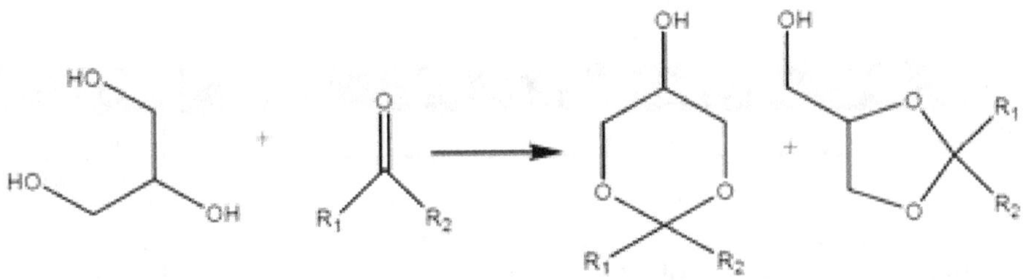

(5)

By partial oxidation of the mixture of primary alcohols from the first ABE fraction containing acetone, ethanol and butanol to aldehydes, a mixture of alcohols and oxo- compounds can be prepared. By using the waste glycerol containing methanol from the biodiesel production (or formaldehyde from the methanol oxidation) can provide a complex reaction mixture which can be condensed into an un-separated multicomponent mixture of various oxygenates with lower oxygen content than the starting glycerol. This mixture does not require complete separation into components or individual compounds to use it as a fuel. The reaction has been studied in the presence of various acidic catalysts as sulfuric acid, sulfonated styrene-divinyl-benzene copolymers and p-toluene-sulfonic acid. All of the catalyst gave similar results, the main product have been the 1,3-dioxolane derivatives. Various other components have also been formed in 1-2 % amount of each. The low-boiling fractions contain mainly the starting alcohols, acetone and dialkoxypropane derivatives, the higher fractions contain mainly 2,2-dimethyl-4-hydroxymethyl-1,3-dioxolane, its mixture with the starting alcohols and the formed dialkoxypropanes. The 2,2-dimethyl-5-hydroxy- 1,3-dioxane has appeared only in the distillation residue because its boiling point is higher than 120 °C. In the acetalization of acetone with glycerol the two possible isomers 1,3- dioxolane or 1,3-dioxane ring containing products can also be formed in the 1,2- or 1,3-type cyclization reactions. The molar ratio of the dioxolane /dioxane and the yields slightly depend on the type of the acidic catalyst. The composition of a typical reaction mixture is illustrated in Table 4. The main product is the 2,2-dimethyl-4-hydroxymethyl-1,3-dioxolane, a smaller amount of the 2,2-dimethyl-5-hydroxy-1,3-dioxane and dialkoxy-propanes are also formed. Two mixed methoxy-group containing acetals are formed, as well. Thus, it seems to be probably that the primarily formed 2,2-dimethoxypropane has reacted with the higher primary alcohols. In order to increase the complexity of the mixture, which is an optimal situation for fuels, the glycerol - MeOH mixture was mixed with the first un-separated fraction of the butanol production which contains EtOH, acetone and BuOH, and reacted with various oxo- compounds prepared from the abovementioned alcohols by oxidation (CH_2O, CH_3CHO and butyraldehyde).

Table 4. The acetals formed in the reaction of ABE solvents and glycerol containing methanol with Varion KSM acidic ion exchanger catalyst at 3 h reflux

Compound	Alcohol	Fraction	Peak area	B.p. range
MeC(OMe)2Me	MeOH	I-IV	1	58-99 °C
MeC(OMe)(OEt)Me	MeOH, EtOH	I-IV	2	58-99 °C
MeC(OEt)2Me	EtOH	III,IV	1	71-99 °C
MeC(OMe)(OBu)Me	MeOH, BuOH	I-IV	1	58-99 °C
2,2-Me_2-4-CH_2OH-1, 3-dioxolane	glycerol	III-VI	88	71-120 °C<
2,2-dimethyl-5-OH-1, 3-dioxane	glycerol	VI	2	120 °C<

It can be seen that from the same molar amounts of the alcohols the acetone prefers the reaction with the glycerol, or the dialkoxy-propanes formed reacts with the glycerol via re- formation of the alcohols.

In this way, the waste stream from ABE and biodiesel production with or without oxidative treatment results an un-separated mixture containing various alcoholic and oxo-components which can react with each other in various water elimination reactions to from a variety of lower oxygen containing acetal/ether type compounds. The formed mixture contains components with a wide boiling range. Table 5 contains the product distribution in a mixture formed in the reaction of 1-1 equivalents of acetone, acetaldehyde, n-butyaldehyde and formaldehyde by 1 equivalent of glycerol and 2-2 equivalents of MeOH, EtOH and BuOH with Varion

KSM sulfonated ion exchanger as catalyst under 3 h reflux. The reaction mixture has been separated into five fractions to study the distribution of each component formed and the starting material in the fractions. Depending on the reaction conditions, molar ratios of each reactant and the catalyst, the product distribution can be varied. Two isomers of 2-alkyl-4-hydroxymethyl dioxolanes are formed which have different boiling points. As an example, the effect of glycerol formal on the properties of the biodiesels can be seen in Table. 6. (Puche, 2009)

Table 5. The identified components of the reaction between biodiesel waste and partially oxidized ABE production waste streams in the presence of Varion KSM catalyst

Compounds	Alcohol	Oxo-reactants	Fraction	Peak area
Dialkoxi-methanes				
$(MeO)_2CH_2$	MeOH	CH_2O	I-II	2
$(EtO)_2CH_2$	EtOH	CH_2O	I-IV	5
$BuOCH_2OMe$	BuOH,MeOH	CH_2O	I-III	15
$BuOCH_2OEt$	BuOH,EtOH	CH_2O	I-IV	14
$(BuO)_2CH_2$	BuOH	CH_2O	II-V	5
Dialkoxyethanes				
$(BuO)(MeO)CHCH_3$	BuOH,MeOH	CH_3CHO	I-IV	1
$(BuO)_2CHCH_3$	BuOH	CH_3CHO	IV-V	2
Dialkoxybutanes				
$(BuO)_2CHCH_2CH_2CH_3$	BuOH	PrCHO	IV-V	2
1,3-Dioxolanes (2 isomers)				
2-Me-4-CH_2OH-1,3-dioxolane	glycerol	CH_3CHO	I-V	6
2-Me-4-CH_2OH-1,3-dioxolane	glycerol	CH_3CHO	V	4
2-Pr-4-CH_2OH-1,3-dioxolane	glycerol	PrCHO	V	12
2-Pr-4-CH_2OH-1,3-dioxolane	glycerol	acetone	IV-V	11
1,3-Dioxanes				
2-Me-5-OH-1,3-dioxane	glycerol	CH_3CHO	IV-V	2
2-Pr-5-OH-1,3-dioxane	glycerol	PrCHO	V	3

Not only acetals, but other ether type components can also be used as fuel blends. The condensation products formed with alcoholic functions can be used for further acetal formation. The dioxolane and dioxane type compounds with alcoholic function groups can be esterified or etherified in a further reaction into other valuable products (Jalinski, 2006).

Table 6. Effect of glycerol formal on properties of methyl ester of rapeseed oil

	RME + glycerol formal				
Glycerol formal content	0	0.5 %	1%	5%	10%
Density, g/cm³	0.8592	0.8620	0.8631	0.8711	0.8802
Freezing point, °C	-7	-16	-21	-21	-21
Viscosity at -10 C°, cSt	Solid	No data	548.2	343.3	No data

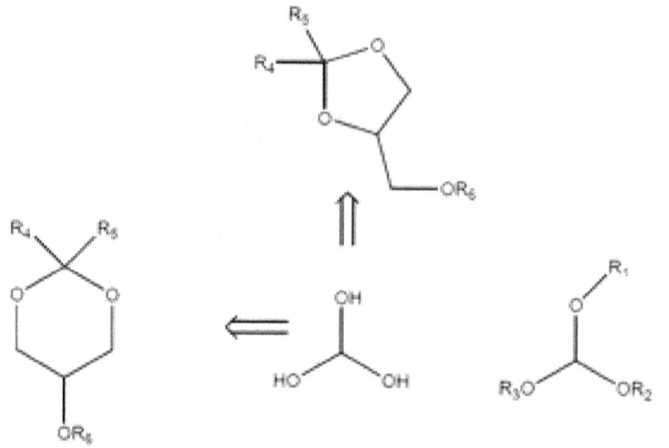

(6)

The general scheme for transformation of glycerol into fuel components with ABE components is given by eqn. (6), where R_1, R_2 and R_3 are Me, Et, Bu, $CH_3C(O)$-, $C_3H_7C(O)$-, R_4 and R_5 are H, Me, Pr, and R_6 means Me, Et, Bu, $CH_3C(O)$-, $C_3H_7C(O)$ or other groups derived from the alcohol-type glycerol condensation products. Transformation of all three hydroxyl groups of the glycerol into alkoxy groups (methoxy, ethoxy or butoxy), or esterifying them with low carbon chain carboxylic acids (acetic acid, butyric acid) decrease the hydrophil nature and oxygen content and increase the combustion heat, the miscibility with fuel. Thus, these compounds are advantageous fuel additives (Mota et al., 2009). Since ethanol, butanol, methanol, acetic and butyric acid are products/by-products and intermediates of the ABE fermentation or biodiesel production, these reactions are candidates for integration into a complex biomass utilization system. The intermediate acetic and butyric acid can also be used as acylation agents for the cyclic acetals, and in this way all product of the ABE fermentation become fuel component. Not only these organic acids but carbonic acid can also acts as acid residue in the esterified products. The carbonate compounds prepared form acetals formed from n-butyraldehyde or acetone and glycerol lowering the soot and the particulate formation during ignition of the diesel fuels (Delfort, 2004). Alkylation or acylation of free hydroxy-groups in 1,3-dioxolane and dioxane type fuel blends increases their solubility with two order of magnitudes (Jalinski, 2006)].

It is obvious, that glycerol which has primary and secondary alcohol functions, and can be condensed with itself to different kind of polyglycerols (Barrault et al., 1998). Polyglycerols can be obtained at high temperature vapor phase reaction over solid catalysts as alkali and alkaline earth metal hydroxides or carbonates, zeolites, La-ion-exchanged zeolites and ion-exchanger resins (Barrault et al., 1998). In the presence of resins, the main product is the diglycerol.

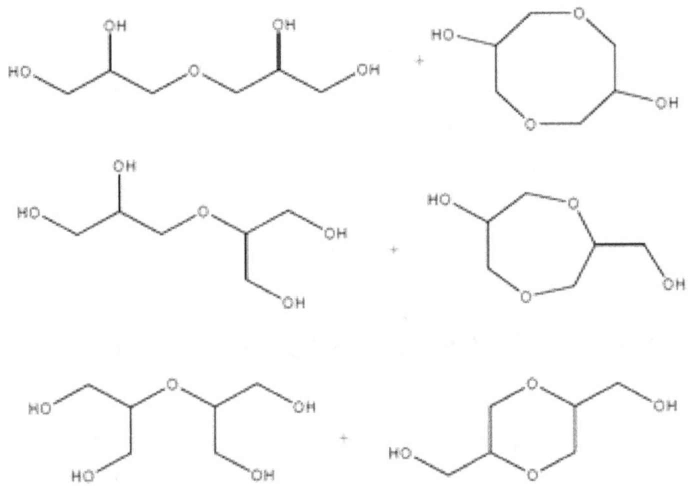

(7)

Since the glycerol has hydroxyl groups with various reactivity, depending on the catalyst and the reaction conditions, various dimers and even more type of oligomers and polymers can be formed. By using these dimers (oligomers) in acetal forming reactions, the complexity of fuel mixture can be further increased.

Not only water elimination, but increasing carbon chain length can decrease the relative oxygen content and increase the combustion heat and improve the fuel properties. Selective etherification of glycerol or the free alcoholic function groups of the condensates formed from the glycerol. The alcohol functions of glycerol or other alcohols formed during polymerization of glycerol or acetal production can easily be alkylated by reaction with isoalkenes (Klepacova et al., 2003). Trans-esterification of crude soya oil with methanol in the presence of NaOH catalyst, then separating the glycerol phase reacted with the mixture in the presence of Amberlyte-15 acidic ion exchanger catalysts for 2 when isobutylene converts the glycerol into ethers. The mixture formed contains 9 % triether, 47 % diether, 21 % mono-ether, 5 % unreacted glycerol, 14 % isobutylene and 4 % methyl esters. By separating and recycling the starting materials and the mono-ethers the residue can be mixed with the ester phase formed in the trans-esterification when a mixture is formed containing 12 % ethers and 88 % methyl esters. Its clouding point is below 0 °C and having a viscosity of 5.94 cSt which is lower with 9 °C and 0.5 cSt, respectively, if this parameters are compared to the ester phase without the addition of glycerol ethers (Barrault et al., 1998).

The oxygenate mixtures produced in the abovementioned ways ensures that a very complex mixture of compounds could be manufactured, in which all components of the ABE fermentation and biodiesel production turn into fuel component. These blending materials have very advantageous properties, decrease the viscosity, decrease the pouring point and soot formation and improve the cetane number. In this way, vegetable oil ester (mainly butyl ester), butanol and acetal or other oxygenate mixture containing biodiesels are formed with much higher production efficiency compared to the classical vegetable oil methyl esters. Thus, our technology can provide an aromatic hydrocarbon-free fuel which can be used even in highly populated large cities. Since biodiesels, fossil diesels and the gasolines can be mixed with pure butanol up to an amount of 40% without influencing the fuel properties, and these oxygenates can also be used around in an amount of 20 %, these new kind of fuel mixtures can provide a solution for the EU demand (incorporation of 20 % biocomponent into fuels until 2020).

5. OTHER ASPECTS OF THE INTEGRATED BIOMASS UTILIZATION SYSTEM

It is an obvious question that which bioalcohol should be used for the replacement of methanol in biodiesel production, or it is worth to change the ethanol blends of fuels to butanol which has much better fuel properties and energy content than the ethanol.

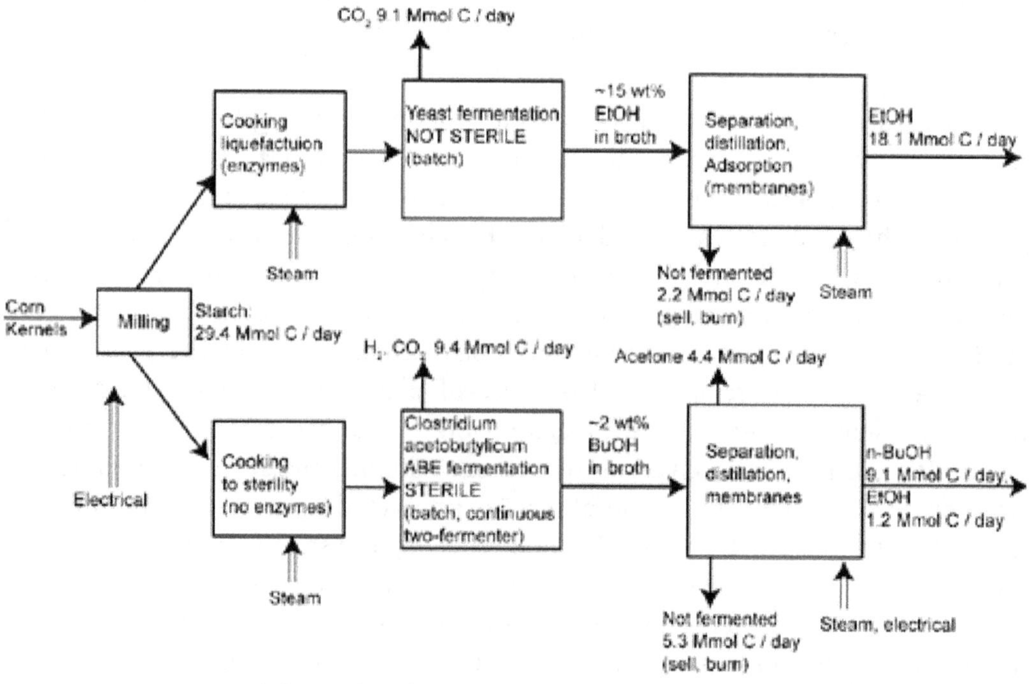

Fig. 3. Comparison of the carbon balances in the fermentation of corn into ethanol (yeast) and ABE (bacteria)

Comparison of technical and economical assessment for corn and switch grass fermented by yeast into ethanol and C. acetobutylicum into butanol showed (Pfromm et al., 2010) that biobutanol production is not competitive with ethanol production. As an example, the carbon balances for corn are illustrated in Fig. 3. However, involving new technologies, new raw materials (e.g. sugar sorghum) and the extractive fermentation processes combined with immobilized cell techniques, and decrease the production cost by means of the new separation technologies, the butanol becomes competitive as blending or reactive component in biofuel production.

5.1 New trends in biobutanol technology

The industrial production of biobutanol has roughly a 100 years history, here only some new trends are reviewed in order to indicate the new perspectives of the biobutanol technology. The so called ABE fermentation produces acetone, butanol and ethanol in a ratio of about 2:1:7 or 3:1:6 depending on the bacteria and the fermentation conditions. A small amount of acetic acid and butyric acid are also formed. The starting materials are C_5 and C_6 sugars, e.g. starch or cellulose hydrolizates, and there are some bacteria strain as well which can utilize cellulose directly. The main problem of the biotechnological butanol production is the toxicity of the solvents formed (mainly the butanol) towards the microorganism (Costa, 1981). The most intensively studied area of the developments is the genetic engineering to produce butanol tolerant bacterium strains or produce less sensitive genetically modified yeast and saving the microorganism from the toxic effects, e.g. by immobilization and capsulation of the bacteria (Park et al., 1989), or by the removal of the accumulated solvents before reaching the toxicity level (Papadopoulos and Linke, 2009; Schmidt et al., 1988;). Combination of the methods provides a good chance to start a continuous ABE fermentation (Hartmeier et al., 1991, Ishii et al., 1985, Kótai and Balogh, 2011). Since the energy demand of the butanol recovery from the dilute solution is one of the main cost factor extraction with a suitable solvent or adsorption on a cheap heterogeneous carrier can be candidates for the development of energy efficient butanol production. Due to the low adsorption capacity of known adsorbents like activated carbon, or the affinity of solid sorbents towards water allowed only utilization at low level. The extraction seemed to be more effective, but the solvents have to meet serious requirements like:

- Non-toxic to the microorganism and high stability,
- High distribution coefficient and selectivity with respect to the product,
- Low viscosity and solubility in the aqueous phase,
- Gravitation separation by density difference,

- Large interfacial tension and low tendency to emulsify the broth,
- High boiling point difference with respect to ABE solvents and low price

By using immobilized microorganism and in situ extraction the continuous production of butanol and ABE solvents can be performed easily, especially, if the integrated biomass utilization system ensures the sugar solution from sorghum processing. Thus, a large amount of ballast materials from corn or starch hydrolysis which increase the dry material content of the mash can be avoided The main problems are;

1. that the solvents which have high selectivity to the ABE solvents are very toxic to the microorganisms,
2. the best distribution coefficients for butanol found among the non-toxic solvents is only 3.5 (oleyl alcohol). In order to apply the high distribution coefficient of a toxic solvent associated with the requirements of the continuous production of butanol, a special system to produce ABE solvents has been developed.

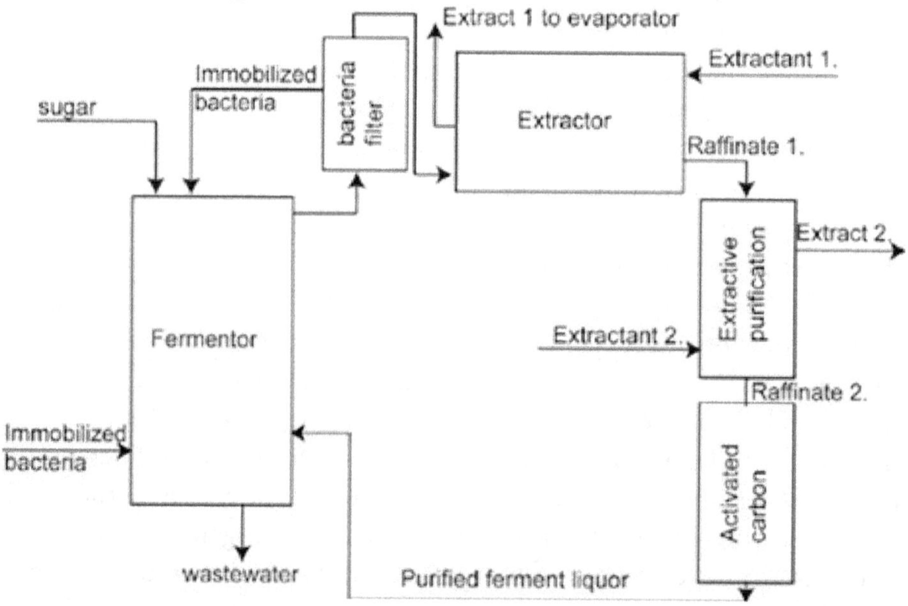

Fig. 4. Flow-sheet of continuous butanol fermentation with outer extraction

Heptanal, which has distribution coefficient for butanol of K_B=11.3 has been selected for the new technique. The toxicity of the heptanal is avoided via a by-pass extraction and a post-extractive purification for removal of traces of extractant from the ferment liquor before recycling it to the fermentation. The flow-sheet of particular method can be seen in Fig. 4. The technology using immobilized bacteria and heptanal as solvent is under development now. After starting the fermentation and before reaching the toxic level of butanol, a part of the ferment liquor is pumped out into a counter-current extractor filled with heptanal. The extraction is a continuous process. It means that the raffinate phase is contacted with another non-toxic solvent and/or a sorbent which removes the toxic solvent from the raffinate before recycling it into the fermentation. It should be noted here, that in case of heptanal the distribution of the ABE solvents can be considered to be advantageous.

Table 7. Distribution coefficients of ABE solvents

Solvent	Heptanal	Oleyl alcohol
Acetone	1.65	0.40
Ethanol	1.01	0.10
Butanol	11.13	3.75

The boiling point of heptanal is higher than the ABE solvents, thus the evaporation of the heptanal is avoided and does not require energy, By using a special fermentor - extractor system (Fig. 5) the extract phase containing the ABE solvents are contacted with another heptanal phase having a smaller volume (1/10-1/5) than the volume of the primary extract. The contact takes place through a special porous wall based on a pumicite - cement composite (Kótai and Balogh, 2011; Kótai et al., 2011.).

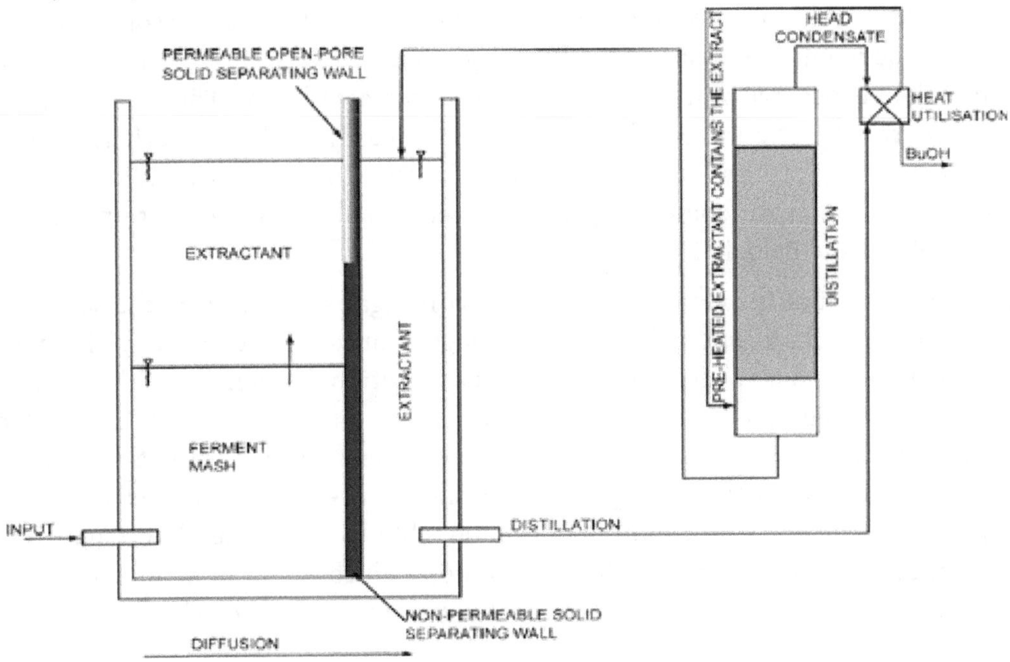

Fig. 5. Fermentor with composite wall

The porous material absorbs the solvents as a sponge, and it acts as a liquid transmitting media. There is no physical mixing only the diffusion controls the distribution of the solvents at the two sides of the wall. In order to keep the concentration difference as driving force, the ABE solvents is produced continuously at the A side (including the extraction step) and the extracted solvents are continuously removed at the B side of the wall, e.g. vacuum distillation, sorption by solid solvents as sulfonated styrene-divinyl-benzene copolymers (Kótai and Balogh, 2011), or with other methods. Due to their higher affinity towards water than the ABE solvents these copolymers ca not remove the AB solvents from the water directly. The Varion KSM sulfonated styrene-divinylbenzene copolymer can absorb the following amount of solvents (Kótai and Balogh, 2011)

Table 8. The adsorbed amount of solvents by solid sorbent based on sulfonated polymer

Solvent	Amount, g/100g	Solvent	Amount, g/100g
Water	87	acetone	40
MeOH	75	heptanal	35
EtOH	73	Oleyl alcohol	35
BuOH	63	Hexane	30

By using waste ion-exchangers, metal-containing activated carbons or metal-free carbons were prepared by leaching of metals from these metal-containing carbons [Kótai and Angyal, 2011], which are candidates as selective spherical sorbents with low hydrodynamic resistance for removing residual extractants and absorb the contaminants occurring in small amount in the ferment liquor. The Fe-containing sorbent are magnetic and could be separated very easily (without filtering) with an electromagnet (Kótai and Angyal, 2011). Elastic waste tires were also sulfonated for the preparation of sulfonated organic polymers to absorb butanol from organic media (Kótai and Balogh, 2011).

The supplement of the continuous reactor with sugar solution from the sorghum processing results a much cleaner technology than the process using the corn mash containing a lot of solid ballast material. Additional advantage of the method is the small volume of the B extract, because the small volume of invaluable solvents (water) needs small volume and investment of the distillation unit. A rough comparison of the energy and material balance shows, that selecting the appropriate bacteria and conditions, it is possible to reach the almost maximal theoretical utilization of glucose (0.37 g butanol/g glucose), so the amount of waste water and the energy to produce a unit of butanol in principle can be decreased to 25-50 %.

Combination of these techniques with new developments in membrane technologies, mainly with selective membrane separation and pervaporation (Liu et al., 2005; Thongsukmak & Sirkar), can provide a feasible butanol production technology.

5.2 Recycling of ash from biomass power plants

Combustion of biomass to produce energy (heat or electricity) always leaves back ash in various amount and composition depending on the biomass used as raw materials in the furnaces. This ash highly alkaline due to its potassium carbonate content (pH >10). Therefore it cannot be used directly as a fertilizer; only very small amount can be applied even for acidic soils. Since the ash contains all microelements and non-volatile elements in the amount absorbed and used by the plants harvested from the given area, their amount and ratio is optimal for the plant. Its insoluble content (e.g. phosphates) can be converted via digestion to utilizable material. In addition, it can be also supplemented with nitrogen fertilizers or other additives. In this way the processed ash becomes a useful material called eco-fertilizer (Angyal et al., 2006). In order to ensure the waste free technological viewpoints, the biomass power plant ash has to transform into fertilizer. It has been known for a long time that the alkaline ash can be neutralized with mineral acids, when potassium containing fertilizers are formed. However, these methods are difficult to perform technically, due to the large volume of gas formed in these reactions. Roughly 150 normal m^3 of carbon dioxide evolves in the reaction of 1 ton of ash with intensive foaming. A biomass power plant with electric capacity of 50 MW produces roughly 50-60 tons ash/day which means producing of > 300 normal m^3 of CO_2 gas/h. Furthermore, a very large volume of dilute fertilizer solution is produced, thus the cost of transportation is very high. The evaporation of the solution is not feasible. In order to solve these problems and ensure recycling the ash components for sustainable biomass production, a new technology has been developed for the neutralization of the biomass power plant ashes. This method ensures complete digestion of potassium and phosphorus content of the ash and decreases the K and P fertilization costs which are essential in the case of plants with fast metabolism such as sugar sorghum or energy grass. The nitrogen supplement is the only one which should be ensured in a usual way with the addition of N fertilizers. In this way the sulfur deficiency of the soils, or extreme sulfur demand as in case of oily plants like rape can also be satisfied (Scherer, 2001).

The new method based on the reaction of sulfuric acid and the biomass ash in a long quasi-closed tube reactor equipped with a screw for moving the reaction mixture. The reaction substantially proceeds during the mixture motion from the one end of the reactor to the other. This technology ensures not only continuous production of neutralized ash, but other important changes also appear in the chemical constitution of the starting ash. Normally, concentrated sulfuric acid does not wet the ash and does not react with it at all, their mixing can be proceed without CO_2 evolution. The reaction starts only in the presence of water. The mortar like mass prepared from the ash/limestone mixture and cc. sulfuric acid are mixed with water and reacted in the tube reactor. The reaction starts only after the dilution of the sulfuric acid. The most advantageous concentration is ~50 % (of sulfuric acid) when the carbonates reacts in a self-sustainable way due to producing water in the neutralization reactions. The key element in the reaction is the quasi-closed tube reactor. During moving the mortar like mass toward the opened end of the tube reactor, the neutralization reaction takes place and CO_2 gas evolves as usual. However, the reaction mass acts as a plug which ensures that the formed CO_2 gas cannot released from the reactor. The in situ evolved CO_2 gas makes micro-bubbles in the material, because of the overpressure of the quasi-closed equipment, and the swelled mass fills out the tube reactor. However, the wall of the reactor and the reaction mass as a plug ensures an overpressure within the reactor. Due to the overpressure no fizzing out occurs within the reactor and the micro-bubbles of the carbon dioxide gas are kept in the mortar-like mass. If the amount of the mass is adjusted to be less in its volume than the volume of the reactor, the mass is blown up due to the evolved gas and fills completely the space within the reactor. The carbon dioxide micro-bubbles have a slight overpressure, thus leaving the tube reactor at the opened end, the gas leaves the semi-solidified mass and the places of the micro-bubbles becomes opened pores. Technologically this method ensures processing of the ash with sulfuric acid in a small volume of continuously operated tube reactor having only a volume of ~3 times larger than the volume of the ash that can be processed (~ 60-80 times larger reactors as used in the classical neutralization methods do not needed (Angyal et al, 2006). Since the formed mortar like mass dries and solidifies easily, after granulating or pelletizing into the usual shape of solid fertilizers, the formed eco- fertilizer can be spread as solid by common facilities.

This method of neutralization has numerous advantages towards the classical neutralization technologies, not only the formation of solid fertilizer instead of dilute liquids, and thus avoiding the high volume expensive reactors during manufacturing, but from chemical viewpoints as well. Normally, the ash formed from straw and energy grass contains a mixed potassium calcium carbonate (Buetschliite), $K_2Ca(CO_3)_2$ as main components, the second most important phase is the KCl, and K_2CO_3 and K_2SO_4 can also be detected by powder X- ray diffraction. Similar amount of magnesium hydroxide and sodium carbonate can also be detected. The ratio of potassium chloride and sulfate depends on the soil composition, and the fertilization and type of fertilizer used (KCl or K_2SO_4) during the production of the wheat of course. Expressing the important metal content in the form oxides are as it follows:

~40 % of K_2O, ~10% CaO, 3.5 % MgO and 2.5 % of Na_2O. The straw contains a lot of chlorides (~7 %), the other anions as sulfate and carbonate expressed in SO_3 and CO_2 are ~10 % and ~20 %, respectively. The potassium-calcium carbonate (or potassium and calcium carbonate as well) easily reacts with diluted (~50 %) sulfuric acid, but not only the expected K_2SO_4 and $CaSO_4$ but their double salts as syngenite ($K_2Ca(SO_4)_2.H_2O$) and polyhalite ($K_2Ca_2Mg(SO_4)_4.2H_2O$) are formed as main products. The syngenite is less soluble (but not completely insoluble) in water and has ion-exchange properties toward ammonium ion, because due to their similar sizes of potassium and ammonium ions they can substitute each other in the structure of this compound.

$$K_2Ca(CO_3)_2 + 2H_2SO_4 = K_2Ca(SO_4)_2.H_2O + H_2O + 2CO_2 \quad (8)$$

The excess of sulfuric acid is neutralized to pH=6 with limestone powder and can be used as simple and general potassium and sulfate fertilizer which has opened pore structures which can absorb water and keep it in the pores, this way increasing the water retaining capacity of the soil. This fertilizer contains soluble phosphates and microelements previously digested by the sulfuric acid treatment. Furthermore, via controlling the amount of the calcium carbonate powder in the last step of the manufacturing, acidic (sub-neutralized), neutral or alkaline (over-neutralized) fertilizers can also be produced. No liquid or solid waste form in this technology, all the used components are built into the structure of the product. Since the water absorbing capacity of these materials, due to the porosity controlled by the synthesis conditions is very high (50-120 % of its mass), it is obvious, that not only water but various aqueous solutions can also be absorbed in these pores. This behavior opens new perspectives, namely absorbing different other fertilizers, e.g. nitrogen fertilizers, insecticides, or any other solutions of important compounds which should be injected into the soils.

The most common nitrogen fertilizer is the ammonium nitrate, however, the metal-doped NH_4NO_3 production has serious problems because the ammonium nitrate is prilled from the melt, and the melted ammonium nitrate is easily exploded due to the catalytic effect of metal compounds. Thus, these metal microelements cannot be added to the melted NH_4NO_3 before prilling. Using ammonium nitrate solutions, the metals compounds can be added in the required amounts without any risk to the porous granulated eco-fertilizer. Although drying after the absorption of the aqueous solutions (e.g NH_4NO_3 solutions) requires extra energy, but the complete process energetically is still more advantageous, due to the following considerations. Ammonium nitrate solution is prepared in an exothermic reaction of a ~65 % of nitric acid with ammonia gas, when an aqueous solution of NH_4NO_3 solution (roughly 80%) is formed. Evaporation of this solution at high temperature leads to the melt of ammonium nitrate which is prilled in the next step of manufacturing. Since we use the NH_4NO_3 solution, which is contacted with the granules, the water removal, without melting of the NH_4NO_3, requires less energy than the final step of the solid NH_4NO_3 manufacturing. This concentrated

NH_4NO_3 solution has acidic character, and easily react with the syngenite and other components of the eco-fertilizer. The concentrated NH_4NO_3 solutions are not only physically absorbed and imbibed in the pores of the eco-fertilizer, but chemically reacts with its components, as well.

$K_2Ca(SO_4)_2.H_2O$ (syngenite) $(NH_4)_2Ca(SO_4)_2.H_2O$(koktait)

The potassium ions can be substituted with ammonium ion with the formation of partially or completely ion-exchanged syngenite-like isomorphous compounds. The completely substituted product is called to be koktaite, $(NH_4)_2Ca(SO_4)_2.H_2O$, which is less soluble in water, thus releases nitrogen slowly into the soil (Angyal et al., 2006; Coates and Woodward, 1988; Von Maessenhausen et al., 1988). The formed potassium nitrate transformed into a solid solution with the excess of the ammonium nitrate, the typical composition of this product was the $K_{0.27}(NH_4)_{0.73}NO_3$. The koktaite and NH_4-syngenites are sparingly soluble in water, thus the ammonium ion concentration liberated in the presence of water is constant at a given temperature and ionic strength of sulfate ion. Since not the full amount of the ammonium ion is liberated, no damages to the plant and losses by washing away, respectively occur even if using in high doses. When the plant absorbs the ammonium-ion from the soil, due to the equilibrium conditions, a part of the solid will dissolve and supply the water with a new amount of ammonium ion. Since the equilibrium concentration is closely constant, the amount of water will control the amount of the released ammonium ion, namely, the release of the ammonium ion from this koktaite type compounds is controlled by raining or irrigating. In drought situation, when there is no absorption of ammonium ion from the soil by the plant, there is no dissolution of ammonium syngenites and releasing ammonium ion which would be decomposed by the soil bacteria as it happens in case of water soluble ammonium ion containing fertilizers. Besides ammonium nitrate, other fertilizer components can also be used to adjust the main element concentrations, such as K, P or N, and to change the available form of these elements in various chemical compounds. The K_2HPO_4 does not react at all with other components of the ash. It is interesting, that ammonium salts as NH_4Cl and $(NH_4)_2SO_4$ cannot transform the syngenite completely into $(NH_4)_2SO_4$, even if the ammonium sulfate is in excess, but in the presence of urea, the transformation is complete. Both KCl and K_2SO_4 decompose the ammonium syngenite, but the mixture of the K_2SO_4 and the NH_4Cl produces $(NH_4)_2Ca(SO_4)_2.H_2O$. Thus, the main factor is probably the ammonium to potassium ion ratio. There is an important difference between the behavior of the potassium sulfate and potassium chloride. The latter compound is more reactive, and KCl, KNO_3 and NH_4MgCl_3 are also formed in its presence. Using various additives not only the ratio of the agriculturally important elements (K, P, N, S) are controlled but their chemical forms can also be altered. Using various kind of soil bacteria and supplementary materials to ensure theirs intensive growing is another possibility for nitrogen-fixation in the treated area. By using the eco-fertilizer technology supplemented with absorbing of aqueous liquids ensures the recycling of the by-product of the biomass power plant providing energy and electricity for the biofuel (biodiesel and biobutanol and supplemented) plants. This approach enables the sustainable development of renewable energy plants, such as sugar sorghum, while simultaneously preserving and enhancing soil quality.

6. CONCISE SUMMARY

Through the judicious selection of biomass from a certain region, the sugar and energy sources, as well as the quantity of vegetable oil produced, can be optimized. To reduce the processing costs of raw materials into sugar-containing mash for fermentation facilities, traditional sugar sources such as corn can be substituted with sugar sorghum, which can be processed similarly to sugarcane. The combustion of residual biomass in power plants or its digestion into biogas is contingent upon the water and protein content of the residue, as well as the heat or energy requirements of fuel-producing (biodiesel, biobutanol, acetals, etc.) or waste processing (fertilizer manufacturing) facilities. Utilizing biomasses with elevated protein and moisture content in biogas facilities is generally more beneficial. The combustion of biogas or its utilization as fuel in gas engines allows for the regulation of heat and electricity production. High cellulose waste can be effectively incinerated in power plants, occasionally following dehydration using low heat value warm water energy streams. By combining these two techniques of energy generation, the byproducts of fuel production can be effectively utilized. The ash and solid leftovers from biomass power plants can be employed as fertilizers by combining them with potassium sulfate or calcium sulfate produced during catalyst recovery (KHSO4 or H2SO4) in biodiesel or acetal facilities. Ultimately, there exist two additional forms of waste. The first is K2SO4 derived from biodiesel technology, while the second is ash resulting from combustion. In addition to integrating energy-producing and consuming facilities and regulating the proportions of raw materials and energy types (heat or electricity), the production technique must also be modified, particularly in biodiesel, biobutanol, and fertilizer plants. This approach can reduce the energy usage of each technological phase.

REFERENCES

Andersen, V. F., Anderson, J. E., Wallington, T. J., Mueller, S. A. & Nielsen, O. J., *Energy & Fuels*, Vol. 24, (2010), pp. 2683-2691. ISSN 0887-0624

Angyal, A., Hujber, O., Kótai, L., Legeza, L. & Sajó, I. E., *HU Patent Appl. 0600390*, 2006.

Arlie, J.-P., *Revue de l' Institute Francais du Petrole*, Vol. 38, No.2. (1983), pp. 251-257. ISSN 0020-2274

Barrault, J., Pouilloux, Y., Vanhove, C., Cottin, K., Abro, S. & Clacens, J. M., *Chemistry and Industry*, Vol. 75, (1998), pp. 13-23. ISSN 0009-3068

Benjumea, P., Agudelo, J. R. & Agudelo, A. F., *Energy & Fuel*, Vol. 25, (2011), pp. 77-85. ISSN 0887-0624

Boocock, D. G. B., Konar, S. K. & Sidi, H., *Journal of the American Oil Chemists Society*, Vol 73 (1996), pp. 1247-1251, ISSN 0003-021X

Boocock, D. G. B., Konar, S. K., Mao, V. & Sidi, H., *Biomass & Bioenergy*, Vol.11, No.1. (1996), pp. 43-50, ISSN: 0961-9534

Boocock, D. G. B., Konar, S. K., Mao, V., Lee, C. & Buligan, S., *Journal of the American Oil Chemists Society*, Vol. 75, (1998), pp. 1167-1172, ISSN 0003-021X

Bruno, T. J., Wolk, A. & Naydich, A., *Energy & Fuel*, 23 (2009), pp. 2295-2306. ISSN 0887- 0624

Coates, R. V. & Woodard, G. D., *Journal of Science of Food & Agriculture*, Vol. 14, No.6, (1963), pp. 398-404, ISSN 0022-5142

Costa, J. M., *Proc. Ann. Biochem. Eng. Symp.* 11th, (1981), pp. 83-90.

Crabbe, E., Nolasco-Hipolito, C., Kobayashi, G., Sonomoto, K. & Ishizaki, A., *Process Biochemistry*, Vol. 37, (2001), pp. 65-71, ISSN 0032-9592.

Delfort, B., Durand, I., Jaecker, A., Lacome, T., Montagne, X. & Fabrice, P., US 7097674, 2004. Di Serio, M., Tesser, R., Pengmei, L. & Santacesaria, E., *Energy & Fuels*, Vol. 22, (2008), pp. 207-217. ISSN 0887-0624

Duck, J. T. & Bruce, C. S., *Journal of Research of the National Bureau of Standards*, (1945), pp.439- 465, ISSN: 0160-1741.

Ferreira, P., Fonseca, I. M., Ramos, A. M., Vital, J. & Castanheiro, J. E., , *Applied Catalysis B. Environmental*, Vol.98, (2010), pp. 94-99, ISSN 0926-3373

Fu B., & Vasudevan, P. T., *Energy & Fuels*, vol. 23, (2009), pp. 4105-4111. ISSN 0887- 0624

Geuens, J., Kremsner, J. M., Nebel, B. A., Schober, S., Dommisse, R. A., Mittelbach, M., Tavernier, S., Kappe, C. O. & Maes, B. U. W., *Energy & Fuels*, vol. 22, (2008), pp. 643- 645. ISSN 0887-0624

Guan, G., Kusakabe, K., Sakurai, N. & Moriyama, K., *Fuel*, Vol. 88, (2009), pp. 81-6, ISSN 0016-2361

Guan, G., Sakurai, N. & Kusakabe, K., *Chemical Engineering Journal*, Vol.146, (2009), pp. 302- 6, ISSN 1385-8947.

Guerro-Perez, M. O., Rosas, J. M., Bedia, J., Rodriguez-Mirasol, J. & Cordero, T., *Recent Patents on Chemical Engineering*, Vol. 2, (2009), pp. 11-21. ISSN 2211-3347

Hartmeier, W., Buecker, C. & Wallrath, J., *Biochem. Bioeng., 2nd Int. Symp., Stuttgart,* (1991), pp. 236-241.

Hill, J., Nelson, E., Tilman, D., Polasky, S. & Tiffany, D., *Proceedings of the National Academy of Sciences* Vol. 103, No. 30, (2006), pp. 11206-11210, ISSN 1091-6490.

Huber, G. W., Iborra, S. & Corma, A., *Chemical Reviews,* Vol. 106, (2006), pp. 4044-4098, ISSN 0009-2665

Ishii, S., Taya, M., Kobayashi, T., *Journal of Chemical Engineering of Japan,* Vol. 18, No.2, (1985), pp. 125-130, ISSN 0021-9592

Ishizaki, A., Michiwaki, S., Crabbe, E., Kobayashi, G., Sonomoto, K. & Yoshino, S., *Journal of Bioscience and Bioengineering,* Vol. 87, No. 3, (1999), pp. 352-356 ISSN 1389-1723

Jalinski, T. J., *WO 2006/084048* A1, 2006

Khelevina, O. G. & Kanyaev, N. P., *Izvestiya Vysshikh Uchebnii Zavedenii, Khimiya i Khimicheskaya Tekhnologiya,* Vol. 11, No.2, (1968), pp. 166-170, ISSN 0579-2991.

Klepacova, K., Mravec, D., Hajekova, E. & Bajus, M., *Petroleum & Coal,* Vol. 45, (2003), pp. 54- 57, ISSN 1335-3055

Kótai, L. & Angyal, A., *Research Report, GOP 1.1.1.,* Axial-Chem Ltd., 2011.

Kótai, L., Angyal, A., Somogyi, I., Bihatsi, L., May, Z., Gömöry, Á. & Tamics, E., *HU 08/00437A,* 2008.

Kótai, L. & Balogh, J., *Research Report, GOP 1.1.1.,* Kemobil Co., 2011.

Kótai, L., Gömöry, Á., Gács, I., Holly, S., Sajó, I. E., Tamics, E., Aradi, T. & Bihátsi, L., *Chemistry Letters,* Vol. 37, No. 10, (2008), pp. 1076-7, ISSN 0366-7022
Kótai, L., Szépvölgyi, J., Tamics, E., *HU 11/00286* A. 2011.

Kótai, L., Tamics, E., Sas, J., Deme, P., Gömöry, Á. & Aradi, T., *HU 06/00886A,* 2006.

Leung, D. Y. C., Wu, X. & Leung, M. K. H., *Applied Energy,* Vol. 87, (2010), pp. 1083-1085, ISSN 0306-2619

Lindquist, C.-J., *WO91/15452,* 1991.

Liu, F., Liu, L., Feng, X., *Separation & Purification Technology,* Vol. 42, (2005), pp. 273-282. ISSN 1383-5866

Maeda, K., Kuramochi, H., Asakuma, Y., Fukui, K., Tsuji,. T., Osako, M. & Sakai, S., *Chemical Engineering Journal,* Vol. 169, (2011), pp. 226-30, ISSN 1385-8947.

Meng, Z., Jiang, J. & Li, X., *Taiyangneng Xuabao,*Vol.30, No.3, (2009), pp. 385-389. ISSN 02540096

Mota, C. J. A., da Silva, C. X. A. & Goncalves, V. L. C., *Quimica Nova* Vol.32, No.3, (2009), pp. 639-648, ISSN 0100-4042

Nimcevic, D., Puntigam, R., Woergetter, M. & Gapes, J. R., *Journal of the American Oil Chemists Society,* Vol. 77, No. 3, (2000), pp. 275-280. ISSN 0003-021X

Papadopoulos, A. I. & Linke, P., *Chemical Engineering and Processing: Process Intensification,* Vol. 48, (2009), pp. 1047-1060, ISSN 0255-2701.

Park, C.-H., Okos, M. R. & Wankat, P. C., *Biotechnology and Bioengineering,* Vol. 34, (1989), pp. 18-29, ISSN 0006-3592

Pimentel, D. & Petzek, T. W., *Natural Resources Research*, Vol. 14, No.1, (2005), pp. 65-76, ISSN 1520-7439.

Pfromm, P. H., Amanor-Boadu, V., Nelson, R., Vadlani, P. & Madl, R., *Biomass and Bioenergy*, Vol. 34, (2010), pp. 515-524, ISSN 0961-9534

Puche, J. D., *US 7637969*B2, 2009.

Scherer, H. W., *European Journal of Agronomy*, Vol. 14, (2001), pp. 81-111, ISSN 1161- 0301

Schmidt, A., Windsperger, A. & Friedl, A., *US 4749495,* 1988.

Schrock, M. D. & Clark, S. J., *Transactions of the ASAE* (1983), pp. 723-727, ISSN 0001- 2351

Silva, P. H. R., Goncalves, V. L. C. & Mota, C. J. A., *Bioresource Technology,* Vol. 101, (2010), 101, pp. 6225-6229, ISSN 0960-8524.

Smith, P. C., O'Neill, B. K., Ngothai, Y. & Nguyen, Q. D, *Energy & Fuels,* Vol. 23, (2009), pp. 3798-3803, ISSN 0887-0624

Soriano, N. U., Jr., Venditti, R. & Argyropoulos, D. S., *Fuel*, Vol. 88, (2009), pp. 560-565, ISSN 0016-2361

Stoldt, S. H. & Dave, H., *EP 0860494* A1, 1998.

Thongsukmak, A. & Sirkar, K. K., *Journal of Membrane Science,* Vol. 302, (2007), pp. 45-58, ISSN 0376-7388.

Von Maessenhausen, W., Czikkely, V. & Jung, J., *US4883530* B2, 1988.

Wahlen, B. D., Barney, B. M. & Seefeldt, L. C., *Energy & Fuels,* Vol. 22 (2008), pp. 4223-28. ISSN 0887-0624.

Welsh, F. W. & Williams, R. E., *Journal of Chemical Technology & Biotechnology*, Vol. 46, No. 3, (1989), pp. 169-78, ISSN 0268-2575

Workman, J.P., Miller, G. L. & Smith, J. L., *Transactions of ASAE*, 1983, 642-645, ISSN 0001-2351

11 Chapter

PRODUCTION OF BIODIESEL VIA IN-SITU SUPERCRITICAL METHANOL TRANSESTERIFICATION

1. INTRODUCTION: AN OVERVIEW

The majority of global energy consumption is sourced from non-renewable fossil fuels, which significantly affect the environment (Warabi et al., 2004). The demand for fossil fuels is escalating rapidly, and it is projected that the remaining global reserves will be depleted by 2020 at the current consumption rate. There is an urgent necessity to pursue alternative fuels to replace diesel due to the progressive depletion of global crude oil stocks. Research is thus directed towards alternate energy. Biomass is one of its candidates, because biomass energy has some advantageous in reproduction, cyclic and carbon neutral properties (Warabi et al., 2004). Biodiesel fuel is one example of biomass energy, and it is generally made of methyl esters of fatty acids produced by the transesterification reaction of triglycerides with methanol with the help of a catalyst (Clark et al., 1984). Alcoholysis of vegetable oils produces fatty acids alkyl esters that are excellent substitutes for conventional fossil diesel fuels (Selmi and Thomas, 1998; De et al., 1999). The viscosity of alkyl esters is nearly twice that of diesel fuel instead of 10–20 times as in the case of neat vegetable oil (Rathore and Madras, 2007).

The use of such edible oil to produce biodiesel is not feasible in view of big gap in the demand and supply of such oils in the country for dietary consumption. Increased pressure to augment the production of edible oils has also put limitations on the use of these oils for production of biodiesel (Sinha et al., 2008). Therefore, biodiesel is actually competing limited land availability with the food industry for the same oil crop. Thus, instead of arable land being utilized to grow food, it is now being used to grow fuel. This will then raise the price of edible oil making the biodiesel produced economically unfeasible as compared to petroleum-derived diesel. In order to overcome this issue, many researchers have begun searching for cheaper and non-edible oils to be used as alternative feedstock for biodiesel production (Kansedo et al., 2009). Few sources have been identified such as waste cooking oil (Wang et al., 2006; Chen et al., 2009) and oils from non-edible oil-producing plants such as *Jatropha curcas* (Heller, 1996; Herrera et al., 2006; Tiwari et al., 2007; Berchmans and Hirata, 2008; Chew, 2009), *Pongamia pinnata* (Meher et al., 2006; Naik et al., 2008; Pradhan et al., 2008), *Calophyllum inophyllum* (Sahoo et al., 2007), cottonseed (Demirbas, 2008; Qian et al., 2008; Rashid et al., 2009), rubber seeds (Ikwuagwu et al., 2000; Ramadhas et al., 2005) and tobacco seeds (Usta, 2005; Veljkovic et al., 2006). Obviously, developing nations have to focus their attention on oils of non-edible nature, which are cheaper (Sinha et al., 2008). In Malaysia, *Jatropha curcas* L. (JCL), could be utilized as a source for production of oil and can be grown in large scale on non-cropped marginal lands and waste lands.

JCL oil is obtained only after going through the following steps: collection of fruit from the trees, separation of seed from the hull, seed drying (Chew, 2009), oil pressing and filtration. Pressing oil from the kernel yields kernel cake (40-50%) and crude oil (50-60%). At present, in the majority of cases oil is generally pressed directly from the seed without separating the kernel and shell. This method produces seed cake (70-75%) and crude oil (25-30%) (Chew, 2009). Much of the un-extractable oil still remains in the seed cake; hence better ways of extracting the oils are needed. Among the extraction techniques reported in the literature include the use of Soxhlet extraction method (Castro and Ayuso, 1998; Ayuso and Castro, 1999; Szentmihalyi et al., 2002; Darcia and Castro, 2004), aqueous enzymatic oil extraction (Rosenthal et al., 1996; Sharma and Gupta, 2006; Jiang et al., 2010) and enzyme assisted three phase partitioning (Shah et al., 2004; Gaur et al., 2007). Some of these extraction methods, however, required a longer extraction time (Chew, 2009). Nowadays, many researchers (Papamichail et al., 2000; King et al., 2001; Cao and Ito 2003; Machmudah et al., 2008) turns to supercritical extraction techniques which is relatively rapid because of the low viscosities and high diffusivities associated with supercritical fluids.

Transesterification is the general term used to describe the important class of organic reactions where an ester is transformed into another ester through interchange of the alkoxy moiety. Several aspects, including the type of catalyst (alkaline, acid or enzyme), alcohol/vegetable oil molar ratio, temperature, purity of the reactants (mainly water content) and free fatty acid content have an influence on the course of the transesterification. In the conventional transesterification of fats and vegetable oils for biodiesel production, free fatty acid and water always produce negative effects, since the presence of free fatty acids and water causes soap formation, consumes catalyst and reduces catalyst effectiveness, all of which result in a low conversion (Demirbas, 2007). In addition to that, more catalyst is required to neutralize free fatty acids of oil with higher free fatty acids content (Kusdiana and Saka, 2004). Thus, the catalytic processes have a high production cost and are energy intensive. One primary problem is due to the vigorous stirring required for the mixing of the two-phase mixture of oil and alcohol. Another problem is the separation of catalyst after the reaction (Madras et al., 2004). Therefore, non-catalytic transesterification has been investigated.

Supercritical fluid extraction using polar solvent such as methanol as an extraction solvent is highly potential extraction technique to be used whereby high yield of oil can be achieved within a shorter time (Hawash et al., 2009). Further, at supercritical state, the solvent solubility increased dramatically, and the extracted oil is relatively low in impurities (Tan et al., 2009). However, there is no details on the maximum crude biodiesel yield can be obtained related to the in-situ supercritical methanol transesterification direct from the seeds.

In situ transesterification differs from the conventional reaction in the sense that the oil- bearing material contacts acidified alcohol directly instead of reacting with purified oil and alcohol. That is, extraction and transesterification of the seed powder proceed within the same process, with alcohol acts as an extracting solvent as well as esterification reagent (Fukuda et. al., 2001). In situ transesterification (Harrington and Evans, 1985; Marinkovic and Tomasevic, 1998; Kildiran et al., 1996; Hass et al., 2004), a biodiesel production method that utilizes the original agricultural products instead of purified oil as the source of triglycerides for direct transesterification, eliminates the costly hexane extraction process and works with virtually any lipid-bearing material. It could reduce the long production system associated with pre-extracted oil and maximize alkyl ester yield. The utilization of reagents and solvents is minimized, hence alleviating concerns over waste disposal. This procedure decreases the final product cost due to a reduced number of unit operations. It is the optimal non-renewable energy source with favorable environmental effects and straightforward extraction.

This work enhances the design, development, and optimization of in-situ supercritical methanol transesterification for biodiesel synthesis using a high-pressure, high-temperature batch reactor system. This study involves the direct production of biodiesel from JCL seeds utilizing methanol under various solvent critical conditions.

2. MATERIALS AND METHODS

2.1 Sample preparation

The *Jatropha curcas* L. (JCL) fruits were obtained with cooperation from the Plantation Unit of Universiti Teknologi MARA Perlis, Malaysia. JCL fruits were cleaned and de-hulled to separate the hull from the seeds. The seeds were then dried in an oven at 105 °C for 35 min (Akbar et al., 2009). The JCL seeds were ground using grinder and sieved through progressively finer screen to obtain particle sizes (d_p) of < 1 mm (Augustus et al., 2002). Sieving was accomplished by shaking the JCL powder in a Endecotts Shaker Model EFL2 for about 30 min and finally stored in a tightly-capped plastic container. The seeds need to be dried and ground in order to remove surface moisture content to obtain constant weight and weaken or rupture the cell walls to release oil for extraction, respectively (Akpan, 2006).

2.2 In-situ supercritical methanol transesterification

A batch type reactor at supercritical methanol was used for in-situ supercritical methanol transesterification of biodiesel from JCL seeds. The in-situ transesterification was carried out at temperatures and pressures ranging from 180 – 300 °C and 6 – 18 MPa, respectively. After a leak-check test, the reactor was pressurized with nitrogen to the desired pressure and heated to reaction temperature at a rate of 5 °C/min. After reaching desired temperature, the reaction was held for periods of 5 – 35 min. A JCL seeds-to-methanol ratio (1:15, 1:20, 1:30, 1:40 and 1:45 w/v) was also investigated. After each reaction, the vessel was removed from the heater and placed into a cold water bath to quench the reaction and depressurized to ambient pressure. The extracted product was discharged from the reactor and was vacuum-filtered on a Buchner funnel and the filter cake was washed with methanol. The extracted products from the in-situ transesterification were allowed to settle and separated into two phases in 500 ml separating funnel. It took about 30 min to separate into two phases, i.e., the top phase consists of the biodiesel (fatty acid methyl ester) and the lower phase consists of the glycerol and other minor components. The schematic diagram of the experimental apparatus of the batch-wise extraction system is shown in Fig. 1.

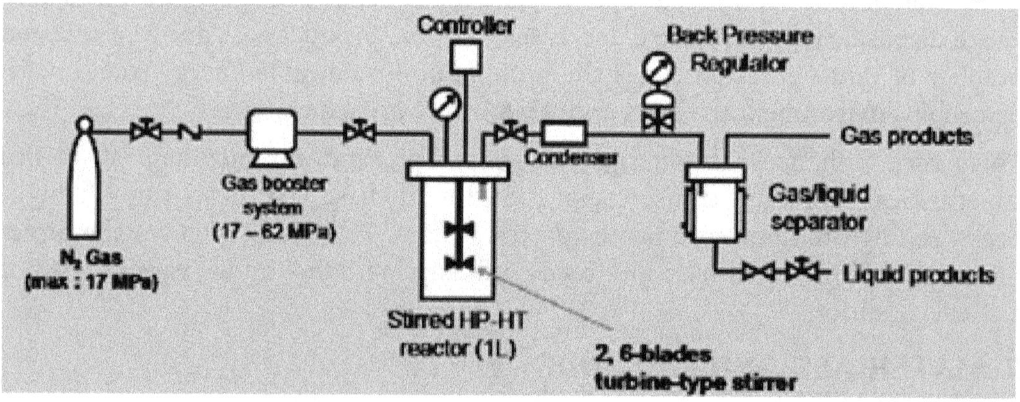

Fig. 1. Schematic diagram of batch-wise extraction system.

2.3 FAMEs analysis

The FAMEs analysis was quantified by Agilent Technologies 6890N with HP-5 5% Phenyl Methyl Siloxane capillary column (30 m by 320 µm by 0.25 mm) and a flame ionization detector. Methyl heptadecanoate (10.0 mg; internal standard) was dissolved in 1 ml hexane to prepare the standard solution. Approximately 100 mg crude methyl ester was dissolved in 1 ml standard solution for GC analysis (Hong, 2009). Approximately 1 µl sample was injected into the GC at an oven temperature of 210 °C with Helium as the carrier gas. The GC oven was programmed at 210 °C, isothermally for 15 min. the FAMEs content was calculated by use of the Equation 1:

$$C = \frac{\sum A - A_{IS}}{A_{IS}} \times \frac{C_{IS} \times V_{IS}}{m} \times 100\% \quad (1)$$

Where:

ΣA = total peak area of methyl ester

A_{IS} = peak area of internal standard (methyl heptadecanoate)

C_{IS} = concentration of the internal standard solution, in mg/ml

V_{IS} = volume of the internal standard solution used, ml

m = mass of the sample, in mg

2.4 Biodiesel properties

The biodiesel was characterized by its density, viscosity, high heating value, cloud and pour points and flash points according to ASTM standards.

3. RESULTS AND DISCUSSION

3.1 Effect of temperature

The effect of temperature on percent of FAMEs yields from JCL seeds were investigated. The parameters were fixed at 12 MPa of pressure, 1:40 (w/v) of seeds-to-methanol ratio, 30 min of reaction time and at varying temperatures of 180, 200, 240, 280 and 300 °C. The results of in-situ supercritical methanol on percent of FAMEs yields from JCL seeds at various temperatures are shown in Table 1. For simplification, the data are also plotted in Fig. 2.

Table 1. In-situ supercritical methanol transesterification[1] results from JCL seeds at various temperatures on percent of FAMEs yield and its contents.

Temperature (°C)	Yields (%)					
	FAMEs	Methyl Palmitate	Methyl Oleate	Methyl Linoleate	Methyl Stearate	Others
180	63.9	10.3	27.9	22.1	3.6	36.1
200	76.0	13.4	34.6	23.2	4.8	24.0
240	90.3	16.2	36.4	31.1	6.6	9.7
280	97.9	18.1	39.5	33.2	7.1	2.1
300	90.9	16.3	36.6	31.3	6.7	9.1

[1]conditions: 12 MPa, 30 min and 1:40 (w/v) seeds-to-methanol ratio.

From Table 1, the results indicate that the percent of FAMEs yields obtained at temperatures of 180 to 300 °C were in the range of 63.9 – 97.9%. The saturated FAMEs content of the seed samples are low, which is between 10.3 – 18.1% for methyl palmitate and 3.6 – 7.1% for methyl stearate. Meanwhile, the content of unsaturated FAMEs, methyl oleate and methyl linoleate are considerably higher at 27.9 – 39.5% and 22.1 – 33.2%, respectively. It should be noted that the critical temperature of methanol is at 240 °C and therefore, the conditions at 180 – 200 °C, 240 and >240 – 300 °C represent subcritical, supercritical and postcritical states of the medium, respectively.

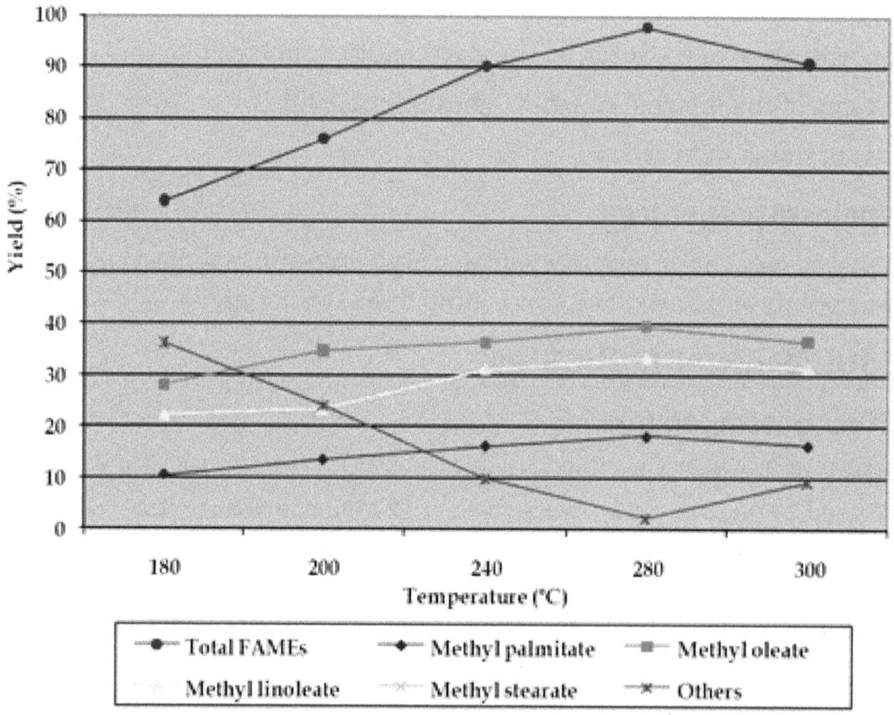

Fig. 2. In-situ supercritical methanol transesterification results from JCL seeds at various temperatures on percent of FAMEs yield and its contents.

At 180 °C, which is the lowest temperature of investigation, low yields of FAMEs (63.9%) were obtained. This observation might be due to the subcritical state of methanol or the instability of the supercritical state of methanol. It was observed that by increasing the reaction temperature to supercritical conditions had a favorable influence on the yield of ester conversion (Demirbas, 2008). Similar results have been reported by Cao et al., (2005), Madras et al., (2004) and Bunyakiat et al., (2006) on soybean oil, sunflower oil and coconut oil, respectively.

Apparently, by increasing the reaction temperature from 200 to 280 °C, the conversion increases significantly with FAMEs yields increased from 76.0 – 97.9%. The higher conversions observed in the supercritical state can be attributed to the formation of a single phase between alcohol and oil (Madras et al., 2004). Under supercritical conditions, the solubility parameter of alcohol reduces and was close to the solubility parameter of oil (Han et al., 2005). According to Petchmala et al.,

(2008), the increase in temperature causes the polarity of methanol to decrease, as a result of the breakdown of the hydrogen bonding of methanol, leading to an increased in the solubility of fatty acids in methanol. The complete solubility occurs as the temperature approaches the mixture critical temperature, at which point the reaction mixture became homogeneous and reaction took place rapidly. In addition, higher temperature contributed to the decomposition of cell walls, and as a result crude biodiesel yield was increased (Machmudah et al., 2007).

At 300 °C, the percent of FAMEs (90.9%) yields were slightly decreased. This observation was due to the decomposition of polyunsaturated methyl esters and unreacted triglycerides in postcritical methanol at severe high temperature (Tan et al., 2009). This finding was further supported by Xin et al., (2008) who suggested that the favorable reaction temperature adopted in supercritical methanol method should be lower than 300 °C. Reaction temperature at above 380 ºC is insuitable for transesterification reaction because the oil and methyl esters tend to decompose at the highest rate. Furthermore, Kusdiana and Saka's (2001) pointed out that saturated and unsaturated FAMEs undergo side reactions such as thermal decomposition and dehydrogenation reactions at temperature >400 °C and >350 °C, respectively. In these experiments, the temperature used was lower than that of Kusdiana and Saka's work and the side reactions did not occur since the temperature was below 300 °C. Furthermore, at 300 °C, a strong burning smell of the extract was detected. Hence, at this point, there is no reason to further increase the extraction temperature beyond 280 °C.

3.2 Effects of pressure

The results of in-situ supercritical methanol transesterification on percent of FAMEs yields from JCL seeds at various pressures are shown in Table 2. For simplification, the data are also being plotted and is shown in Fig. 3. The temperature was fixed at 280 °C based on the maximized yield conditions from the previous experiment.

Table 2. In-situ supercritical methanol transesterification1 results from JCL seeds at various pressures on FAMEs yield and its contents.

Pressure (MPa)	Yields (%)					
	FAMEs	Methyl	Methyl Oleate	Methyl Linoleate	Methyl Stearate	Others
6	80.6	13.1	41.0	20.4	6.1	19.4
8	95.6	15.7	47.1	26.3	6.5	4.4
12	97.9	18.1	39.5	33.2	7.1	2.1
16	93.5	16.0	38.4	32.1	7.0	6.5
18	92.5	16.0	38.6	31.1	6.8	7.5

[1]conditions: 280 °C, 30 min and 1:40 (w/v) seeds-to-methanol ratio.

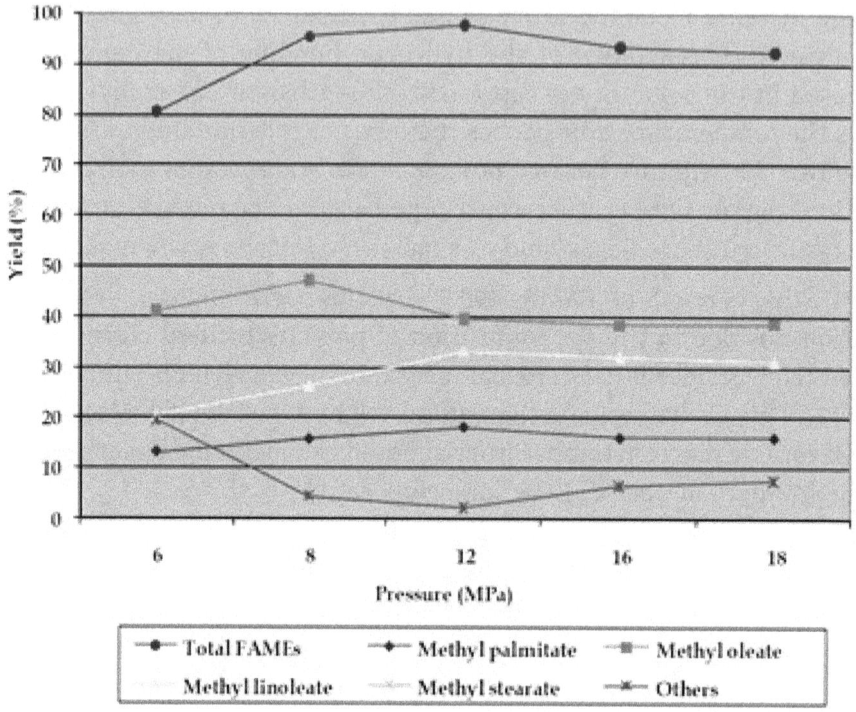

Fig. 3. In-situ supercritical methanol transesterification results from JCL seeds at various pressures on FAMEs yield and its contents.

From Table 2, the results indicated that the percent of FAMEs yields obtained at temperatures of 280 °C and pressures of 6-18 MPa was in the range of 80.6 – 97.9% with maximum yields at 12 MPa. The saturated FAMEs content of the seed samples are low, which is between 13.1 – 18.1% for methyl palmitate and 6.1 – 7.1% for methyl stearate. Meanwhile, the content unsaturated FAMEs, methyl oleate and methyl linoleate are considerably higher at 38.6 – 47.1% and 20.4 – 33.2%, respectively.

At the lowest pressure of 6 MPa, FAMEs yields are only 80.6%, but increases to 97.9% when the pressure are increased to 12 MPa. The high FAMEs yields achieved at 12 MPa, which is slightly above the critical pressure of methanol (8.09 MPa), might be due to the increase in solvent power of methanol with increasing pressure.

Further, increasing the pressure to 18 MPa, the FAMEs yield decreases slightly to 92.5%. After the pressure increased to a specific level, the increase of pressure does not cause an obvious improvement in the FAME yield (He et al., 2007). This phenomenon might be due to the maximum solubility and/or hydrogen donor ability of the solvent that has been achieved regardless of high pressure employed.

As the pressure of the system increased, the solubility parameter of the methanol decreased and is close to the solubility parameter of the oil, thus forming a single phase between the alcohol and the oil. Based on these results, it can be seen that the fact that both temperature and pressure play an important role that contributes to high extraction yield, with the later being more prominent. Based on these results, it can be seen that the fact that both temperature and pressure play an important role that contributes to high yield, with the later being more prominent.

3.3 Effects of reaction time

Table 3 and Fig. 4 shows the effect of reaction time on percent of FAMEs yields from JCL seeds using in-situ supercritical methanol transesterification. The reaction conditions were fixed based on maximum yields at optimized conditions discussed previously, i.e. 280 °C of temperature and 12.7 MPa of pressure.

Table 3. In-situ supercritical methanol transesterification[1] results of JCL seeds at various reaction times on percent of FAMEs yield and its contents.

Pressure (MPa)	Yields (%)					
	FAMEs	Methyl	Methyl Oleate	Methyl Linoleate	Methyl Stearate	Others
5	88.4	15.0	38.5	28.6	28.6	11.6
10	94.2	16.6	40.5	30.6	6.5	5.2
20	96.0	17.2	39.6	32.0	7.2	4.0
30	97.9	18.1	39.5	33.2	7.1	2.1
35	93.1	16.6	38.8	30.8	6.9	6.9

[1]conditions: 280 °C, 12.7 MPa and 1:40 (w/v) seeds-to-methanol ratio.

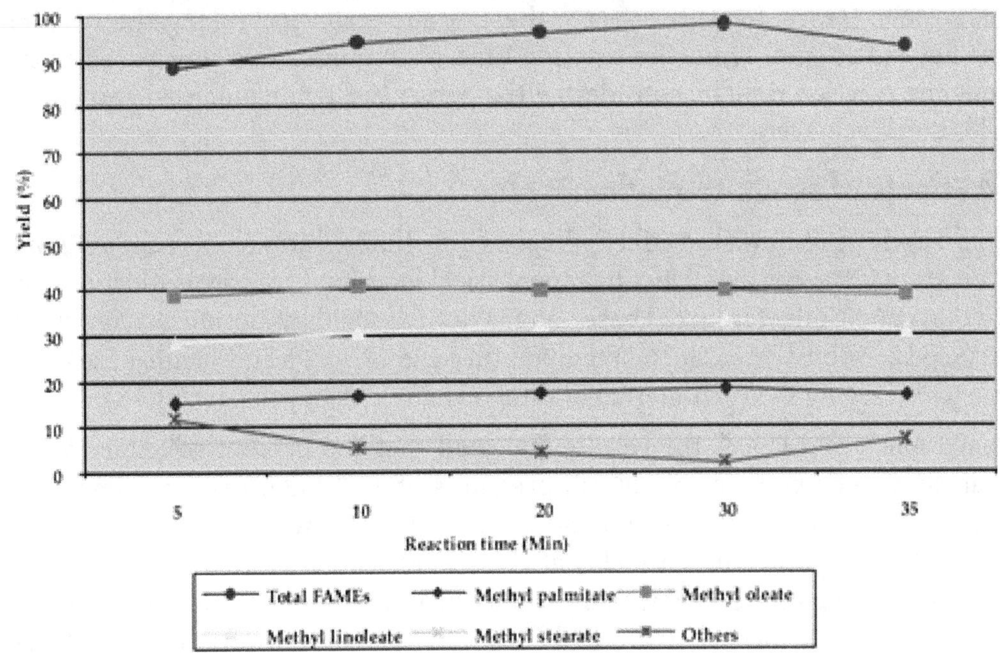

Fig. 4. In-situ supercritical methanol transesterification results from JCL seeds at various reaction times on percent of FAMEs yield and its contents.

From Table 3 and Fig. 4, the results indicated that the percent of FAMEs yields obtained at temperatures of 280 °C, pressures of 12.7 MPa, seeds-to-methanol ratio of 1:40 (w/v) and reaction time of 5 – 35 min was in the range of 88.4 – 97.9% with maximum yields at 30 min.

The saturated FAMEs content of the seed samples are low, which is between 15.0 – 18.1% for methyl palmitate and 6.3 – 7.2% for methyl stearate. Meanwhile, the content unsaturated FAMEs, methyl oleate and methyl linoleate are considerably higher at 38.5 – 39.5% and 28.6 – 33.2%, respectively.

From the results, it can be seen that the percent of FAMEs yields were only 88.4% at 5 min of reaction time. According to Saka and Kusdiana (2001), in the common method, the reaction is initially slow because of the two-phase nature of the methanol/oil system, and slows even further because of the polarity problem even with the help of an acid or an alkali catalyst. However, as described in this work, supercritical method can readily solve these problems because of the supercritical temperature and pressure employed. It can be seen that the conversion was increased in the reaction time ranges between 5 and 30 min with the percent of FAMEs yields showed a slight increase in the range of 88.4 – 97.9%.

Further, the results indicated that an extension of the reaction time from 30 to 35 min had leads to a reduction in the FAMEs yield (93.1%). This is because longer reaction time enhanced the hydrolysis of esters (reverse reaction of transesterification), resulted in loss of esters as well as causing more fatty acids to form soap (Eevera et al., 2009). Hence, for this process, there is no reason to prolong the reaction time beyond 30 min. Thus, the reaction time of 30 min can be considered as the economic reaction time by considering the percent of crude biodiesel and FAMEs yields being achieved.

3.4 Effects of seeds-to-methanol ratio

Table 4 and Fig. 5 shows the effect of seeds-to-methanol ratio on percent of FAMEs yields from JCL seeds using in-situ supercritical methanol transesterification. The reaction conditions were fixed based on maximized yields at optimized conditions discussed previously, i.e. 280 °C of temperature and 12.7 MPa and 30 min of reaction time with varying seeds-to-methanol ratio of 1:20, 1:30 and 1:40 (w/v).

From Table 4 and Fig. 5, the results indicated that the percent of FAMEs yields obtained at temperatures of 280 °C, pressures of 12.7 MPa, reaction time of 30 min and at various seeds- to-methanol ratio (1:15 – 1:45 w/v) was in the range of 89.0 – 97.9%, with maximum yields at 1:40 (w/v). The saturated FAMEs content of the seed samples are low, which is between 15.2 – 18.1% for methyl palmitate and 6.6 – 7.5% for methyl stearate. Meanwhile, the content unsaturated FAMEs, methyl oleate and methyl linoleate are considerably higher at 37.4 – 39.5% and 29.8 – 33.2%, respectively.

Table 4. In-situ supercritical methanol transesterification1 results of JCL seeds at various seed-to-methanol ratios on percent of FAMEs yield and its contents.

Seed-to-methanol ratio (w/v)	Yields (%)					
	FAMEs	Methyl	Methyl Oleate	Methyl Linoleate	Methyl Stearate	Others
1:15	89.0	15.2	37.4	29.8	6.6	11.0
1:20	94.4	16.9	38.8	31.6	7.1	5.6
1:30	95.9	17.5	38.6	32.3	7.5	4.1
1:40	97.9	18.1	39.5	33.2	7.1	2.1
1:45	97.0	17.4	40.1	32.3	7.2	3.0

1 conditions: 280 °C, 12.7 MPa, 30 min reaction time.

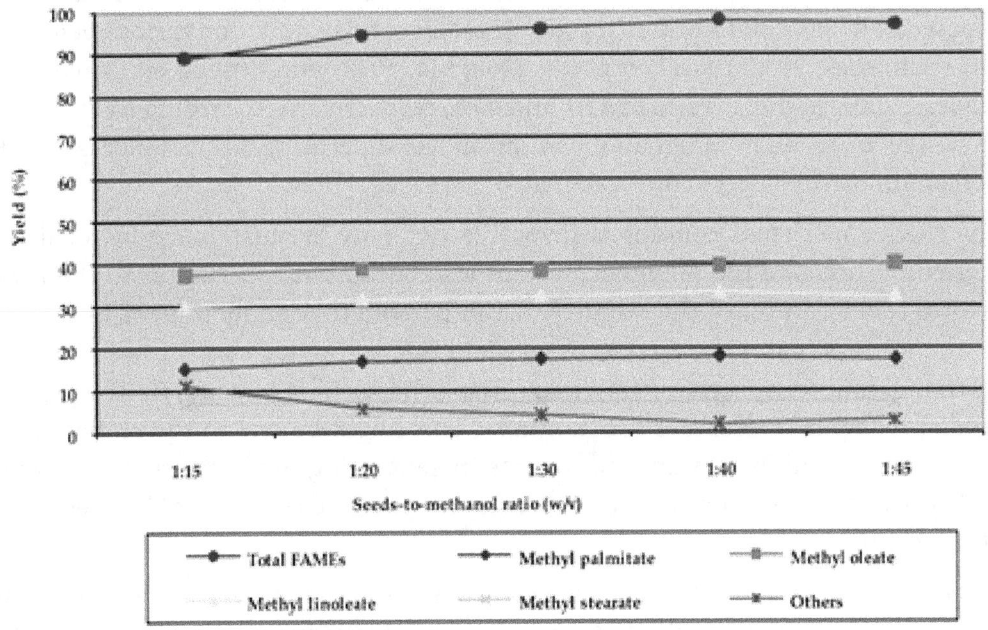

Fig. 5. In-situ supercritical methanol transesterification results of JCL seeds at various seeds- to-methanol ratios on percent of FAMEs yield and FAMEs contents.

From Table 4 and Fig. 5, the results indicated that the percent of FAMEs yields obtained at temperatures of 280 °C, pressures of 12.7 MPa, reaction time of 30 min and at various seeds- to-methanol ratio (1:15 - 1:45 w/v) was in the range of 89.0 - 97.9%, with maximum yields at 1:40 (w/v). The saturated FAMEs content of the seed samples are low, which is between 15.2 – 18.1% for methyl palmitate and 6.6 - 7.5% for methyl stearate. Meanwhile, the content unsaturated FAMEs, methyl oleate and methyl linoleate are considerably higher at 37.4 - 39.5% and 29.8 - 33.2%, respectively.

Obviously, at the lowest seeds-to-methanol ratio of 1:15 (w/v), the percent of FAMEs yields was relatively low (89.0%) and increased with increasing seeds-to-methanol ratio. When the methanol content in the supercritical fluids increased, the percent conversion of methyl ester also increased. The higher methanol content is favorable not only because more molecules of methanol surround the oil molecules but also because it contributes to the lower critical temperature of the mixture. Maximum percent of crude biodiesel and FAMEs yields were obtained at a 1:40 (w/v) of seeds-to-methanol ratio. This is a significant difference from conventional catalytic reaction for which at least 1 h of reaction time is needed to attain the same yield. In this reaction, an excess of methanol was used in order to shift the equilibrium in the direction of the products (Demirbas, 2007). Kusdiana and Saka (2001) have suggested that higher molar ratios of methanol to oil also result in a more efficient transesterification reaction. The results obtained shows good agreement with previous work, where maximum conversion was obtained for rapeseed oil (Saka and Kusdiana, 2001) at molar ratio of 42:1, for various vegetable oils (Demirbas, 2002; Diasakou et al., 1998; Ma, 1998) and linseed oil (Varma and Madras, 2007) at molar ratio of 41:1 and 40:1, respectively. According to Bunyakiat et al., (2006), when the methanol content in the supercritical fluids increased, the percent of methyl esters conversion also increased.

The higher methanol content is favorable not only because more molecules of methanol surround the oil molecules but also because it contributes to the lower critical temperature of the mixture. It can be seen that an increment in seed-to-methanol ratio can enhance biodiesel yield due to higher contact area between methanol and triglycerides. However, when the ratio is beyond 40, the yield of biodiesel begins to decrease substantially. This might be due to the restriction of the reaction equilibrium and difficulties in separating excessive methanol from methyl esters and glycerol, which subsequently lowered the yield of biodiesel (Tan et al., 2009).

Moreover, it was observed that for high seeds-to-methanol ratio added the set up required longer time for the subsequent separation stage since separation of the FAMEs layer from the organic layer becomes more difficult with the addition of a large amount of methanol. This is due to the fact that methanol, with one polar hydroxyl group, can work as an emulsifier that enhances emulsion. Operating beyond the optimal value, the ester yield would not be increased but will result in additional cost for methanol recovery (Eevera et al., 2009). Therefore, increasing the seeds-to-methanol ratio is another important parameter affecting the FAMEs yield. This report is in line with the results of many investigations based on neat vegetable oils (Freedman et al., 1984; Zhang et al., 2003; Leung et al. 2006, Eevera et al., 2009).

3.5 Biodiesel characterization

The biodiesel obtained through the one-step supercritical methanol extraction and transesterification in-situ process in this experiment was dark yellow in color. Compositions of samples were analyzed by GC. Figure 6 shows the total ion current chromatogram of the biodiesel. Furthermore, Table 5 shows the names, structure and compositions of *Jatropha curcas* L. FAMEs.

Fig. 6 depicts the gas chromatographic evaluation of the FAMEs produced over the course of reaction. The methyl esters analyzed by GC appear in the retention time of less than 15 min in the chromatograms. The weight percentages were similar for all of the variables condition; temperature, pressure, reaction time and seeds-to-methanol ratio of in-situ transesterification, as suggested by Carrapiso et al., (2000) that transesterification was random. The average saturated FAMEs content of the seed samples are low: 18.1% for methyl palmitate (C17:0) and 7.1% for methyl stearate (C19:0). The average content of the unsaturated FAMEs, methyl oleate (C19:1) and methyl linoleate (C19:2) are considerably higher at 39.5 and 33.2%, respectively which are comparable to the fatty acid composition in crude JCL oil feedstock. Depending on the origin, either oleic or linoleic acid content is higher. In this case, the seed oil belongs to the oleic or linoleic acid group, to which similar to the majority of vegetable oils (Carrapiso et al., 2000).

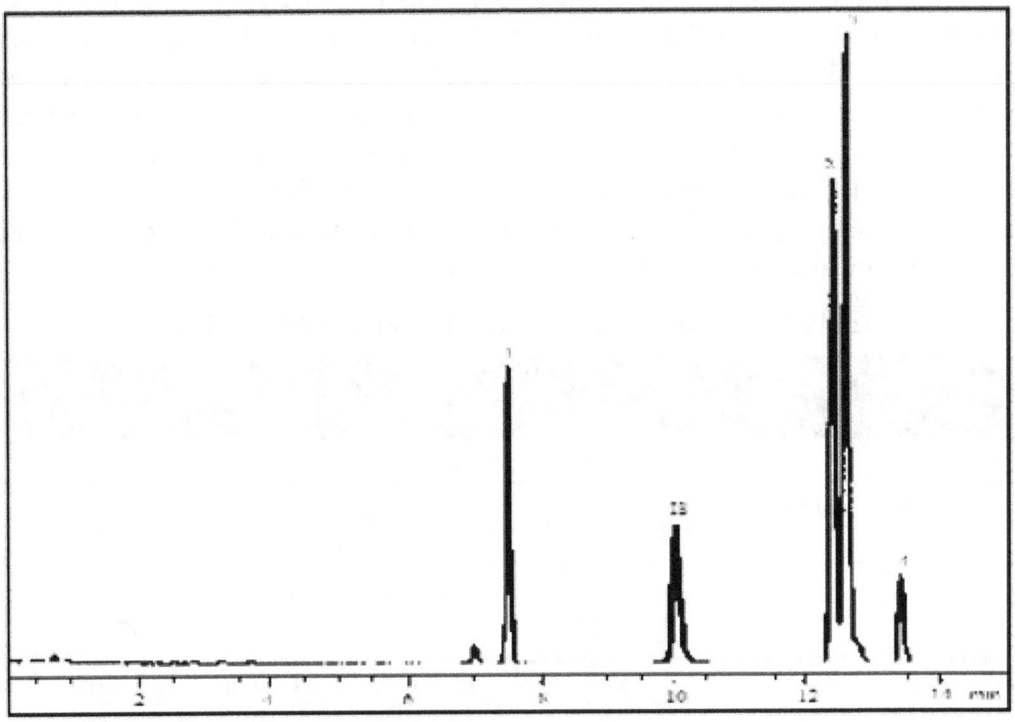

IS: Internal standard (Methyl heptadecanoate)

Fig. 6. Total ion current chromatogram of the biodiesel.

Table 5. Names, structure and compositions of *Jatropha curcas* L. FAMEs.

Peak No.	Name	Wt%
1	Methyl Palmitate	18.46
2	Methyl Oleate	40.41
3	Methyl Linoleate	33.91
4	Methyl stearate	7.22

3.5.1 Biodiesel characterization

Vegetable oil methyl esters, commonly referred to as "biodiesel" are prominent candidates as alternative Diesel fuels. Biodiesel is technically competitive with or offers technical advantages compared to conventional petroleum Diesel fuel. The vegetable oils, as alternative engine fuels, are all extremely viscous with viscosities ranging from 10 to 20 times greater than that of petroleum Diesel fuel (Demirbas, 2003). The purpose of the transesterification process is to lower the viscosity of the oil. In this study, in-situ supercritical methanol transesterification for production of biodiesel from Jatropha curcas L. (JCL) seeds was generate via 1000 ml high-temperature high-pressure batch-wise reactor system in an absence of catalyst. The reaction conditions were conducted at 280 °C of temperature, 12.7 MPa of pressure, 30 min of reaction time and 1:40 of seeds-to-methanol ratio at 450 rpm of stirring rate. Samples of the biodiesel obtained from the in-situ experiment were determined using reference methods published by American Society for Testing and Materials (ASTM) D6751. In order to ensure that it can be used in diesel engine without any modification, the properties of biodiesel produced from this in-situ transesterification reaction was comparable with fuel properties of No. 2 Diesel. Fuel properties of No. 2 Diesel, JCL biodiesel and ASTM D6751 derived biodiesel standards is shown in Table 6 for comparison.

Table 6. Fuel properties of No. 2 Diesel and JCL biodiesel.

Properties	No.2 Diesela	JCL biodieselb	ASTM D6751a	JCL Biodiesel (This study)
Specific gravity	0.85	0.86 to 0.87	0.87 to 0.90	0.87
Kinematic viscosity @ 40 °C (cSt)	1.9 to 4.1	4.23 to 5.65	1.9 to 6	5.27
Cloud point (°C)	-19 to -8	8 to 10.2	Report	-2.06
Pour point (°C)	-34 to -10	4.2 to 6	-15 to 10	0
Flash point (°C)	51 to 85	130 to 192	130 min	100
Calorific value (MJ/kg)	45.0 to 45.3	38.5-42.7	41.0	39.3

[a]Demirbas, (2008); Encinar, (2005); Vyas, (2009)

[b]Ghadge and Rehman, (2005); Vyas, (2009); Sahoo and Das, (2009)

The properties of biodiesel produced from this in-situ supercritical methanol transesterification were comparable with fuel properties of commercial No. 2 Diesel. It was found that specific gravity of JCL biodiesel was 0.87 g/cm^3 and it falls between the ASTM D6751 ranges. Fuel injection equipment operates on a volume metering system, hence a higher density for biodiesel results in the delivery of a slightly greater mass of fuel (Demirbas, 2005). The kinematic viscosity was 5.27 cSt. Among the general parameters for biodiesel the viscosity of FAMEs can go very high levels and hence it is important to control it within an acceptable level to avoid negative impacts on fuel injector's system performance (Murugesan et al., 2009). The flash point was determined to be at 100 °C. Since biodiesel has a higher flash point than diesel, it is a safer fuel than diesel. Addition of a small quantity of biodiesel with diesel increases the flash point of diesel which can result in improved fire safety for transport purpose (Lu et al., 2009) and it is safer to store biodiesel-diesel blends in comparison to diesel alone (Sahoo et al., 2009). Meanwhile, the pour point was measured to be 0 °C which was slightly higher than that of No. 2 Diesel fuel. This might be due to the presence of wax, which begins to crystallize with the decrease in temperature. This finding was agreed with Vyas et al., (2009) and Raheman and Ghadge, (2007). The problems of higher pour point of JCL biodiesel could be overcome by blending with diesel. The cloud point was reported to be -2.06 °C. The cloud point depends upon the feedstock used and must be taken into consideration if the fuel is to be used in cold environments (Fernando et al., 2007). The calorific value of JCL biodiesel was 39.3 MJ/kg, which was almost 88% of the calorific value of diesel (44.8 MJ/kg). The lower calorific value of JCL is because of the presence of oxygen in the molecular structure, which is confirmed by elemental analysis also. Furthermore, the presence of oxygen in the biodiesel helps for complete combustion of fuel in the engine. These findings were also agreed by Sinha et al., (2008). Therefore, they could be excellent substitutes and blends of No. 2 diesel fuel.

4. CONCISE SUMMARY

The data indicate that temperature is a crucial factor in this in-situ process. With the rise in temperature, the yields of crude biodiesel and fatty acid methyl esters (FAMEs) likewise escalated.. The crude biodiesel and FAMEs of the yields reached a maximum (59.9 and 97.9, respectively) at 280 °C and then decreased with increasing temperature. The loss was caused by thermal decomposition, dehydrogenation and other side reactions. For the effect of pressure, the crude biodiesel and FAMEs yield increased with increasing pressure. Above 12 MPa, no improvement of both yields was observed. The optimum pressure was thus fixed at 12.7 MPa in this experiment. For the effect of reaction time, it can be seen that the conversion was increased in the reaction time ranges between 5 and 30 min, and thereafter reduced as a representative of the equilibrium conversion. The excess reaction time did not promote the conversion but favors the reverse reaction of transesterification which resulted in a reduction in the ester yield. The optimal FAMEs yield was found to be 97.9% in 30 min. For the effect of seeds-to-methanol ratio, the maximum crude biodiesel and FAMEs yields were obtained at a 1:40 of seeds-to-methanol ratio. It can be seen that an increment in seed-to- methanol ratio can enhance biodiesel yield due to higher contact area between methanol and triglycerides. However, when the ratio is beyond 40, the yield of biodiesel begins to decrease substantially.

The advantage of this technology is that it necessitates only a singular process, so circumventing the conventional oil extraction procedure. Furthermore, due to the non-catalytic approach, the purification of products following the transesterification reaction is far simpler than the conventional method. Consequently, this novel approach provides an alternate method for the direct conversion of fruits into methyl esters through a more streamlined and abbreviated production process.

5. REFERENCES

Ayuso, L. E. G. & Castro M. D. L. (1999). A multivariate study of the performance of a microwave-assissted Soxhlet extractor for olive seeds. *Analytica Chimica Acta*, Vol. 382(3), pp. 309-316.

Berchmans, H. J. & Hirata, S. (2008). Biodiesel production from crude *Jatropha curcas* L. seed oil with a high content of free fatty acids. *Bioresource Technology*, Vol. 99, pp. 1716– 1721.

Bunyakiat, K.; Makmee, S.; Sawangkeaw, R. & Ngamprasertsith. (2006). Continous production of biodiesel via transesterification from vegetable oils in supercritical methanol. *Energy & Fuel*, Vol. 20, pp. 812-817.

Cao, W.; Han, H. & Zhang, J. (2005). Preparation of biodiesel from soybean oil using supercritical methanol and co-solvent. *Fuel*, Vol. 84(4), pp. 347-351.

Cao, X. & Ito, Y. (2003). Supercritical fluid extraction of grape seed oil and subsequent separation of free fatty acids by high-speed counter-current chromatography. *Journal of Chromatography A*, Vol. 1021, pp. 117–124.

Carrapiso, A. I. and García, C. (2000). Development in lipid analysis: some new extraction techniques and in situ transesterification. *Lipids*, Vol. 35, pp. 1167–1177.

Castro, M. D. L. & Ayuso, L. E. G. (1998). Soxhlet extraction of solid materials: Soxhlet extraction of solid materials: anoutdated technique with a promising innovative future. *Analytica Chimica Acta*, Vol. 369(1-2), pp. 1-10.

Chen, W. H.; Chen, C. H.; Chang, C. M. J.; Chiu, Y. H. & Hsiang, D. (2009). Supercritical carbon dioxide extraction of triglycerides from *Jatropha curcas* L. seeds. *Journal of Supercritical Fluids*, Vol. 51, pp. 174–180.

Chew, C. S. (2009). *Jatropha curcas.* L.: Development of a new oil crop for biofuel (Summary). New and Renewable Energy Group, Strategy and Industry Research Unit, The Institute of Energy Economics, Japan (IEEJ).

Clark, S. J.; Wagner, L.; Schrock, M. D. & Pinnaar, P. G. (1984). Methyl and ethyl esters as renewable fuels for diesel engines. *Journal of American Oil Chemists' Society,* Vol. 61, pp. 1632–1638.

Darcia, J. L. L. & Castro, M. D. L. (2004). Ultrasound-assisted Soxhlet extraction: An expeditive approach for solid sample treatment – Application to the extraction of total fat from oleaginous seeds. *Journal of Chromatography A,* Vol. 1034 (1-2), pp. 237- 242.

De, B. K.; Bhattacharyya, D. K. & Bandhu, C. (1999). Enzymatic synthesis of fatty alcohol esters by alcoholysis. *Journal of American Oil Chemists' Society,* Vol. 76, pp. 451-453.

Demirbas, A. (2003). Biodiesel fuels from vegetable oils via catalytic and non-catalytic supercritical alcohol transesterifications and other methods: a survey. *Energy Convers Mgmt, Vol.* 44, pp. 2093–109.

Demirbas, A. (2005). Biodiesel production from vegetable oils via catalytic and non catalytic supercritical methanol transesterification methods. *Progress in Energy and Combustion Science, Vol.* 31, pp. 466-487.

Demirbas, A. (2007). Biodiesel from sunflower oil in supercritical methanol with calcium oxide. *Energy Conversion and Management,* Vol. 48(3), pp. 937-941.

Demirbas, A. (2008). Studies On cottonseed oil biodiesel prepared in non-catalytic SCF conditions. *Bioresource Technology,* Vol. 99(5), pp. 1125-1130.

Diasakou, M.; Louloudi, A. & Papayannakos, N. (1998). Kinetics of the non-catalytic transesterification of soybean oil. *Fuel,* Vol. 77(12), pp. 1297-1302.

Eevera, T.; Rajendran, K. & Saradha, S. (2009). Biodiesel production process optimization and characterization to assess the suitability of the product for varied environmental conditions. *Renewable Energy,* Vol. 34(3), pp. 762-765.

Encinar, J. M.; Gonzalez, J. F. & Reinares, A. R. (2005). Biodiesel from Used Frying Oil. Variables Affecting the Yields and Characteristics of the Biodiesel. *Industrial & Engineering Chemistry Research,* Vol. 44(15), pp. 5491–5499.

Fernando, S.; Karra, P.; Hernandez, R. & Jha, S. K. (2007). Effect of incompletely converted soybean oil on biodiesel quality. *Energy,* Vol. 32, pp. 844-851.

Freedman, B.; Pyrde, E. H. & Mounts, T. L. (1984). Variables affecting the yield of fatty esters from transesterified vegetable oils. *Journal of American Oil Chemists' Society,* Vol. 61(10), pp. 1638–1643.

Fukuda, H.; Kondo, A. & Noda, H. (2001). Biodiesel fuel production by transesterification of oils. *Journal of Bioscience and Bioengineering*, Vol. 92(5), pp. 405-416.

Gaur, R.; Sharma, A.; Khare, S. K. & Gupta, M. N. (2007). A novel process for extraction of edible oils: Enzyme assisted three phase portioning (EATPP). *Bioresource Technology*, Vol. 98(3), pp. 696-699.

Ghadge, S. V. & Raheman, H. (2005). Biodiesel production from mahua (Madhuca indica) oil having high free fatty acids. *Biomass and Bioenergy*, Vol. 28, pp. 601–605.

Han, H.; Cao, W. & Zhang, J. (2005). Preparation of biodiesel from soybean oil using supercritical methanol and Co_2 as co-solvent. *Process Biochemistry, Vol.* 40, pp. 3148- 3151.

Harrington, K. J. & Evans, C. D. A. (1985). Transesterification in situ of sunflower seed oil. *Industrial & Engineering Chemistry Product Research & Development,* Vol. 62, pp. 314– 318.

Hass, M. J.; Scott, K. M.; Marmer, W. N. & Foglia, T. A. (2004). In situ alkaline transesterification: an effective method for the production of fatty acid esters from vegetable oils, *Journal of American Oil Chemists' Society*, Vol. 81, pp. 83–89.

Hawash, S.; Kamal, N.; Zaher, F.; Kenawi, O. & El Dewani, G. (2009). Biodiesel fuel from *Jatropha* oil via non-catalytic supercritical methanol transesterification. *Fuel*, Vol. 88(3), pp. 579-582.

He, H.; Wang, T. & Zhu, S. (2007). Continuous production of biodisel fuel from vegetable oil using supercritical methanol process. *Fuel, Vol.* 86, pp. 442-447.

Heller, J. (1996). *Physic nut. Jatropha curcas L. Promoting the conservation and use of underutilized and neglected crops.* Institute of Plant Genetics and Crop Plant Research, Gatersleben/ International Plant Genetic Resources Institute, Rome.

Herrera, J. M.; Siddhuraju, P.; Francis, G., Ortiz.; G. D. V. & Becker, K. (2006). Chemical composition, toxic/antimetabolic constituents, and effects of different treatments on their levels, in four provenances of *Jatropha curcas* L. from Mexico. *Food Chemistry*, Vol. 96(1), pp. 80–89.

Ikwuagwu, O. E.; Ononogbu, I. C. & Njoku, O. U. (2000). Production of biodiesel using rubber [*Hevea brasiliensis (Kunth. Muell.)*] seed oil. *Industrial Crops and Products*, Vol. 12(1), pp. 57-62.

Jiang, L.; Hua, D.; Wang, Z. & Xu, S. (2009). Aqueous enzymatic extraction of peanut oil and protein hydrolysates. *Food and Bioproducts Processing*, Vol. 88 (2-3), pp. 2010.

Kansedo, J.; Lee, K. T. & Bhatia, S. (2009). *Cerbera odollam* (sea mango) oil as promising non-edible feedstock for biodiesel production. *Fuel*, Vol. 88(6), pp. 1148-1150.

Kildiran, G.; Yucel, S. O. & Turkay, S. (1996). In-situ alcoholysis of soybean oil, *Journal of American Oil Chemists' Society*, Vol. 73, pp. 225–228.

King, J. W.; Mohamed, A.; Taylor, S. L.; Mebrahtu, T. & Paul, C. (2001). Supercritical fluid extraction of *Vernonia galamensis* seeds. *Industrial Crops and Products*, Vol. 14, pp. 241–249.

Kusdiana, D. & Saka, S. (2001). Kinetics of transesterification in rapeseed oil to biodiesel fuel as treated in Supercrititcal methanol. *Fuel*, Vol. 80, pp. 693-698.

Kusdiana, D. & Saka, S. (2004). Effects of water on biodiesel fuel production by supercritical methanol treatment. *Bioresource Technology*, Vol. 91, pp. 289–295.

Leung, D. Y. C.; Koo, B. C. P. & Guo, Y. (2006). Degradation of biodiesel under different storage conditions. *Bioresource Technology*, Vol. 97(2), pp. 250-256.

Lu, H.; Liu, Y.; Yang, Y.; Chen, M. & Liang, B. (2009). Production of biodiesel from *Jatropha curcas* L. oil. *Computers and Chemical Engineering*, Vol. 33 (5), pp. 1091- 1096.

Ma, F.; Clements, L. D. & Hanna, M. A. (1998). The effects of catalysts, free fatty acids and water on transesterification of beef tallow. *Transactions of the American Society of Agricultural Engineers, Vol.* 41, pp. 1261–1264.

Machmudah, S.; Kawahito, Y.; Sasaki, M. & Goto, M. (2007). Supercritical CO_2 extraction of rosehip seed oil: Fatty acids composition and process optimization. *Journal of Supercritical Fluids*, Vol. 41(3), pp. 421–428.

Machmudah, S.; Kondo, M.; Sasaki, M.; Goto, M.; Munemasa, J. & Yamagata, M. (2008). Pressure effect in supercritical CO_2 extraction of plant seeds. *Journal of Supercritical Fluids*, Vol. 44, pp. 301–307.

Madras, G.; Kolluru, C. & Kumar, R. (2004). Synthesis of biodiesel in supercritical fluids. *Fuel, Vol.* 83, pp. 2029-2033.

Marinkovic, S. S. &Tomasevic, A. (1998). Transesterification of sunflower oil in situ. *Fuel*, Vol. 77(12), pp. 1389-1391.

Meher L. C.; Kulkarni M. G.; Dalai A. K. & Naik, S. N. (2006). Transesterification of karanja (Pongamia pinnata) oil by solid basic catalysts. *European Journal of Lipid Science & Technology*, Vol. 108, pp. 389–397.

Murugesan, A.; Umarani, C.; Chinnusamy, T. R.; Krishnan, M.; Subramanian, R. & Neduzchezhain, N. (2009). Production and analysis of bio-diesel from non- edible oils—A review. *Renewable and Sustainable Energy Reviews*, Vol. 13, pp. 825–834.

Naik, M.; Meher, L. C.; Naik, S. N. & Das, L. M. (2008). Production of biodiesel from high free fatty acid Karanja (*Pongamia pinnata*) oil. *Biomass and Bioenergy*, Vol. 32(4), pp. 354-357.

Papamichail, I.; Louli, V. & Magoulas, K. (2000). Supercritical fluid extraction of celery seed oil. *Journal of Supercritical Fluids*, Vol. 18, pp. 213–226.

Petchmala, A.; Yujaroen, D.; Shotipruk, A.; Goto, M. & Sasaki, M. (2008). Production methyl esters from palm fatty acids in supercritical methanol. Chiang Mai Journal *of Science,* Vol. 35(1), pp. 23-28.

Pradhan, R. C.; Naik, S. N.; Bhatnagar, N. & Swain, S. K. (2008). Moisture-dependent physical properties of Karanja (*Pongamia pinnata*) kernel. *Industrial Crops and Products*, Vol. 28(2), pp. 155-161.

Qian, J.; Wang, F.; Liu, S. & Yun, Z. (2008). In situ alkaline transesterification of cottonseed oil for production of biodiesel and nontoxic cottonseed meal. *Bioresource Technology*, Vol. 99(18), pp. 9009-9012.

Raheman, H. & Ghadge, S. V. (2007). Performance of compression ignition engine with mahua (*Madhuca indica*) biodiesel. *Fuel*, Vol. 86(16), pp. 2568-2573.

Ramadhas, A. S.; Jayaraj, S. & Muraleedharan, C. (2005). Biodiesel production from high FFA rubber seed oil. *Fuel*, Vol. 84(4), pp. 335-340.

Rashid, U.; Anwar, F. & Knothe, G. (2009). Evaluation of biodiesel obtained from cottonseed oil. *Fuel Processing Technology*, Vol. 90(9), pp. 1157-1163.

Rathore, V. & Madras, G. (2007). Synthesis of biodiesel from edible and non-edible oils in supercritical alcohols and enzymatic synthesis in supercritical carbon dioxide. *Fuel*, Vol. 86, pp. 2650–2659.

Rosenthal, A.; Pyle, D. L. & Niranjan, K. (1996). Aqueous and enzymatic processes edible oil extraction for edible oil extraction. *Enzyme and Microbial Technology*, Vol. 19, pp. 402-420.

Sahoo, P. K. & Das, L. M. (2009). Process optimization for biodiesel production from *Jatropha*, karanja and *Polanga* oils. *Fuel*, Vol. 88(9), pp. 1588-1594.

Sahoo, P. K.; Das, L. M.; Babu, M. K. G. & Naik, S. N. (2007). Biodiesel development from high acid value polanga seed oil and performance evaluation in a CI engine. *Fuel*, Vol. 86(3), pp. 448-454.

Saka, S. & Kusdiana, D. (2001). Biodiesel fuel from rapeseed oil as prepared in supercritical methanol. *Fuel,* Vol. 80, pp. 225-231.

Selmi, B. & Thomas, D. (1998). Immobilized lipase-catalyzed ethanolysis of sunflower oil in a solvent-free medium. *Journal of American Oil Chemists' Society,* Vol. 75, pp. 691-695.

Sharma, A. & Gupta, M. N. (2006). Ultrasonic pre-irradiation effect upon aqueous enzymatic oil extraction from almond and apricot seeds. *Ultrasonics Sonochemistry*, Vol. 13(6), pp. 529-534.

Sinha, S.; Agarwal, A. K. & Garg, S. (2008). Biodiesel development from rice bran oil: Transesterification process optimization and fuel characterization. *Energy Conversion and Management*, Vol. 49(5), pp. 1248-1257.

Tan, K. T.; Lee,K. T. & Mohamed, A. R. (2009). Production of FAME by palm oil transesterification via supercritical methanol technology. *Biomass and Bioenergy*, Vol. 33(8), pp. 1096-1099.

Tiwari, A. K.; Kumar, A. & Raheman, H. (2007). Biodiesel production from *Jatropha* oil (*Jatropha Curcas)* with high free fatty acids: an optimized process", *Biomass and Bioenergy*, Vol. 31, pp. 569-575.

Usta, N. (2005). An experimental study on performance and exhaust emissions of a diesel engine fuelled with tobacco seed oil methyl ester. *Energy Conversions and management*, Vol. 46(15-16), pp. 2373-2386.

Varma, M. N. & Madras, G. (2007). Synthesis of biodiesel from castor oil and linseed oilin supercritical fluids, *Industrial & Engineering Chemistry Research,* Vol. 46, pp. 1–6.

Veljkovic, V. B.; Lakicevic, S. H.; Stamenkowc, O. S.,;Todorovic, Z. B. & Lazic, M. L. (2006). Biodiesel production from tobacco (*Nicotianna tabacum* L.) seed oil with a high content of free fatty acids. *Fuel*, Vol. 85(17-18), pp. 2671-2675.

Vyas, A. P.; Subrahmanyam, N. & Patel, A. (2009). Production of biodiesel through transesterification of *Jatropha* oil using KNO_3/Al_2O_3 solid catalyst. *Fuel*, Vol. 88(4), pp. 625-628.

Wang, Y.; Ou, S.; Liu, P.; Xue, F. & Tang, S. (2006). Comparison of two different processes to synthesize biodiesel by waste cooking oil. *Journal of Molecular Catalysis A: Chemical*, Vol. 252(1-2), pp. 107-112.

Warabi, Y.; Kusdiana, D. & Saka, S. (2004). Reactivity of triglycerides and fatty acids of rapeseed oil in supercritical alcohols. *Bioresource Technology,*Vol. 9, pp. 283- 287.

Xin, J.; Imahara, H. & Saka, S. (2008). Oxidation stability of biodiesel fuel as prepared by supercritical methanol. *Fuel*, Vol. 87(10-11), pp. 1807-1813.

Zhang, Y.; Dube, M. A.; McLean, D. D. & Kates, M. (2003). Biodiesel production from waste cooking oil: 1. Process design and technological assessment. *Bioresource Technology*, Vol. 89(1), pp. 1-16.

12 Chapter

TRANSESTERIFICATION IN SUPERCRITICAL CONDITIONS.

1. INTRODUCTION: AN OVERVIEW

Supercritical transesterification, the process of biodiesel production under supercritical conditions, is a catalyst-free chemical reaction involving triglycerides, the primary constituents of vegetable oils and/or animal fats, and low molecular weight alcohols, such as methanol and ethanol, conducted at temperatures and pressures exceeding the critical point of the mixture. Figure 1 illustrates the complete transesterification reaction.

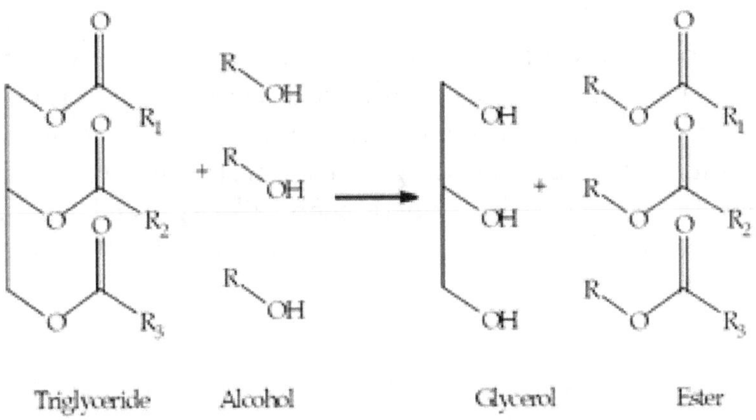

Fig. 1. The overall transesterification reaction (R is a small alkyl group, R_1, R_2 and R_3 are a fatty acid chain)

The reaction mechanism for supercritical transesterification has been proposed to be somewhat alike the acidic-catalyzed reaction as described in Section 1.2.

Since the actual feedstocks are not composed solely of triglycerides, especially the low-grade feedstocks, but are also contaminated with water and free fatty acids, some side reactions also take place under supercritical conditions (see Section 1.3). For example, the esterification of free fatty acids with alcohols increases the fatty acid alkyl ester content in the biodiesel product, while the thermal cracking of unsaturated fatty acids decreases the esters content.

The earlier research on supercritical transesterification mostly employed methanol as the reacting medium and reacting alcohol at the same time due to the fact that it has the lowest critical point and the highest activity (Warabi et al., 2004). Ethanol is also an interesting candidate because it can be industrially produced from renewable sources in many countries nowadays. However, other supercritical mediums, such as methyl acetate (Saka & Isayama, 2009) and dimethyl carbonate (Ilham & Saka, 2009; Tan et al., 2010b), have also been used to produce biodiesel, but these are not be described in this chapter since the chemical reaction involved is not a transesterification reaction.

1.1 The definition of supercritical transesterification

A pure substance ordinarily exists in a solid, liquid or gaseous state, depending on the temperature and pressure. For example, methanol is in a liquid state at ambient temperature (and pressure) and changes to a gaseous state above its boiling point. A gaseous substance can be compressed to a liquid state when a pressure above the boiling point is applied. Until the critical temperature is reached, a gaseous substance cannot be compressed to the liquid state. In the same manner, a compressed liquid substance cannot be heated to a gaseous state at its critical pressure.

Above its inherent critical temperature and pressure, the substance becomes a supercritical fluid, which is a non-condensable dense fluid. In the supercritical state, the density is generally in a range between 20 – 50% of that in the liquid state and the viscosity is close to that in the gaseous state. In other words, the molecules in the supercritical fluid have high kinetic energy like a gas and high density like a liquid. Therefore, the chemical reactivity can be enhanced in this state.

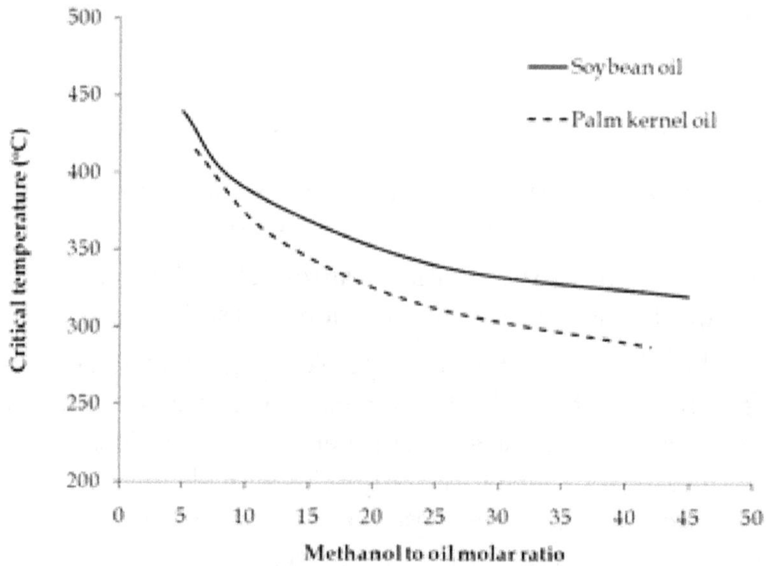

Fig. 2. The estimated critical temperature of soybean oil-methanol (Hegel et al., 2008) and palm kernel oil-methanol (Bunyakiat et al., 2006) mixtures.

The critical point of any transesterification reaction mixture is mostly calculated by the critical properties of the alcohols and the vegetable oils and/or animal fats. However, the critical properties of vegetable oils and/or animal fats cannot be experimentally measured because they thermally decompose before the critical point is reached. In addition, the molecular structure of vegetable oils and/or animal fats is impossible to know because the exact distribution of the fatty acids chain in triglycerides mixture is unknown.

Therefore to estimate the critical properties of vegetable oils and/or animal fats, their molecular is assumed to be a simple triglyceride (tripalmitin, triolein, etc.) or pseudo- triglycerides (Espinosa et al., 2002), with the proportion of such different simple triglycerides reflecting the actual overall fatty acid composition in the feedstock. The type of simple triglycerides or the pseudo-triglycerides are thus defined by their actual fatty acid profile in the vegetable oils and/or animal fats. For instance, soybean oil has linoleic acid as the major fatty acid, and so it is usually assumed to be trilinolein. Next, the critical properties of the simple triglycerides or the pseudo-triglycerides are estimated by the Fedor and Lydersen group contribution method (Poling et al., 2001), or similar. After the critical properties of each of the triglycerides are estimated, the critical point of the mixture can be estimated by mixing rules, such as the Lorentz-Berthelot-type (Bunyakiat et al., 2006) and the group-contribution with associated mixing rules (Hegel et al., 2008).

For example, the critical temperature and pressure of soybean oil-methanol and palm kernel oil-methanol mixtures are illustrated in Figs. 2 and 3, respectively.

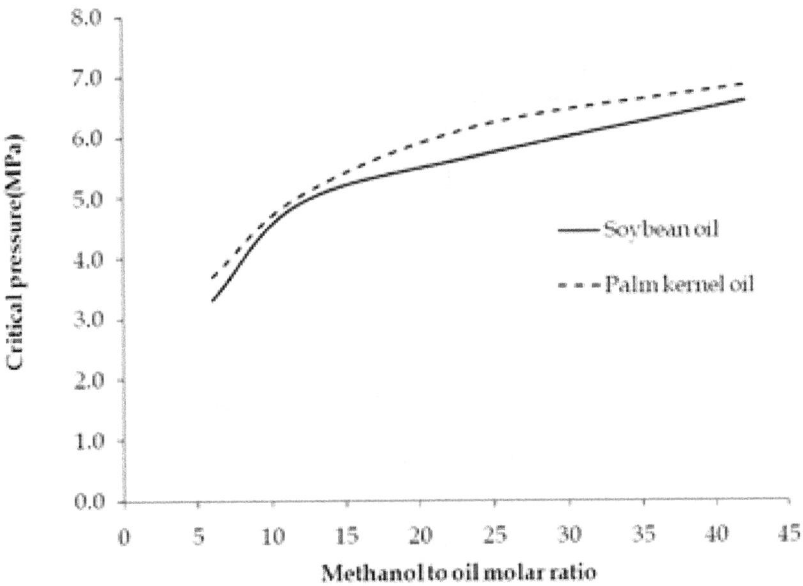

Fig. 3. The estimated critical pressure of soybean oil-methanol (Hegel et al., 2008) and palm kernel oil-methanol (Bunyakiat et al., 2006) mixtures.

From Figs. 3 and 4, it is clear that the critical point of the reaction mixture depends on the alcohol to oil molar ratio, so the selected alcohol to oil molar ratio will reflect the operating temperature and pressure, as described in Section 2.1. For a high transesterification conversion (triglyceride to alkyl ester) at a constant methanol to oil molar ratio, the operating temperature and pressure have to be approximately 1.5- to 2.0-fold over the critical point of the reaction mixture. For example, the optimal conditions at a methanol to oil molar ratio of 42:1 is 350 °C and 20 MPa, respectively. Therefore, the definition of supercritical conditions is the temperature and pressure above the critical point of the reaction mixture, which is calculated from the critical properties of the vegetable oils and/or animal fats and the alcohols.

1.2 The reaction mechanism of supercritical transesterification

Below the critical point of mixture, transesterification can take place in the presence of acidic or basic catalysts. Thus, the reaction mechanisms of transesterification are divided into acid and base catalyzed paths, as summarized elsewhere (Meher et al., 2006). The reaction mechanism of supercritical transesterification is somewhat similar to the acid catalyzed path in that the hydrogen bond of the alcohol is weakened at high temperatures (Hoffmann & Conradi, 1998). However, whilst the acid-catalyzed transesterification reaction is much slower than the base- catalyzed one at ambient conditions, the supercritical transesterification is much faster and achieves complete conversion of triglycerides to esters rapidly because the chemical kinetics are dramatically accelerated under supercritical conditions.

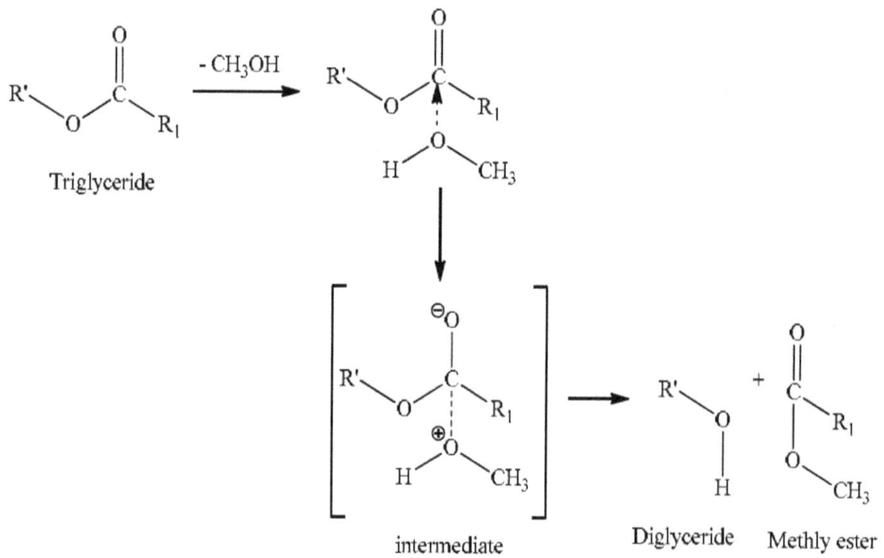

Fig. 4. The proposed reaction mechanism of transesterification in supercritical methanol (R' is a diglyceride group and R_1 is a fatty acid chain).

The reaction mechanism of supercritical transesterification, as shown in Fig. 4, was proposed by the analogues between the hydrolysis of esters in supercritical water (Krammer & Vogel, 2000) and the transesterification of triglycerides in supercritical methanol (Kusdiana & Saka, 2004b). It is assumed that the alcohol molecule (in this case methanol) directly attacks the carbonyl carbon of the triglyceride because the hydrogen bond energy is lowered; which would allow the alcohol to be a free monomer. In the case of methanol, the transesterification is completed via transfer of a methoxide moiety, whereby fatty acid methyl esters and diglycerides are formed. Consequently, the diglyceride reacts with other methanol molecules in a similar way to form the methyl ester and monoglyceride, the later of which is further converted to methyl ester and glycerol in the last step. The same process is applicable to other primary alkyl alcohols, such as ethanol.

1.3 The side reactions in the supercritical conditions

Firstly, the hydrolysis reaction of alkyl esters and triglyceride can take place at over 210 °C (Khuwijitjaru et al., 2004) and over 300 °C (W. King et al., 1999), respectively, in present of water and pressure above 20 MPa. The overall hydrolysis reaction of triglycerides and alkyl esters is shown in Figs. 5 and 6, respectively, with the principal products of the hydrolysis reaction being the respective fatty acids which are subsequently converted to alkyl esters under the supercritical condition by the esterification, as illustrated in Fig. 7.

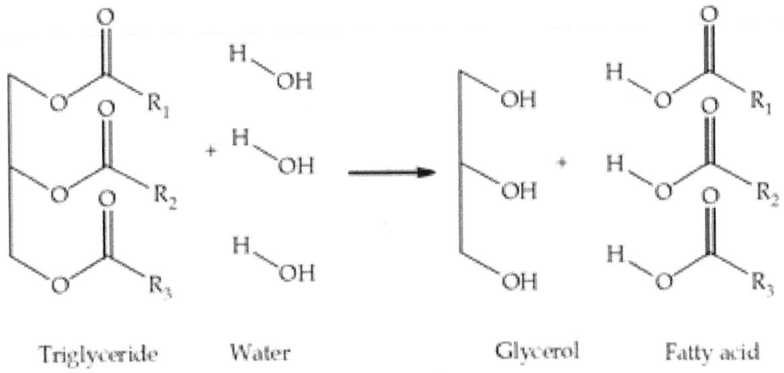

Fig. 5. The overall hydrolysis reaction of triglycerides under supercritical conditions (R is an alkyl group and R_1, R_2 and R_3 are fatty acid chains)

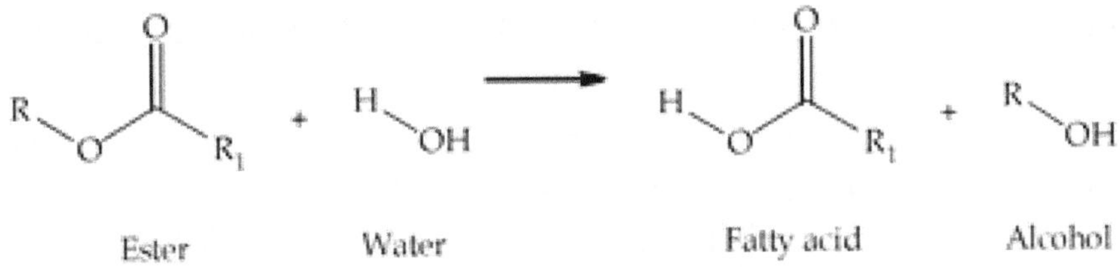

Fig. 6. The hydrolysis reaction of alkyl esters under supercritical conditions (R is an alkyl group and R_1 is a fatty acid chain)

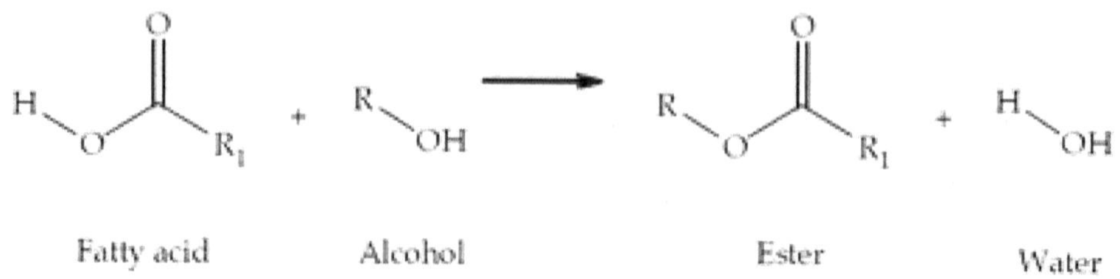

Fig. 7. The esterification reaction under supercritical conditions (R is an alkyl group and R_1 is a fatty acid chain)

In actuality, the hydrolysis reaction or the presence of water and free fatty acids do not affect the final alkyl ester content obtained for supercritical transesterification (Kusdiana & Saka, 2004b), because the alcohols have a much higher reactivity than water at the optimal point of supercritical transesterification. For example, the chemical rate constant for the transesterification of rapeseed oil is approximately 7-fold higher than that for the hydrolysis of soybean oil (Khuwijitjaru et al., 2004; Kusdiana & Saka, 2001).

Secondly, the thermal cracking of unsaturated fatty acids, especially the polyunsaturated fatty acids, occurs at temperatures over 300 °C and reaction times over 15 min (Quesada- Medina & Olivares-Carrillo, 2011). An example of the thermal cracking of a palmitic, oleic and linoleic acid based triglyceride is illustrated in Fig. 8.

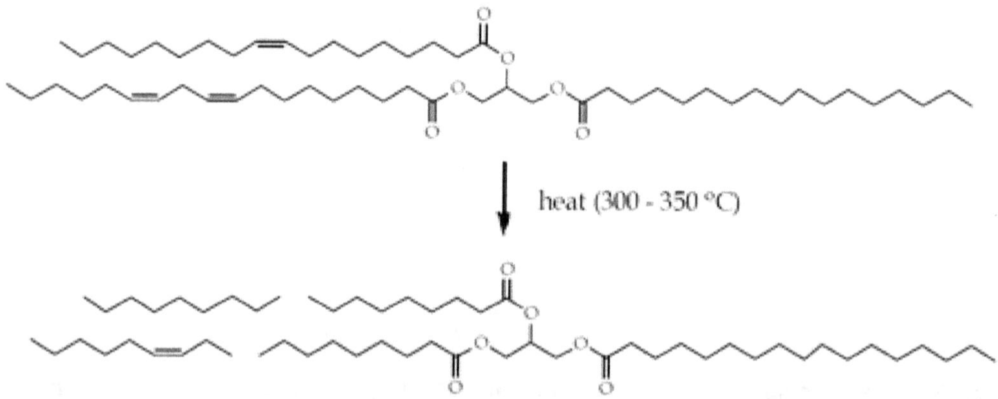

Fig. 8. The thermal cracking reaction of a triglyceride under supercritical conditions at a temperature range of 300 - 350 °C and a reaction time of over 15 min.

The triglyceride product from the thermal cracking reaction can be transesterified afterwards under supercritical conditions to alkyl esters. However, these alkyl esters are not the fatty acid alkyl esters of the common fatty acids in vegetable oils and/or animal fats. Therefore, the thermal cracking reaction reduces the acceptable primary alkyl ester content, as quantified and defined by the international standard for biodiesel (EN14103), and this is especially the case for the oils with a high polyunsaturated fatty acid content, such as soybean and sunflower oil.

In addition, triglycerides are decomposed to fatty acids and some gaseous products within the temperature range of 350 - 450 °C, as shown in Fig. 9 (Lima et al., 2004; Marulanda et al., 2009). In the same manner, with thermal cracking at 300 - 350 °C, the fatty acids product can be esterified under supercritical conditions, but the alkyl ester content is also decreased. However, the small hydrocarbon molecules of the thermal cracking products could improve some fuel properties of biodiesel, such as viscosity, density and cold flow properties.

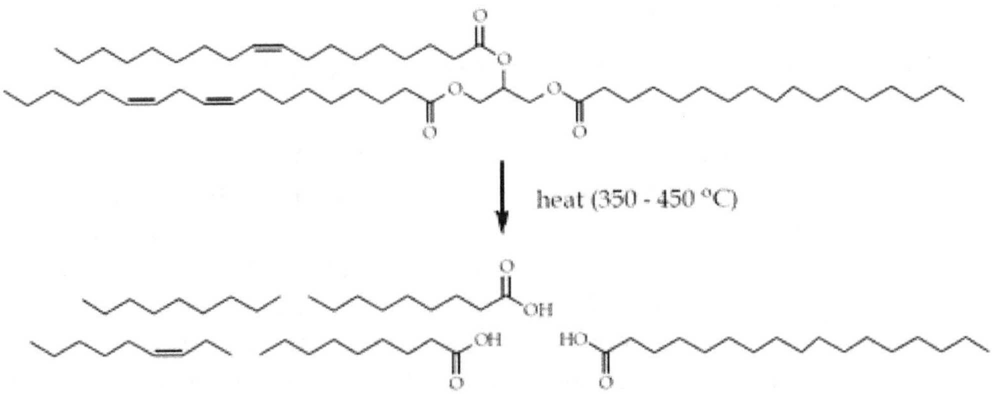

Fig. 9. The thermal cracking reaction of triglycerides under supercritical conditions at temperature range of 350 - 450 °C

1.4 The original reaction parameters and optimal conditions in the early scientific articles

The reaction parameters that were typically investigated in supercritical transesterification reactions are the temperature, pressure, alcohol to oil molar ratio and reaction time in batch and continuous reactors, and are summarized in Table 1.

The extent of the reaction was reported in terms of the % alkyl ester content and the % conversion of triglycerides. The % alkyl esters content refers to the alkyl esters of the common fatty acids in the vegetable or animal oils/fats that can be identified by different analytical techniques, while the % triglycerides conversion implies the remaining triglyceride reactant that is converted to fuels. Note that the % alkyl esters content refers to the specified esters, which must not be less than 96.5%, in the International standard (EN14214) for biodiesel fuel. It should also be noted that a high alkyl esters content infers a high triglyceride conversion, but, in contrast, a high triglyceride conversion does not have to infer a high alkyl esters content because the triglycerides could have been converted by the side reactions to other products.

According to Table 1, the original optimal conditions, defined as yielding the highest extent of reaction as over 90% conversion or over 96% alkyl esters content, were within 300 – 350 ºC, 20 – 35 MPa, an alcohol to oil molar ratio of 40:1 – 42:1 and a reaction time of 5 – 30 min, for both methanol and ethanol. These parameters are referred to as the original supercritical transesterification parameters, and have been employed to study the effects of each parameter, the chemical kinetics, the phase behavior and the economical feasibility of the process. Since the original parameters are elevated conditions, innovative techniques are proposed to reduce these original parameters.

2. PROCESS OVERVIEW

2.1 The effects of reaction parameters on the % conversion in supercritical transesterification

Among the general operating parameters mentioned previously (temperature, pressure, alcohol to oil molar ratio and reaction time), the reaction temperature is the most decisive parameter for indicating the extent of the reaction. This is as a result of the accelerated chemical kinetics and changes to the alcohol's properties. For example, the rate constant of supercritical transesterification is dramatically enhanced some 7-fold as the temperature is increased from 210 to 280 ºC at 28.0 MPa and a 42:1 methanol to oil molar ratio (He et al., 2007a), whilst the degree of hydrogen bonding also suddenly drops as the temperature is increased from 200 to 300 ºC at 30.0 MPa (Hoffmann & Conradi, 1998). However, where a maximum alkyl ester content is required, that is for biodiesel production, the higher operating temperatures cause a negative effect on the proportion of alkyl esters obtained in the product due to the thermal cracking reaction. Indeed, the thermal cracking is the chemical limitation of supercritical transesterification, and this is discussed in Section 2.4.2.

The reaction pressure also has a significant effect on the efficiency of the supercritical transesterification reaction below 20.0 MPa, but the effects tend to be negligible above 25.0 MPa (He et al., 2007a; He et al., 2007b), due to the fact that increasing the reaction pressure simultaneously increases both the density of the reaction mixture (Velez et al., 2010) and the degree of hydrogen bonding (Hoffmann & Conradi, 1998) at an otherwise constant temperature and alcohol to oil molar ratio. The transesterification conversion is enhanced with an increased reaction mixture density, due to the resulting increased volumetric concentration of alcohols and the residence time in a tubular reactor, which is commonly used to investigate the effect of pressure. On the other hand, the increasing degree of hydrogen bonding or alcohol cluster size weakens the nucleophilic strength of supercritical alcohols and so the reactivity of supercritical alcohol is reduced with increasing pressure. Thus, the desirable pressure for supercritical transesterification is in the range of 20.0 – 35.0 MPa.

Table 1. The original reaction parameters and optimal conditions of supercritical transesterification for various oil types and alcohols.

Oil type	Alcohol	T (oC)	P (MPa)	Alcohol: oil (mol: mol)	Reaction time (min)	Reactor (size / type) a	Extent of reactionb	Ref
Coconut & Palm kernel	Methanol	350	19	42:1	7 - 15	251-mL / CT	95% MEC	(Bunyakiat et al., 2006)
Hazelnut kernel & Cottonseed	Methanol	350	NR	41:1	5	100-mL / Batch	95% MEC	(Demirbas, 2002)
Palm and Groundnut	Methanol	400	20	50:1	30	11-mL / Batch	95% Con.	(Rathore & Madras, 2007)
Palm kernel	Methanol	350	20	42:1	30	250-mL / Batch	95% MEC	(Sawangkeaw et al., 2007)
Rapeseed	Methanol	350	45	42:1	4	5-mL / Batch	98% MEC	(Saka & Kusdiana, 2001)
Rapeseed	Methanol	350	20	42:1	30	200-mL / CT	87% MEC	(Minami & Saka, 2006)
Soybean	Methanol	350	20	42:1	30	250-mL / Batch	95% MEC	(Yin et al., 2008a)
Sunflower	Methanol & Ethanol	400	20	40:1	30	8-mL / Batch	97% Con.	(Madras et al., 2004)
Castor	Ethanol	300	20	40:1	NR	42-mL / CT	75% EEC	(Vieitez et al., 2011)
Soybean	Ethanol	350	20	40:1	15	42-mL / CT	80% Con.	(Silva et al., 2007)
Sunflower	Ethanol	280	NR	40:1	5	100-mL / Batch	80% EEC	(Balat, 2008)

From Table 1, the original alcohol to oil molar ratio for supercritical transesterification is in range of 40:1 – 42:1. The alcohol to oil molar ratio affects the supercritical transesterification efficiency strongly below 24:1, but its effect is then reduced with increasing alcohol to oil molar rations to plateau at over a 50:1 alcohol to oil molar ratio for methanol (He et al., 2007a) or 70:1 for ethanol (Silva et al., 2007) at 330 ºC and 20.0 MP. This is likely to be due to the fact that the operating temperature and pressure are much higher than the critical point of the reaction mixture, and are still located in the supercritical region.

As mentioned in Section 1.1, the critical point of the reaction mixture decreases with increasing alcohol to oil molar ratios. Thus, the optimal reaction temperature and pressure at a high alcohol to oil molar ratio is always milder than that at a low molar ratio. Nonetheless, the large amount of alcohol not only increases the required reactor volume but importantly it consumes a large amount of energy to heat the reactant and also to subsequently recover the excess alcohol. The energy for recycling excess alcohol might be minimized by a low temperature separation process, such as the use of a medium-pressure flash drum (Diaz et al., 2009), whereas the additional energy for heating the excess reactant alcohol cannot be avoided. Therefore, the use of assisting techniques, as described in Sections 3.2 – 3.4, have been introduced to decrease the alcohol to oil molar ratio whilst maintaining the transesterification conversion efficiency (and so the fatty acid alkyl ester content).

The effect of the reaction time on the transesterification conversion follows the general rate law. For example, the alkyl ester content increases gradually with reaction time and then remains constant after the optimal point (maximum conversion to alkyl esters) is reached. The optimal reaction time for supercritical transesterification at around 300 – 350 ºC varied between 4 to 30 minutes, depending upon the reactor size and type. Since the effect of residence time is directly related to the chemical kinetics of transesterification, the optimal reaction time at low temperatures is longer than that at high temperatures.

2.2 The chemical kinetics and phase behavior in supercritical transesterification

2.2.1 The chemical kinetics of supercritical transesterification

The chemical kinetics of supercritical transesterification is divided into three regions, that of the slow (<280 ºC), transition (280 - 330 ºC) and fast (>330 ºC) regions, and usually follows the first-order rate law with respect to the triglyceride concentration alone (He et al., 2007a; Kusdiana & Saka, 2001; Minami & Saka, 2006). Here, the reaction mechanism is merged into one overall step and the concentrations of all intermediates (mono- and diglycerides) are ignored. However, the first-order kinetic model is only suitable for a high alcohol to oil molar ratio, due to the insignificant changes in the alcohol concentration, but this increasingly becomes untrue as the alcohol to oil molar ratio decreases.

For the first-order model, the rate constants for each vegetable oil have a different temperature sensitivity, as noticed by the slope of Arrhenius' plot (Sawangkeaw et al., 2010). For example, the rate constants of rapeseed and soybean oil depend more strongly on the temperature than that for sunflower, palm and groundnut oils. The rate constants of saturated triglycerides were found to be faster than unsaturated triglycerides and slow down with increasing levels of double bonds in the triglyceride molecule (Rathore & Madras, 2007; Varma & Madras, 2006). However, saturated fatty acids have a slightly lower reactivity than the unsaturated fatty acids (Warabi et al., 2004).

On the other hand, a second-order kinetic model with respect to both the triglycerides and alcohol concentrations has also been proposed (Diasakou et al., 1998; Song et al., 2008). This divides the transesterification reaction into three steps; the reaction between a triglyceride and an alcohol that generates a diglyceride and an alkyl ester, the diglyceride and alcohol and finally the monoglyceride and alcohol. The concentration of the intermediates is then taken into account and the rate constants are found by mathematical model fitting. Thus, although more complex, the second-order kinetic model is more appropriate than the first- order model for reactions involving alcohol to oil molar ratios below 24:1.

2.2.2. The phase behavior of reactants in supercritical transesterification

The fact that the required optimal operating parameters can become milder with the addition of co-solvents has spurred much interest in the phase behavior of reactants during supercritical transesterification. The phase behavior of soybean oil-methanol with propane as a co-solvent was reported first (Hegel et al., 2007), followed by that for soybean oil- methanol and soybean oil-ethanol with carbon dioxide as the co-solvent (Anitescu et al., 2008). The study of the phase behavior of supercritical transesterification, when performed in a high-pressure view cell, revealed that the liquid - liquid (LL) alcohol - triglycerides mixture transforms to a vapor – liquid - liquid (VLL) phase equilibrium. The VLL equilibrium consists of two immiscible liquid phases (triglycerides and alcohol) and a vapor phase which mainly contains alcohol. Then, the VLL equilibrium changes to a vapor - liquid (VL) phase as a result of the triglycerides dissolving into the supercritical alcohol phase. Finally, the VL equilibrium merges to a one-phase supercritical at nearly the estimated critical point of mixture.

The transition temperature of the VLL to VL equilibriums decreases with increasing methanol to oil molar ratios (Anitescu et al., 2008; Hegel et al., 2007). For example, the reaction mixture of soybean oil and methanol are partially miscible up to temperatures close to 350 ºC at a methanol to oil molar ratio of 24:1, while the two liquid phases of soybean oil and methanol become completely miscible at 180 ºC and 157 ºC with a methanol to oil molar ratio of 40:1 and 65:1, respectively. For a soybean oil-ethanol mixture, it becomes a VL equilibrium at a lower temperature than that for the soybean oil-methanol due to the higher solubility of soybean-oil in ethanol than in methanol (Anitescu et al., 2008).

The transition from a VL system to a one-phase supercritical system was observed near the estimated critical temperature of the mixture, as described in Section 1.1.

At a methanol to oil molar ratio of 24:1, the critical temperature of the soybean oil-methanol mixture was 377 ºC where the transition temperature was reported to be higher than 350 ºC (Anitescu et al., 2008). Moreover, the transition temperature of the two-phase VL to a one-phase supercritical could be reduced by the addition of gaseous co-solvents, such as carbon dioxide and propane. For instance, the addition of 24% by weight of propane decreased the transition temperature of soybean oil-methanol, at a methanol to oil molar ratio of 65:1, from 315 ºC to 243 ºC (Hegel et al., 2007). Therefore, the addition of gaseous co-solvents is able to reduce the original severe conditions due to their ability to lower the transition temperature from a VL system to a one-phase supercritical system.

2.3 The advantages and drawbacks of supercritical transesterification

Novel solid heterogeneous catalysts that catalyze the transesterification on acidic or basic surfaces instead of in solution have been proposed to overcome the drawbacks of the conventional homogeneous catalytic method, which in part are the same as the supercritical transesterification. The enzyme catalysts typically also allow for a very high selectivity on the alkyl ester products (Helwani et al., 2009; Lene et al., 2009).

2.3.1 The advantages of supercritical transesterification

The advantages of supercritical transesterification over the conventional homogeneous catalytic method are feedstock flexibility, higher production efficiency and it is more environmentally friendly. The feedstock quality is far less influential under supercritical conditions than with the heterogeneous catalytic method, whilst supercritical transesterification has a similar advantage with respect to the product separation as the novel catalytic methods, but it has a higher production efficiency than both novel catalytic methods. The feedstock flexibility is the most important advantage to consider for biodiesel production methods because the resultant biodiesel price strongly depends on the feedstock price (Kulkarni & Dalai, 2006; Lam et al., 2010). The free fatty acids and moisture in low- grade feedstocks and hydrated ethanol pose a negative effect on the basic homogeneous and heterogeneous catalytic methods. Whereas, free fatty acid levels and moisture contents in the feedstock do not significantly affect supercritical transesterification with methanol or ethanol. Therefore, supercritical transesterification is more suitable for use with the low- grade and/or the hydrated ethanol feedstocks (Demirbas, 2009; Gui et al., 2009; Kusdiana & Saka, 2004b; Vieitez et al., 2011). For example, the in-situ transesterification of wet algal biomass in supercritical ethanol gave a 100% alkyl ester yield (Levine et al., 2010).

Supercritical transesterification has a better production efficiency than the conventional catalytic method because it requires a smaller number of processing steps. For instance, the feedstock pretreatment to remove moisture and free fatty acids, and the post-production product treatment steps, such as neutralization, washing and drying, are not necessary. In addition, the rate of reaction under supercritical conditions is significantly faster than the conventional catalytic method, so that the supercritical transesterification requires a smaller reactor size for a given production output.

With respect to environmental aspects, supercritical transesterification does not require any catalysts or chemicals, whilst the waste from the pretreatment and post-treatment steps are also reduced, since those steps are not necessary, leading to the generation of insignificant waste levels. However, the distillation process to recover the excess alcohol requires a large amount of energy which reduces the environmentally friendly advantage of the process (Kiwjaroun et al., 2009). Thus to maintain an environmentally friendly advantage, low- energy separation methods, such as medium pressure flash drum, must be applied to recover the excess alcohol (Diaz et al., 2009).

2.3.2 The drawbacks of supercritical transesterification

The original parameters to achieve a high transesterification conversion were a high temperature (330 – 350 ºC), high pressure (19 – 35 MPa) and high alcohol to oil molar ratio (1:40 – 1:42). Indeed, the high temperature and pressure requires both an expensive reactor and a sophisticate energy and safety management policy. As a result of the high alcohol to oil molar ratio a large energy consumption in the reactants pre-heating and recycling steps is required. Moreover, the high amount of alcohol in the biodiesel product retards the biodiesel-glycerol phase separation. Therefore, the use of those original parameters results in high capital costs, especially for the reactor and pump, being somewhat higher than the novel catalytic methods.

To increase the technical and economical feasibility of supercritical transesterification, further studies are required to reduce the energy consumption and operating parameters of this process. For example, the integration of a heating and cooling system can improve (reduce) the energy demand. The experimental techniques that have demonstrated the ability to lower the original parameters for supercritical transesterification are illustrated in sections 3.2 – 3.4.

2.4 The economical feasibilities and chemical limitations of supercritical transesterification

The economical feasibilities of supercritical transesterification, compared with the conventional homogeneous catalytic methods, have been studied by computer simulation (van Kasteren & Nisworo, 2007). These studies usually employed the original parameters for transesterification in supercritical methanol (350 ºC,

20.0 MPa and 1:42 methanol to oil molar ratio) and the general parameters for the conventional catalytic methods, such as a reaction at 60 ºC, 0.1 MPa and a methanol to oil molar ratio of 9:1. With respect to chemical limitations, supercritical transesterification is limited by the operating temperature due to the thermal cracking of the unsaturated fatty acids.

2.4.1 The economical feasibilities of supercritical transesterification

Supercritical transesterification with the original reacting parameters is economically competitive compared to the conventional catalytic method especially when low-grade feedstocks are employed (van Kasteren & Nisworo, 2007). As expected, the supercritical transesterification has a larger capital cost, due to the required reacting and pumping systems, than the conventional catalytic method, but has no additional capital and operating costs on feedstock pre-treatment, product post-treatment and waste management.

For a better economic feasibility, research into ways to reduce the high operating conditions and lower energy consumption are warranted. For example, supercritical transesterification is not economically feasible when the heating and cooling integration is not employed (Marchetti & Errazu, 2008), while it is a feasible method when heat integration and the presence of catalysts are applied (D'Ippolito et al., 2006; Glišic et al., 2009; van Kasteren & Nisworo, 2007). The addition of calcium oxide as a solid catalyst and the reduction of the alcohol to oil molar ratio significantly decreased the total energy demand and improved the economic feasibility as well (Glišic & Skala, 2009). Furthermore, the addition of propane as co-solvent also enhanced the economic feasibility of supercritical transesterification with methanol (van Kasteren & Nisworo, 2007). However, additional feasibility studies on other assisting techniques that lower the original parameters (see Sections 3.2 - 3.4) are still required.

2.4.2 The chemical limitation of supercritical transesterification

To fulfill the international standard of biodiesel (EN14214), which requires over 96.5% esters content, thermal cracking of polyunsaturated fatty acids is a serious obstacle. At over 300 ºC and a reaction time of over 15 min, the methyl linoleate content in biodiesel decreases by approximately 10% compared with the level in the feedstock (Quesada-Medina & Olivares- Carrillo, 2011). Whereas, the % recovery of biodiesel samples which are prepared from various vegetable oils remains constant after exposure with supercritical methanol at 270 ºC over 40 min (Imahara et al., 2008). Therefore, the 96.5% alkyl esters content requirement for biodiesel cannot be achieved when an operating temperature of over 300 ºC and a reaction time of over 15 min are employed.

To prevent this thermal degradation, the suggested temperature for supercritical transesterification is below 300 ºC, and preferably 270 ºC. However, the required reaction time to nearly complete transesterification conversion at a 42:1 alcohol to oil molar ratio is then significantly longer at more than 90 min (Minami & Saka, 2006). This prolonged reaction time might cause a decline in the production efficiency obtained by supercritical transesterification, but it could be shortened by the use of assisting methods, as discussed in Sections 3.2 – 3.4.

On the other hand, the gradual heating technique in a tubular reactor has been demonstrated to avoid the thermal cracking of unsaturated fatty acids and shorten the reaction time at the same time (He et al., 2007b). For instance, when the reaction mixture is heated in a tubular reactor gradually from 100 ºC at the inlet to 320 ºC at the outlet, the biodiesel product obtained after 25 min of reaction time has an over 96% methyl ester content (He et al., 2007b).

3. PROCESS IMPROVEMENTS

The process improvements to the supercritical transesterification can be divided into three routes; the addition of the co-solvents, the use of catalysts and process modifications. The general goal, to reduce the original parameters altogether, is the most challenging aspect for supercritical transesterification. The reduced parameters are 270 – 300 °C, 15 – 20 MPa and an alcohol to oil molar ratio of 24:1 – 35:1.

3.1 The chronological development of supercritical transesterification

In 1998, non-catalytic transesterification of soybean oil at the near-critical point of methanol (230 ºC, 6.2 MPa and a 27:1 methanol to oil molar ratio) was invented as an alternative method to produce biodiesel, but this method obtained only an 85% methyl ester content after over 10 hours (Diasakou et al., 1998). In 2001, the pioneering transesterification of rapeseed oil in supercritical methanol at 350 ºC, 45 MPa and a 42:1 methanol to oil molar ratio, attaining a high methyl ester content (98%) after only 4 min was reported (Kusdiana & Saka, 2001; Saka & Kusdiana, 2001). Transesterification in supercritical methanol has evolved continuously since 2001.

In 2002, the transesterification of cottonseed, hazelnut kernel, poppy seed, safflower and sunflower derived oils in supercritical methanol were evaluated, with a nearly complete transesterification reaction being found for all of the vegetable oils (Demirbas, 2002). Meanwhile, the effect of water and free fatty acids (Kusdiana & Saka, 2004b), the catalytic effect of a metal reactor for supercritical transesterification with methanol (Dasari et al., 2003; Kusdiana & Saka, 2004a) and the reactivity of supercritical alcohols were all reported (Warabi et al., 2004). In 2004, the first supercritical transesterification of sunflower oil with ethanol and supercritical carbon dioxide in the presence of a lipase enzyme were investigated in a batch reactor (Madras et al., 2004). However, during 2001 – 2005, the maximum alkyl ester contents were generally observed at nearly the same reaction conditions as that reported earlier by the Japanese pioneers (Kusdiana & Saka, 2001; Saka & Kusdiana, 2001).

In 2005, carbon dioxide and propane were introduced as co-solvents to obtain milder operating parameters for the supercritical transesterification with methanol (Cao et al., 2005; Han et al., 2005). Then, the two-step supercritical process (Minami & Saka, 2006) was demonstrated to reduce those operating parameters. In the following years, various catalysts were employed to assist the supercritical transesterification to achieve the maximum alkyl esters content but at milder operating conditions (Demirbas, 2007; Wang et al., 2008; Wang et al., 2007; Wang & Yang, 2007; Yin et al., 2008b). The continuous production of biodiesel in supercritical methanol was reported in 2006 (Bunyakiat et al., 2006) (Minami & Saka, 2006) and 2007 (He et al., 2007b). Therefore, the research focus on the reduction of the elevated operating conditions and continuous process has been ongoing since 2005.

In 2007, the gradual heating technique was introduced to limit or prevent thermal cracking of the unsaturated fatty acids and so prevent the reduction in the final methyl esters content obtained (He et al., 2007b). At the same time, the effect of using co-solvents to reduce the viscosity of vegetable oils was successfully investigated (Sawangkeaw et al., 2007). Supercritical transesterification in ethanol was studied in a continuous reactor in 2008 (Vieitez et al., 2008). In 2009, carbon dioxide was applied to supercritical transesterification with ethanol to reduce the operating conditions (Bertoldi et al., 2009). From 2007 to 2010, numerous additional studies, such as vapor-liquid equilibria of binary systems (Anitescu et al., 2008; Fang et al., 2008; Shimoyama et al., 2008; Shimoyama et al., 2009; Tang et al., 2006), phase behavior of the reaction mixture (Glišic & Skala, 2010; Hegel et al., 2008; Hegel et al., 2007), thermal stability of unsaturated fatty acids in supercritical methanol (Imahara et al., 2008) and process simulation and economic analysis (Busto et al., 2006; D'Ippolito et al., 2006; Deshpande et al., 2010; Diaz et al., 2009; van Kasteren & Nisworo, 2007) were reported, leading to a better understanding of the supercritical transesterification process.

3.2 The addition of co-solvents

The co-solvents that have been used in supercritical transesterification are liquid co- solvents, such as hexane and tetrahydrofuran (THF), and gaseous co-solvents, such as propane, carbon dioxide (CO_2) and nitrogen (N_2). Both types of co-solvents have different purposes and advantages that will be presented accordingly.

The liquid co-solvents are added into the supercritical transesterification reaction to reduce the viscosity of the vegetable oils, which might otherwise pose some pumping problems in a continuous process (Sawangkeaw et al., 2007). Since hexane is the conventional solvent for vegetable oil extraction, it could be possible to combine the supercritical transesterification after the extraction process using hexane for both. Additionally, THF improves the solubility of alcohols in the triglyceride and so forms a single phase mixture, allowing a single high- pressure pump to be employed to feed the reaction mixture into the reactor. A small amount of liquid co-solvent, up to ~20% (v/v) of hexane in vegetable oil, neither affects the transesterification conversion nor lowers the original operating parameters. Whereas, an excess amount of hexane shows a negative effect on the final obtained alkyl esters content due to dilution and obstruction of the reactants (Tan et al., 2010a).

The addition of gaseous co-solvents to the supercritical transesterification reaction aims to reduce the original operating parameters. Due to the fact that the critical properties of gaseous co-solvents are much lower than alcohol and triglycerides, the addition of a small amount of gaseous co-solvents dramatically decreases the critical point of the reaction mixture allowing the use of milder operating parameters. For example, 0.10 mole of CO_2 or

0.05 mole of propane per mole of methanol lowers the reaction temperature and methanol to oil molar ratio to 280 °C and 1:24, respectively (Cao et al., 2005; Han et al., 2005). Furthermore, it was reported that the addition of N_2 improved the oxidation stability and reduced the total glycerol content in the biodiesel product (Imahara et al., 2009). Gaseous co- solvents have the advantage of easier separation from the product than the liquid co- solvents. For instance, they can be separated from the biodiesel product by expansion without using additional energy at the end of the transesterification process, unlike the liquid co-solvents that typically need to be recovered by distillation.

3.3 The use of catalysts

The homogeneous acidic and basic catalysts, such as H_3PO_4, NaOH and KOH, have been applied to supercritical transesterification to obtain milder operating conditions (Wang et al., 2008; Wang et al., 2007; Yin et al., 2008b). However, despite the milder operating conditions and faster rate of reaction obtained compared to the catalyst-free process, the addition of homogeneous catalysts is not an interesting idea because the problem of subsequent catalyst separation and waste management still remain, the same situation as with the conventional homogeneous catalytic process. The use of solid heterogeneous catalysts might enhance the technical and economical feasibility of using supercritical transesterification as a result of the ease of separation of the catalysts. However, the acidic and basic heterogeneous catalysts have different characteristics and advantages, as will be discussed below.

The acidic heterogeneous catalysts, such as WO_3/ZrO_2, zirconia-alumina, sulfated tin oxide and Mg–Al–CO_3 hydrotalcites, have been evaluated in the supercritical transesterification process (Helwani et al., 2009). However, despite the presence of the catalysts, the chemical kinetics of the acidic heterogeneous catalysts at atmospheric pressure were slower than the catalyst-free process. For example, the transesterification of soybean oil in supercritical methanol at 250 °C and a 40:1 methanol to oil molar ratio in the presence of WO_3/ZrO_2 as catalyst still takes 20 hours to attain a 90% conversion level (Furuta et al., 2004). However, the acidic catalysts are less sensitive to moisture and free fatty acid content than the basic catalysts and so they could be appropriate for low-grade feedstocks.

Alternatively, basic heterogeneous catalysts, such as CaO (Demirbas, 2007) MgO (Demirbas, 2008) and nano-MgO (Wang & Yang, 2007), have been applied to supercritical transesterification to reduce the original operating conditions. These catalysts have the ability to catalyze the transesterification reaction at the boiling point of alcohols and are stable at supercritical conditions. As expected, the rate of reaction at the supercritical conditions is faster than that at lower temperatures. For example, the CaO catalyst takes over 180 min to reach over 95% conversion at 65 °C (Liu et al., 2008), but only 10 min to reach complete conversion at 250 °C (Demirbas, 2007). Unfortunately, the basic catalysts can be poisoned by the presence of water and free fatty acids. Therefore, further studies on using low-grade feedstocks with basic heterogeneous catalysts are still required.

3.4 The process modifications

The two-step process is based on firstly a hydrolysis reaction in subcritical water to obtain fatty acid products and then secondly the transesterification and esterification reactions in supercritical alcohol to form the alkyl esters product. The two-step process reduces the optimal operating parameters successfully since the hydrolysis and esterification reactions reach complete conversion at a lower temperature than the transesterification reaction does (Minami & Saka, 2006). Nonetheless, the two-step process is more complicated than the single-step process. For example, the process has high-pressure reactors that connect in series with a high-pressure water-glycerol-fatty acid phase separator. Furthermore, the glycerol-water stream, which is contaminated by trace amounts of fatty acids, requires more separation units. Although a distillation tower is the simplest separation unit for handling the glycerol-water stream, it consumes a large amount of energy to operate.

The high-temperature process involves increasing the operating temperature to 400 to 450 °C (Marulanda et al., 2009; Marulanda et al., 2010), so that the operating pressure, methanol to oil molar ratio and reaction time for complete conversion are reduced to 10.0 MPa, 6:1 and 4 min, respectively. As expected, the unsaturated fatty acids are partially consumed by thermal degradation but the oxidation resistance or storage stability of the product might be enhanced. Under these conditions it was reported that triglyceride and glycerol convert to oxygenate liquid fuel with a conversion of up to 99.5%. The glycerol dehydration both increases the fuel yield by up to 10% and reduces the amount of glycerol by-products (Aimaretti et al., 2009). By using the high-temperature process, the simultaneous conversion of triglyceride, free fatty acids and glycerol to liquid fuel is an alternative option that will increase the feasibility and profitability of supercritical transesterification.

4. PROCESS PROSPECTIVE

In this section, the process prospective is split into two on the basis of the operating temperature since the temperature is the key parameter and chemical limitation for supercritical transesterification. The low-temperature approach aims to produce biodiesel that fulfills the 96.5% alkyl esters content requirement for biodiesel, while the high- temperature approach proposes an alternative method to synthesize the biofuel from a triglyceride-base biomass in supercritical conditions.

4.1 The low-temperature approach

The term "Low-temperature approach" defines supercritical transesterification within a temperature range of 270 – 300 ºC so as to avoid the thermal degradation of unsaturated fatty acids and to maximize the alkyl esters content in the product. Without the assistance of any co-solvent, catalyst or other process modification techniques, the low-temperature approach employs a high pressure, a high alcohol to oil molar ratio and a long reaction time to achieve the >96.5% alkyl esters content required for biodiesel composition by the international standard. However, with the assisting techniques, as mentioned in Sections 3.2 – 3.4, the optimal conditions of low-temperature approach generally involve 20 – 30 MPa, an alcohol to oil molar ratio of 24:1 and a reaction time over 30 min. The biodiesel product, which typically exceeds the 96.5% alkyl esters content of the international standard for biodiesel (EN14214), can be used as biodiesel.

For future research involving the low-temperature approach, the use of low-grade feedstocks and/or heterogeneous catalysts are very interesting topics. Alternatively, studies on scale up continuous reactors which are more suitable for an industrial scale are required. These have been successfully evaluated in lab-scale tubular reactors (Bunyakiat et al., 2006; He et al., 2007b; Minami & Saka, 2006), but an evaluation on a scaled-up reactor is presently lacking. An optimal reaction time to achieve over 96.5% alkyl esters content is the most important finding for the low-temperature approach studies because it corresponds with reactor sizing and reflects on the economical feasibility.

4.2 The high-temperature approach

The high-temperature approach uses supercritical transesterification at temperatures over 400 ºC, as described in Section 3.4. Even though the mono-alkyl esters content in the product from the high-temperature process is always lower than the biodiesel specification value of 96.5%, it can be proposed as an alternative biofuel that would require further studies on engine testing and fuel properties itself. Improved fuel properties, such as the viscosity and density of the biofuel product, from the high- temperature approach have been proposed (Marulanda et al., 2009). Furthermore, the operating temperature and pressure used in the high-temperature approach are close to those for catalytic hydrocracking in conventional petroleum refining, so it has a high possibility that it can be realized in an industrial scale.

Since the high-temperature approach, as recently initiated, has evaluated the triglycerides found in soybean oil (Anitescu et al., 2008) and chicken fat (Marulanda et al., 2009; Marulanda et al., 2010) only, then additional research into other triglycerides are needed. In addition, studies on the economical feasibility and environmental impact are also required. Indeed, the complete fuel properties need examining along with engine testing for the biofuel product for the high-temperature approach (Basha et al., 2009). On the other hand, the fine studies on the reactions pathways and/or chemical kinetics are also attractive works to better understand the high-temperature approach.

5. CONCISE SUMMARY

Supercritical transesterification is an advantageous technique for producing biodiesel in an environmentally sustainable manner due to its flexibility in feedstock, production efficiency, and ecological benefits. For comprehensive information, review articles on supercritical transesterification using methanol (de Boer & Bahri, 2011; Sawangkeaw et al., 2010), or ethanol (Balat, 2008; Pinnarat & Savage, 2008), as well as other supercritical technologies (Lee & Saka, 2010; Tan & Lee, 2011), are accessible in other sources.

Despite the expansion of the information base regarding this process over the past decade, further efforts are necessary for a comprehensive understanding. Despite its benefit of feedstock versatility, there has been no research on the utilization of low-grade feedstocks in supercritical transformation. Additionally, prospective investigations for both low-temperature and high-temperature methods, as previously indicated, are necessary to achieve supercritical transesterification on an industrial scale.

REFERENCES

Aimaretti, N., Manuale, D.L., Mazzieri, V.M., Vera, C.R. & Yori, J.C. (2009). Batch Study of Glycerol Decomposition in One-Stage Supercritical Production of Biodiesel. *Energy and Fuels*, Vol.23, No.2, pp. (1076-1080), ISSN 0887-0624

Anitescu, G., Deshpande, A. & Tavlarides, L.L. (2008). Integrated technology for supercritical biodiesel production and power cogeneration. *Energy and Fuels*, Vol.22, No.2, pp. (1391-1399), ISSN 0887-0624

Balat, M. (2008). Biodiesel fuel production from vegetable oils via supercritical ethanol transesterification. *Energy Sources, Part A: Recovery, Utilization and Environmental Effects*, Vol.30, No.5, pp. (429-440), ISSN 1556-7036

Basha, S.A., Gopal, K.R. & Jebaraj, S. (2009). A review on biodiesel production, combustion, emissions and performance. *Renewable and Sustainable Energy Reviews*, Vol.13, No.6- 7, pp. (1628-1634), ISSN 1364-0321

Bertoldi, C., da Silva, C., Bernardon, J.P., Corazza, M.L., Filho, L.C., Oliveira, J.V. & Corazza, F.C. (2009). Continuous Production of Biodiesel from Soybean Oil in Supercritical Ethanol and Carbon Dioxide as Cosolvent. *Energy and Fuels*, Vol.23, No.10, pp. (5165-5172), ISSN 0887-0624

Bunyakiat, K., Makmee, S., Sawangkeaw, R. & Ngamprasertsith, S. (2006). Continuous Production of Biodiesel via Transesterification from Vegetable Oils in Supercritical Methanol. *Energy and Fuels*, Vol.20, No.2, pp. (812-817), ISSN 0887-0624

Busto, M., D'Ippolito, S.A., Yori, J.C., Iturria, M.E., Pieck, C.L., Grau, J.M. & Vera, C.R. (2006). Influence of the Axial Dispersion on the Performance of Tubular Reactors during the Noncatalytic Supercritical Transesterification of Triglycerides. *Energy and Fuels*, Vol.20, No.6, pp. (2642-2647), ISSN 0887-0624

Cao, W., Han, H. & Zhang, J. (2005). Preparation of biodiesel from soybean oil using supercritical methanol and co-solvent. *Fuel*, Vol.84, No.4, pp. (347-351), ISSN 0016- 2361

D'Ippolito, S.A., Yori, J.C., Iturria, M.E., Pieck, C.L. & Vera, C.R. (2006). Analysis of a Two- Step, Noncatalytic, Supercritical Biodiesel Production Process with Heat Recovery. *Energy and Fuels*, Vol.21, No.1, pp. (339-346), ISSN 0887-0624

Dasari, M., Goff, M. & Suppes, G. (2003). Noncatalytic alcoholysis kinetics of soybean oil. *Journal of the American Oil Chemists' Society*, Vol.80, No.2, pp. (189-192), ISSN 0003- 021X

de Boer, K. & Bahri, P.A. (2011). Supercritical methanol for fatty acid methyl ester production: A review. *Biomass and Bioenergy*, Vol.35, No.3, pp. (983-991), ISSN 0961- 9534

Demirbas, A. (2002). Biodiesel from vegetable oils via transesterification in supercritical methanol. *Energy Conversion and Management*, Vol.43, No.17, pp. (2349-2356), ISSN 0196-8904

Demirbas, A. (2007). Biodiesel from sunflower oil in supercritical methanol with calcium oxide. *Energy Conversion and Management*, Vol.48, No.3, pp. (937-941), ISSN 0196- 8904

Demirbas, A. (2008). Biodiesel from vegetable oils with MgO catalytic transesterification in supercritical methanol. *Energy Sources, Part A: Recovery, Utilization and Environmental Effects*, Vol.30, No.17, pp. (1645-1651), ISSN 1556-7036

Demirbas, A. (2009). Biodiesel from waste cooking oil via base-catalytic and supercritical methanol transesterification. *Energy Conversion and Management*, Vol.50, No.4, pp. (923-927), ISSN 0196-8904

Deshpande, A., Anitescu, G., Rice, P.A. & Tavlarides, L.L. (2010). Supercritical biodiesel production and power cogeneration: Technical and economic feasibilities. *Bioresource Technology*, Vol.101, No.6, pp. (1834-1843), ISSN 0960-8524

Diasakou, M., Louloudi, A. & Papayannakos, N. (1998). Kinetics of the non-catalytic transesterification of soybean oil. *Fuel*, Vol.77, No.12, pp. (1297-1302), ISSN 0016- 2361

Diaz, M.S., Espinosa, S. & Brignole, E.A. (2009). Model-Based Cost Minimization in Noncatalytic Biodiesel Production Plants. *Energy and Fuels*, Vol.23, No.11, pp. (5587- 5595), ISSN 0887-0624

Espinosa, S., Fornari, T., Bottini, S.B. & Brignole, E.A. (2002). Phase equilibria in mixtures of fatty oils and derivatives with near critical fluids using the GC-EOS model. *The Journal of Supercritical Fluids*, Vol.23, No.2, pp. (91-102), ISSN 0896-8446

Fang, T., Shimoyama, Y., Abeta, T., Iwai, Y., Sasaki, M. & Goto, M. (2008). Phase equilibria for the mixtures of supercritical methanol + C18 methyl esters and supercritical methanol + α-tocopherol. *The Journal of Supercritical Fluids*, Vol.47, No.2, pp. (140- 146), ISSN 0896-8446

Furuta, S., Matsuhashi, H. & Arata, K. (2004). Biodiesel fuel production with solid superacid catalysis in fixed bed reactor under atmospheric pressure. *Catalysis Communications*, Vol.5, No.12, pp. (721-723), ISSN

Glišic, S., Lukic, I. & Skala, D. (2009). Biodiesel synthesis at high pressure and temperature: Analysis of energy consumption on industrial scale. *Bioresource Technology*, Vol.100, No.24, pp. (6347-6354), ISSN 0960-8524

Glišic, S. & Skala, D. (2009). The problems in design and detailed analyses of energy consumption for biodiesel synthesis at supercritical conditions. *The Journal of Supercritical Fluids*, Vol.49, No.2, pp. (293-301), ISSN 0896-8446

Glišic, S.B. & Skala, D.U. (2010). Phase transition at subcritical and supercritical conditions of triglycerides methanolysis. *The Journal of Supercritical Fluids*, Vol.54, No.1, pp. (71- 80), ISSN 0896-8446

Gui, M.M., Lee, K.T. & Bhatia, S. (2009). Supercritical ethanol technology for the production of biodiesel: Process optimization studies. *The Journal of Supercritical Fluids*, Vol.49, No.2, pp. (286-292), ISSN 0896-8446

Han, H., Cao, W. & Zhang, J. (2005). Preparation of biodiesel from soybean oil using supercritical methanol and CO_2 as co-solvent. *Process Biochemistry*, Vol.40, No.9, pp. (3148-3151), ISSN 1359-5113

He, H., Sun, S., Wang, T. & Zhu, S. (2007a). Transesterification Kinetics of Soybean Oil for Production of Biodiesel in Supercritical Methanol. *Journal of the American Oil Chemists' Society*, Vol.84, No.4, pp. (399-404), ISSN 0003-021X

He, H., Wang, T. & Zhu, S. (2007b). Continuous production of biodiesel fuel from vegetable oil using supercritical methanol process. *Fuel*, Vol.86, No.3, pp. (442-447), ISSN 0016-2361

Hegel, P., Andreatta, A., Pereda, S., Bottini, S. & Brignole, E.A. (2008). High pressure phase equilibria of supercritical alcohols with triglycerides, fatty esters and cosolvents. *Fluid Phase Equilibria*, Vol.266, No.1-2, pp. (31-37), ISSN 0378-3812

Hegel, P., Mabe, G., Pereda, S. & Brignole, E.A. (2007). Phase Transitions in a Biodiesel Reactor Using Supercritical Methanol. *Industrial and Engineering Chemistry Research*, Vol.46, No.19, pp. (6360-6365), ISSN 0888-5885

Helwani, Z., Othman, M.R., Aziz, N., Kim, J. & Fernando, W.J.N. (2009). Solid heterogeneous catalysts for transesterification of triglycerides with methanol: A review. *Applied Catalysis A: General*, Vol.363, No.1-2, pp. (1-10), ISSN 0926-860X

Hoffmann, M.M. & Conradi, M.S. (1998). Are There Hydrogen Bonds in Supercritical Methanol and Ethanol? *The Journal of Physical Chemistry B*, Vol.102, No.1, pp. (263- 271), ISSN 1520-6106

Ilham, Z. & Saka, S. (2009). Dimethyl carbonate as potential reactant in non-catalytic biodiesel production by supercritical method. *Bioresource Technology*, Vol.100, No.5, pp. (1793-1796), ISSN 0960-8524

Imahara, H., Minami, E., Hari, S. & Saka, S. (2008). Thermal stability of biodiesel in supercritical methanol. *Fuel*, Vol.87, No.1, pp. (1-6), ISSN 0016-2361

Imahara, H., Xin, J. & Saka, S. (2009). Effect of CO_2/N_2 addition to supercritical methanol on reactivities and fuel qualities in biodiesel production. *Fuel*, Vol.88, No.7, pp. (1329- 1332), ISSN 0016-2361

Khuwijitjaru, P., Fujii, T., Adachi, S., Kimura, Y. & Matsuno, R. (2004). Kinetics on the hydrolysis of fatty acid esters in subcritical water. *Chemical Engineering Journal*, Vol.99, No.1, pp. (1-4), ISSN 1385-8947

Kiwjaroun, C., Tubtimdee, C. & Piumsomboon, P. (2009). LCA studies comparing biodiesel synthesized by conventional and supercritical methanol methods. *Journal of Cleaner Production*, Vol.17, No.2, pp. (143-153), ISSN 0959-6526

Krammer, P. & Vogel, H. (2000). Hydrolysis of esters in subcritical and supercritical water. *The Journal of Supercritical Fluids*, Vol.16, No.3, pp. (189-206), ISSN 0896-8446 Kulkarni, M.G. & Dalai, A.K. (2006). Waste Cooking Oil An Economical Source for Biodiesel:‰ A Review. *Industrial and Engineering Chemistry Research*, Vol.45, No.9, pp. (2901-2913), ISSN 0888-5885

Kusdiana, D. & Saka, S. (2001). Kinetics of transesterification in rapeseed oil to biodiesel fuel as treated in supercritical methanol. *Fuel*, Vol.80, No.5, pp. (693-698), ISSN 0016- 2361

Kusdiana, D. & Saka, S. (2004a). Catalytic effect of metal reactor in transesterification of vegetable oil. *Journal of the American Oil Chemists' Society*, Vol.81, No.1, pp. (103- 104), ISSN 0003-021X

Kusdiana, D. & Saka, S. (2004b). Effects of water on biodiesel fuel production by supercritical methanol treatment. *Bioresource Technology*, Vol.91, No.3, pp. (289-295), ISSN 0960-8524

Lam, M.K., Lee, K.T. & Mohamed, A.R. (2010). Homogeneous, heterogeneous and enzymatic catalysis for transesterification of high free fatty acid oil (waste cooking oil) to biodiesel: A review. *Biotechnology Advances*, Vol.28, No.4, pp. (500-518), ISSN 0734-9750

Lee, J.-S. & Saka, S. (2010). Biodiesel production by heterogeneous catalysts and supercritical technologies. *Bioresource Technology*, Vol.101, No.19, pp. (7191-7200), ISSN 0960- 8524

Lene, F., Knud, V.C. & Birgir, N. (2009). A review of the current state of biodiesel production using enzymatic transesterification. *Biotechnology and Bioengineering*, Vol.102, No.5, pp. (1298-1315), ISSN 1097-0290

Levine, R.B., Pinnarat, T. & Savage, P.E. (2010). Biodiesel Production from Wet Algal Biomass through in Situ Lipid Hydrolysis and Supercritical Transesterification. *Energy and Fuels*, Vol.24, No.9, pp. (5235-5243), ISSN 0887-0624

Lima, D.G., Soares, V.C.D., Ribeiro, E.B., Carvalho, D.A., Cardoso, É.C.V., Rassi, F.C., Mundim, K.C., Rubim, J.C. & Suarez, P.A.Z. (2004). Diesel-like fuel obtained by pyrolysis of vegetable oils. *Journal of Analytical and Applied Pyrolysis*, Vol.71, No.2, pp. (987-996), ISSN 0165-2370

Liu, X., He, H., Wang, Y., Zhu, S. & Piao, X. (2008). Transesterification of soybean oil to biodiesel using CaO as a solid base catalyst. *Fuel*, Vol.87, No.2, pp. (216-221), ISSN 0016-2361

Madras, G., Kolluru, C. & Kumar, R. (2004). Synthesis of biodiesel in supercritical fluids. *Fuel*, Vol.83, No.14-15, pp. (2029-2033), ISSN 0016-2361

Marchetti, J.M. & Errazu, A.F. (2008). Technoeconomic study of supercritical biodiesel production plant. *Energy Conversion and Management*, Vol.49, No.8, pp. (2160-2164), ISSN 0196-8904

Marulanda, V.F., Anitescu, G. & Tavlarides, L.L. (2009). Biodiesel Fuels through a Continuous Flow Process of Chicken Fat Supercritical Transesterification. *Energy and Fuels*, Vol.24, No.1, pp. (253-260), ISSN 0887-0624

Marulanda, V.F., Anitescu, G. & Tavlarides, L.L. (2010). Investigations on supercritical transesterification of chicken fat for biodiesel production from low-cost lipid feedstocks. *The Journal of Supercritical Fluids*, Vol.54, No.1, pp. (53-60), ISSN 0896- 8446

Meher, L.C., Vidya Sagar, D. & Naik, S.N. (2006). Technical aspects of biodiesel production by transesterification--a review. *Renewable and Sustainable Energy Reviews*, Vol.10, No.3, pp. (248-268), ISSN 1364-0321

Minami, E. & Saka, S. (2006). Kinetics of hydrolysis and methyl esterification for biodiesel production in two-step supercritical methanol process. *Fuel*, Vol.85, No.17-18, pp. (2479-2483), ISSN 0016-2361

Pinnarat, T. & Savage, P.E. (2008). Assessment of noncatalytic biodiesel synthesis using supercritical reaction conditions. *Industrial and Engineering Chemistry Research*, Vol.47, No.18, pp. (6801-6808), ISSN 0888-5885

Poling, B.E., Prausnitz, J.M. & O'connell, J.P. (2001). *The properties of gases and liquids.* (5 ed). McGraw-Hill, ISBN 7011-6822, New York.

Quesada-Medina, J. & Olivares-Carrillo, P. (2011). Evidence of thermal decomposition of fatty acid methyl esters during the synthesis of biodiesel with supercritical methanol. *The Journal of Supercritical Fluids*, Vol.56, No.1, pp. (56-63), ISSN 0896- 8446

Rathore, V. & Madras, G. (2007). Synthesis of biodiesel from edible and non-edible oils in supercritical alcohols and enzymatic synthesis in supercritical carbon dioxide. *Fuel*, Vol.86, No.17-18, pp. (2650-2659), ISSN 0016-2361

Saka, S. & Isayama, Y. (2009). A new process for catalyst-free production of biodiesel using supercritical methyl acetate. *Fuel*, Vol.88, No.7, pp. (1307-1313), ISSN 0016-2361

Saka, S. & Kusdiana, D. (2001). Biodiesel fuel from rapeseed oil as prepared in supercritical methanol. *Fuel*, Vol.80, No.2, pp. (225-231), ISSN 0016-2361

Sawangkeaw, R., Bunyakiat, K. & Ngamprasertsith, S. (2007). Effect of co-solvents on production of biodiesel via transesterification in supercritical methanol. *Green Chemistry*, Vol.9, No.6, pp. (679-685), ISSN 1463-9262

Sawangkeaw, R., Bunyakiat, K. & Ngamprasertsith, S. (2010). A review of laboratory-scale research on lipid conversion to biodiesel with supercritical methanol (2001-2009). *The Journal of Supercritical Fluids*, Vol.55, No.1, pp. (1-13), ISSN 0896-8446

Shimoyama, Y., Abeta, T. & Iwai, Y. (2008). Prediction of vapor-liquid equilibria for supercritical alcohol + fatty acid ester systems by SRK equation of state with Wong- Sandler mixing rule based on COSMO theory. *The Journal of Supercritical Fluids*, Vol.46, No.1, pp. (4-9), ISSN 0896-8446

Shimoyama, Y., Abeta, T., Zhao, L. & Iwai, Y. (2009). Measurement and calculation of vapor- liquid equilibria for methanol + glycerol and ethanol + glycerol systems at 493- 573 K. *Fluid Phase Equilibria*, Vol.284, No.1, pp. (64-69), ISSN 0378-3812

Silva, C., Weschenfelder, T.A., Rovani, S., Corazza, F.C., Corazza, M.L., Dariva, C. & Oliveira, J.V. (2007). Continuous Production of Fatty Acid Ethyl Esters from Soybean Oil in Compressed Ethanol. *Industrial and Engineering Chemistry Research*, Vol.46, No.16, pp. (5304-5309), ISSN 0888-5885

Song, E.-S., Lim, J.-w., Lee, H.-S. & Lee, Y.-W. (2008). Transesterification of RBD palm oil using supercritical methanol. *The Journal of Supercritical Fluids*, Vol.44, No.3, pp. (356-363), ISSN 0896-8446

Tan, K.T. & Lee, K.T. (2011). A review on supercritical fluids (SCF) technology in sustainable biodiesel production: Potential and challenges. *Renewable and Sustainable Energy Reviews*, Vol.15, No.5, pp. (2452-2456), ISSN 1364-0321

Tan, K.T., Lee, K.T. & Mohamed, A.R. (2010a). Effects of free fatty acids, water content and co-solvent on biodiesel production by supercritical methanol reaction. *The Journal of Supercritical Fluids*, Vol.53, No.1-3, pp. (88-91), ISSN 0896-8446

Tan, K.T., Lee, K.T. & Mohamed, A.R. (2010b). Optimization of supercritical dimethyl carbonate (SCDMC) technology for the production of biodiesel and value-added glycerol carbonate. *Fuel*, Vol.89, No.12, pp. (3833-3839), ISSN 0016-2361

Tang, Z., Du, Z., Min, E., Gao, L., Jiang, T. & Han, B. (2006). Phase equilibria of methanol- triolein system at elevated temperature and pressure. *Fluid Phase Equilibria*, Vol.239, No.1, pp. (8-11), ISSN 0378-3812

van Kasteren, J.M.N. & Nisworo, A.P. (2007). A process model to estimate the cost of industrial scale biodiesel production from waste cooking oil by supercritical transesterification. *Resources, Conservation and Recycling*, Vol.50, No.4, pp. (442-458), ISSN 0921-3449

Varma, M.N. & Madras, G. (2006). Synthesis of Biodiesel from Castor Oil and Linseed Oil in Supercritical Fluids. *Industrial and Engineering Chemistry Research*, Vol.46, No.1, pp. (1-6), ISSN 0888-5885

Velez, A., Hegel, P., Mabe, G. & Brignole, E.A. (2010). Density and Conversion in Biodiesel Production with Supercritical Methanol. *Industrial and Engineering Chemistry Research*, Vol.49, No.16, pp. (7666-7670), ISSN 0888-5885

Vieitez, I., da Silva, C., Borges, G.R., Corazza, F.C., Oliveira, J.V., Grompone, M.A. & Jachmanián, I. (2008). Continuous Production of Soybean Biodiesel in Supercritical Ethanol-Water Mixtures. *Energy and Fuels*, pp., ISSN 0887-0624

Vieitez, I., Pardo, M.J., da Silva, C., Bertoldi, C., de Castilhos, F., Oliveira, J.V., Grompone, M.A. & Jachmanián, I. (2011). Continuous synthesis of castor oil ethyl esters under supercritical ethanol. *The Journal of Supercritical Fluids*, Vol.56, No.3, pp. (271-276), ISSN 0896-8446

W. King, J., L. Holliday, R. & R. List, G. (1999). Hydrolysis of soybean oil in a subcritical water flow reactor. *Green Chemistry*, Vol.1, No.6, pp. (261-264), ISSN 1463-9262

Wang, C.-W., Zhou, J.-F., Chen, W., Wang, W.-G., Wu, Y.-X., Zhang, J.-F., Chi, R.-A. & Ying,

W.-Y. (2008). Effect of weak acids as a catalyst on the transesterification of soybean oil in supercritical methanol. *Energy and Fuels*, Vol.22, No.5, pp. (3479-3483), ISSN 0887-0624

Wang, L., He, H., Xie, Z., Yang, J. & Zhu, S. (2007). Transesterification of the crude oil of rapeseed with NaOH in supercritical and subcritical methanol. *Fuel Processing Technology*, Vol.88, No.5, pp. (477-481), ISSN 0378-3820

Wang, L. & Yang, J. (2007). Transesterification of soybean oil with nano-MgO or not in supercritical and subcritical methanol. *Fuel*, Vol.86, No.3, pp. (328-333), ISSN 0016- 2361

Warabi, Y., Kusdiana, D. & Saka, S. (2004). Reactivity of triglycerides and fatty acids of rapeseed oil in supercritical alcohols. *Bioresource Technology*, Vol.91, No.3, pp. (283- 287), ISSN 0960-8524

Yin, J.-Z., Xiao, M. & Song, J.-B. (2008a). Biodiesel from soybean oil in supercritical methanol with co-solvent. *Energy Conversion and Management*, Vol.49, No.5, pp. (908-912), ISSN 0196-8904

Yin, J.-Z., Xiao, M., Wang, A.-Q. & Xiu, Z.-L. (2008b). Synthesis of biodiesel from soybean oil by coupling catalysis with subcritical methanol. *Energy Conversion and Management*, Vol.49, No.12, pp. (3512-3516), ISSN 0196-8904

13
Chapter

ALTERNATIVE TECHNIQUES FOR THE PRODUCTION OF FATTY ACID ALKYL ESTERS: MICROWAVES, RADIO FREQUENCY, AND ULTRASOUND

1. INTRODUCTION: AN OVERVIEW

The manufacturing of biodiesel is a contemporary and technologically advanced field for researchers, gaining significance daily due to rising petroleum prices and environmental benefits. (Mustafa, 2011).

Biodiesel is a mixture of mono-alkyl esters of long chain fatty acids, is an alternative fuel made from renewable sources as vegetable oils and animal fats. It is biodegradable, non- toxic, show low emission profiles and also is beneficial environmentally. (Fangrui and Milford, 1999).

Biodiesel is quite similar to petroleum-derived diesel in its main characteristics such as cetane number, energy content, viscosity, and phase changes. Biodiesel contains no petroleum products, but it is compatible with conventional diesel and can be blended in any proportion with fossil-based diesel to create a stable biodiesel blend. Therefore, biodiesel has become one of the most common biofuels in the world (Lin et al., 2011). There are four primary techniques for biodiesel production: direct use and blending of raw oils, micro- emulsions, thermal cracking and trans-esterification (Siddiquee and Rohani, 2011).

Direct use of vegetable oil and animal fats as combustible fuel is not suitable due to their high kinematic viscosity and low volatility. Furthermore, its long term use posed serious problems such as deposition, ring sticking and injector chocking in engine. Microemulsions with alcohols have been prepared to overcome the problem of high viscosity of vegetable oils. Another alternative way to produce biodiesel is through thermal cracking or pyrolysis. However, this process is rather complicated to operate and produce side products that have not commercial value. The most commonly used method for biodiesel production is trans- esterification (also known as alcoholysis) reaction in presence of a catalyst. Trans- esterification is the process of exchanging the alkoxy group of an ester compound with another alcohol (Lam et al., 2010).

Esterification is the sub category of trans-esterification. This requires two reactants, carboxylic acids (fatty acids) and alcohols. Esterification reactions are acid-catalyzed and proceed slowly in the absence of strong acids such as sulfuric, phosphoric, sulfonic-organic acids and hydrochloric acid (Vyas et al., 2010).

The fatty acid methyl esters (FAME) are more used because of its facility of production, however, presents operating problems at low temperatures for its high content of saturated fractions that crystallize and can block the filters of the engines. One of the alternatives to reduce the flow properties at low temperatures (FPLT) of methyl esters specially the obtained from oil palm is use alkyl esters, obtained through of trans-esterification with branched alcohols, that prevent the agglomeration and formation of crystals of these methyl esters.

Alkyl esters can be produced through trans-esterification of triglycerides, which are separated by immiscibility and higher density. (Marchetti et al., 2007; Ma and Hanna, 1999; Vicente et al., 2004)

Very few studies have been made with the aim to obtain alkyl esters and all are obtained by homogeneous catalysis (Lee et al., 1995). Yields of these reactions are very low by the high steric hindering that presenting the branched alcohols. To increase the conversion, in this work, we propose use assisted reactions by alternative methods.

The synthesis of fatty acid alkyl esters through alternative methods, including electromagnetic radiation (microwave and radio frequency) and ultrasound, provides a rapid and straightforward approach to this valuable biofuel. This process features a brief reaction time, a low reactant ratio, operational simplicity, a significant decrease in by-product formation, and diminished energy consumption. This work proposes a review of the pertinent features of production optimization, intrinsic effects, and the most significant factors in the synthesis and characterisation of fatty acid alkyl esters (biodiesel) utilizing different methods: Microwaves, Radio Frequency, and Ultrasound.

2. FATTY ACID ALKYLESTERS PRODUCTION ASSISTED BY MICROWAVES

Electromagnetic radiation (EMR) is a form of energy exhibiting wave like behaviour as it travels through space. EMR has both electric and magnetic field components, which oscillate in phase perpendicular to each other and perpendicular to the direction of energy propagation. Electromagnetic radiation is classified according to the frequency of its wave. In order of increasing frequency and decreasing wavelength, these are radio waves, microwaves, infrared radiation, visible light, ultraviolet radiation, X-rays and gamma rays (Serway and Jewett, 2004).

Microwaves belong to the portion of the electromagnetic spectrum with wavelengths from 1 mm to 1 m with corresponding frequencies between 300 MHz and 300 GHz.

Within this portion of the electromagnetic spectrum there are frequencies that are used for cellular phones, radar, and television satellite communications. For microwave heating, two frequencies, reserved by the Federal Communications Commission (FCC) for industrial, scientific, and medical (ISM) purposes are commonly used for microwave heating. The two most commonly used frequencies are 0.915 and 2.45 GHz. Recently, microwave furnaces that allow processing at variable frequencies from 0.9 to 18 GHz have been developed for material processing (Thostenson and Chou, 1999). Microwave radiation was discovered as a heating method in 1946, with the first commercial domestic microwaves being introduced in the 1950s. The first commercial microwave for laboratory utilization was recognized in 1978 (Gedye et al., 1986; Giguere et al., 1986).

Over the last decade, microwave dielectric heating as an environmentally benign process has developed into a highly valuable technique, offering an efficient alternative energy source for numerous chemical reactions and processes. It has many advantages compared to conventional oil-bath heating, such as non-contact heating, energy transfer instead of heat transfer, higher heating rate, rapid start-up and stopping of the heating, uniform heating with minimal thermal gradients, selective heating properties, reverse thermal effects (heating starting from the interior of the material body), energy savings and higher yields in shorter reaction time (Tierney and Lidstrom, 2005). Microwave heating is based dielectric heating, the ability of some polar liquids and solids to absorb and convert microwave energy into heat. In this context, a significant property is the mobility of the dipoles by either ionic conduction or dipolar polarization and the ability to orient them according to the direction of the electric field. The orientation of the dipoles changes with the magnitude and the direction of the electric field. Molecules that have a permanent dipole moment are able to align themselves through rotation, completely or at least partly, with the direction of the field. Therefore, energy is lost in the form of heat through molecular friction and dielectric loss (Loupy, 2002). The amount of heat produced by this process is directly related to the capability of the matrix to align itself with the frequency of the applied electric field. If the dipole does not have enough time to realign, or reorients too rapidly with the applied field, no heating occurs (Kappe, 2004).

The production of biodiesel via the conventional heating system appears to be inefficient due to the fact that the heat energy is transferred to the reactants through conduction, convection and radiation from the surface of the reactor. Hence, conventional heating requires longer reaction time and a larger amount of heat energy to obtain a satisfactory biodiesel. The replacement of conventional heating by microwave radiation for the transesterification process is expected to shorten the reaction time due to the transfer of heat directly to the reactants. The microwave radiation during the transesterification process is expected to create (i) an alignment of polar molecules such as alcohols with a continuously changing magnetic field generated by microwaves and (ii) molecular friction due to which heat will be generated (Yaakob et al., 2009).

The involvement of such heterogeneous catalytic systems under microwave conditions represents an innovative approach with processing advantages. These solid-state catalysts find scope in the context of green chemistry development as they are active in solvent free or dry media synthesis, with potential advantages in terms of separation, recovery post- reaction and recycling assays. The creation of hot spots, specific under MW conditions, is typically utilized for energy saving as improved yields and selectivities are recorded after shorten reaction times at lower nominal temperatures. These hot spots may induce a re- organization of the catalyst under microwave conditions and are probably responsible for reaction rates and selectivity enhancement (compared to conventional heating at the same nominal temperature) (Richela et al., 2011)

2.1 Esterification reactions assisted by MW

The esterification reaction is a slow equilibrium, and can be catalyzed by Brønsted acids such as sulfuric acid. The main problem is the generation of highly acidic waste causing a serious environmental problem, and to reduce this problem have been used alternative heterogeneous catalysts and microwaves as a heating source to promote and increase the yielding. Algunos catalizadores empleados son: scandium triflate and bismuth triflate (Socha and Sello, 2010), sulfated zirconia (Kim et al., 2011a), niobium oxide (Melo et al., 2010), entre otros.

The temperature presented a pronounced effect on the conversion, following an exponential dependence. The results for a distinct molar ratio of alcohol/fatty acid indicated that the increase of this parameter lead to a decrease on the reaction conversion. In general, the esterification reaction under microwave irradiation yielded similar results to those obtained with the conventional heating but with very fast heating rates (Melo et al., 2009). The pulsed microwaves with repetitive strong power could enhance the efficiency of biodiesel production relative to the use of continuous microwave with mild power (Kim et al., 2011b). Electric energy consumption for the microwave heating in this accelerated esterification was only 67% of estimated minimum heat energy demand because of significantly reduced reaction time (Kim et al., 2011a).

For oils with a high content of free fatty acid FFA as palm oil, has been proposed obtain alkyl ester from crude palm oil (CPO), using microwaves like heating source, in a process of two stages by means of homogeneous and heterogeneous catalysis; the first stage (esterification), was made using sulfuric acid and Dowex

50X2, Amberlyst 15 and Amberlite IR-120 resin catalysts, to diminish the acid value of the oil, avoiding the soap formation and facilitating the separation of the phases. In these works has been reported the obtaining of alkyl ester using alcohols non-conventional such as: ethanol (EtOH) (Suppalakpanya et al., 2010a, 2010b), isopropyl (IsoprOH), isobutyl (IsobuOH), 2-butyl (2- BuOH) and Isopentyl (IsopentOH) alcohols (Mazo and Rios, 2010a; Mazo and Rios, 2010b), where was found that that the acidity order obtained for the catalysts is Dowex < Amberlite < Amberlyst, and the order for the alcohols: Methanol < isopropyl alcohol < isobutyl alcohol < 2-butyl alcohol < isopentyl alcohol, because Dowex microreticular resin presents the lowest divinylbenzene (2%), which has a lower cross-linking that produces a high expansion of the resin in a polar medium, and the resin can expand their pores up to 400%, enabling the income of the voluminous substrate (FFA) and its protonation. Amberlyst 15 macroreticular resin is activated due to its surface area, and the protons located on the outer surface seem that catalyse the esterification because the interiors are inaccessible due to high cross-linking. The reaction is favoured with the increasing of polarity of solvents.

Table 1 shows the work carried out for bio-diesel production by esterification of FFA under different conditions using microwave irradiation.

Table 1. Microwave assisted esterification.

FFA	Catalyst	Catalyst amount (%)	Alcohol	Oil to alcohol molar ratio	Microwave reaction conditions	Ester conver-sion (%)	Ref.
Oleic Linoleic	- -	- -	MeOH EtOH MeOH	1:10	Synthos 3000- Anton Paar. 1400W. 30min, 200°C	51.8 31.5 49.6	(Melo et al., 2009)
Oleic	Sulfated Zirconia	5 wt	MeOH	1:20	Experimental MW heating system 20min, 60°C	90.0	(Kim et al., 2011a)
Oleic	Amberlyst 15 dry	10 wt	MeOH	1:20	Experimental MW heating system. Pulsed MW. 15min, 60°C	66.1	(Kim et al., 2011b)
Oleic	Niobium Oxide Sulfated Zirconia	5 wt	MeOH	1:10	Synthos 3000- Anton Paar. 1400W. 20min, 200°C	68.0 68.7	(Melo et al., 2010)

Linoleic Linoleic Oleic Oleic Myristic Myristic Palmitic Palmitic	Sc(OTf)3 Bi(OTf)3 Sc(OTf)3 Bi(OTf)3 Sc(OTf)3 Bi(OTf)3 Sc(OTf)3 Bi(OTf)3	1%mol	MeOH	48 eq	Biotage MW reactor. 1min, 150°C	97.0 98.0 100.0 88.0 98.0 90.0 100.0 99.0	(Socha and Sello, 2010)
FFA Palm Oil	H2SO4	2.5%wt Oil	MeOH IsoprOH IsoBuOH 2-BuOH Isopen-tOH	1:8	Domestic MW 1000W 60 min, 60°C 60 min, 75°C 60 min, 105°C 60 min, 90°C 60 min, 115°C	99.8 99.8 96.2 95.5 90.8	(Mazo and Rios, 2010a)
FFA Palm Oil	H2SO4	4% wt FFA	EtOH	1:24	Domestic MW 800W 60 min, 70°C	87.7	(Suppalak-panya et al., 2010a and b)
FFA Palm Oil	Dowex 50X2	10%wt Oil	MeOH IsoprOH IsoBuOH 2-BuOH Isopen-tOH	1:20	Domestic MW 1000W 60 min, 60°C 60 min, 75°C 60 min, 105°C 60 min, 90°C 60 min, 115°C	95.6 86.2 82.8 78.5 77.7	(Mazo and Rios, 2010b)
FFA Palm Oil	Amberlite IR120	10%wt Oil	MeOH IsoprOH IsoBuOH 2-BuOH Isopen-tOH	1:20	Domestic MW 1000W 60 min, 60°C 60 min, 75°C 60 min, 105°C 60 min, 90°C 60 min, 115°C	91.3 85.1 81.4 74.8 74.1	(Mazo and Rios, 2010b)
FFA Palm Oil	Amber-lyst15	10%wt Oil	MeOH IsoprOH IsoBuOH 2-BuOH Isopen-tOH	1:20	Domestic MW 1000W 60 min, 60°C 60 min, 75°C 60 min, 105°C 60 min, 90°C 60 min, 115°C	91.4 84.7 80.6 66.8 73.5	(Mazo and Rios, 2010b)

2.2 Transesterification reactions assisted by MW

Vegetable oils are becoming a promising alternative to diesel fuel because they are renewable in nature and can be produced locally and in environmentally friendly ways. Edible vegetable oils such as canola and soybean oil in the USA, palm oil in Malaysia, rapeseed oil in Europe and corn oil have been used for biodiesel production and found to be good diesel substitutes. Non-edible vegetable oils, such as Pongamia pinnata (Karanja or Honge), Jatropha curcas (Jatropha or Ratanjyote), Madhuca iondica (Mahua) and Castor Oil have also been found to be suitable for biodiesel production (Yusuf et al., 2011).

Transesterification (also called alcoholysis) is the reaction of a fat or oil with an alcohol (with or without catalyst) to form esters and glycerol. Since the reaction is reversible, excess alcohol is used to shift the equilibrium to the product side (Fangrui and Milford, 1999). Under Transesterification reaction with alcohol the first step is the conversion of triglycerides to diglycerides, which is followed by the subsequent conversion of higher glycerides to lower glycerides and then to glycerol, yielding one methyl ester molecule from each glyceride at each step (Hideki et al., 2001).

Several examples of microwave irradiated transesterification methods have been reported using homogenous alkali catalyst (Kumar et al., 2011; Azcan and Danisman, 2008), acid catalyst (Mazo and Rios, 2010a) and heterogeneous alkali catalyst (Patil et al., 2011), heterogeneous acid catalyst (Yuan et al., 2009) and enzymatic (Yu et al., 2010).

Microwave synthesis is not easily scalable from laboratory small-scale synthesis to industrial multi kilogram production. The most significant limitation of the scale up of this technology is the penetration depth of microwave radiation into the absorbing materials, which is only a few centimeters, depending on their dielectric properties. The safety aspect is another reason for rejecting microwave reactors in industry (Groisman and Aharon, 2008).

The preparation of biodiesel using a scientific microwave apparatus offers a fast, easy route to this valuable biofuel with advantages of a short reaction time, a low oil/methanol ratio, and an ease of operation. The methodology allows for the reaction to be run under atmospheric conditions; it is complete in a matter of a few minutes and can be performed on batch scales up to 3 kg of oil at a time (Leadbeater and Stencel, 2006).

The continuous-flow preparation of biodiesel using a commercially available scientific microwave apparatus offers a fast, easy route to this valuable biofuel. The methodology allows for the reaction to be run under atmospheric conditions and performed at flow rates of up to 7.2 L/min using a 4 L reaction vessel. Energy consumption calculations suggest that the continuous-flow microwave methodology for the transesterification reaction is more energy-efficient than using a conventional heated apparatus (Barnard et al., 2007).

Few studies report the use of alcohols different to methanol. Alcohols more used are ethanol and butanol, and the latter is a versatile and sustainable platform chemical that can be produced from a variety of waste biomass sources. The emergence of new technologies for the production of fuels and chemicals from butanol will allow it to be a significant component of a necessarily dynamic and multifaceted solution to the current global energy crisis. Recent work has shown that butanol is a potential gasoline replacement that can also be blended in significant quantities with conventional diesel fuel (Harvey and Meylemans, 2011).

Table 2 shows the work carried out for bio-diesel production from various feedstocks, catalysis and alcohols under different conditions using microwave irradiation.

Table 2. Microwave assisted transesterification.

Oil	Catalyst	Catalyst amount (%wt)	Alcohol	Oil to alcohol molar ratio	Microwave reaction conditions	Ester conversion (%)	Ref.
Castor	50% H_2SO_4/C	5	MeOH	01:12	MAS-1 Shanghai Sineo MW 65°C, 60 min	94	(Yuan et al., 2009)
Castor	SiO_2/50%H_2SO_4 SiO_2/50%H_2SO_4 Al2O3/50%KOH	10	MeOH EtOH MeOH	01:06	Domestic MW 540W 60°C, 30 min 60°C, 20 min 60°C, 5 min	95 95 95	(Perin et al., 2008)
Jatropha	KOH	1.5	MeOH	01:07.5	Start Synth-Milestone 1200W 65°C, 2 min	99	(Shakinaz et al., 2010)
Waste frying	SrO $Sr(OH)_2$	1.5	MeOH	1:4 wt	Domestic MW 900W 60°C, 40 s	99 97	(Koberg et al., 2011)
Pongamia Pinnata	NaOH KOH	0.5 1.0	MeOH	01:06	Start Synth-Milestone 1200W 60°C, 7 min	95.3 96	(Kumar et al., 2011)
Rapeseed	NaOH KOH	1	MeOH	-	Start Synth-Milestone 1200W 60°C, 5 min	91.7 90.8	(Azcan and Danisman, 2008)
Soybean	Nano CaO	3.0	MeOH	01:07	ETHOS900 Milestone 900W 65°C, 60 min	96.6	(Hsiao et al., 2011)

Oil	Catalyst	Catalyst amount (%wt)	Alcohol	Oil to alcohol molar ratio	Microwave reaction conditions	Ester conversion (%)	Ref.
Soybean	Novozym 435	3	MeOH	01:6	MCR-3 Shanghai JieSi 800W 40°C, 12h	94	Yu et al., 2010)
Canola	$ZnO/La_2O_2CO_3$	1	MeOH	1:1 wt	Biotage MW reactor. <100°C, 5 min	>95	(Jin et al., 2011)
Camelina	BaO SrO	1.5 2	MeOH	01:09 01:12	Domestic MW 800W 100°C, 4 min 60°C, 4 min	95 78	(Patil et al., 2011)
Camelina	NaOH KOH BaO SrO $BaCl_2$/AA SrCl2/AA	0.5 1 1.5 2 2 2	MeOH	01:09	Domestic MW 800W 60°C, 60 s 60°C, 60 s 60°C, 4 min 60°C, 4 min 60°C, 5 min 60°C, 5 min	95 85 95 95 27 20	(Patil et al., 2010)
Safflower	NaOH	1	MeOH	01:10	Start labstation-Milestone 60°C, 6 min	98.4	(Duz et al., 2011)
Soybean Rice Bran	NaOH	0.6	EtOH	01:05	ETHOS E-Milestone 73°C, 10 min	99.25 99.34	(Terigar et al., 2010)
Rapeseed	KOH	1	ButOH	01:04	MARS CEM Corp. 117°C, 30 min	100	(Geuens et al., 2008)
Soybean	H_2SO_4 KOH	5 1	ButOH ButOH	01:06	CEM Discover 300W CEM MARS 1600W 100°C, 15 min 120°C, 1 min	93 93	(Leadbeater et al., 2008)
Jatropha Waste frying	NaOH	1	MeOH	01:12	MW650 Aurora Instruments MW discovery 65°C, 7 min	89.7 88.63	(Yaakob et al., 2009)
Palm	H_2SO_4	3	MeOH IsoprOH IsoBuOH 2-BuOH IsopentOH	01:30	Domestic MW 1000W 60°C, 5h 75°C, 5h 105°C, 5h 90°C, 5h 115°C, 5h	49.4 62.39 67.39 62.39 75	(Mazo and Rios, 2010a)

Oil	Catalyst	Catalyst amount (%wt)	Alcohol	Oil to alcohol molar ratio	Microwave reaction conditions	Ester conversion (%)	Ref.
Palm	$NaOCH_3$	0.9	MeOH IsoprOH IsoBuOH 2-BuOH IsopentOH	01:27	Domestic MW 1000W 60°C, 1h 75°C, 1h 105°C, 1h 90°C, 1h 115°C, 1h	99.9 99.87 88.39 83.19 81.63	(Mazo and Rios, 2010a)
Palm	K2CO3	3	MeOH IsoprOH IsoBuOH 2-BuOH IsopentOH	01:20	Domestic MW 1000W 60°C, 3h 75°C, 3h 105°C, 3h 90°C, 3h 115°C, 3h	8.63 49.51 67.59 52 54.59	(Mazo and Rios, 2010b)
Palm	KOH	1.5	EtOH	01:04	Domestic MW 800W 70°C, 5 min	97.4	(Suppalak-panya et al., 2010a)

2.3 Optimization production biodiesel under MW irradiation

Some examples about the obtaining of biodiesel making a response surface methodology (RSM) was used to analyze the influence of the process variables (oil to methanol ratio, catalyst concentration, and reaction time) on the fatty acid methyl ester conversion, are shown in Table 3, where is confirmed that the microwave energy has a significant effect on esterification and transesterification reactions.

Table 3. Recent examples of optimization of reaction conditions a for production of biodiesel from various feedstocks using response surface methodology

Oil	Catalyst	Catalyst amount (%wt)	Alcohol	Oil to alcohol molar ratio	Microwave reaction conditions	Ester conversion (%)	Ref.
Dry algae	KOH	2.0	MeOH	1:12 (wt/vol)	Domestic MW 800W 60°C, 4min	64.18	(Patil et al., 2011)
Macau-ba	Novozyme 435 Lipozyme IM	2.5	EtOH	1:9	Synthos 3000-Anton Paar. 1400W. 30°C, 15min	45.2 22.9	(Nogue-ira et al., 2010)
Pon-gamia pinnata	Esterifica-tion: H2SO4 Transester-ification: KOH	3.73 1.33	MeOH MeOH	33.83% (w/w) 33.4% (w/w)	Domestic MW 800W 60°C, 190s 60°C, 150s	87.4 89.9	(Ven-katesh et al., 2011)

3. FATTY ACID ALKYLESTERS PRODUCTION ASSISTED BY RADIO FREQUENCY

Radio waves, whose wavelengths range from more than 104 m to about 0.1 m, are the result of charges accelerating through conducting wires. They are generated by such electronic devices as LC oscillators and are used in radio and television communication systems (Serway and Jewett, 2004).

Radio frequency (RF) heating is a promising dielectric heating technology which provides fast heat generation through a direct interaction between an RF electromagnetic field and the object being heated (Piyasena et al., 2003). Compared to microwave heating, a popular dielectric heating technology, RF heating systems are simpler to configure and have a higher conversion efficiency of electricity to electromagnetic power (Wang et al., 2003). Moreover, RF energy has deeper penetration into a wide array of materials than microwave energy, increasing feasibility of RF heating for industrial scale applications.

Very few publications have been obtained by this alternative heating method, which use a RF heating apparatus (SO6B; Strayfield Fastran, UK). The distance between the two electrodes was fixed at 15 cm. A 150-mL conical flask coupled with a water-cooling reflux condenser was used as a reactor. Schematic diagram and photograph are shown in Fig. 1.

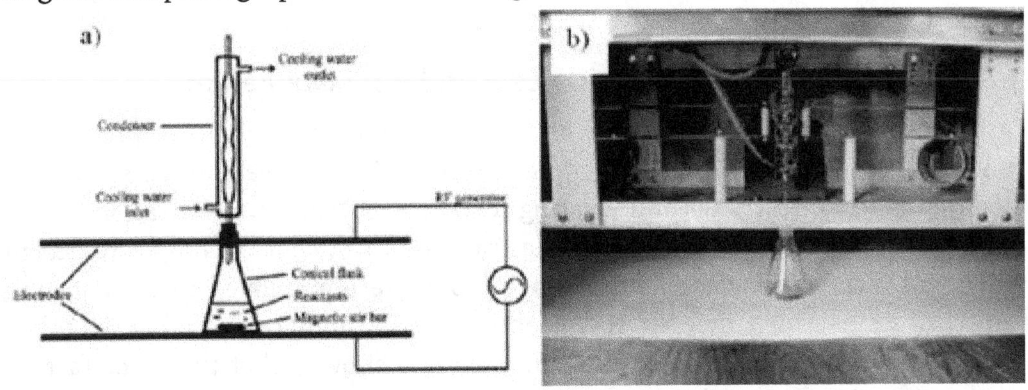

Fig. 1. a) Schematic diagram of RF heating apparatus (Lui et al., 2010) and b) Photograph of RF heating apparatus (Lui et al., 2008).

Applications to obtaining biodiese using different oils, reaction conditions and catalysts are described below:

3.1 Esterification reactions assisted by RF

Efficient biodiesel conversion from waste cooking oil with high free fatty acids (FFAs) was achieved via a two-stage procedure (an acid-catalyzed esterification followed by an alkali- catalyzed transesterification) assisted by radio frequency (RF) heating. In the first stage, with only 8-min RF heating the acid number of the waste cooking oil was reduced from 68.2 to 1.64 mg KOH/g by reacting with 3.0% H_2SO_4 (w/w, based on oil) and 0.8:1 methanol (weight ratio to waste oil). Then, in the second stage, the esterification product (primarily consisting of triglycerides and fatty acid methyl esters) reacted with 0.91% NaOH (w/w, based on triglycerides) and 14.2:1 methanol (molar ratio to triglycerides) under RF heating for 5 min, and an overall conversion rate of 98.8 ± 0.1% was achieved. Response surface methodology was employed to evaluate the effects of RF heating time, H_2SO_4 dose and methanol/oil weight ratio on the acid-catalyzed esterification. A significant positive interaction between RF heating time and H_2SO_4 concentration on the esterification was observed (Lui et al., 2010).

3.2 Transesterification reactions assisted by RF

Efficient biodiesel production from beef tallow was achieved with radio frequency (RF) heating. A conversion rate of 96.3% was obtained with a NaOH concentration of 0.6% (based on tallow), an RF heating for 5 min, and a methanol/tallow molar ratio of 9:1. Response surface methodology was employed to evaluate the influence of NaOH dose, RF heating time, and methanol/tallow ratio. The alkaline concentration showed the largest positive impact on the conversion rate. Similar fast conversion from canola oil to biodiesel was achieved in our previous work, indicating that RF heating, as an accelerating technique for biodiesel production, had a large applying area (Lui et al., 2011).

3.3 Optimization production biodiesel under RF irradiation

Fast transesterification of canola oil and methanol for biodiesel production was achieved using radio frequency (RF) heating. The conversion rate of oil to biodiesel reached 97.3% with RF heating for 3 min, a NaOH concentration (based on oil) of 1.0%, and a methanol/oil molar ratio of 9:1. A central composite design (CCD) and response surface methodology (RSM) were employed to evaluate the impact of RF heating time, NaOH concentration, and molar ratio of methanol to oil on conversion efficiency. Experimental results showed that the three factors all significantly affected the conversion rate. NaOH concentration had the largest influence, with the effect being more pronounced at lower (0.2-0.6%, based on weight of oil) concentration. No evident interaction among the three factors was observed. RF heating efficiency was primarily related to the amount of NaOH and methanol. The scale of the experiment was increased by five times (from 20 to 100 g oil per batch) without decrease of the conversion rate, indicating the scale-up potential of RF heating for biodiesel production (Lui et al., 2008).

4. FATTY ACID ALKYLESTERS PRODUCTION ASSISTED BY ULTRASOUND

Sound waves are the most common example of longitudinal waves. They travel through any material medium with a speed that depends on the properties of the medium. As the waves travel through air, the elements of air vibrate to produce changes in density and pressure along the direction of motion of the wave. If the source of the sound waves vibrates sinusoidally, the pressure variations are also sinusoidal (Serway and Jewett, 2004).

Sound waves are divided into three categories that cover different frequency ranges (1) Audible waves lie within the range of sensitivity of the human ear. They can be generated in a variety of ways, such as by musical instruments, human voices, or loudspeakers. (2) Infrasonic waves have frequencies below the audible range. Elephants can use infrasonic waves to communicate with each other, even when separated by many kilometers. (3) Ultrasonic waves have frequencies above the audible range. You may have used a "silent" whistle to retrieve your dog. The ultrasonic sound it emits is easily heard by dogs, although humans cannot detect it at all. Ultrasonic waves are also used in medical imaging (Serway and Jewett, 2004).

Sonochemistry is a branch of chemical research dealing with the chemical effects and applications of ultrasonic waves, that is, sound with frequencies above 20 kHz that lie beyond the upper limit of human hearing. The development of ultrasound in organic synthesis began on 1930 when Richards and Loomis, 1927, applied ultrasound (100-500 KHz) in organic synthesis for determine the effect on the solubility of gases for first time. Developments were very slow, then Luche and Damiano, 1980, reported metal activation reactions using ultrasound probes. Thereafter, reaction systems using US to speed up chemical reactions have been developed.

A low frequency ultrasonic irradiation could be useful for transesterification of triglyceride with alcohol. Ultrasonication provides the mechanical energy for mixing and the required activation energy for initiating the transesterification reaction (Singh et al., 2007). Ultrasonication increases the chemical reaction speed and yield of the transesterification of vegetable oils and animal fats into biodiesel. Ultrasonic assisted transesterification method presents advantages such as shorter reaction time and less energy consumption than the conventional mechanical stirring method, efficient molar ratio of methanol to TG, and simplicity (Ji et al., 2006; Siatis et al., 2006).

Many researchers have tried to solve the mass-transfer limitation problem in biodiesel synthesis using ultrasonic cavitation and hydrodynamic cavitation. Cavitation has been shown to efficiently speed up the transesterification reaction because it simultaneously supplies heating as well as the stirring effect as a result of jet formation on bubble collapse. Cavitation is basically the formation, growth, and implosive collapse of gas or vapour filled microbubbles and can be induced acoustically (using ultrasound) or hydrodynamically in a body of liquid. The collapse of these bubbles lead to local transient high temperatures (g 5000 K) and pressures (g 1000 atm), resulting in the generation of highly reactive species, such as OH•, HO_2•, and H• radicals in water. Cavitation effects also increase the mass and heat transfers in a medium and accelerate the reaction rates and yields (Mahamuni and Adewuyi, 2009).

Main factors that vary the yielding in the production of biodiesel using US are:

Effect of Ultrasonic Frequency on Biodiesel Yield. The frequency of the ultrasound has a significant effect on the cavitation process because it alters the critical size of the cavitation bubble, which in turn changes the intensity of the collapse of the cavitation bubbles.

Effect of Ultrasonic Power on Biodiesel Yield. It is well-known that as the ultrasonic power increases, the size of the cavitation bubbles increase leading to more intense collapse of bubble, which causes better emulsion formation of oil and methanol resulting into higher interfacial surface area for mass transfer and hence the higher biodiesel yield. The BD yield increased with increasing ultrasonic power from 150 to 450 W, but the ME content decreased at ultrasonic powers over 450 W. This is due to the decrease of the real irradiation time caused by the increase in the pulse interval required for tuning the temperature due to the extension of the irradiation power (Lee et al., 2011).

Effect of Catalyst Loading. As the amount of KOH increases, the concentration of methoxide anions, which are responsible for nucleophilic attack on the triglyceride molecules to produce biodiesel, also increase, resulting in higher biodiesel yield.

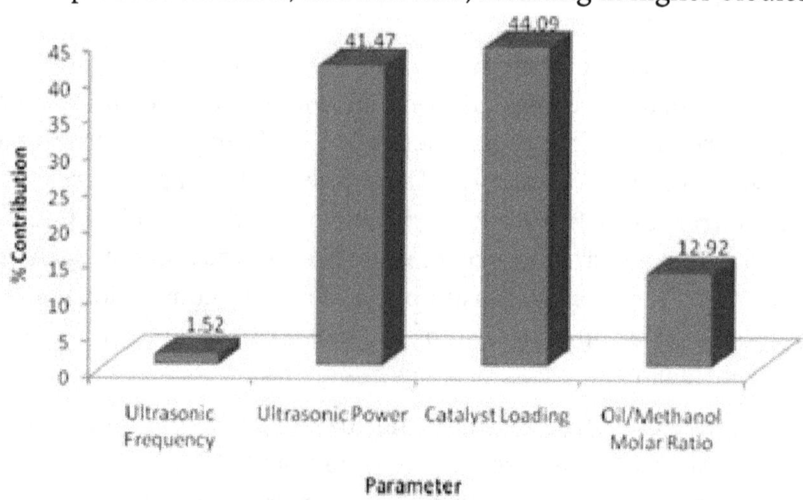

Fig. 2. Percentage contribution of the factors contributing to the production of biodiesel by ultrasound

Effect of Oil/Methanol Molar Ratio. As oil and methanol are not miscible into each other, they form a heterogeneous reaction mixture and mass transfer between these two phases becomes important for the transesterification reaction. The presence of ultrasound can help increase the mass transfer between the two phases by the formation of a fine emulsion, which increases the interfacial area between the two phases. Ultrasound can also increase the mass transfer coefficient due to the presence of acoustic streaming and jet formations at the end of cavitation bubble collapse near the phase boundary between oil and methanol phases.

As shown in Fig. 2, the factors with more contribution to the production of biodiesel are ultrasonic power and catalyst loading, then oil/methanol molar ratio and finally, the frequency.

4.1 Esterification reactions assisted by ultrasound

Not found reports of this methodology for the esterification reaction in the obtaining of biodiesel.

4.2 Transesterification reactions assisted by ultrasound

The works that use US in the transesterificatication reaction for the obtaining biodiesel use edible oils as soybean, triolein, palm, canola, fish and coconut. Also, use no edible oil as Jatropha, homogeneous basic catalysts as KOH, and heterogeneous basic catalysts as CaO, SrO, BaO, Na/SiO_2 and Novozym435 enzymes and lipase. Table 8 shows the work carried out for bio-diesel production from various feedstocks under different conditions using ultrasound irradiation.

The equipments used are conformed by transducer, cleaner and probe used in batch processes. In recent years, chemistry in flowing systems has become more prominent as a method of carrying out chemical transformations, ranging in scale from microchemistry up to kilogram-scale processes. Compared to classic batch ultrasound reactors, flow reactors stand out for their greater efficiency and flexibility as well as lower energy consumption. Cintas et al., 2010, developed a new ultrasonic flow reactor, a pilot system well suited for reaction scale up. This was applied to the transesterification of soybean oil with methanol for biodiesel production. This reaction is mass-transfer-limited initially because the two reactants are immiscible with each other, then because the glycerol phase separates together with most of the catalyst (Na or K methoxide). In our reactor a mixture of oil (1.6 L), methanol and sodium methoxide 30% in methanol (wt/wt ratio 80:19.5:0.5, respectively) was fully transesterified at about 45 C in 1 h (21.5 kHz, 600 W, flow rate 55 mL/min). The same result could be achieved together with a considerable reduction in energy consumption, by a two-step procedure: first a conventional heating under mechanical stirring (30 min at 45°C), followed by ultrasound irradiation at the same temperature (35 min, 600

W, flow rate 55 mL/min). Our studies confirmed that high-throughput ultrasound applications definitively require flow reactors (Cintas et al., 2010). The detailed scheme of the system is showed in Fig. 3.

Table 4. Ultrasound assisted transesterification

Oil	Catalyst	Catalyst amount (%wt)	Alcohol	Oil to Alcohol molar ratio	Ultrasonic reaction condition	Source of ultrasound	Ester conversion (%)	Ref
Soybean	KOH	0.5	MeOH	1:6	611kHz, 139W, 26°C, 30min	Multifrequency transducer UES300C sonochemist	90	(Mahamuni and Adewuyi, 2009)
Triolein	KOH	1.0	MeOH	1:6	40kHz, 1200W, 25°C, 30min	Honda electronic cleaner	99	(Hanh et al., 2008)
Soybean	Novozym 435	6.0	MeOH	1:6	40kHz, 500W (50%), 40°C, 4h 0.5%v/v tert-amyl alcohol/oil	Ultrasonic bath KQ500DV Kunshan	96	(Yu et al., 2010)
Soybean	$NaOCH_3$	20g (30% in MeOH)	MeOH	1.6L:80g MeOH	21.5kHz,600W, 45°C, 1h Flow 55mL/min	3 transducer (21.5kHz)	90	(Cintas et al., 2010)
Jatropha	Na/SiO_2	3.0	MeOH	1:9	24kHz, 200W, 15min	UP200S ultrasonic Gmblt	98.53	(Kumar et al., 2010a)
Canola Soybean Corn	KOH	1.0	MeOH	1:6	450W, 55°C, 30min	Probe type VCX-600	98 97 95	(Lee et al., 2011)
Jatropha	Lipase Chromobacterium viscosum	5.0	MeOH		0.7s, 100W/m3, 30min	UP200S Hielscher ultrasonic Gmblt	84.5	(Kumar et al., 2011)
Palm	KOH	20.0	MeOH		Petroleum ether Ethyl methyl ketone 47kHz, 340W, 60°C, 2h	Water bath Bransonic cleaner	75.2 60	(Boey et al., 2011)
Palm	CaO SrO BaO	3.0	MeOH		20kHz, 200W (50%), 65°C, 60min	Transducer and probe	77.3 95.2 95.2	(Mootabadi et al., 2010)
Waste frying	KOH	First stage 0.7 Second stage 0.3	MeOH		20kHz, 25°C, 5min	Horn transducer	81 99	(Thanh et al., 2010)
Fish	$NaOC_2H_5$	0.8	EtOH		35kHz, 20kHz, 20°C, 30min	Bath Probe	95 95	(Armenta et al., 2007)
Coconut	KOH	0.75	EtOH		24kHz, 200W, 7min	UP200S Hielscher ultrasonic Gmblt	98	(Kumar et al., 2010b)

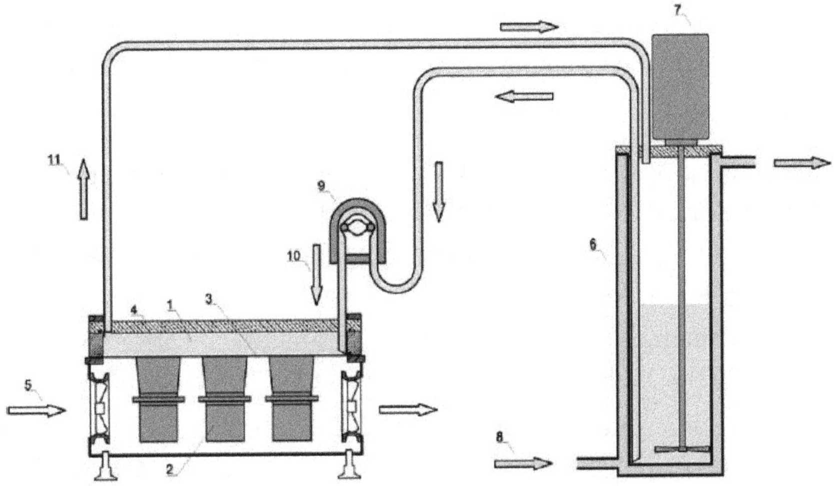

1. Sonication chamber
2. US transducers
3. Titanium plate
4. Screwed cover
5. Air cooling
6. Tank
7. Mechanical stirrer
8. Thermostatting liquid
9. Peristaltic pump
10. Inlet
11. Outlet

Fig. 3. Detailed scheme of the system for biodiesel production (Cintas et al., 2010).

4.3 Optimization production biodiesel under ultrasound

This paper utilizes the Taguchi optimization methodology (L9 orthogonal array) to optimize various parameters for the ultrasound-assisted, KOH-catalyzed transesterification of soybean oil with methanol. The statistical tool used in the Taguchi method to analyze the results is the analysis of variance (ANOVA). The optimum conditions are determined to be 581 kHz, 143 W, 0.75% (w/w) KOH loading at 1:6 oil/methanol molar ratio, resulting in more than 92.5% biodiesel yield in less than 30 min. Confirmation experiments have been performed to prove the effectiveness of the Taguchi technique after the optimum levels of process parameters are determined. (Mahamuni et al., 2010).

5. FUTURE DIRECTIONS

Most reports focus on the production of first-generation biofuels. These synthesis approaches aim to produce second and third generation fuels to advance sustainable chemistry and utilize renewable raw materials that do not compete with food sources. The integration of biotechnological procedures with these novel technologies would enhance existing methods, decreasing duration and augmenting production.

The creation of additives that enhance the characteristics of biodiesel would facilitate an enhancement in the cold flow qualities of biodiesel derived from oils such as palm, achievable through these technologies. Efforts should focus on advancing these technologies at both pilot and industrial scales, emphasizing continuous operations, minimal energy usage, and economic viability.

6. REFERENCES

Armenta, R.E.; Vinatoru, M.; Burja, A.M.; Kralovec, J.A.; Barrow, C.J. (2007) Fish Oil Transesterification of Fish Oil to Produce Fatty Acid Ethyl Esters Using Ultrasonic Energy. J Am Oil Chem Soc, Vol.84, No.11, (November 2007), pp.1045–1052 ISSN 0003-021X

Azcan, N. & Danisman, A. Microwave assisted transesterification of rapeseed oil. Fuel, Vol.87, No. 10-11, (August 2008), pp.1781–1788. ISSN: 0016-2361

Bandow, H. (2010) A two-step continuous ultrasound assisted production of biodiesel fuel from waste cooking oils: A practical and economical approach to produce high quality biodiesel fuel. Bioresource Technology, Vol.101, No.14, (July 2010), pp.5394– 5401, ISSN 0960-8524

Barnard, T.M.; Leadbeater, N.E.; Boucher, M.B.; Stencel, L.M. & Wilhite, B.A.(2007) Continuous-Flow Preparation of Biodiesel Using Microwave Heating. Energy & Fuels, Vol.21, No.3, pp.1777-1781, ISSN: 0887-0624

Boey, P.L; Ganesan, S.; Maniam, G.P. & Ali, D.M.H. (2011) Ultrasound aided in situ transesterification of crude palm oil adsorbed on spent bleaching clay. Energy Conversion and Management, Vol.52, No.5, (May 2011), pp.2081–2084, ISSN: 0196- 8904

Cintas, P.; Mantegna, S.; Calcio, E. & Cravotto, G. (2010) A new pilot flow reactor for high- intensity ultrasound irradiation. Application to the synthesis of biodiesel. Ultrasonics Sonochemistry, Vol.17, No.6 (August 2010) 985–989. ISSN: 1350-4177

Dharmendra Kumar, Gajendra Kumar, Poonam, C.P. Singh. (2010) Fast, easy ethanolysis of coconut oil for biodiesel production assisted by ultrasonication. Ultrasonics Sonochemistry, Vol.17, No.3, (March 2010), pp. 555–559. ISSN: 1350-4177

Duz, M.Z.; Saydut, A. & Ozturk, G. (2011) Alkali catalyzed transesterification of safflower seed oil assisted by microwave irradiation. Fuel Processing Technology, Vol.92, No.3, (March 2011), pp.308–313, ISSN: 0378-3820

Fangrui, Ma.; & Milford, A. (1999) Biodiesel Production: a Review. Bioresource Technology, Vol.70, No.1, (October 1999), pp. 1-15, ISSN 0960-8524

Gedye, R.; Smith, F.; Westaway K.; Ali, H.; Baldisera, L.; Laberge, L. & Rousell, J. (1986) The use of microwave ovens for rapid organic synthesis. Tetrahedron Lett, Vol.27, No.3, pp.279–82, ISSN: 0040-4039

Geuens, J.; Kremsner, J.M.; Nebel, B.A.; Schober, S.; Dommisse, R.A.; Mittelbach, M.; Tavernier, S.; Kappe, O.; & Maes, B. (2008) Microwave-Assisted Catalyst-Free Transesterification of Triglycerides with 1-Butanol under Supercritical Conditions. Energy & Fuels, Vol.22, No.1, pp.643–645, ISSN: 0887-0624

Giguere, R.J.; Bray, T.L.; Duncan, S.M. & Majetich, G. (1986) Application of commercial microwave ovens to organic synthesis. Tetrahedron Lett, Vol.27, No.41, pp.4945–58, ISSN: 0040-4039

Groisman, Y. & Gedanken, A. (2008) Continuous flow, circulating microwave system and its application in nanoparticle fabrication and biodiesel synthesis. J Phys Chem C, Vol.112, No.24, pp.8802-8808, ISSN: 1932-7447.

Hanh, H.D.; The Dong, N.; Starvarache, C.; Okitsu, K.; Maeda, Y. & Nishimura, R. (2008) Methanolysis of triolein by low frequency ultrasonic irradiation. Energy Conversion and Management, Vol.49, No.2, (February 2008), pp. 276–280, ISSN: 0196-8904

Harvey, B.G & Meylemans, H.A. (2011) The role of butanol in the development of sustainable fuel technologies. J Chem Technol Biotechnol, Vol.86, No.1, (January 2011), pp.2–9, ISSN: 0268-2575

Hideki, F.; Akihiko, K. & Hideo, N. (2001) Biodiesel fuel production by transesterification of oils. J Biosci Bioeng, Vol.92, No.5, pp.405-416, ISSN: 1389-1723

Hsiao, M.C; Lin, C.C. & Chang, Y.H. Microwave irradiation-assisted transesterification of soybean oil to biodiesel catalyzed by nanopowder calcium oxide. Fuel, Vol.90, No.3, (March 2011), pp.1963–1967, ISSN: 0016-2361

Jianbing, J.; Jianli, W.; Yongchao, L.; Yunliang, Y. & Zhichao, X. (2006) Preparation of biodiesel with the help of ultrasonic and hydrodynamic cavitation. Ultrasonics, Vol.44, No.1, (December 2006), pp.e411-e414, ISSN: 0041-624X

Jin, L.; Zhang, Y.; Dombrowski, J.P.; Chen, C.; Pravatas, A.; Xu,L.; Perkins, C. & Suib, S.L. (2011) ZnO/La2O2CO3 layered composite: A new heterogeneous catalyst for the efficient ultra-fast microwave biofuel production. Applied Catalysis B: Environmental. In press. ISSN: 0926-3373

Kappe CO. (2004) Controlled microwave heating in modern organic synthesis. Angewandte Chemie International Edition, Vol.43, No.46, (November 2004), pp.6250–84, ISSN: 1521-3773

Kim, D.; Choi, J.; Kim, G.J.; Seol, S.K.; Ha, Y.C.; Vijayan, M.; Jung, S.; Kim, B.H.; Lee, D. & Park, S.S. (2011) Microwave-accelerated energy-efficient esterification of free fatty acid with a heterogeneous catalyst. Bioresource Technology, Vol.102, No.3, (February 2011), pp. 3639–3641, ISSN 0960-8524

Kim, D.; Choi, J.; Kim, G.; Seol, S.K. & Jung, S. (2011) Accelerated esterification of free fatty acid using pulsed microwaves. Bioresource Technology, In press, ISSN 0960-8524

Koberg, M.; Abu-Much, R. & Gedanken, A. (2011) Optimization of biodiesel production from soybean and wastes of cooked oil: Combining dielectric microwave irradiation and a SrO catalyst. Bioresource Technology, Vol.102, No.2, (January 2011), pp. 1073–1078, ISSN 0960-8524

Kumar, S.A.; Sandun, F. & Hernandez, R. (2007) Base-catalyzed fast transesterification of soybean oil using ultrasonication. Energy & Fuel, Vol.21, No.2, pp.1161-1164, ISSN: 0887-0624

Kumar, R.; Kumar, G.R. & Chandrashekar, N. (2011) Microwave assisted alkali-catalyzed transesterification of Pongamia pinnata seed oil for biodiesel production. Bioresource Technology, Vol.102, No.11, (June 2011), pp. 6617-6620, ISSN 0960-8524

Kumar, D.; Kumar, G.; Poonam, & Singh, P. (2010) Ultrasonic-assisted transesterification of Jatropha curcus oil using solid catalyst, Na/SiO2. Ultrasonics Sonochemistry, Vol.17, No.5, (June 2010), pp.839–844, ISSN: 1350-4177

Kumar, G.; Kumar, D.; Poonam, Johari, R. & Singh, C.P. (2011) Enzymatic transesterification of Jatropha curcas oil assisted by ultrasonication. Ultrasonics Sonochemistry, In press, ISSN: 1350-4177

Lam, M.K.; Lee, K.T. & Mohamed, A.R. (2010) Homogeneous, heterogeneous and enzymatic catalysis for transesterification of high free fatty acid oil (waste cooking oil) to biodiesel: A review. Biotechnology Advances, Vol.28, No.4 , (July-August 2010), pp. 500–518, ISSN 0734-9750

Leadbeater, N.E.; Barnard, M.; & Stencel, M.L. (2008) Batch and Continuous-Flow Preparation of Biodiesel Derived from Butanol and Facilitated by Microwave Heating. Energy & Fuels, Vol.22, No.3, pp.2005–2008, ISSN: 0887-0624

Leadbeater, N.E. & Stencel, L.M. (2006) Fast, Easy Preparation of Biodiesel Using Microwave Heating. Energy & Fuels, Vol.20, No.5, pp.2281-2283, ISSN: 0887-0624

Lee, S.B.; Lee, J.D. & Hong, K. (2011) Ultrasonic energy effect on vegetable oil based biodiesel synthetic process. Journal of Industrial and Engineering Chemistry, Vol.17, No.1, (January 2011), pp.138–143 ISSN: 1226-086X

Lee, I.; Johnson, L.A. & Hammond, E. (1995) Use of branched-chain esters to reduce the crystallization temperature of biodiesel. J Am Oil Chem Soc, Vol.72, No.10, (October 1995), pp.1155-1160 ISSN 0003-021X

Lin, L.; Cunshan, Z.; Vittayapadung, S.; Xiangqian, S. & Mingdong, D. (2011) Opportunities and challenges for biodiesel fuel. Applied Energy, Vol.88, No.6, (June 2011), pp.1020–1031. ISSN: 0306-2619

Liu, S.; Wang, Y.; McDonald, T. & Taylor, S.E. (2008) Efficient Production of Biodiesel Using Radio Frequency Heating. Energy & Fuels, Vol.22, No.3, pp.2116–2120, ISSN: 0887- 0624

Liu, S.; McDonald, T. & Wang, Y. (2010) Producing biodiesel from high free fatty acids waste cooking oil assisted by radio frequency heating. Fuel, Vol.89, No.10, (October 2010), pp.2735–2740 ISSN: 0016-2361

Liu, S.; Wang, Y.; Oh, J-H. & Herring, J.L. (2011) Fast biodiesel production from beef tallow with radio frequency heating. Renewable Energy, Vol.36, No.3, (March 2011), pp.1003-1007, ISSN: 0960-1481

Loupy, A. (2002) Microwave in organic synthesis, Wiley- VCH Verlag GmbH & Co. KGaA. ISBN:3-527-30514-9, Weinheim, Germany.

Luche, J.L. and Damiano, J.C. (1980) Ultrasounds in Organic Syntheses. 1. Effect on the Formation of Lithium Organometallic Reagents. J. Am. Chem. Soc, Vol.102, No.27, pp.7926-7927, ISSN: 0002-7863

Mahamuni, N.N. & Adewuyi, Y.G. (2010) Application of Taguchi Method to Investigate the Effects of Process Parameters on the Transesterification of Soybean Oil Using High Frequency Ultrasound. Energy & Fuels, Vol.24, No.3, pp.2120–2126, ISSN: 0887- 0624

Mahamuni, N.N & Adewuyi, Y.G. (2009) Optimization of the Synthesis of Biodiesel via Ultrasound-Enhanced Base-Catalyzed Transesterification of Soybean Oil Using a Multifrequency Ultrasonic Reactor. Energy & Fuels, Vol.23, No.5, pp.2757–2766, ISSN: 0887-0624

Marchetti, J.; Miguel, V. & Errazua, A. (2007) Possible methods for biodiesel production. Renewable and Sustainable Energy Reviews, Vol.11, No.6, (August 2007), pp.1300– 1311. ISSN: 1364-0321

Mazo, P., Rios, L. (2010) Esterification and transesterification assisted by microwave of crude palm oil. Homogeneous catalysis. Latin American Applied Research, Vol.40, No.4, p 337-342, ISSN 0327-0793

Mazo, P., Rios, L. (2010) Esterification and transesterification assisted by microwave of crude palm oil. Heterogeneous catalysis. Latin American Applied Research, Vol.40, No.4, p 343-349, ISSN 0327-0793

Melo, C.A.; Albuquerque, C.; Carneiro, J.; Dariva, C.; Fortuny, M.; Santos, A.; Egues, S. & Ramos, A. (2010) Solid-Acid-Catalyzed Esterification of Oleic Acid Assisted by Microwave Heating. Ind. Eng. Chem. Res, Vol.49, No.23, pp.12135–12139, ISSN: 0888-5885.

Melo, C.A.; Albuquerque, C.E.R.; Fortuny, M.; Dariva, C.; Egues, S.; Santos, A.F. & Ramos, A.L.D. (2009) Use of Microwave Irradiation in the Noncatalytic Esterification of C18 Fatty Acids. Energy & Fuels, Vol.23, No.1, pp.580–585, ISSN: 0887-0624

Mootabadi, H.; Salamatinia, B.; Bhatia, S. & Abdullah, A.Z. (2010) Ultrasonic-assisted biodiesel production process from palm oil using alkaline earth metal oxides as the heterogeneous catalysts. Fuel, Vol.89, No.8, (August 2010), pp.1818–1825, ISSN: 0016-2361

Mustafa, B. (2011) Potential alternatives to edible oils for biodiesel production – A review of current work. Energy Conversion and Management, Vol.52, No.2, (February 2011), pp.1479–1492, ISSN: 0196-8904

Nogueira, B.M; Carretoni, C.; Cruza, R.; Freitas, S.; Melo, P.A.; Costa-Félix,R.; Pinto, J. & Nele, M. (2010) Microwave activation of enzymatic catalysts for biodiesel production. Journal of Molecular Catalysis B: Enzymatic, Vol.67, No.1-1, (November 2010), pp.117–121, ISSN: 1381-1177

Patil,P.; Gude, V.G.; Pinappu, S. & Deng, S. (2011) Transesterification kinetics of Camelina sativa oil on metal oxide catalysts under conventional and microwave heating conditions. Chemical Engineering Journal, Vol.168, No.3, (April 2011), pp.1296– 1300. ISSN: 1385-8947

Patil, P.D.; Gude, V.G.; Camacho,L.M. & Deng, S. (2010) Microwave-Assisted Catalytic Transesterification of Camelina Sativa Oil. Energy & Fuels, Vol.24, No.2, pp. 1298– 1304, ISSN: 0887-0624

Patil, P.D.; Gude, V.G.; Mannarswamy, A.; Cooke, P.; Munson-McGee, S.; Nirmalakhandan, N.; Lammers, P. & Deng, S. (2011) Optimization of microwave-assisted transesterification of dry algal biomass using response surface methodology. Bioresource Technology, Vol.102, No.2, (January 2011), pp.1399–1405, ISSN 0960- 8524

Perin, G.; Alvaro, G.; Westphal, E.; Viana, L.H.; Jacob, R.G.; Lenardao, E.J. & D'Oca, M.G.M. (2008) Transesterification of castor oil assisted by microwave irradiation. Fuel, Vol.87, No.12, (September 2008), pp.2838–2841, ISSN: 0016-2361

Piyasena, P.; Dussault, C.; Koutchma, T.; Ramaswamy, H.S. & Awuah, G.B. (2003) Radio frequency heating of foods: principles, applications and related properties–a review. Crit Rev Food Sci Nutr, Vol.43, No.6, pp.587–606, ISSN: 1040-8398

Raymond A. S. & Jewett, J.W. (2004) Physics for Scientists and Engineers, Thomson Brooks/Cole, 6th edition, ISBN: 0534408427, Washigton, USA.

Richards, W.T. & Loomis, A.L. (1927) The chemical effects of high frequency sound waves. J. Am. Chem. Soc, 1927, Vol.49, No.12, pp.3086-3088. ISSN: 0002-7863

Richela, A.; Laurenta, P.; Wathelet, B.; Wathelet, J-P. & Paquot, M. (2011) Current perspectives on microwave- enhanced reactions of monosaccharides promoted by heterogeneous catalysts. Catalysis Today, In press, ISSN: 0920-5861

Shakinaz A.E.S.; Refaat, A.A. & Shakinaz T.E.S. (2010) Production of biodiesel using the microwave technique. Journal of Advanced Research, Vol.1, No.4, (October 2010), pp. 309–314, ISSN: 2090-1232.

Siatis, N.G.; Kimbaris, A.C.; Pappas, C.S.; Tarantilis, P.A. & Polissiou, M.G. (2006). Improvement of biodiesel production based on the application of ultrasound: monitoring of the procedure by FTIR spectroscopy. J Am Oil Chem Soc, Vol.83, No.1, (January 2006), pp.53-57 ISSN 0003-021X

Siddiquee, M.N. & Rohani, S. (2011) Lipid extraction and biodiesel production from municipal sewage sludges: A review. Renewable and Sustainable Energy Reviews, Vol.15, No.2, (February 2011), pp.1067–1072 ISSN: 1364-0321

Socha, A. & Sello, J.K. (2010) Efficient conversion of triacylglycerols and fatty acids to biodiesel in a microwave reactor using metal triflate catalysts. Org. Biomol. Chem, Vol.8, No.20, pp.4753–4756, ISSN. 1477-0520

Suppalakpanya, K.; Ratanawilai, S.B. & Tongurai, C. (2010) Production of ethyl ester from esterified crude palm oil by microwave with dry washing by bleaching earth. Applied Energy, Vol.87, No.7, (July 2010), pp. 2356–2359, ISSN: 0306-2619

Suppalakpanya, K.; Ratanawilai, S.B. & Tongurai, C. Production of ethyl ester from crude palm oil by two-step reaction with a microwave system. Fuel, Vol.89, No.8, (August 2010), pp. 2140–2144, ISSN: 0016-2361

Terigar, B.G.; Balasubramanian, S.; Lima, M. & Boldor, D. (2010) Transesterification of Soybean and Rice Bran Oil with Ethanol in a Continuous-Flow Microwave-Assisted System: Yields, Quality, and Reaction Kinetics. Energy & Fuels, Vol.24, No.12, pp.6609–6615, ISSN: 0887-0624

Thostenson, E.T. & Chou, T.W. (1999) Microwave processing: fundamentals and applications. Composites Part A: Applied Science and Manufacturing, Vol.30, No.9, (September 1999), pp.1055–1071

Tierney, J.P. & Lidstrom, P. (2005) Microwave assisted organic synthesis, Blackwell Publishing Ltd, ISBN 0-8493-2371-1, Oxford, UK.

Venkatesh, H.; Regupathi, I. & Saidutta, M.B. Optimization of two step karanja biodiesel synthesis under microwave irradiation. Fuel Processing Technology, Vol.92, No.1, (January 2011), pp.100–105, ISSN: 0378-3820

Vicente, G.; Martínez, M. & Aracil, J. (2004) Integrated biodiesel production: a comparison of different homogeneous catalysts systems. Bioresource Technology, Vol.92, No.3, (May 2004), pp. 297-305, ISSN 0960-8524

Vyas, A.P.; Verma, J.L.; Subrahmanyam, N. A review on FAME production processes. Fuel, Vol.89, No.1, (January 2010), pp.1–9, ISSN: 0016-2361

Wang Y, Wig TD, Tang J, Hallberg LM. (2003) Dielectric properties of foods relevant to RF and microwave pasteurization and sterilization. J Food Eng, Vol.57, No.3, (May 2003), pp.257–68, ISSN: 0260-8774

Yaakob, Z.; Ong, B.H.; Kumar, M.N.S. & Kamarudin, S. K. (2009) Microwave-assisted transesterification of jatropha and waste frying palm oil. International Journal of Sustainable Energy, Vol.28, No.4, pp.195-201, ISSN: 1478-646X

Yuan, H.; Yang, B.L. & Zhu, G.L. (2009) Synthesis of Biodiesel Using Microwave Absorption Catalysts. Energy & Fuels, Vol.23, No.1, pp.548–552, ISSN: 0887-0624

Yu, D.; Tian, L.; Wu, H.; Wang, S.; Wang, Y.; Ma, D. & Fang, X. Ultrasonic irradiation with vibration for biodiesel production from soybean oil by Novozym 435. Process Biochemistry, Vol.45, No.4, (April 2010), pp.519–525. ISSN: 1359-5113

Yu, D.; Tian, L.; Ma, D.; Wu, H.; Wang, Z.; Wang, L. & Fang, X. (2010) Microwave-assisted fatty acid methyl ester production from soybean oil by Novozym 435. Green Chem, Vol.12, No.5, pp.844–850. ISSN: 1463-9262

Yusuf, N.N.A.N.; Kamarudin, S.K. & Yaakub, Z. (2011) Overview on the current trends in biodiesel production. Energy Conversion and Management, Vol.52, No7, (July 2011), pp.2741–2751, ISSN: 0196-8904

Chapter 14

REACTIVE DISTILLATION FOR THE SYNTHESIS AND CHARACTERIZATION OF BIODIESEL VIA TRANSESTERIFICATION

1. INTRODUCTION: AN OVERVIEW

Escalating global fuel prices, increasing energy demand, and apprehensions regarding climate change are the principal causes fueling a resurgence of interest in renewable energy sources and bioenergy. Currently, global energy demand has risen substantially due to industrialization and population growth. Consequently, the existing finite reservoirs will soon be exhausted at the present pace of usage. The research in energy concentrates on identifying an alternate energy source to petroleum-derived diesel.

India imported roughly two-thirds of its petroleum needs last year, incurring a cost of around Rs. 80,000 crores in foreign money. A mere 5% substitution of petroleum gasoline with biofuel might benefit India and conserve Rs. 4000 crores annually in foreign cash. It is imperative to investigate alternatives to petroleum fuels to mitigate the import expenditure. Biodiesel serves as an appropriate alternative to petroleum-based diesel. It is biodegradable, nearly devoid of sulfur, and a renewable fuel, although its production methods are not environmentally sustainable. This alternative fuel comprises methyl or ethyl esters, produced through the transesterification of triglycerides (TG) or the esterification of free fatty acids (FFAs). Biodiesel fuel has gained appeal due to its environmental advantages, as it is derived from renewable biomass sources such as plants, vegetable oils, and animal fats.

Presently, the majority of biodiesel is produced through the transesterification of consumable resources, including animal fats, vegetable oils, and waste cooking oils, utilizing alkaline catalytic conditions. Nonetheless, the substantial utilization of catalysts, the generation of soaps, and the diminished yields render biodiesel now more costly than petroleum-derived fuel. Moreover, the plants utilized for the production of vegetable oils sequester more CO2 from the environment than the quantity emitted during the combustion of these oils. [1].

The three fundamental methods for biodiesel synthesis from oils and fats are base-catalyzed transesterification of the oil, direct acid-catalyzed transesterification of the oil, and the conversion of the oil to its fatty acids followed by the generation of biodiesel. Among these three ways, the predominant generation of biodiesel occurs through the base-catalyzed reaction pathway.

The stoichiometric equation for transesterification reaction [9] in general can be represented as follows:

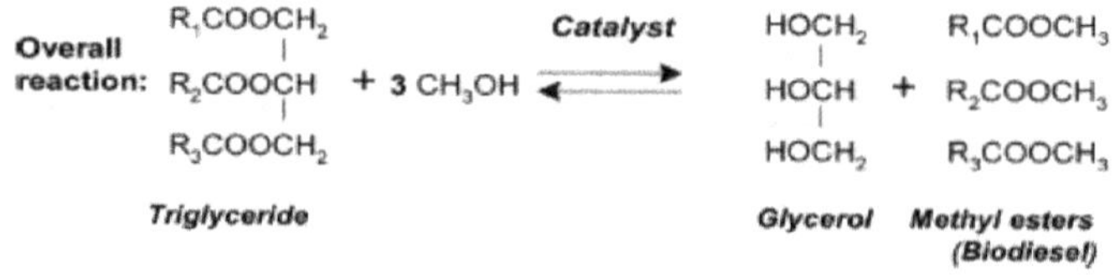

2. BIODIESEL SCENARIOS WORLDWIDE

Table 1. World biodiesel productions by region and selected countries 2005-2009 (Thousand barrels per day)

Sr.No	Region/Country	2005	2006	2007	2008	2009
1	North America	6.1	17.1	33.7	45.9	35.2
2	United States	5.9	16.3	32.0	44.1	32.9
3	Central and south America	0.5	2.2	15.2	38.6	57.9
4	Brazil	0.0	1.2	7.0	20.1	27.7
5	Europe	68.1	113.2	137.5	155.0	172.6
6	France	8.4	11.6	18.7	34.4	41.1
7	Germany	39.0	70.4	78.3	61.7	51.2
8	Italy	7.7	11.6	9.2	13.1	13.1
9	Eurasia	0.3	0.3	0.7	2.5	3.8
10	Lithuania	0.1	0.2	0.5	1.3	1.9
11	Asia and Oceania	2.2	9.1	15.8	28.8	38.5
12	China	0.8	4.0	6.0	8.0	8.0
13	India	0.2	0.4	0.2	0.2	0.4
14	Korea South	0.2	0.6	1.7	3.2	5.0
15	Malaysia	0.0	1.1	2.5	4.5	5.7
16	Thailand	0.4	0.4	1.2	7.7	10.5
	WORLD	77.2	142.0	202.9	270.9	308.2

Source- U.S. Energy Information Administration, International Energy Statistics, Biofuels Production

3. REACTIVE DISTILLATION

Reactive distillation is a chemical unit operation in which chemical reaction and product separation occurs simultaneously in one unit. Reactive distillation column consists of a reactive section in the middle with non-reactive rectifying and stripping sections at the top and bottom.

Let us begin by considering a reversible reaction scheme where A and B react to give C and D. The boiling point of the components follows the sequence A, C, D and B. The traditional flow sheet for this process consists of a reactor followed by a sequence of distillation columns. The mixture of A and B is fed to the reactor, where the reaction takes place in the presence of a catalyst and reaches equilibrium. A distillation train is required to produce pure products C and D. The unreacted components, A and B, are recycled back to the reactor.

The Reactive distillation technology offers many benefits as well as restrictions over the conventional process of reaction followed by distillation or other separation approaches. Reducing capital cost, higher conversion, improving selectivity, lower energy consumption, the reduction or elimination of solvents in the process and voidance of azeotropes are a few of the potential advantages offered by Reactive distillation. Conversion can be increased far beyond what is expected by the equilibrium due to the continuous removal of reaction products from the reactive zone. This helps to reduce capital and investment costs and may be important for sustainable development due to a lower consumption of resources. [7]

The fig.1 represents the general configuration of reactive distillation.

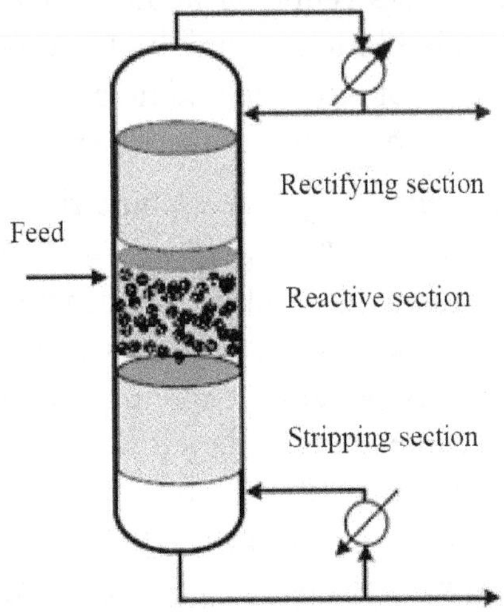

Fig. 1. The general configuration of Reactive Distillation

Based on the applied separation technology, reactive distillation, reactive extraction, reactive adsorption and other combined processes have been distinguished. The combined simultaneous performance of chemical reaction and a multi-component distillation process is an alternative, which has been increasingly used for the large-scale production of relevant chemicals. The use of reactive distillation process can have several advantages such as reduced downstream processing, utilization of heat of reaction for evaporation of liquid phase, simple temperature control of reactor, possibility of influencing chemical equilibria by removal of products and limitations imposed by azeotropic mixture. Several commercially important processes in reactive distillation have been identified in some recent reviews. [7]

Reactive distillation has been successfully applied for the etherification reaction to produce fuel ethers such as methyl tert-butyl ether (MTBE), tert-amyl methyl ether(TAME) and ethyl tertbutyl ether (ETBE). These have been the model reactions for the studies in reactive distillation in the last two decades. A small number of industrial applications of reactive distillation have been around for many decades. Low chemical equilibrium constants can be overcome and high conversions achieved by the removal of products from the location where the reaction is occurring. [6]

It may be advantageous for liquid-phase reaction systems when the reaction must be carried out with a large excess of one or more of the reactants, when a reaction can be driven to completion by removal of one or more of the products as they are formed, or when the product recovery or by-product recycle scheme is complicated or made infeasible by azeotrope formation. Novel processes were proposed based on catalytic reactive distillation and reactive absorption to biodiesel production from esterification and transesterification reactions. The major benefits of this approach were: investment costs reducing about 45% energy savings compared to conventional reactive distillation, very high conversions, increased unit productivity, no excess of alcohol required and no catalyst neutralization step The advantage of reactive distillation can be summarized as follows [3]

a. Simplification: From design view point the combinations of reaction system and separation system can lead to significant capital saving.
b. Improved conversion of reactant approaches 100%. This increase in conversion gives a benefit in reduced recycle costs.
c. Improved selectivity: where, removing one of the products from the reaction mixture or maintaining a low concentration of one of the reagents can lead to reduction of the rates of side reactions and hence improved selectivity for the desired products.
d. Significantly reduced catalyst requirement for the same degree of conversion.
e. Avoidance of azeotropes: RD is particularly advantageous when the reactor product is a mixture of species that can form several azeotropes with each other. RD conditions can allow the azeotropes to be “reacted away” in a single vessel.
f. There is a reduced by-product formation.

g. Heat integration benefits: If the reaction is exothermic, the heat of reaction can be used to provide the heat of vaporization and reduce the reboiler duty.
h. Removal of the product from a system at equilibrium will cause more products to form. Therefore reactive distillation is capable to increase the conversion of equilibrium limited reaction.

Biodiesel production by reactive distillation

As the reaction and separation occurs simultaneously in the same unit in reactive distillation, it is attractive in those systems where certain chemical and phase equilibrium conditions exist. Because there are many types of reactions, there are many types of reactive distillation columns. In this section we describe the ideal classical situation, which will serve to outline the basics of reactive distillation. Consider the system in which the chemical reaction involves two reactants (A and B) producing two products (C and D). The reaction takes place in the liquid phase and is reversible.

$$A+B \leftrightarrow C+D$$

The number of the separation steps depends on the number of products, catalysts, solvents as well as reactants which are not converted. The main objective functions to increase process economics are selectivity as well as reaction yield what influences the reactor design.

Usually, each unit operation is typically performed in individual items of equipment, which, when arranged together in sequence, make up the complete process plant. As reaction and separation stages are carried out in discrete equipment units, equipment and energy costs are added up from these major steps. However, this historical view of plant design is now being challenged by seeking for combination of two or more unit operations into the one plant unit [4].

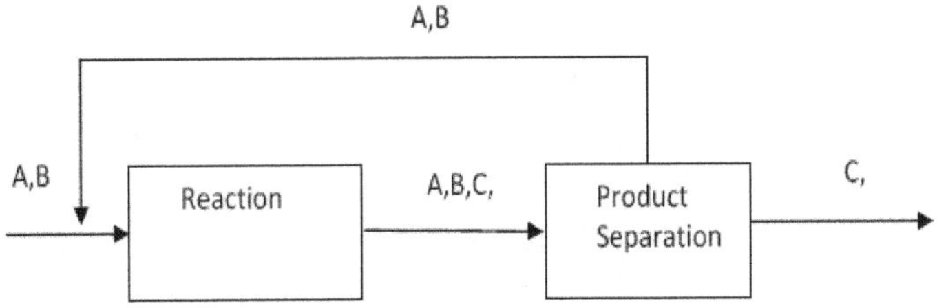

Fig. 2. Standard process scheme for reversible reactions in which the conversion is limited by the chemical equilibrium [9]

For reactive distillation to work, we should be able to remove the products from the reactants by distillation. This implies that the products should be lighter and/or heavier than the reactants. In terms of the relative volatilities of the four components, an ideal case is when one product is the lightest and the other product is the heaviest, with the reactants being the intermediate boiling components.

$$\alpha C > \alpha D > \alpha D$$

The most obvious way to improve the reaction yield in an integrated unit is a continuous separation of one product out of the reaction zone. This allows for getting a 100% conversion in case of reversible reactions [9].

$$A+B \leftrightarrow C+D$$

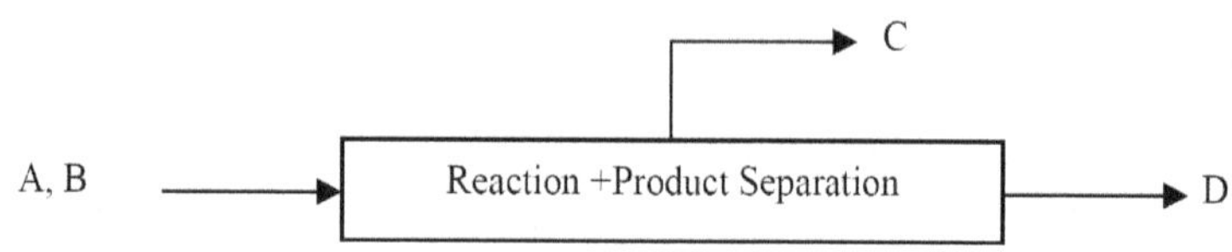

Fig. 3. Complete conversions of reactants in case of equilibrium reaction [7]

Figure 4 presents the flow sheet of this ideal reactive distillation column. In this situation the lighter reactant A is fed into the lower section of the column but not at the very bottom. The heavier reactant B is fed into the upper section of the column but not at the very top. The middle of the column is the reactive section and contains number of reaction trays. The vapor flow rates through the reaction section change from tray to tray because of the heat of the reaction. As component A flows up the column, it reacts with descending B. Very light product C is quickly removed in the vapor phase from the reaction zone and flows up the column. Likewise, very heavy product D is quickly removed in the liquid phase and flows down the column. The section of the column above where the fresh feed of B is introduced (the rectifying section with NR trays) separates light product C from all of the heavier components, so a distillate is produced that is fairly pure product C.

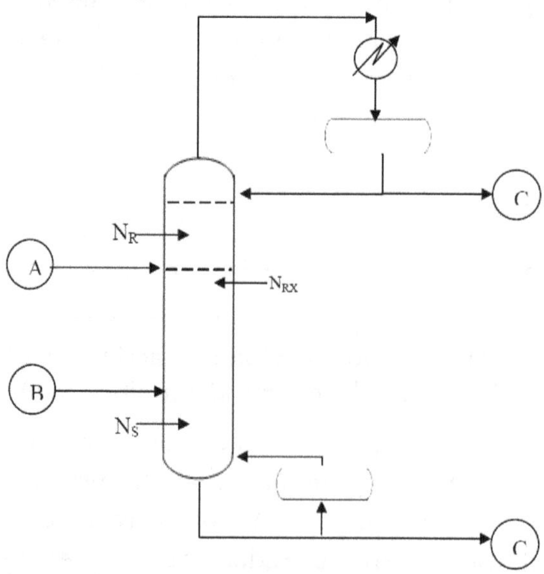

Fig. 4. Flow sheet of ideal reactive distillation column

The section of the column below where the fresh feed of A is introduced (the stripping section with NS trays) separates heavy product D from all of the lighter components, so a bottom is produced that is fairly pure product D. The reflux flow rate and the reboiler heat input can be manipulated to maintain these product purities. The specific numerical case has 30 total trays, consisting of 10 stripping trays, 10 reactive trays, and 10 rectifying trays. Trays are numbered from the bottom. Note that the concentrations of the reactants peak at their respective feed trays. The purities of the two products are both 95 mol%, with B the major impurity in the bottoms and A the major impurity in the distillate [7].

Reactive distillation column must be adjusted to achieve these specifications while optimizing some objective function such as total annual cost (TAC). These design degrees of freedom include pressure, reactive tray holdup, number of reactive trays, location of reactant feed streams, number of stripping trays, number of rectifying trays, reflux ratio, and reboiler heat input [9].

Tray holdup is another design aspect of reactive distillation that is different from conventional. Holdup has no effect on the steady-state design of a conventional column. It certainly affects dynamics but not steady-state design. Column diameter is determined from maximum vapor loading correlations after vapor rates have been determined that achieve the desired separation. Typical design specifications are the concentration of the heavy key component in the distillate and the concentration of the light key component in the bottoms.

However, holdup is very important in reactive distillation because reaction rates directly depend on holdup (or the amount of catalyst) on each tray. This means that the holdup must be known before the column can be designed and before the column diameter is known. As a result, the design procedure for reactive distillation is iterative. A tray holdup is assumed and the column is designed to achieve the desired conversion and product purities. The diameter of the column is calculated from maximum vapor-loading correlations. Then the required height of liquid on the reactive trays to give the assumed tray holdup is calculated. Liquid heights greater than 10–15 cm are undesirable because of hydraulic pressure drop limitations. Thus, if the calculated liquid height is too large, a new and smaller tray holdup is assumed and the design calculations repeated. An alternative, which may be more expensive in terms of capital cost, is to make the column diameter larger than that required by vapor loading [9].

4.CASESTUDY-TRANSESTERIFICATIONBYREACTIVEDISTILLATION FOR SYNTHESIS AND CHARACTERIZATION OF BIODIESEL

4.1 Materials and methods

Materials:

a. Oil Feed stocks:

In this study, three commercially available feed stocks of vegetable oils are used. They are

1. Castor seed oil
2. Cottonseed oil
3. Coconut oil

Table 2. Physical Properties of Vegetable Oil Feed stocks Used For Transesterification

Sample	Kinematic Viscosity, cst (mm2/s)	Density (Kg/m3 at 288K)	Flash point oC	Pour point oC	Saponification value
Castor oil	115 (at 60°C)	938	229	-33	182
Coconut oil	24.85 (at 40 °C)	907	225	20	191.1
Cottonseed oil	35.42 (at 40 °C)	904	15	-15.5	192

b. Methanol:

Methanol (Merck) of 99.5% purity (density: 0.785 g/mL at 30°C) was used in this transesterification process.

c. Catalyst:

In this study the catalysts used are:

1. Homogeneous base catalysts (KOH & NaOH)
2. Heterogeneous solid acid catalysts (Amberlyst 15)

The two homogeneous basic catalysts (KOH & NaOH) used for reactive distillation were purchased from local Chemical store at Amravati. M.S.The heterogeneous catalyst used for transesterification Amberlyst BD15 was purchased from Dayo Scientific Laboratory, Nashik Road, Nashik, M.S. India.

Amberlyst-15:

Amberlyst 15 wet is a macro reticular, strongly acidic, polymeric catalyst. Its continuous open pore structure makes it an excellent heterogeneous acid catalyst for a wide variety of organic reactions. Amberlyst 15 is extremely resistant to mechanical and thermal shocks. It also possesses greater resistance to oxidants such as chloride, oxygen and chromates than most other polymeric catalyst. It can use directly in the aqueous system or in organic medium after conditioning with a water miscible solvent. Amberlyst 15 has optimal balance of surface area, acid capacity and pore diameter to make it the catalyst of choice for esterification reactions.

Table 3. Characteristics of Amberlyst-15 catalyst

Physical forms	Opaque beads
Ionic form as shipped	Hydrogen
Total exchange capacity	≥1.7 eq /L
Moisture holding capacity	52-57%
Harmonic mean size	600-850 μm
Fine contents	< 0.355 mm :1.0%
Coarse beads	> 1.180 mm :5.0%
Average pore diameter	24 nm
Surface area	45 m^2 / gm
Shrinkage water to methanol	4.0%

4.2 Transesterification

Transesterification also called alcoholysis is the most common way to produce biodiesel. This involves a catalyzed chemical reaction between vegetable oil and an alcohol to yield fatty acid alkyl esters (i.e., biodiesel) and glycerol. Transesterification is the displacement of alcohol from an ester by another alcohol in a process similar to hydrolysis, except that an alcohol is employed instead of water. Triglycerides, as the main component of vegetable oil, consist of three long chain fatty acids esterified to a glycerol backbone. When triglycerides react with an alcohol (e.g., methanol), the three fatty acid chains are released from the glycerol skeleton and combine with the alcohol to yield fatty acid alkyl esters (e.g., fatty acid methyl esters or biodiesel). Glycerol is produced as a by-product.

The mechanism of transesterification can be represented as follows:

$$\begin{matrix} CH_2-OOC-R_1 \\ CH-OOC-R_2 \\ CH_2-OOC-R_1 \end{matrix} + 3R'OH \rightleftharpoons \begin{matrix} R_2-OOC-R' \\ R_2-OOC-R' \\ R_3-OOC-R' \end{matrix} + \begin{matrix} CH_2-OH \\ CH_2-OH \\ CH_2-OH \end{matrix} \qquad (1)$$

$$\text{Triglyceride} + \text{Alcohol} \xrightleftharpoons{\text{Catalyst}} \text{Esters} + \text{Glycerol}$$

4.2.1 Transesterification of vegetables oils

In the transesterification of different types of oils, triglycerides react with an alcohol, generally methanol or ethanol, to produce esters and glycerin. To make it possible, a catalyst is added to the reaction.The overall process is normally a sequence of three consecutive steps, which are reversible reactions. In the first step, from triglycerides diglyceride is obtained, from diglyceride monoglyceride is produced and in the last step, from monoglycerides glycerin is obtained. In all these reactions esters are produced. The stoichiometric relation between alcohol and the oil is 3:1. However, an excess of alcohol is usually more appropriate to improve the reaction towards the desired product:

$$\begin{gathered} \text{Triglyceride (TG)} + R'OH \underset{k_2}{\xrightarrow{k_1}} \text{Diglycerides (DG)} + R'COOR_1 \\ \text{Diglycerides (DG)} + R'OH \underset{k_4}{\xrightarrow{k_3}} \text{Monoglycerides (MG)} + R'COOR_2 \\ \text{Monoglycerides (MG)} + R'OH \underset{k_6}{\xrightarrow{k_5}} \text{Glycerin (GL)} + R'COOR_3 \end{gathered} \quad (2)$$

Startup Procedures of transesterification using reactive distillation:

To start of each experiment, approximate 2 L of oil and 250 mL of methanol were injected into the column. The reboiler heater was set to 120°C and allowed to heat for approximately 1.5 hours till the temperature of the top column reached 62°C.

Steady-operation:

The inputs, both oil at 55°C and methanol at 30○C, were pumped into a short tube mixer to mix the oil with the methanol/catalyst solution. Then the reactant mixture at 62°C was entered to the top of the RD column. In the RD column, triglyceride in the reactant mixture further reacted with the present methanol. The product mixture was withdrawn from the reboiler section and sent to a glycerol ester separator, where the glycerol and esters were separated by gravity in a continuous mode. Every hour, samples were collected from reboiler to analyze the biodiesel composition and methanol content.

In this experimentation reaction parameters has been optimized and an optimized process has been investigated for biodiesel production by transesterification of vegetable oil using reactive distillation technique

Calculations:

The ester content (C) expressed as a fraction in percent, is calculated using the following formula:

$$C = \frac{\left(\sum A\right) - A_{EI}}{A_{EI}} \times \frac{C_{EI} \times V_{EI}}{m} \times 100\% \quad (4)$$

ΣA = the total peak area from the FAME $C_{14:0}$ to $C_{24:1}$

A_{EI} = the peak area of methyl heptadecanoate

C_{EI} = the concentration , in mg/ml of the methyl heptadecanoate solution

V_{EI} = the volume, in ml of the methyl heptadecanoate solution

m = the mass, in mg of the sample

5. EXPERIMENTAL SETUP

The system consists of a reactive distillation column fed at the top with the initial reactive solution (oil, alcohol, catalyst). This solution slowly travels down between the plates. When the solution exits the column; the alcohol that has not reacted is recuperated by evaporation. Then, the vapors are re-circulated in the reactive distillation column in the upward direction passing through the plates. As the vapors travel through, interactions between the gaseous alcohol and the liquid solution occur. This then would increase the effective oil to alcohol ratio up to 20:1 (He, Singh et al.2006), thus shifting the reaction equilibrium to the product side and therefore increasing the reaction efficiency. Finally, once the alcohol vapors have reached the top of the reactive distillation column, they are condensed through a condenser allowing the remaining alcohol fraction to re-enter the system. The experimental setup is shown in fig.5 below.

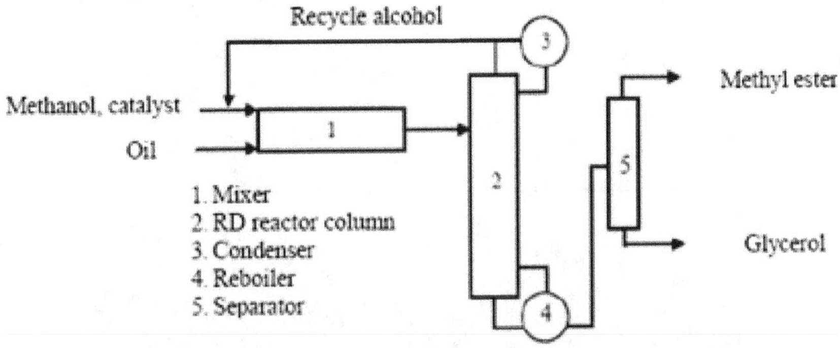

Fig. 5. Schematic of Reactive distillation column for biodiesel

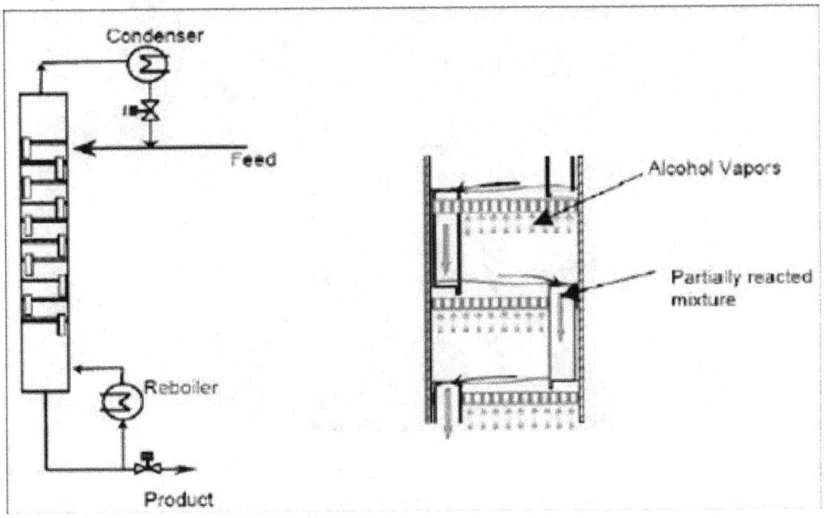

Singh, Thompson Et Al. 2004; Thompson and He 2007

Fig. 6. Operation in Reactive Distillation column

In the present experimental study, packed bed Lab-scale reactive distillation column is designed and constructed. This column made up of glass (Inner dia: 30mm, Height of column: 210mm) has been used. The RD column packings used were glass packing. The feed reactants entering into the column were distributed over the packings by the use of distributor plates. The process parameters studied here are alcohol-to-oil ratio.{3:1, 4:1 and 9:1, Optimum methanol-to-oil molar ratio = 4:1},Flow rates of reactants {2, 4, 6 ml/min, Optimum flow rate = 4ml/min},Reaction time {Residence time of 2min, 3min., 6min, optimum residence time = 3min},Temperature {55, 60, 65 °C, Optimum temperature = 65 °C}.

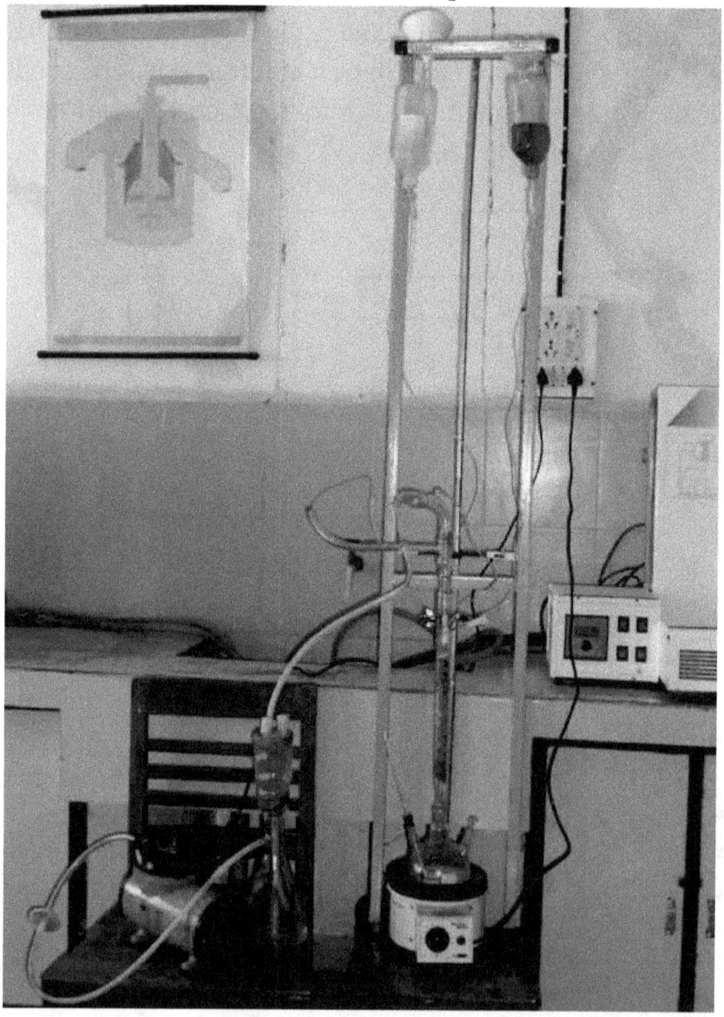

Fig. 7. a) View of experimental lab apparatus of Reactive distillation

The RD reactor consists of perforated plates or packed sections. For packed columns the packing holds certain amount of reacting liquid in it, forming mini-reactors. Un-reacted alcohol is vaporized from the reboiler, flows upward constantly, and bubbles through the liquid in the packing, which provides a uniform mixing. The thru-vapor is condensed at the top of the RD column and refluxes partially back to the column and the rest combines with the feeding stream

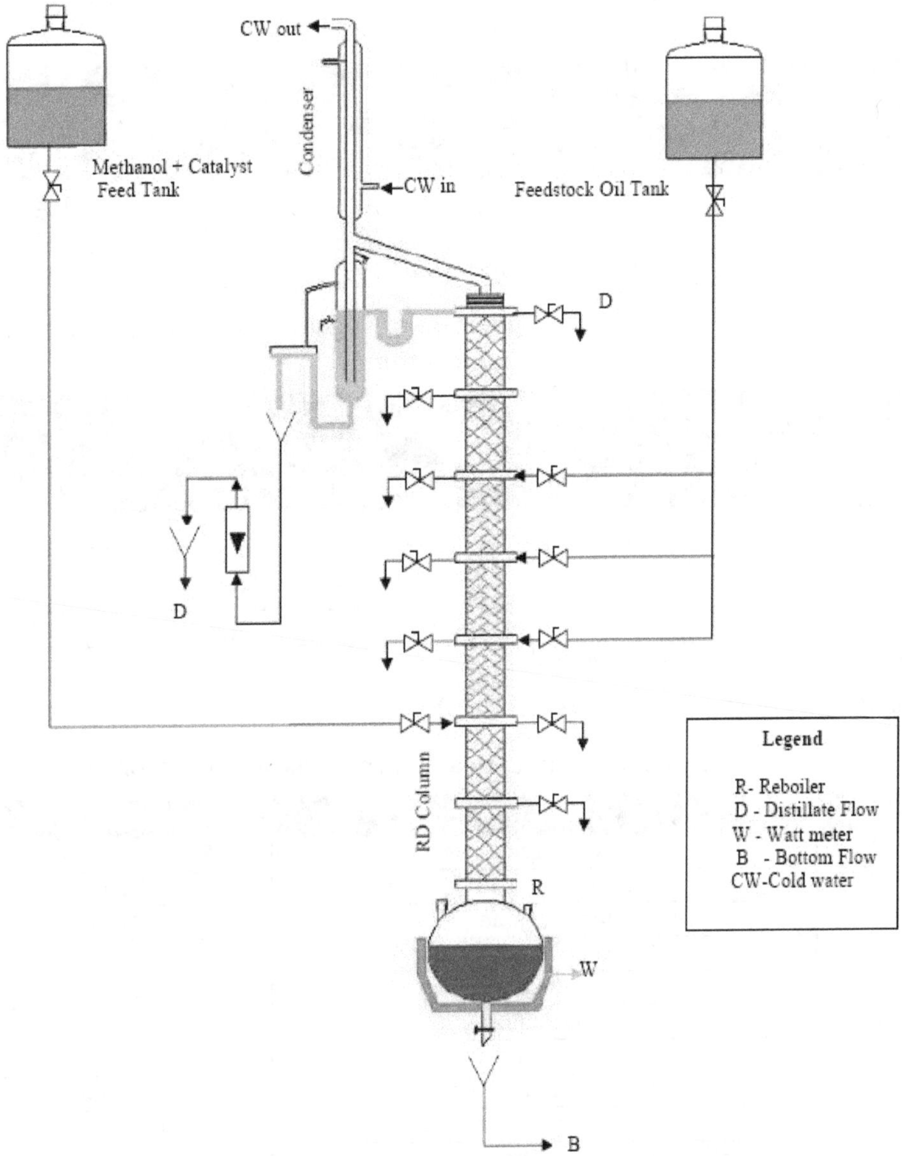

Fig. 7. b) Schematic Diagram of Experimental Setup of Continuous Reactive Distillation Column for biodiesel

In this study, three non edible vegetable oils namely, castor seed oil, coconut oil and cottonseed oil were used one by one for transesterification.

5.1 Physical and chemical characteristics of the feedstock vegetable oils used for the production of biodiesel

Table 4. Physical Properties of Vegetable Oil Feed stocks Used for Transesterification:

Sample	Kinematic Viscosity, cst(mm2/s)	Density (Kg/m3 at 288K)	Flash point °C	Pour point °C	Saponification value
Castor oil	115 (at 60oC)	938.8	229	-33	182
Cottonseed oil	35.42 (at 40 oC)	904	15	-15.5	192
Coconut oil	24.85 (at 40 oC)	907	225	20	191.1

Table 5. Table Fatty acids composition of vegetable oils samples under consideration

Oil Sample	Palmitic acid (16:0)	16:1 (Palmit-oleic)	Stearic acid(18:0)	Oleic acid (18:1)	Linoleic acid (18:2)	Linolenic acid (18:3)	Other
Castor oil[a]	1.1	0	3.1	4.9	1.3	0	89.6
Cottonseed oil	28.7	0	0.9	13.0	57.4	0	0
Coconut oil	9.7	0.1	3.0	6.9	2.2	0	65.7

Castor oil contains 89.6% ricinoloic acid

Physical and chemical characteristics of castor oil

Table 6. Physical and chemical characteristics of castor oil

Sr no.	Parameters	Values
1	Appearance	Pale dark yellow
2	Density (at 15oC)	938.8 kg/m3
3	Iodine value	82 -90
4	Saponification value	182
5	Flash Point	229 oC
6	Pour point	-33 oC
7	Acid value	2.0 max
8	Moisture and Volatiles	0.50% max
9	Specific gravity at 20oC	0.954 – 0.967
10	Kinematic Viscosity ,cst(mm2/s)	115 (at 60oC)
	Fatty acids content (%)	
11	Ricinoleic acid	89.6
12	Linoleic acid	4.2%
13	Oleic acid	3%
14	Stearic acid	1%
15	Palmitic acid	1%
16	Linolenic acid	0.3%

Physical and chemical characteristics of cottonseed oil

Its fatty acid profile generally consists of 70% unsaturated fatty acids including 18% monounsaturated (oleic), 52% polyunsaturated (linoleic) and 26% saturated (primarily palmitic and stearic).

Table 7. Physical and chemical characteristics of cottonseed oil

Sr. no.	Parameters	Values
1	Appearance	Golden yellow
2	Density	904
3	Kinematic viscosity(at 40°C)	35.42 cSt
4	Saponification value	192
5	Flash point	15°C
6	Pour point	-15.5 °C
7	Acid value	0.6518 mg of KOH/gm of oil
8	Free fatty acids	0.3258%
9	Molecular weight	863
10	Specific gravity at 20°C	0.9406 gm/ml

Table 8. Fatty acids composition of cottonseed oil

Sr no.	Parameters	Values
1	Palmitic acid	23%
2	Oleic acid	18%
3	Linoleic acid	52%
4	Aleic acid	19%
5	Alpha lenoleic acid	1%
6	Stearic acid	3%

Physical and chemical characteristics of coconut oild

Table 9. Physical and chemical characteristics of coconut oil

Sr no.	Parameters	Values
1	Melting point(°C)	24
2	Moisture %	<0.1
3	Kinematic viscosity(mm^2/s)	24.85 (at 40 °C)
4	Density (at 15°C)	907 kg/m^3)
3	Iodine value(cgI_2/g)	12-15
5	Saponification value	191.1
6	Flash Point	225 °C
7	Pour point	20 °C
8	Total phenolics(mg/kg)	620
Fatty acids content (%)		
9	Saturates	92.0
10	Monosaturates	6.0
11	Polyunsaturates	2.0

5.2 Continuous transesterification by reactive distillation for synthesis of biodiesel

The process parameters studied here are alcohol-to-oil ratio.{3:1, 6:1 and 9:1, Optimum methanol-to-oil molar ratio =6:1},Reaction time {Residence time of 2min, 3min., 6min, optimum residence time = 3min},Temperature {55, 60, 65oC, Optimum temperature = 65oC }, catalyst loading (1,1.5 and 2 % by wt of oil).

5.2.1 Effect of methanol to oil molar ratio on methyl ester conversion

Feedstock oil was held in a separate heated reservoir maintained at 50°C.The methanol-to-oil molar ratios used were 3.0, 6.0 and 9.0. From several trials, it was found that an overall flow rate of 5-6 ml/min with the column temperature at 64°C provided residence time of about 6min without any significant operational difficulties. The column temperature was maintained by controlling the reboiler heat input. Temperatures above 64°C caused excessive entrainment and a reduction in methanol concentrations in the liquid phase. In preparation for each trial, stock alcoholic KOH was prepared on a stirring plate at a ratio that corresponded to 1, 1.5 and 2 % KOH w/w of oil for each given methanol-to-oil molar ratio, and placed in a holding reservoir next to the RD column. Optimum reaction time in biodiesel formation (1min in prereactor +5min in RD column=6min.). Reaction time by using RD column is 20 times shorter than that in typical batch processes. Also productivity of RD reactor system is 6 to 10 times higher than that of batch and existing continuous flow processes.

The main process parameters examined in this study were as shown below: For individual oils (Castor, Cottonseed and Coconut oil) under consideration

Table 10. (a) Effect of Methanol to oil Molar ratio on methyl esters conversion

Methanol/oil molar Ratio (mol/mol)	Methyl esters Conversion (%), Castor oil	Methyl esters Conversion (%), Cottonseed oil	Methyl esters Conversion (%), Coconut oil
3	56	68	55
6	68	72	88
9	72	74	89

Temperature = 64oC, Flow rate =6ml/min, Reaction time = 6min., Catalyst (KOH) =1% by wt. of oil)

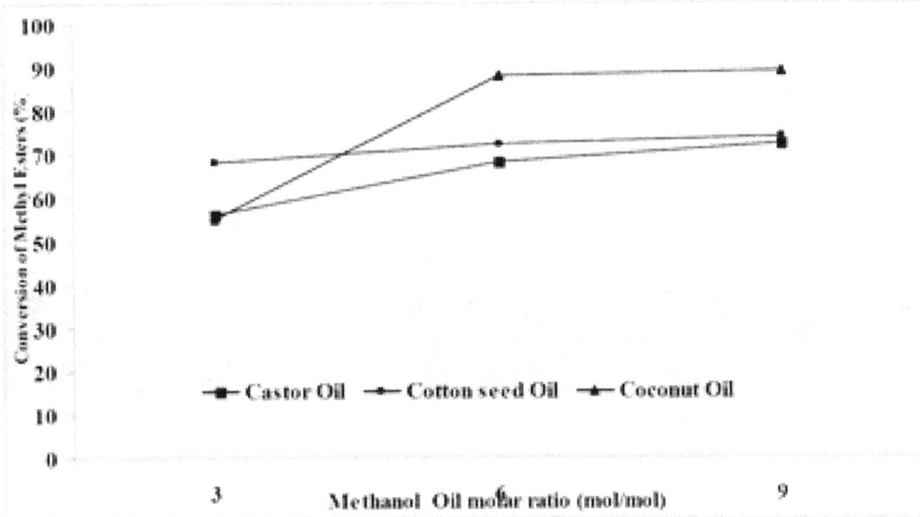

Optimum Molar ratio of Methanol- to- oil= 6:1

Fig. 8. Effect of Methanol to oil Molar ratio on methyl ester conversion

Table 10. (b) Effect of methanol-to-mixed oil feed molar ratio on methyl ester conversion

Reaction Time(min.)	Methanol/oil molar Ratio (mol/mol)=3	Methanol/oil molar Ratio (mol/mol)=6	Methanol/oil molar Ratio (mol/mol)=9
2	68	72	55
6	89	96	88
8	91	94	89

Reaction temperature =60oC, Catalyst concentration =1 wt%, Methanol to mixed oil molar ratio = 3:1, 6:1, 9:1

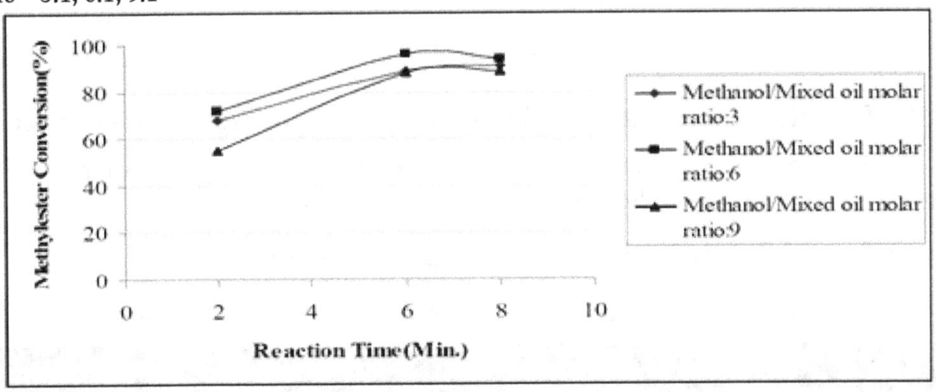

Fig. 9. Effect of Methanol to Mixed oil Molar ratio on methyl ester conversion

The three vegetable oil feedstock's under consideration were mixed and three mixed feed oils were prepared for the experimental runs. The effect of Methanol to Mixed oil Molar ratio on methyl ester conversion was observed as shown in fig.9. The conversion of methyl esters was found to increase with increase in molar ratio during initial reaction time. Also the highest conversion o e 6min in RD column. Thus it can be concluded that the mixed oils can be used for synthesis of biodiesel. This would help in reduction in overall cost of biodiesel synthesis by cutting down the cost of expensive oil by replacing the portion of expensive oils by cheaper oils or the oils which are easily available in abundance.

Table 10. (c) Effect of reaction time using different mixed oils on methyl ester conversion

Reaction Time(min.)	Methyl ester conversion (%) Mixed oil 1	Methyl ester conversion (%) Mixed oil 2	Methyl ester conversion (%) Mixed oil 3
2	65	77	78
6	89	92	95
8	90	93	95

Reaction temperature =60°C, Catalyst concentration =1 wt%, Methanol to mixed oil molar ratio = 6:1, Mixed oil 1= 50% Castor oil+50% Cottonseed oil, Mixed oil 2 = 50% Castor oil+ 50% Coconut oil, Mixed oil 3 = 50% Coconut oil + 50% Cottonseed oil

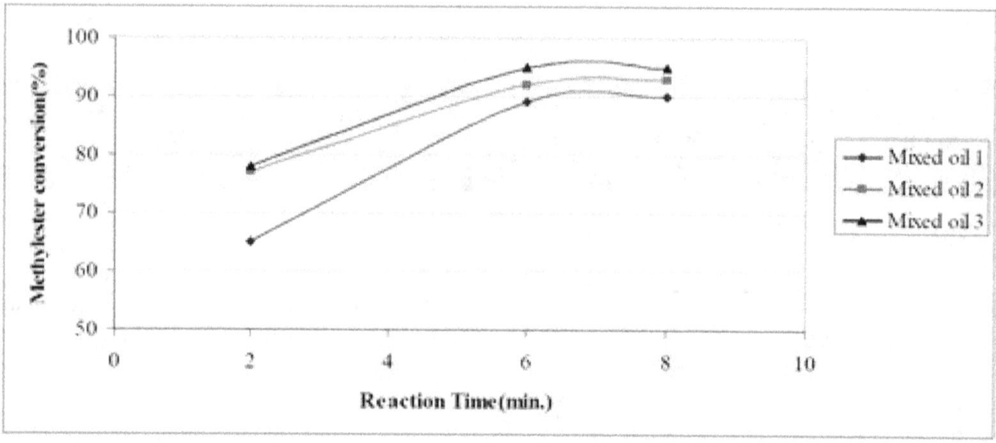

Fig. 10. Effect of reaction time using different mixed oils on methyl ester conversion

It was observed that mixed oil 3 i.e. 50% Coconut oil + 50% Cottonseed oil showed the maximum methyl ester conversion.

5.2.2 Effect of flow rates on methyl ester conversion

Table 11. Effect of Flow rates on methyl ester conversion

Flow rate, ml/min	Methyl esters Conversion (%) Castor oil	Methyl ester conversion (%) Cottonseed oil	Methyl ester Conversion (%) Coconut oil
5	90	92	90
6	94	96	93
7	94	95	93

Molar ratio of Methanol to Oil = 6:1, Reaction time = 6min., Catalyst (KOH) =1% (by wt. of oil), Temperature = 60°C

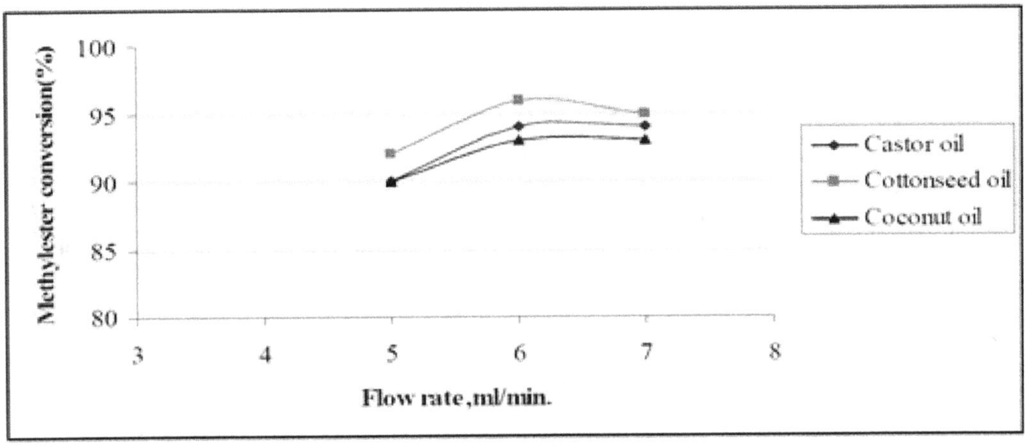

Optimum flow rate = 6ml/min

Fig. 11. Effect of Flow rates on methyl ester conversion

The feed stream flow rates for the test run were chosen carefully in order to avoid any column flooding problem. The flow rate, which is inversely related to retention time, is used as an experimental factor to interpret the reaction conversion with the liquid retention time. The flow rate achieved in the experimental runs varied from 5 to 7 mL/min. The retention time varied from about 4 to 8 min. These values may not be the actual reaction time because of some reaction that takes place in the reboiler. Since the concentrations of methanol and catalyst were very small in the liquid phase of the reboiler, it was not possible to determine the actual retention time of reactants and catalyst in the reboiler. The effect of flow rate was mainly on the production of methyl ester of the reactor. The % weight of methyl ester decreased while the flow rate increased since the retention time is less. For the RD operation of setup in this study, the feed flow rate should not be higher 6mL/min in order to avoid flooding in column and this rate were considered as optimum range of operation.

5.2.3 Effect of reaction time on methyl ester conversion

Table 12. Effect of Reaction time on methyl ester conversion

Reaction Time (min.)	Methyl esters Conversion (%) , Castor oil	Methyl esters Conversion (%), Cottonseed oil	Methyl esters Conversion (%), Coconut oil
4	92	95	89
6	94	96	93
8	94	96	93

Molar ratio of Methanol to Oil = 6:1, Catalyst (KOH) = 1% (by wt. of oil), Flow rate = 6ml/min, Temperature = 60°C

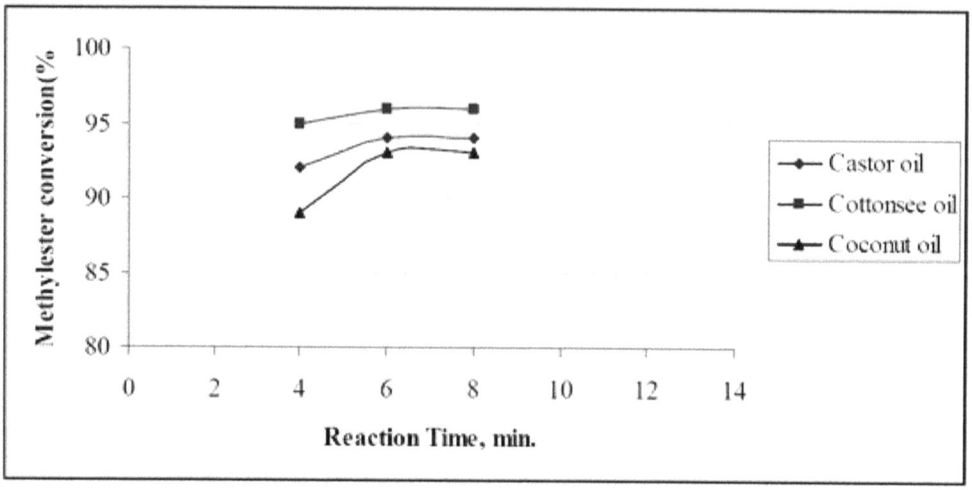

Optimum Reaction time = 6min.

Fig. 12. Effect of reaction time on methyl ester conversio

The rate of transesterification depends on the time of reaction as shown in fig.12 The reaction was slow during the first few minutes due to time taken for mixing and dispersion of methanol with the triglycerides in the oil. However the rate of reaction increased steadily from 6 min of the reaction. The residence time of the reactants in the RD Column was 6min. at which the highest ME conversion was achieved. Thus the reaction time clearly influences the conversion rate of methyl esters.

5.2.4 Effect of catalyst loading on methyl esters conversion

The type of catalyst and the amount of catalyst has a great impact on formation of biodiesel. The adequate catalyst loading is necessary to obtain the maximum conversion of triglycerides to methyl esters. In the above experimentation, different types of catalysts with different catalyst loadings were utilized for transesterification reaction and their effects were studied. In the course of the tests, it is observed that addition of excess amount of catalyst, give rise to the formation of an emulsion, which has increased the viscosity and led to the formation of gel.

a. Effect of catalyst (KOH) loading on methyl ester conversion

Table 13. Effect of catalysts loadings on methyl ester conversion

KOH Catalyst loading (wt.%)	Methyl esters Conversion (%), Castor oil	Methyl esters Conversion (%), Cottonseed oil	Methyl esters Conversion (%), Coconut oil
1	80	92	95
1.5	78	93	96
2	60	93	96

Methanol to oil molar ratio = 6, Temperature =60°C, Catalyst loadings used for experimentation = 1% , 1.5% and 2% KOH by wt of oil

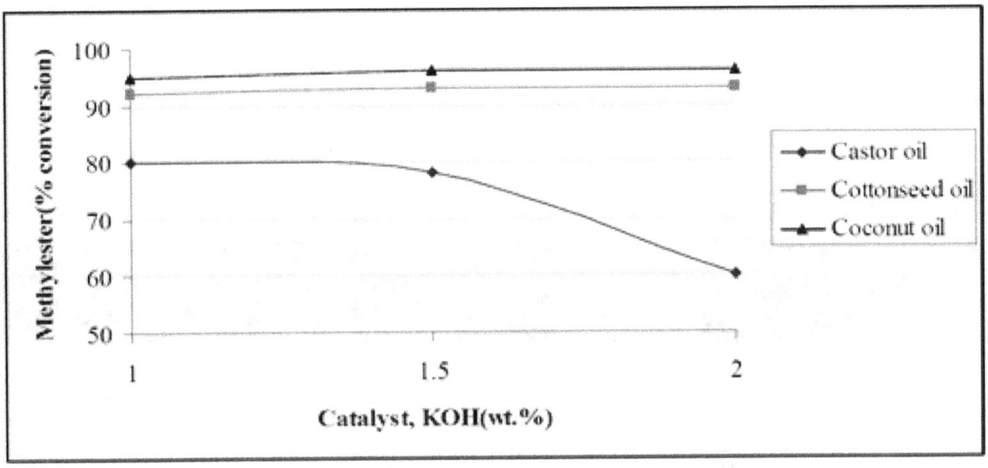

Fig. 13. Effect of catalyst (KOH) loading on Methyl ester Conversiond

b. Effect of catalyst(NaOH) loading on methyl ester conversion

Table 14. Effect of Catalyst (NaOH) loadings on Methyl ester Conversion

NaOH Catalyst loading (wt.%)	Methyl esters Conversion (%), Castor oil	Methyl esters Conversion (%), Cottonseed oil	Methyl esters Conversion (%), Coconut oil
0.5	72	89	94
1	80	80	95
1.5	78	91	96
2	60	91	97

Methanol to oil molar ratio= 6, Temperature =60oC, Catalyst loadings used for experimentation = 0.5%, 1%, 1.5% and 2% NaOH

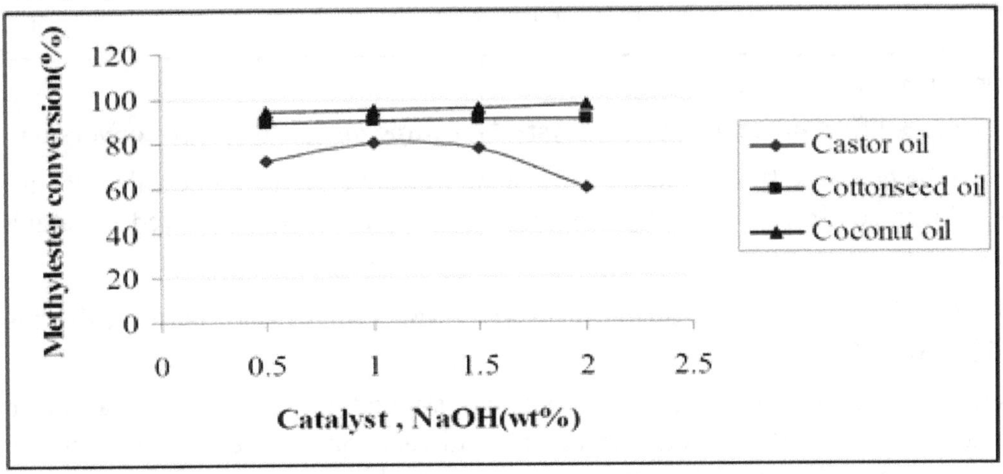

Fig. 14. Effect of catalyst (NaOH) loading on Methyl ester Conversion

It can be seen that the conversion of triglyceride to methyl ester increased as the catalyst loading increased from 0.5 to 1 wt.%. However, with the increase of the catalyst loading from 1% to 2 wt.%, the conversion of triglyceride to methyl ester decreased. Since the conversions of triglycerides to methyl esters did not change significantly for the catalyst loadings from 1wt.%, the optimum catalyst loading for this reaction was 1wt.%.

c. Effect of catalyst (Amberlyst-15) loading on Methyl ester Conversion

Table 15. Effect of catalyst (Amberlyst-15) loading on Methyl ester Conversion

Amberlyst-15 Catalyst loading (wt.%)	Methyl esters Conversion (%), Castor oil	Methyl esters Conversion (%), Cottonseed oil	Methyl esters Conversion (%), Coconut oil
3	89	90	88
4.5	77	80	76
6	56	65	60

Methanol to oil molar ratio= 6, Reaction temperature =60oC, Reaction time = 1.5 hrs

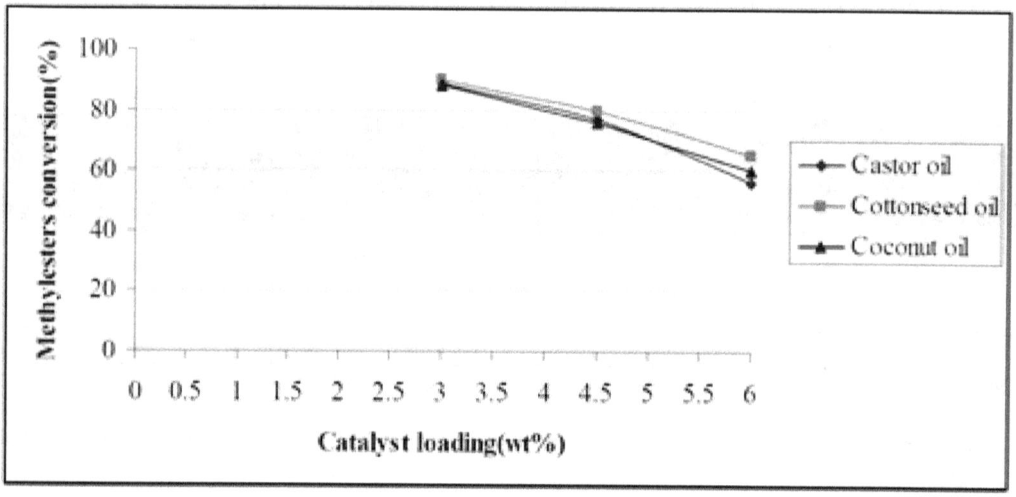

Optimum catalyst loading = 3wt%

Fig. 15. Effect of catalyst (Amberlyst-15) loading on Methyl ester Conversion

It can be observed that there was no rise in ME conversion when catalyst loading was increased from 3 to 6 wt%. The highest conversion was achieved at catalyst loading of 3 wt%.

5.2.5 Effect of reaction time on methyl ester conversion using amberlyst -15 catalyst

Amberlyst-15 a solid acid catalysts been studied for transesterification for production of methyl esters. However, mild reaction conditions are necessary to avoid degradation of the catalyst. At a relatively low temperature (60 °C), the conversion of castor, cottonseed and coconut oil was reported to be only between 2% to 5%, when carrying out the reaction at atmospheric pressure and a 6:1 methanol-to-oil initial molar ratio.

Table 16. Effect of Reaction time on Methyl ester Conversion using Amberlyst -15 Catalyst

Reaction Time(hrs)	Methyl esters Conversion (%), Castor oil	Methyl esters Conversion (%), Cottonseed oil	Methyl esters Conversion (%), Coconut oil
0.5	60	65	55
1	76	79	74
1.5	80	85	82
3	72	80	76
3.5	65	62	60
4	56	50	52

Methanol to oil molar ratio= 6, Reaction temperature =60oC, Catalyst loading=3wt.% Amberlyst-15

During the initial 30 minutes reaction time the percent conversion was less. But as the reaction time progressed the conversion of triglycerides to methyl esters was increased till further one hour. After 1.5 hrs of reaction time, there was no significant rise in conversion. But the ME conversion decreased after 1.5 hrs. So, 1.5 hrs is obtained as the optimum reaction time for this study.

Fig. 16. Effect of reaction time on Methyl ester Conversion using Amberlyst -15 Catalyst

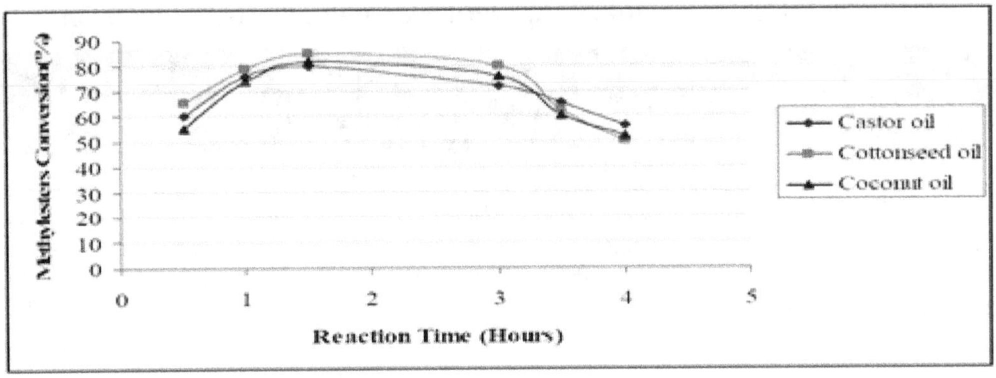

Optimum reaction time =1.5 hours

5.2.6 Effect of methanol to oil molar ratio on methyl ester conversion using Amberlyst-15 catalyst

Table 17. Effect of methanol to oil molar ratio on Methyl ester conversion using Amberlyst- 15 catalyst

Methanol-t-oil molar ratio	Methyl esters Conversion (%), Castor oil	Methyl esters Conversion (%), Cottonseed oil	Methyl esters Conversion (%), Coconut oil
6	89	92	90
9	76	84	77
12	44	73	39

Reaction time = 1.5 hrs, Reaction temperature =60oC, Catalyst loading=3wt. % Amberlyst-15

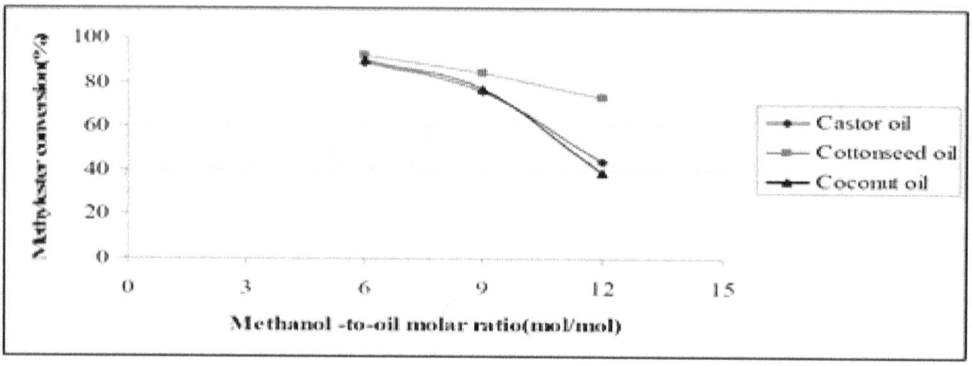

Optimum methanol to oil molar ratio = 6:1

Fig. 17. Effect of methanol to oil molar ratio on methyl ester conversion using Amberlyst-15 catalyst

It can be observed that there was no rise in ME conversion when methanol to oil molar ratio was increased from 6:1 to 12:1 .The highest conversion was achieved at methanol to oil molar ratio of 6:1.So, it is considered as optimum methanol to oil molar ratio.

5.2.7 Evaluation of column operating conditions

Effect of reaction temperature on methyl ester conversion

Table 18. Effect of reaction temperature on methyl ester conversion

Temperature (oC)	Methyl esters Conversion (%), Castor oil	Methyl esters Conversion (%), Cottonseed oil	Methyl esters Conversion (%), Coconut oil
55	89	84	90
60	92	96	96
65	94	97	96

Molar ratio of Methanol to Oil = 6:1, Catalyst (KOH) = 1% (by wt. of oil), Flow rate = 6ml/min, reaction time = 6min

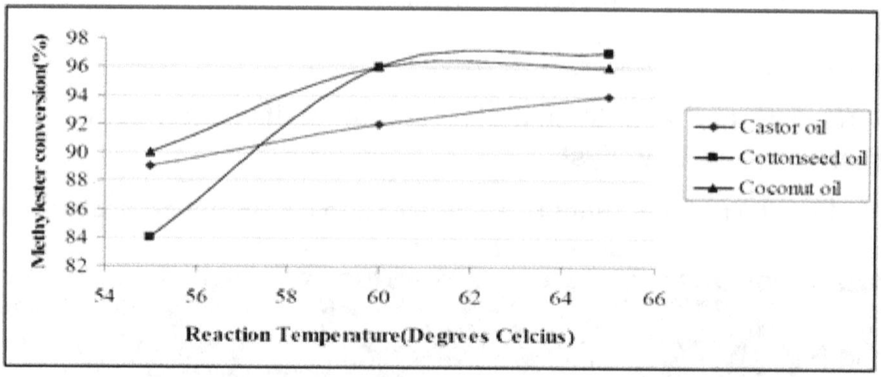

Optimum reaction Temperature = 60°C

Fig. 18. Effect of Temperature on Methyl ester conversion (%)

The effect of temperature on methyl ester conversion was studied by trans-esterification at different temperature i.e. 55 °C, 60 °C and 65 °C. It was observed that the conversion increased with increase in temperature from 55 °C to 60 °C . But there was no significant rise in conversion after 60 °C upto 65 °C. So, the optimum value of temperature for this transesterification reaction was considered as 60 °C.

5.2.8 Effect of reboiler temperature on methyl ester content of the product

The function of reboiler is to vaporize the residual methanol present in the liquid reaching the bottom of the column. At steady state, the boiling-up rate of methanol is determined by the heat load on the reboiler, heat transfer efficiency and the amount of methanol in the reboiler. Methanol boils at 64.7oC, however, according to the experiments, sufficient methanol vapors were generated only with reboiler temperature higher than 90oC. Depending upon the methanol concentrations, therefore, reboiler temperature in the experimental design varied from 80oC to 120oC in order to produce smooth and consistent methanol vapor flow rates. It was found that the lower reboiler temperatures are favorable for better reactor performance. A possible reason is that with higher operating temperatures, the rates of soap formation increase more rapidly than that of transesterification.

Table 19. Effect of reboiler temperature on methyl ester content of the product

Time (hrs)	%Methyl ester at 120oC	%Methyl ester at 100oC	%Methyl ester at 90oC	%Methyl ester (at 85oC
0	0	0	0	0
1	25	34	10	33
2	44	55	28	63
3	56	67	49	68
4	62	74	68	70
5	66	78	80	72
6	68	80	87	72
8	68	83	90	74
10	66	86	91	75
12	66	88	93	75
14	67	87	92	73
15	67	87	92	74

Molar ratio of Methanol to Oil = 6:1, Flow rate =6ml/min, Reaction time = 6min., Catalyst (KOH) =1% by wt. of oil)

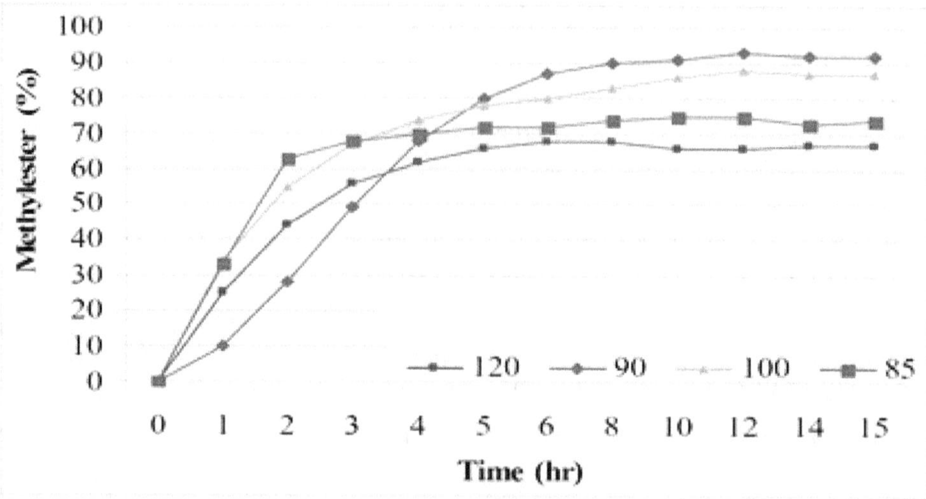

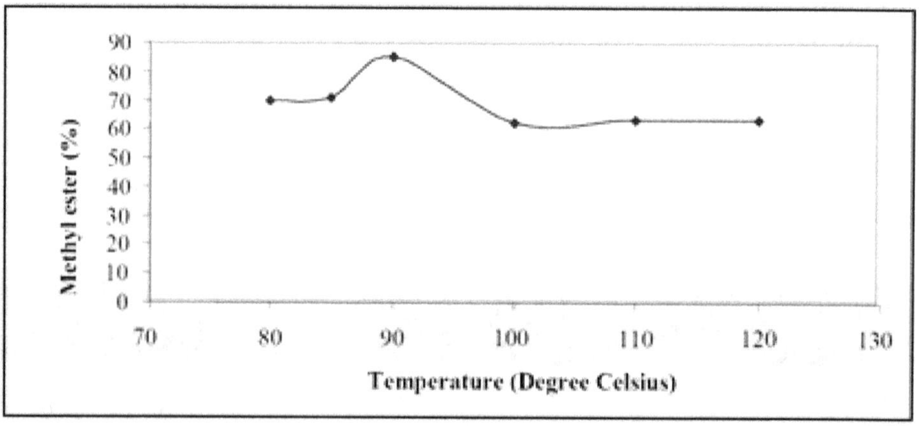

Fig. 19. The effect of the reboiler temperature on methyl esters content in product

5.2.9 Biodiesel yield

The biodiesel yield was determined indirectly from the triglyceride as discussed in the materials and methods section. The percentage composition of the triglyceride consumed was assumed to be converted to methyl ester. Therefore, the amount of biodiesel produced was calculated from the difference in composition of triglyceride in the original oil and final biodiesel.

6. CONCISE SUMMARY

This study analysis yields the following conclusions:

Reactive distillation serves as a techno-economic method for the synthesis of biodiesel from vegetable oil via the transesterification process. The proposed approach can significantly enhance the economics of contemporary biodiesel synthesis and decrease the number of downstream processes.

Using reactive distillation for synthesis of methyl ester has several advantages such as high unit productivity, up to 6–10 times higher than of the current process, lower excess alcohol requirements, reduced capital and operating costs, due to less units and lower energy consumption, Sulfur-free fuel, since solid acids do not leach into the product, no waste streams because no salts are produced. Also there is significant reduction in reaction time as well as in number of equipment units compared to that in conventional batch and continuous transesterification processes.

The feasibility of using mixture of two different vegetable oils was tested and it was found that coconut oil and cottonseed oil if mixed in equal proportion and used for transesterification reaction, the methyl esters conversion as high as 95% can be achieved.

From the experimental results the optimization of parameters obtained was as follows: Transesterification reaction is affected by methanol to oil molar ratio, catalyst loading, reaction temperature, flow rate of reactant streams, reaction time, mixing intensity. In the present study reaction was carried out using three values of Methanol to oil molar ratios (3:1, 6:1 and 9:1) for individual oils (castor, cottonseed and coconut oil) and KOH catalyst for synthesis of biodiesel the highest ME conversions obtained for castor, cottonseed and coconut oil transesterification were 68%, 72% and 88% respectively. The highest ME conversion was obtained for coconut oil in this case. Whereas by using the heterogeneous catalyst Amberlyst -15, the highest ME conversions obtained for castor, cottonseed and coconut oil transesterification were 89%, 92% and 90% respectively for methanol to oil molar ratio of 6:1.

Also the possibility of using mixed oils was investigated by using three seed oils in 3 different proportions, such as mixed oil 1 (50% castor oil+50% cottonseed oil), mixed oil 2 (50% castor oil+50% coconut oil) and mixed oil 3(50% coconut oil+50% cottonseed oil).The highest ME conversion (95%) was obtained for mixed oil 3 at methanol to oil molar ratio of 6:1.Thus the option of using mixed oils is feasible in case of scarcity of one of particular feed stock oil in particular region. Also the application can be further investigated to reduce the cost of production of biodiesel by using cheaper feed stocks in more proportion in mixed oil feed.

The flow rate of reactants has an impact on reaction rate of biodiesel production. Out of 5ml/min, 6ml/min and 7ml/min flow rates of reactants to the RD column transesterification reaction. The highest ME conversion (96%) was obtained for cottonseed oil at reactants flow rate of 6ml/min.

For a practical and economic feasible transesterification process, it is necessary to limit the reaction time at a certain period. Longer reaction time could also permit reversible transesterification reaction to occur, which eventually could reduce the yield of fatty acid alkyl esters. Thus, optimization of reaction time is also necessary. In this study, the reaction time was varied from 4 min, 6min and 8min. It was observed that the highest ME conversion (96%) was obtained for cottonseed oil at 6min reaction time. Whereas for heterogeneous catalyst, reaction times of 0.5,1, 1.5, 3, 3.5 and 4 hrs using catalyst Amberlyst-15 for Methanol to oil molar ratio 6:1, Reaction temperature of 60oC, Catalyst loading of 3wt.% were used and the highest ME conversion (85%) was obtained for cottonseed oil after 1.5 hrs reaction time. There was no significant rise in conversion rate after 1.5 hrs.

The catalyst plays an important role in transesterification reaction. The type and quantity of catalyst usually depend upon the quality of feed stock and method applied for transesterification. Three values of catalyst loadings of 1, 1.5 and 2 wt% KOH were used for Methanol to oil molar ratio 6:1 at 60°C .The highest ME conversion (96%) was observed for coconut oil at 1.5 wt% KOH catalyst loading.

Similarly the second homogeneous catalyst NaOH also resulted in the same conversion for the same experimental conditions. Whereas the heterogeneous catalyst Amberlyst-15 was used in three catalyst loadings of 3, 4.5 and 6 wt% Amberlyst-15 for Methanol to oil molar ratio of 6:1, Reaction temperature of 60°C, Reaction time of 1.5 hrs, the highest ME conversion (90%) was obtained for cottonseed oil at 3wt% Amberlyst-15 catalyst loading.

The transesterification was carried out at reaction temperatures of 55, 60 and 65°C for individual oils-castor, cottonseed and coconut oil and the highest ME conversions obtained for castor, cottonseed and coconut oil transesterification were 92%, 96% and 96% respectively for 60°C temperature using KOH catalyst. For homogeneous catalyst, moderate reaction temperature is enough to commence the reaction whereas for heterogeneous catalyst the operating temperature varies depending upon activation energy and conditions to produce the high yield of methyl esters. For NaOH catalyst the same optimum value of 60 °C temperature was obtained in batch transesterification process with maximum ME conversion of 96%.

The sufficient methanol vapors were generated only with reboiler temperature higher than 90°C. Depending upon the methanol concentrations, therefore, reboiler temperature in the experimental design was varied from 80°C to 120°C in order to produce smooth and consistent methanol vapor flow rates. It was found that the lower reboiler temperatures are favorable for better reactor performance. A possible reason is that with higher operating temperatures, the rates of soap formation increase more rapidly than that of transesterification.

It was found that cottonseed oil resulted into maximum yield of biodiesel. Usually crude cottonseed oil contains palmitic acid (22- 26%), oleic acid (15-20%), linoleic acid (49- 58%) and approximately 10% mixture of arachidic acid, behenic acid and lignoceric acid, as well as about 1% sterculic and malvalic acids. In this study, the used crude cottonseed oil contained 24.60% of palmitic acid, 17.09% of oleic acid, and 50.50% of linoleic acid. Since higher amount of free fatty acids (FFA) (>1% w/w) in the feedstock can directly react with the alkaline catalyst to form soaps, which are subject to form stable emulsions and thus prevent separation of the biodiesel from the glycerol fraction and decrease the yield, it is better to select reactant oils with low FFA content or to remove FFA from the oil to an acceptable level before the reaction. Nevertheless, the FFA (calculated as oleic acid) content of the crude cottonseed oil used in this experiment was only 0.8%, which was in an allowed level for being directly used for reaction with the alkaline catalyst to produce biodiesel.

The castor oil showed yield of biodiesel less than that of cottonseed oil due to its high acid value. High acid value leads to neutralization of part of catalyst present thus producing soaps within the reaction medium which reduce mass transfer during the reaction. Also in basic medium, the hydroxyl group at C-12 of ricinoleic acid is converted into an alcohoxide derivative that can compete with the generation of methoxide species and compromise the conversion reaction. The Coconut oil contains approximately 92.1% saturated fatty acids, 6.2% monounsaturated fatty acids, 1.6% polyunsaturated fatty acids. Different fatty acids in coconut oil range from C6 - C18 carbon atom chains. It contains Lauric acid(over 50%) higher in composition So, it was concluded that cottonseed oil was the most feasible feedstock among the three vegetable oil feed stocks under consideration for this study.

The manufacture of biodiesel via transesterification using reactive distillation is both technically and economically viable, and scaling it up to an industrial level is advisable to satisfy future energy demands.

REFERENCES

Chuaohuymak Pojanalai, Sookkumnerd Terasut, Kinetics of Homogeneous Transesterification Reaction of Palm oil and Methanol, Technology p. 1-6,2005.

Demirbas Ayhan, Biodiesel A Realistic Fuel Alternative. Trabzon: Springer-Verlag London Limited, 2008.

Drapcho, Caye, Nghiem Phu Nhuan and Terry H.Walker. Biofuels Engineering Process Technology. New York Chicago San Fr: The McGraw-Hill Companies, Inc, 2008.

Kalayasiri, P., Jayashke, N. and Krisnangkura, K. "Survey of seed oils for use as diesel fuels." Journal of American Oil Chemical Society (1996): 73:471–474.

Noshadi I, A Review of Biodiesel Production Via Reactive Distillation AICHE Conference, Spring Session, 2011

Richardson, Colson and. Particle Technology and Separation Processes. Newcastle: Library of Congress Cataloguing in Publication Data, 1991.

Sharma Y.C, Singh Bhaskar, A hybrid feedstock for a very efficient preparation of biodiesel, Technology, Volume, Pages 1267-1273, 2010

Sundmatcher, Kai and Achim Kienle. Reactive Distillation Status and Future Directions.

Mhar. Achim Kienle Kai Sundmatcher. Megdeburg: Wiley-VCH, 2002.

Viswanathan B. and A.V.Ramaswamy, 'Selection Of Solid Heterogeneous Catalysts For Transesterification Reaction', National Centre for Catalysis Research, Indian Institute of Technology, Madras, Chennai 600 036

Zhang Y, Dube M A, McLean D D, Kates D, analysis, Bioresource Technology 90 (3) p. 229- 240, 2003

http://www.crnindia.com

http://www.castoroil.in

http://www.mcxindia.com

http://www.biodiesel.org

http://www.coconutboard.nic.in

http://www.biodiesel.com

http://www.teriin.org

http://www.indexmundi.com

15 Chapter

GAS-LIQUID PROCESS, THERMODYNAMIC PROPERTIES (19 BLENDS), EFFICIENCY AND ENVIRONMENTAL IMPLICATIONS, SEM PARTICULATE MATTER ASSESSMENT, AND ON-ROAD BUS TRIAL OF AN ESTABLISHED NOX-REDUCING BIODIESEL

1. INTRODUCTION: AN OVERVIEW

Biodiesel has garnered global interest as a renewable fuel and blending agent, noted for its reduced greenhouse gas emissions, but with a marginal increase in NOx emissions (Michael & Robert, 1998) in comparison to petroleum diesel. Vegetable oils (Srivastava & Prasad,2000; Prasad & Mohan,2003) namley soybean, sunflower, cottonseed and rapeseed have been examined for fuel without/ with a small modification in the engine. A number of problems, mainly high viscosity, are associated with vegetable oils when directly used as fuel in the CI engines (Agarwal, 1998; Sinha & Misra , 1997; Roger & Jaiduk 1985). It is difficult to reduce particulate matter (PM) and oxides of nitrogen (NO_x) (Mohamad and et al., 2002) simultaneously owing the trade-off between NO_x and PM. Moreover, methyl esters of vegetable oils are sulphur free and possess good lubricating properties (De-Gang et al., 2005). Depending upon the climate and the soil conditions, different countries looking for different type of vegetable oil (Goering et al., 1982; Fernando et al., 2003; Antolin et al., 2002; Freedam et al., 1986; Noureddini & Zhu, 1997; Mohamad et al., 2002; Yi-Hsu & Shaik, 2005; Sukumar et al., 2005) used for the biodiesel production; soybean oil in US, rapeseed oil in Europe, palm oil in Malaysia and Indonesia, and coconut oil in the Philippine are being considered (Barnwal & Sharma , 2005).

In India, out of more than 125 million tons (Arumugam et al., 2003) of rice production, about 6 million tons of rice bran and 1 million ton of RBOBD is produced annually (Table 1). General characteristics of refined rice bran oil are as follows: sp gr, 0.916 kg/m^3; ref index 1.47; Cloud index, 17; iodine value, 99-108; saponification value, 180-190; unsopanifiable matter, 3.5(%); smoke point, 213 °C; and fire point,

352 °C. General properties of vegetable oil based biodiesel (Table 2) show many variations that might be due to the conversion to biodiesel through different raw materials and different processes.

This chapter delineates the procedure for the synthesis of rice bran oil biodiesel (RBOBD), its composition and physicochemical features, engine test outcomes, scanning electron microscope (SEM) imagery, particulate matter in exhaust emissions, and reductions in emissions..

Table 1. Annual production (metric million tons) of rice, rice bran and oil in the world.

Country	Rice	Rice bran	Oil
China	181	14.5	2.47
India	137	6.8	1.02
Indonesia	50	4.0	0.68
Bangladesh	38	3.0	0.51
Vietnam	32	2.6	0.44
Thailand	24	1.9	0.32
Myanmar	20	1.6	0.27
Philippines	13	1.0	0.17
Japan	11	0.9	0.15
Brazil	10	0.8	0.14

Table 2. Properties of Biodiesel from vegetable oils.

Oil Name	ρ (kg/m3)	μ (mm2/s)	Cetane No:	Calorific value (MJ/kg)	Flash point (°C)
Pea	0.883	4.9	54	33.6	176
Soya	0.885	4.5	45	33.5	178
Bab	0.875	3.6	63	31.8	117
Palm	0.880	5.7	62	33.5	164
Sun	0.860	4.6	49	33.5	183
Diesel	0.855	3.06	50	43.8	76
B20	0.859	3.2	51	43.2	128
Range	0.85-0.88	3.2-5	45-62	32-44	76-183

2. PROCESS DESCRIPTION

Oil having specific gravity in the range of 0.85 – 0.96 and iodine value not exceeding 208 is heated to a temperature not exceeding 120°C for not less than 2hrs to adjust the moiture content at a level not exceeding 0.5% and is transesterified using 8 to 42% w/w, of alcohol of general formula R-OH, where R represents (C_nH_{2n+1}), n being any integer between 1 and 5, by known method in presence of not more than 0.5% w/w, of a known catalyst, at a temperature higher than the boiling point of the alcohol but not exceeding 215°C for not less than 30 minutes under continuous turbulent condition at rpm in the range of 100-150 to get a mixture of ester and glycerol. The Reynolds number (N_{Re}) is maintained at not less than 4000 irrespective of the type of the reactor. The mixture of ester and glycerol is subjected to separation by known method for a period of not less than 4 hrs and the top layer ester is purified by conventional method for a period of not less than 8hrs. The process of separation as well as purification is repeated for not less than three times in succession to get biodiesel.

Fig. 1. Lab scale experimental setup

In lab scale experimental setup Fig.1, RBO was taken in the continuous stirred tank glass reactor (1 l) with reflex condenser, temperature control and agitation control setup. In another reactor, NaOH (50 g) was dissolved in methanol (300 ml). This solution was added slowly at the reactor maintained at 65-70 °C for 150 min. Then the entire mixture kept in the separating funnel. The top layer, biodiesel, is taken for the removal of methanol in the ROTO vacuum distiller. Then the methyl ester washed of distilled water (1 l) in the same reactor for 30 min. After washing, top layer in the separating funnel has to be washed with saline water for two times. Finally, clear biodiesel was kept in the oven for 4 h at 100°C. The ready to use biodiesel few samples shown in Fig.2.

Fig. 2. Ready to use biodiesel samples

In the bench scale level, Rice Bran Oil (RBO) experiments were carried out with standardized process conditions in high-pressure Parr Reactor (Fig.3.) inbuilt sophisticated controlling systems of reactor (20 l). Rice Bran Oil Biodiesel RBOBD (>150 l) was produced.

Fig. 3. Bench Scale lab Parr Reactor

In each lot, biodiesel sample has been analyzed for the conversion, fuel properties and composition. Quality consistency conformed by C_{13} and Proton of JEOL ECA 500 MHz NMR analysis and the composition by GCMS. All chemicals used were of LR/AR grade. A typical NMR spectrum show in Fig.4.

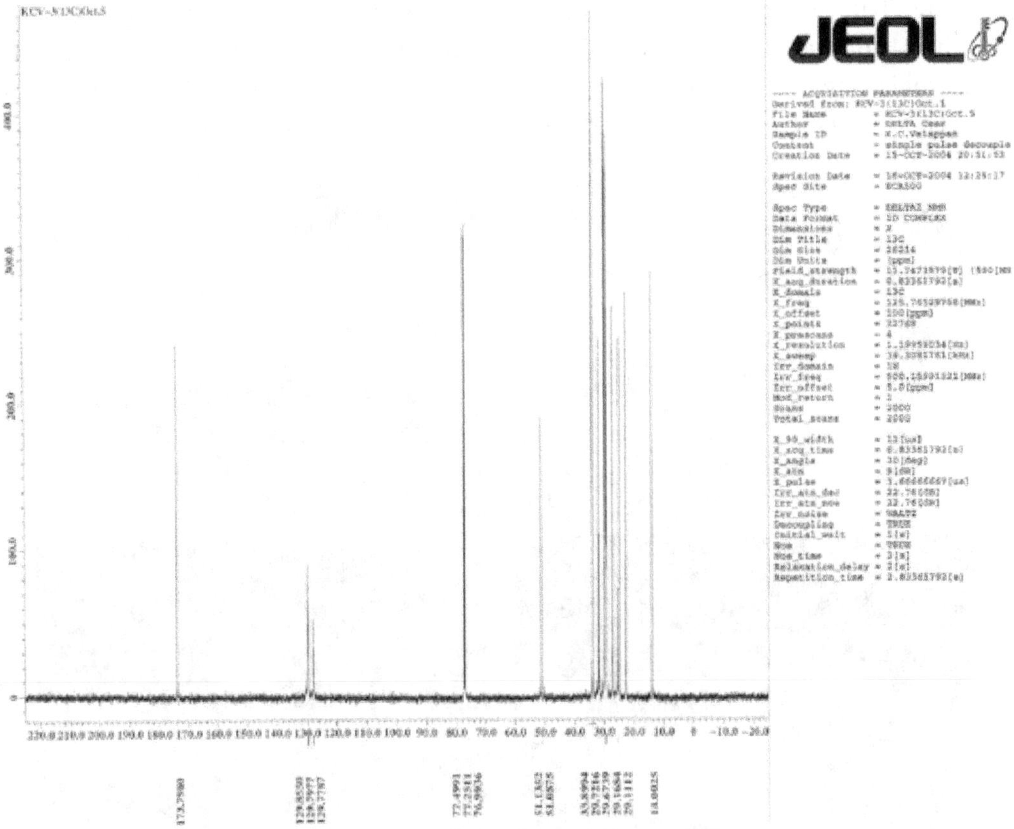

Fig. 4. A Typical C13, NMR Spectrum

The brief process description has been followed at the Pilot-scale preparation of biodiesel (Fig. 5. (a)), which was used for on-road trails from rice bran oil is following.

Rice bran oil is filtered to remove any impurities. 69 lit. of moisture free refined oil is taken in a Pilot Plant scale reactor (Fig. 5. (b)) of capacity 120 lit. Fitted with a reflux condenser and heated with agitation to 65ºC. Then 345 gms of sodium hydroxide, 20.7 lit. of methanol are mixed separately and the mixture is slowly added to oil at 65ºC.

The reaction mixture is mixed well, temperature is maintained at 65-70ºC throughout the reaction and the reaction time is 150 min. When the reaction is complete, the contents are allowed to cool and transferred to a separating tank. After overnight settling, the mixture gets separated into two layers due to density difference.

The bottom layer-Glycerol is separated. The top layer - biodiesel is distilled at 65ºC to recover unreacted alcohol. Then the methyl ester is washed for 30 minutes at 50ºC with equal volumes of 0.1% dil. acetic acid to remove any traces of un reacted alkali. In case of emulsion formation after washing, saline water is used for second washing. The pH of the ester layer is adjusted to neutral while washing. After washing, the layers are allowed to settle for 30 min. The top layer is separated and biodiesel is dried in a pan drier for 2 hrs at 110ºC. Then it is filtered to separate any traces of impurities. The final ready to use biodiesel product is found to be 60 lit.

Fig. 5. (a) Pilot-scale preparation of biodiesel (Fig. 5. (b))

Few thousand liters of Biodiesel produced in the pilot level which is used as feul in the on- Road bus trails. More than 26000 km exprimental trials were carried out in the Metropolitan Transport Corporation (MTC) buses in Chennai, Government of Tamil Nadu. Few clipings of MTC bus trails are shown in Fig.6. Initialy four buses have been taken for on-road trials in a single root but fuelled with different biodiesl percentage namely, B5, B10,B20 and B50. Then all the buses fuelled with 100% Biodiesel. The MTC, government of Tamil Nadu, has submitted the officeal report about the on –raod trials. The Fig. 7 showing the highligts signed by the MTC highre officails of the report in the reginal language namely TAMIL and Fig 8. Showing its translation in English.

3. ENGINE TESTING AND EXHAUST GAS ANALYSIS

RBOBD was tested in Kirloskar four stroke, single cylinder, water cooled, direct injection IC engine (Fig.9) with following parameters: bore, 80 mm; stroke, 110 mm; swept volume, 553 cm^3; clearance volume, 36.87 cm^3; compression ratio, 16.5:1; rated output, 3.7 kW at 1500 rpm; rated speed, 1500 rpm; injection pressure, 240 bar; fuel injection timing, 24 BTDC; type of combustion chamber, hemispherical open; lubricating oil, SAE 40; connecting rod length, 235 mm; valve diam, 33.7 mm; and maximum valve lift, 10.2 mm.

Fig. 5. (b) Pilot Plant scale reactor

Fig. 6. Few clipings of MTC bus trials

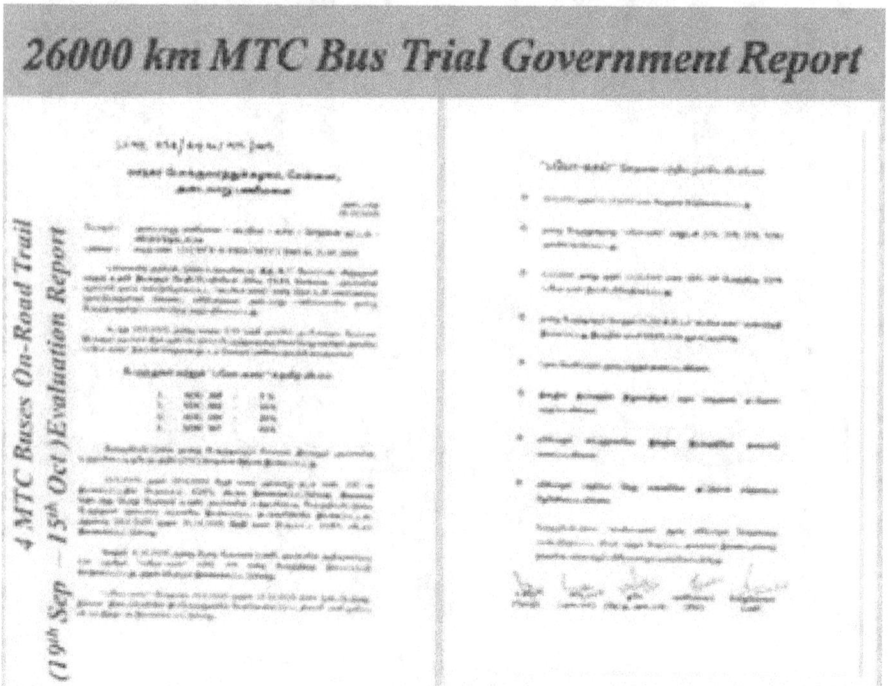

26000 km MTC Bus Trial Government Report

4 MTC Buses On-Road Trail (19th Sep – 15th Oct) Evaluation Report

Fig. 7. Showing the highligts signed by the MTC highre officails

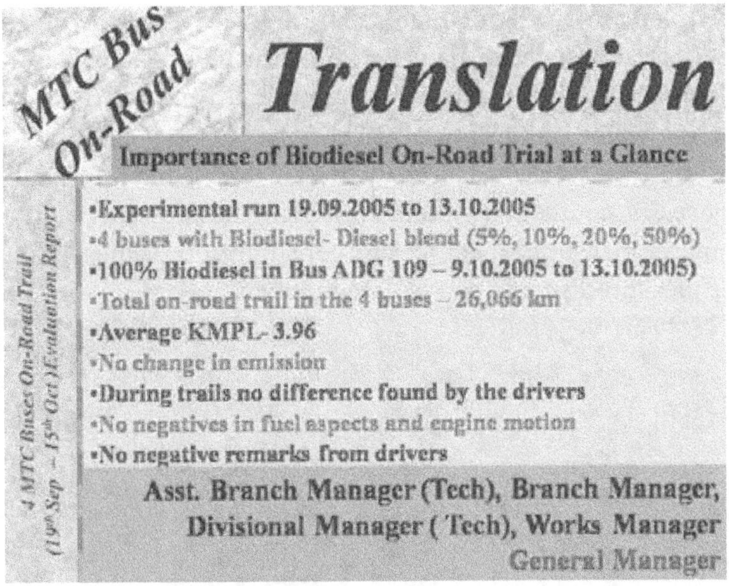

Fig. 8. Highlights Translation in English

Fig. 9. A Test engine

DYNALOG, PCI 1050 system has been used for digital data acquisition during the engine trial. Online engine calibration (Fig.10) with a special software namey "Engine-soft". The very brief specifications are Number of channels (16); Resolution (12- bit A/D); Input range (± 10 V, ± 5V, 0 -10 V); Accuracy (0.025%) and Conversion time (8 μs).

Engine was coupled to a swinging field separating exciting type DC generator and loaded by electrical resistance bank to apply various load. An iron-constantan thermocouple measured exhaust gas temperature and mercury thermometer measured cooling water temperature. Carbon monoxide (CO), nitrous oxide (NO_x) and hydrocarbons (HC) were measured by DELTA 1600-L and MRU OPTRANS 1600, a fully microprocessor controlled system employing nondestructive IR technique. A U-tube manometer measured specific fuel consumption. TI diesel tune, 114-smoke density tester measured smoke particulate number.

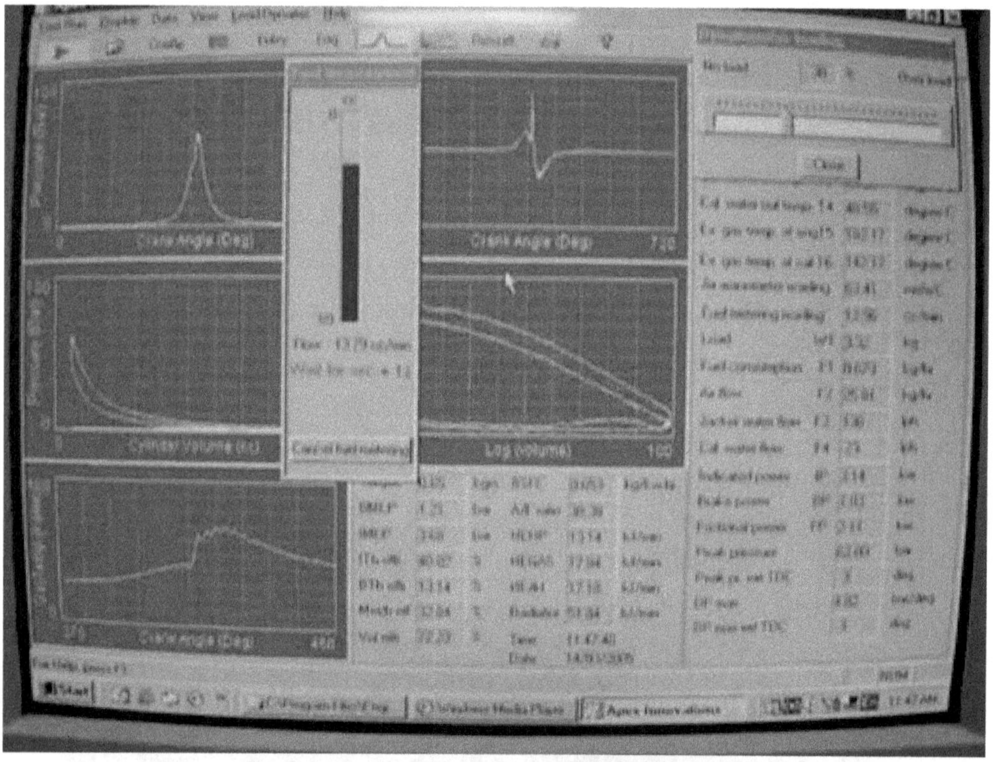

Fig. 10. Engine Calibration with "Engine-Soft"

The engine was started on neat diesel fuel and warmed up till liquid cooling water temperature was stabilized. During the performance of each trail, data were collected on time taken for 10 ml of fuel, load, exhaust gas temperature, cooling water inlet and outlet temperature, CO, CO_2, O_2, HC, NOx, smoke and sound. Graphical comparisons are described in the results and discussion. Smoke samples were collected in a white filter paper; this was taken for Scanning Electron Microscope (SEM) analysis to find the size of the particulate matter and to visualize the quantity of agglomeration. The SEM image is shown in Fig 10. Based on the data, specific fuel consumption, indicative thermal efficiency, brake thermal efficiency, mechanical efficiency and total fuel consumption were estimated. Similar procedures were repeated for RBOBD.

4. RESULTS AND DISCUSSION

4.1 Process conditions and compositions

RBOBD contains (GC-MS) esters of following acids: palmitic, 16; stearic, 2; oleic, 42; linoleic, 38; linolenic, 1.4; and arachidic, 0.6%. Quality consistency was conformed by C_{13} and Proton of JEOL ECA 500 MHz NMR. Physico-chemical characteristics of RBOBD and its 19 blends (Table 3) show that most of the parameters comply with international standards of biodiesel. An NMR spectrum is already shown in Fig 4.

4.2 Comparison of Brake Power and Specific Fuel Consumption (SFC)

SFC of diesel, RBOBD and its various blends at different load (0-3.78 kW) were estimated and graphical representaion is shown in Fig 11. In comparison to diesel, a slight increase (10-15%) of SFC was found for RBOBD, B40, B50, B60 and B80 throughout all loads. At the maximum load (3.78 kW), SFC of B60 was found higher in comparison to the other blends. In particular to B20, the result shows that SFC was lower than diesel and other RBOBD and its blends in all the loads. The maximum increase (11.6%) was found at load 1.89 kW.

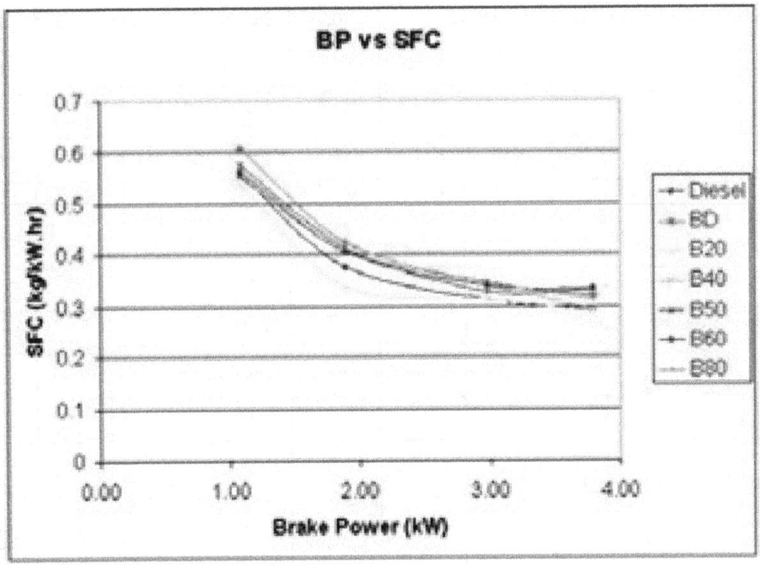

Fig. 11. Comparison of brake power and specific fuel consumption

4.3 Comparison of Brake Power and Fuel Consumption Time (FCT)

FCT of RBOBD and its various blends have been found less than the FCT of diesel, graphical representaion is shown in Fig 12. Slight decrease (5-10 %) of FCT was found for all fuels. Maximum decrease of FCT (12.5 %) was found at the brake power of 3.78 kW for B50 and B60. But, in particular, for B20, there was slight increase of FCT for the entire range of brake power. Maximum increase of FCT (12 %) was at 1.89 kW and minimum (3 %) at 3.78 kW.

Table 3. Physico-chemical characteristics of RBOBD and its 19 blends

Parameters	BD5	BD10	BD15	BD20	BD25	BD30	BD35	BD40	BD45	BD50
Acid value	0.27	0.3	0.31	0.49	0.5	0.57	3.49	0.74	0.54	0.87
Ash Content	0.0005	0.0006	0.0008	0.0010	0.0013	0.0051	0.0068	0.0051	0.0034	0.0038
Calcium	Nil	Nil	Nil	Nil	Nil	Nil	Nil	Nil	Nil	Nil
Carbon	82.27	82.17	82.07	81.97	81.90	81.85	81.80	81.90	81.98	82.01
Carbon residue (%)	0.00	0.0026	0.0030	0.0036	0.0041	0.0071	0.0096	0.011	0.01	0.012
Cetane Number	49	50	49	49	48	48	49	48	50	49
Cloud Point (C)	22	20	20	14	16	15	21	23	24	19
Density @ 15 C	0.8288	0.8335	0.8351	0.8391	0.8414	0.8437	0.8475	0.8520	0.8556	0.8588
Distillation 85	330	333	334	338	342	343	345	343	343	348
Distillation 95	344	347	347	350	356	358	360	358	360	361
Ester content	11.6	20.5	29.3	39.3	49.87	62.6	73.4	85.7	96.2	106.1
Flash Point	40	40	42	42	42	42	44	44	46	44
Free Glycerol	0.008	0.0091	0.010	0.011	0.009	0.012	0.014	0.013	0.016	0.015
G.C.V	10850	10720	10600	10610	10530	10500	10390	10300	10270	10210
Hydrogen	12.62	12.54	12.60	12.58	12.63	12.60	12.59	12.62	12.65	12.56
Iodine Value	10.2	14.1	17.9	23.3	28.7	32.5	35.5	42.3	46.2	49.5
N.C.V.	10181	10055	9932	9943	9860	9832	9723	9631	9600	9544
Nitrogen	0.031	0.030	0.030	0.034	0.031	0.033	0.034	0.032	0.030	0.032
Oxygen	5.063	5.245	5.285	5.402	5.426	5.505	5.564	5.436	5.33	5.308
Phosphorus	0.0006	0.00071	0.010	0.0015	0.0023	0.0029	0.0034	0.0043	0.0051	0.0058
Potassium	2.0	1.9	1.8	1.7	1.8	1.65	1.70	1.90	1.80	1.6
Pour point C	-22	-20	-20	-18	-15	-15	-16	15	-14	-13
Sodium	2.1	2.0	1.8	1.9	1.6	1.8	1.50	1.50	1.60	1.4
Sulphated Ash	0.0010	0.0013	0.0015	0.0022	0.0031	0.0068	0.0094	0.0064	0.0050	0.0064
Sulphur	0.016	0.015	0.015	0.014	0.013	0.012	0.012	0.012	0.010	0.009
Sulphur	0.016	0.015	0.015	0.014	0.013	0.012	0.012	0.012	0.010	0.009
Total contamination	0.096	0.009	0.011	0.011	0.011	0.012	0.009	0.0098	0.011	0.013
Total Glycerol	0.013	0.018	0.019	0.022	0.025	0.033	0.036	0.043	0.051	0.054
Viscosity @ 40 C	2.6	2.7	2.9	3.0	3.1	3.4	3.4	3.5	3.7	3.9
Water & sediments	0.022	0.026	0.048	0.029	0.024	0.027	0.030	0.029	0.029	0.031
Water content	0.021	0.024	0.045	0.026	0.022	0.022	0.026	0.027	0.027	0.0281
Acid value	1.08	0.97	1.07	1.08	1.2	1.27	1.4	1.41	1.43	1.54

Parameters	BD5	BD10	BD15	BD20	BD25	BD30	BD35	BD40	BD45	**BD50**
Ash Content	0.0037	0.0047	0.0045	0.0039	0.0076	0.0080	0.0081	0.0086	0.0093	0.0096
Calcium	Nil	Nil	Nil	Nil	Nil	Nil	-	-	-	-
Carbon	83.17	85.10	83.01	82.96	82.90	82.79	82.80	82.70	82.56	82.42
Carbon residue (%)	0.02	0.021	0.031	0.019	0.021	0.016	0.025	0.029	0.029	0.036
Cetane Number	49	50	48	49	48	47	47	46	46	45
Cloud Point Deg.C	9	9	12	14	14	13	16	14	19	23
Density @ 15 Deg.C	0.8625	0.8635	0.8666	0.8704	0.8748	0.8794	0.8814	0.8854	0.8889	0.8901
Distillation 85	342	326	340	342	344	345	344	343	347	348
Distillation 95	350	342	350	352	356	357	356	355	359	361
Ester content	111.7	120.2	130.3	132.2	150.5	156.2	164.9	164.9	181.1	192
Flash Point	43	46	64	60	64	70	68	90	90	124
Free Glycerol	0.016	0.015	0.018	0.017	0.019	0.018	0.020	0.021	0.024	0.025
G.C.V	10190	10090	10050	9970	9920	9860	9750	9690	9610	9810
Hydrogen	12.84	12.94	12.89	12.80	12.76	12.80	12.70	12.69	12.72	12.60
Iodine Value	51.3	56.2	57.2	63.2	67.7	72.7	77.3	80.7	85.6	86.4
N.C.V.	9509	9364	9367	9242	9284	9241	9187	9077	9016	8942
Nitrogen	0.030	0.032	0.031	0.029	0.028	0.028	0.029	0.029	0.031	0.030
Oxygen	3.936	3.904	3.904	4.188	4.271	4.362	4.451	4.563	4.672	4.933
Phosphorus	0.062	0.064	0.0068	0.0078	0.0076	0.0084	0.0083	0.0074	0.0095	0.0097
Potassium	1.2	2.0	2.1	2.1	2.0	2.2	2.9	2.2	2.1	1.9
Pour point Deg.C	-13	-13	-12	-12	-11	-11	-10	-10	-9	-8
Sodium	3.1	3.6	3.6	3.5	3.4	3.7	3.4	3.0	2.9	2.2
Sulphated Ash	0.0073	0.0080	0.0064	0.0057	0.0081	0.0061	0.0069	0.0053	0.012	0.013
Sulphur	0.024	0.024	0.023	0.023	0.021	0.020	0.020	0.018	0.017	0.017
Sulphur	0.024	0.024	0.023	0.023	0.021	0.020	0.020	0.018	0.017	0.017
Total contamination	0.015	0.018	0.022	0.021	0.024	0.025	0.029	0.030	0.035	0.044
Total Glycerol	0.064	0.069	0.070	0.074	0.075	0.083	0.088	0.096	0.099	0.10
Viscosity @ 40 Deg.C	4.0	4.2	4.40	4.60	4.8	5.0	5.2	5.5	5.7	6.0
Water & sediments	0.014	0.018	0.021	0.021	0.0034	0.024	0.028	0.031	0.035	0.041
Water content	0.012	0.017	0.020	0.020	0.021	0.022	0.028	0.029	0.035	0.041

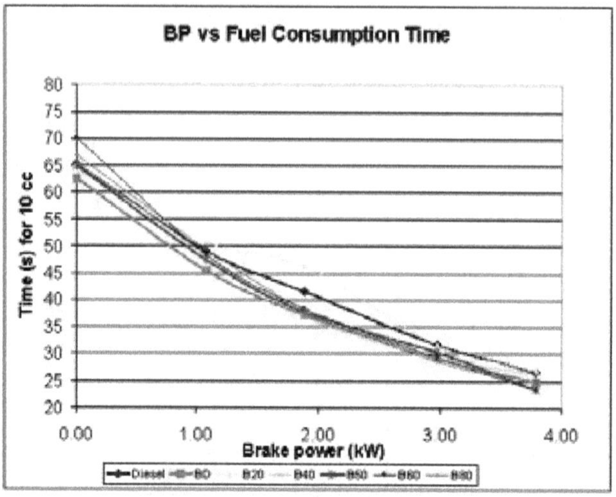

Fig. 12. Comparison of brake power and fuel consumption time

4.4 Comparison of Brake Power and Total Fuel Consumption (TFC)

TFC increased with increase of brake power, graphical representaion is shown in Fig 13. A maximum increase (13%) was at the load 1.89 kW. TFC's of RBOBD blends, B40, B50, B60 and B80, are higher (5-10%) than the TFC of diesel. But B20's TFC is slightly lesser than diesel and all the other RBOBD blends from the minimum load to the maximum load. Maximum TFC decrease (10%) was observed for B20 at 1.89 kW. Overall trend shows that the percentage decrease in TFC is inversely proportional to the brake power. At the maximum load, the increasing order of TFC is B20, Diesel, B40, B80, RBOBD and B60.

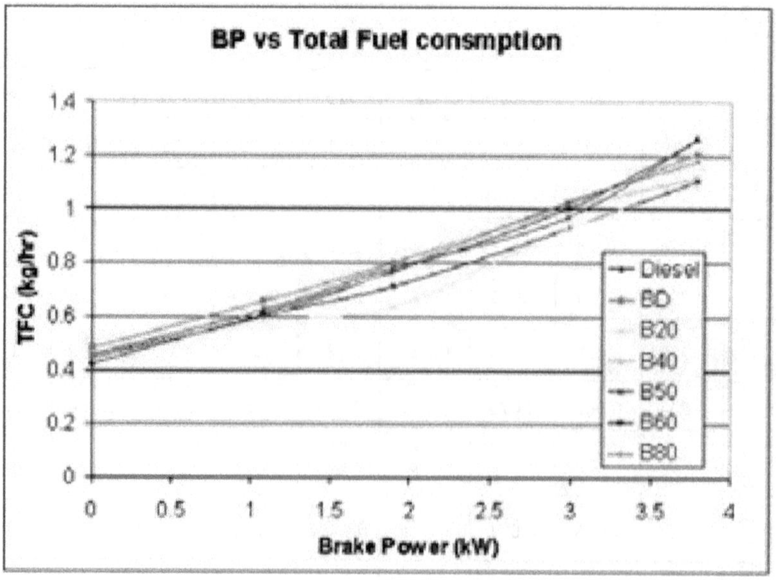

Fig. 13. Comparison of brake power and total fuel consumption

4.5 Comparison of Brake Power and Exhaust Gas Temperature (EGT)

EGT increases with increase of brake power, graphical representaion is shown in Fig 14.

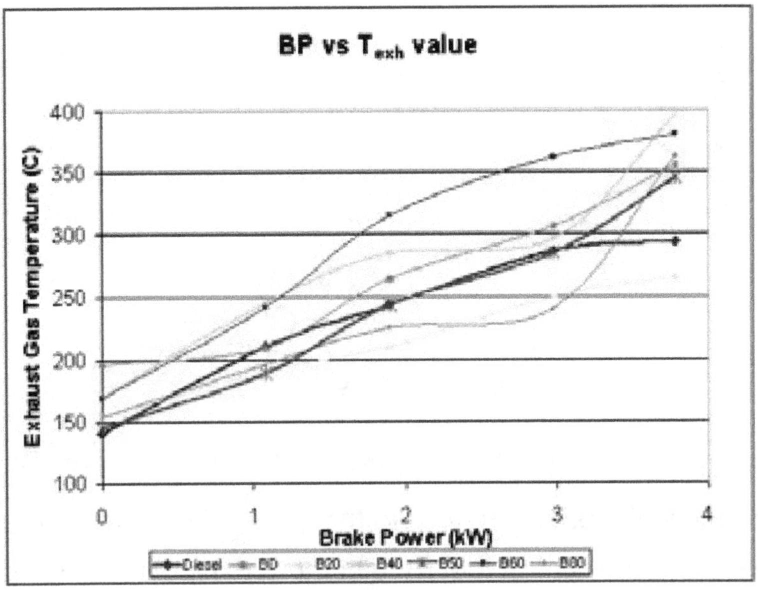

Fig. 14. Comparison of brake power and exhaust gas Temperatur

In comparison with EGT of diesel in each load, EGT of RBOBD and all the blends were higher. The highest value of EGT (395°C) was found with B40 at the maximum load of 3.78 kW, whereas corresponding value of normal diesel was 294°C only. Percentage increase of EGT of RBOBD decreased with the increase of load. Maximum increase (40%) of EGT was found at lower load of zero brake power. EGT of B40, B50, B60 and B80 were 350-400°C at the maximum load. The percentage increase (20-40%) of EGT of B60 was higher than all the loads. EGT of B20 was found to be slightly lower than EGT of normal diesel in all loads (0-3.78 kW). The minimum EGT decrease (9.5%) and maximum decrease (16.3%) of RBOBD and blends were found at 1.1 Kw and 2.98 kW respectively as compared to diesel. EGT of B20, B50 and B80 were found to be lower than EGT of diesel at 0-2.97 kW.

4.6 Comparison of Brake Power and Brake Thermal Efficiency (BTE)

BTE increases with increase of load, graphical representaion is shown in Fig 15.

BTE of RBOBD was less (5%) than diesel with respect to all loads. All other RBOBD blends (B40, B50, B60 and B80) were within 5 % only. Maximum reduction (20%) of BTE was found at 1.89 kW and minimum increase (7%) at 2.9765 kW. BTE of B20 was found higher than BTE of normal diesel in all loads. At maximum load (3.78 kW), BTE for B20 (29.7%), B40 (28.6%), B50 (25.6%) and B60 & B80 (25.5%) are higher than BTE of diesel.

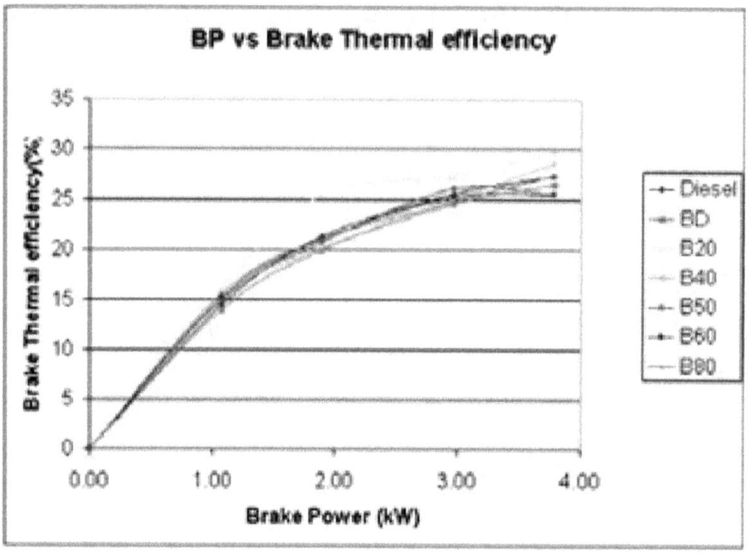

Fig. 15. Comparison of brake power and brake thermal efficiency

4.7 Comparison of Brake Power and Indicative Thermal Efficiency (ITE)

ITE of RBOBD, B50, B60 and B80 were found lower than ITE of diesel, graphical representaion is shown in Fig 16 . ITE of B20 and B40 were slightly more than ITE of diesel. Maximum ITE (57.9%) was found for B20 at 1.89 kW. Reduction (5-15%) was observed in ITE of various blends. But the ITE of all fuels shows that values are almost steady throughout the entire brake power.

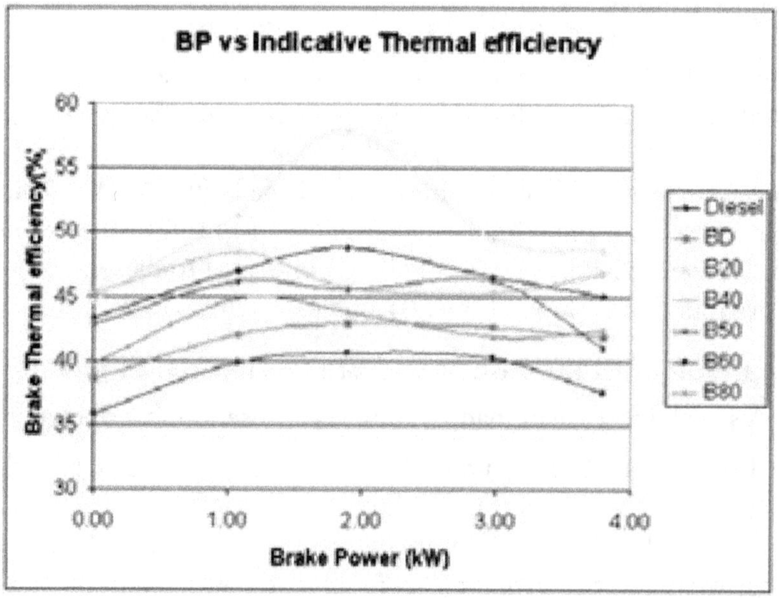

Fig. 16. Comparison of brake power and indicative thermal efficiency

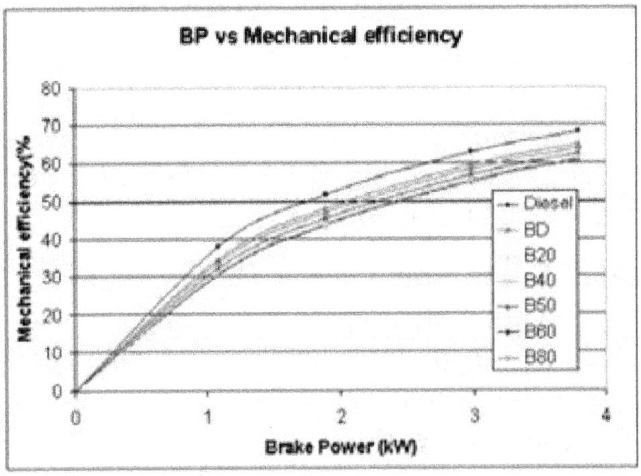

Fig. 17. Comparison of brake power and mechanical efficiency

4.8 Comparison of Brake Power And Mechanical Efficiency (ME)

ME of the engine run with RBOBD and various blends increases with increase of brake power in comparison with normal diesel, graphical representaion is shown in Fig 17. ME of RBOBD was less (5%) than diesel with respect to all loads. There was a slight increase of ME for all RBOBD blends in the order of B40, B50, B20, B80 and B60. Highest ME (68%) observed for B60 was at the maximum load. Overall trend shows that percentage increase of ME was decreased with increase of load. The result of B60 shows that the minimum increase (12%) was found at maximum load and maximum increase (24%) at minimum load.

4.9 Comparison of Brake Power and Hydrocarbons (HC)

HC increased with increase of brake power, graphical representaion is shown in Fig 18. RBOBD and its five blends showed lower HC (50-100%) than diesel. Reduction of HC (60%) of RBOBD and its blends are at the maximum load; B50 shows higher HC reduction than other blends at all loads.

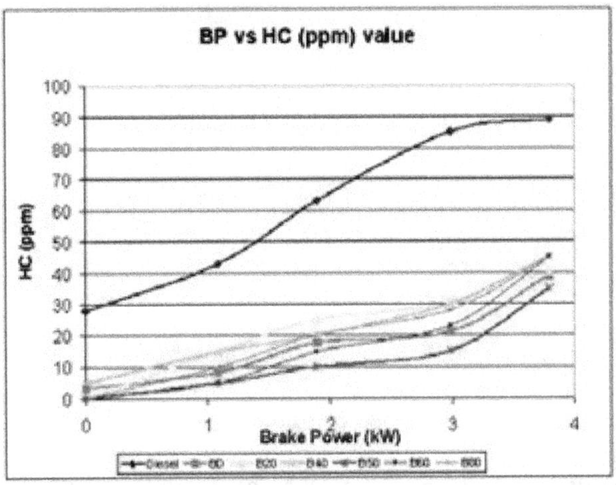

Fig. 18. Comparison of brake power and hydrocarbons

4.10 Comparison of Brake Power and CO emission

CO emission increased with increase of brake power, graphical representaion is shown in Fig 19. There was decrease of CO emission (50-80%) of RBOBD and its all five blends in comparison with CO emission of diesel. Reduction (>50%) of CO emission was found at 2.97 kW. Within blends, B20 shows lower CO emission (70-80%), which decreased with increase of load.

4.11 Comparison of Brake Power and CO2 emission

CO_2 emission increased with increase of brake bower, graphical representaion is shown in Fig 20. There was slight increase in CO_2 emission of RBOBD and its blends as compared to diesel. More variation of percentage increase was found within all RBOBD blends at the load 1.89 kW. The overall trend shows that the CO_2 emissions are similar to diesel at each load.

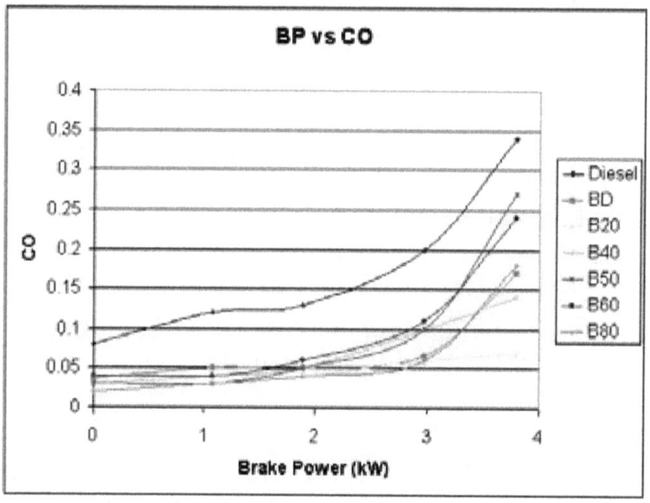

Fig. 19. Comparison of brake power and carbon monoxide emission

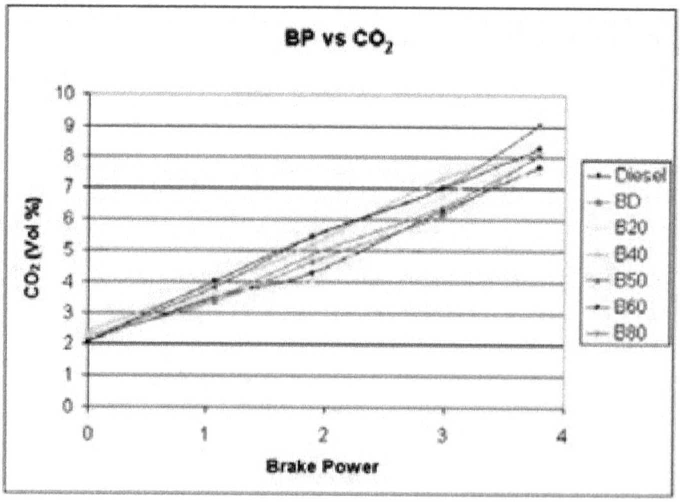

Fig. 20. Comparison of brake power and carbon dioxide emission

4.12 Comparison of Brake Power and NOx emission

NO_x increased with increase of brake power, graphical representaion is shown in Fig 21. There was a reduction (10-55%) of NO_x of RBOBD and its blends in comparison with NO_x values of diesel in each load. The trend shows that at minimum load, percentage reduction was maximum and at the maximum load, the percentage reduction of NO_x was minimum. The percentage reduction of NO_x decreased with increase of brake power. NO_x values at maximum load (3.78 kW) were found to be: diesel, 942; B80, 858; B50, 782; RBOBD, 753; B20, 677; and B60, 660 ppm. At 3.78 kW, maximum reduction (28%) was found for B20 and minimum (8.9%) for B80.

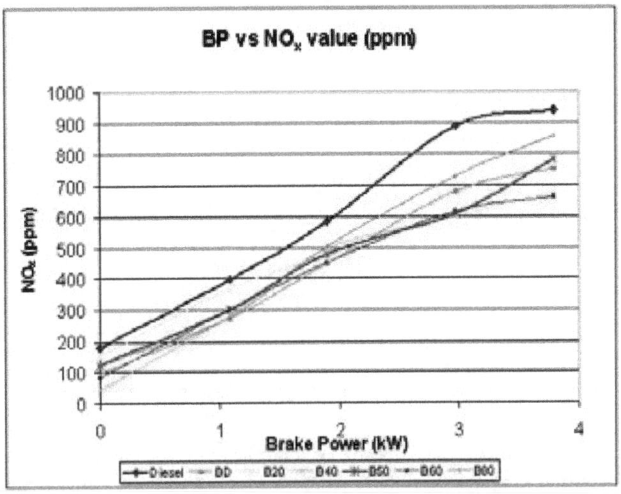

Fig. 21. Comparison of brake power and nitrogen oxides emission

4.13 Comparison of Brake Power and O_2

O_2 decreased with increase of brake bower, graphical representaion is shown in Fig 22. Deviations (5-10%) were found for RBOBD and its blends. At maximum load, O_2 (6.5) in B50 was less (25%) than O_2 (9.1) of diesel.

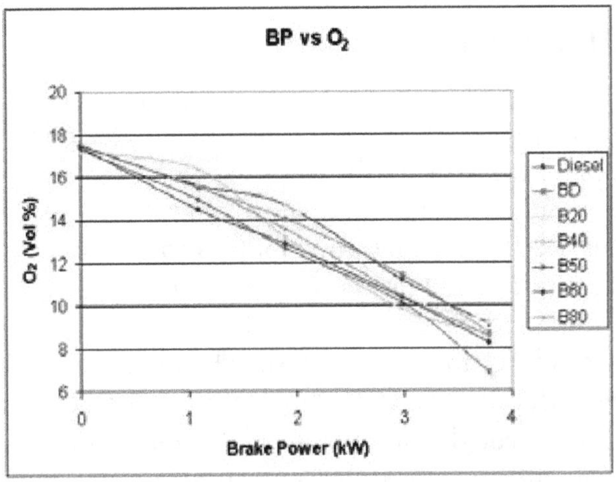

Fig. 22. Comparison of brake power and Oxygen

4.14 Comparison of Brake Power and sound

Sound or noise increased with increase of load, graphical representaion is shown in Fig 23. Sound values of RBOBD and its blends are found lower (15-30%) than the sound values of diesel throughout the brake power. Within comparison of RBOBD and its blends, there was not much change in sound in all the loads. The minimum decrease (13.6%) was observed at the minimum load, and the maximum decrease (30%) at the maximum load (3.78 kW). At the higher load, sound reduction (21-30%) for RBOBD and all of its blends compared to diesel.

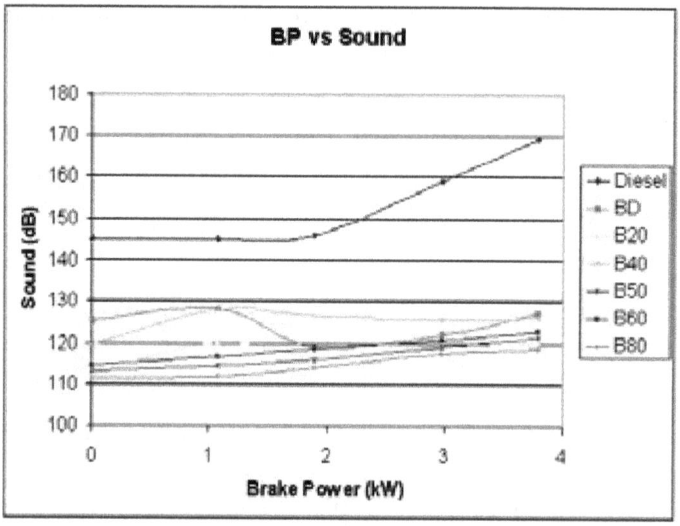

Fig. 23. Comparison of brake power and sound

4.15 Comparison of SEM image

During the engine trails, smoke was collected in the white filter paper. Few samples at the maximum load were taken for SEM analysis at different resolutions. SEM shows much reduction in the particulate matter in the biodiesel as compared to diesel (Fig. 24). The particle size in the smoke of RBOBD is less than 0.5 μm.

5. CONCISE SUMMARY

This is the most ambitious and successful program for the development of alternative energy technologies, which is a significant worldwide priority and will benefit the environment. These are the essential elements for energy security and yield beneficial economic, social, and environmental effects. Conventional fossil fuel energy sources are the primary contributors to climate change; biodiesel, in contrast, reduces emissions and promotes a sustainable society. The manufacture of biodiesel with this method could reduce imports of fossil fuels and reliance, thereby liberating funds for investment in social and economic development. This is the process innovation for the less NO_x emitting Biodiesel. An international Patent has been filed through the Intellectual Property Management Division of Council of Scientific and Industrial Research (CSIR), New Delhi in the year 2003.

(Ref No. IMPD. 0290 NF 2003, PCT / IB 03/ 05349 (21.11.2003, US &JP 20050108927 and World Intellectual Property Organization-WO/2005/052103)., Patent granted Firest in Singapore, Patent Number 119411 on 31st May 2006 and then Australia, patent Number-2003282270 and Sri Lanka Patent Number - 13950. In India, this is the first technological breakthrough in On-Road trials fuelled with biodiesel were successfully carried out in state government metropolitan Transport corporation (MTC) buses more than 26000 km without any engine modifications. There are 15% increases in the KMPL found after the trials. The MTC bus trials prove that this Biodiesel is substitute for diesel in the real conditions. On-Road trials fuelled with 100% Biodiesel (free biodiesel samples) by owner-driven in different vehicle models, namely, *Toyota Qualis, Mitsubishi Lancer, Bolero, Hyundai Acent etc.*

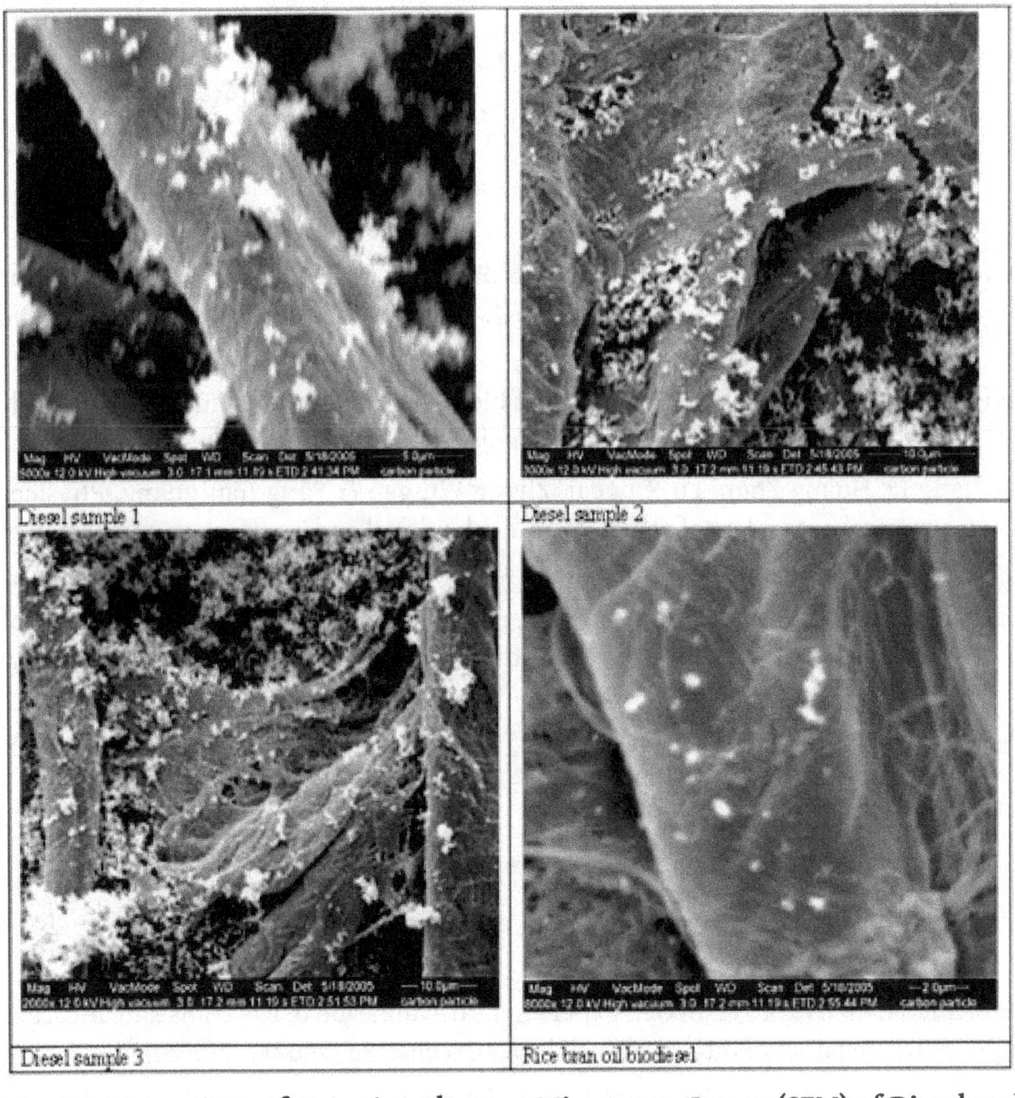

Fig. 24. Comparison of Scanning Electron Microscope Image (SEM) of Diesel and RBOBD

RBOBD was obtained by the optimized process conditions of transesterification and purification. RBOBD (> 150 l) is produced in Bench-Scale Parr reactor. RBOBD showed 15% increase in SFC, 25% increase in BTE, less exhaust gas temperature, 12% increase in FCT, more ITE and ME, 60% reduction in HC, more than 75% reduction in CO, minimum of 10% reduction in NO_x and 30 % sound reduction. The combustion of RBOBD and its blends produces fewer pollutants than diesel. Scanning Electron Microscopy (SEM) reveals a decrease in particulate matter, with solid particle dimensions measuring less than 0.5 µm. This study illustrates that RBOBD serves as an eco-friendly replacement fuel for diesel without requiring any modifications to the engine.

REFERENCES

Agarwal A K, Vegetable oils verses diesel fuels: Development and use of biodiesel in compression ignition engine, *TIDE*, 8 (1998) 191-204.

Antolin G, Tinaut F V, Briceno Y, Castano V, Perez C & Ramirez A I, Optimization of biodiesel production by sunflower oil transesterification, *Biores Technol*, 83 (2002) 111-114.

Arumugam C, Kumar S, Rajam L & Sundarresan A, Integrated technology for refining rice bran oil and by-products recovery, *6th Nat Semin on Rice Bran Oil*, Lucknow, 2003.

Barnwal B K & Sharma M P, Prospects of biodiesel production from vegetable oils in India, *Renewable & Sustainable Energy Rev*, 9 (2005) 363-378.

De-Gang Li, Huang Zhen, Lu Xingcai, Zhang Wu-gao & Yang Jian-guang, Physico-chemical properties of ethanol-diesel blend fuel and its effect on performance and emissions of diesel engines, *Renewable Energy*, 30 (2005) 967-976.

Fernando Neto da Silva, Antonio S P & Jorge R T, Technical feasibility assessment of oleic sunflower methyl ester Utilisation in diesel bus engine, *Energy Convers & Manage*, 44 (2003) 2857-2878.

Freedam B, Butterfield R O & Pryde E H, Transesterification kinetics of soyabean oil, *J Am Oil Chem Soc*, 63 (1986) 1375-1380.

Goering C E, Schwab A, Doughherty M, Pryde M & Heakin A, Fuel properties of eleven vegetable oils, *Trans ASAE*, 85 (1982)1472-1483.

Michael S G & Robert L M, Combustion of fat and vegetable oil derived fuels in diesel engines, *Prog Energy Combust Sci*, 24(1998) 125-164.

Mohamad I, Al-Widyan & Ali O A, Experimental evaluation of the transesterification of waste palm oil into biodiesel, *Biores Technol*, 85 (2002) 253-256.

Mohamad I, Al-widyan, Ghassan T & Moh'd A Q, Utilization of ethyl ester of waste oils as fuel in diesel engines, *Fuel Processing Technol*, 76 (2002) 91-103.

Noureddini H & Zhu D, Kinetics of transesterification of soybean oil, *J Am Chem Soc*, 74 (1997) 1457-1463.

Prasad Rao G A & Rama Mohan P, Effect of supercharging on performance of a DI diesel engine with cotton seed oil, *Int J Energy Conserv & Manage*, 44 937-940.

Roger A K & Jaiduk J O, A rapid engine test to measure injector fouling in diesel engine using vegetable oil fuels, *J Am Oil Chem Soc*, 62 (1985) 1563-1564.

Sinha S & Misra N C, Diesel fuel alternative from vegetable oils, *Chem Engg World*, 32 (1997) 77-80.

Srivastava A & Prasad R, Triglycerides-based diesel fuels, *Renewable & Sustainable Energy Rev*, 4 (2000) 111-133.

Sukumar P, Vedaraman N, Rambrahamam B V & Nagarajan G, Mahua (*Madhuca indica*) seed oil: A source of renewable energy in India, *J Sci Ind Res*, 64 (2005) 890-896.

Yi-Hsu J & Shaik R V, Rice bran oil as potential resource for biodiesel: A review, *J Sci Ind Res*, 64 (2005) 866-882.

16 Chapter

PRODUCTION OF BIODIESEL UTILIZING SOLID CATALYST

1. INTRODUCTION: AN OVERVIEW

Biodiesel is often synthesized through the transesterification of vegetable oils or animal fats using chemical catalysts, particularly in the presence of strong acidic or basic solutions, including hydrochloric acid, sulfuric acid, sodium hydroxide, sodium methoxide, and potassium hydroxide. Homogeneous alkali catalysts can efficiently convert triglycerides into their respective fatty acid methyl esters (FAMEs) with high yield, reduced time, and minimal cost. Nonetheless, isolating the catalyst from the product mixture for recycling presents significant technological challenges. Subsequent to the reaction, the catalyst must be neutralized or extracted using a substantial volume of hot water, resulting in a significant quantity of industrial wastewater.

Common plant oils, including soybean oil, rapeseed oil, and palm oil, constitute the primary edible oils. They are unsuitable as raw materials, especially in developing nations, because to restricted availability and elevated costs. Consequently, inexpensive lipids, including non-consumable oils (e.g., Jatropha oil), animal fats, and waste oils, are utilized as optimal feedstocks. Such oils typically contain water and free fatty acids (FFAs) that will produce soap when homogeneous base catalysts are employed. Conversely, homogeneous acid catalysts are corrosive to equipment. Solid heterogeneous catalysts are employed to address these issues, as they are non-corrosive, non-toxic, and readily separable for recycling. The reusability of heterogeneous catalysts enables continuous fixed-bed operation. This continuous method can reduce product separation and purification expenses, rendering it competitively competitive with conventional petroleum-based diesel fuel. This chapter delineates solid heterogeneous catalysts utilized in biodiesel generation and their conventional catalytic mechanisms..

2. HETEROGENEOUS SOLID CATALYSTS

In laboratory-scale experiment, heterogeneous processes could be run in a continuous mode with a packed-bed continuous flow reactor. Heterogeneous catalysts were easily separated from the products, water-washing process and neutralization steps were avoided. Contaminated water from this process was greatly reduced, and the sewage treatment fees were also minimized. New types of heterogeneous catalysts have mushroomed and developed in recent years.

2.1 Heterogeneous acid catalysts

Acid catalysts can simultaneously catalyze both esterification and transesterification, showing a much higher tolerance to FFAs and water than basic homogeneous catalysts (e.g., NaOH and KOH). Homogeneous acid-catalyzed reaction is about 4000 times slower than the homogeneous base-catalyzed reaction (Lotero et al., 2005). Heterogeneous acid catalysts performed less activety, but they are favorable for low-qualified oil feedstocks with high FFAs. Now, synthetic solid acids have already amounted to hundreds of species, most of them can be used in esterification and transesterification reactions. Solid acids keep stable activity in conversion of low-qualified oils or fats to biodiesel. Currently developed solid acid catalysts are introduced in the following sections: cation exchange resin (i.e. Amberlyst-15 and NR50), mineral salts (i.e. ferric sulfate, zirconium sulfate, alum phosphate and zirconium tungsten), supported solid acid and heteropolyacid catalysts.

2.1.1 Ion-exchange resins

Ion-exchange resins are widely used in important industrial processes for both separation and reaction applications. They are less expensive than lipase and supercritical methanol. Ion exchange resins also help to separate biodiesel and glycerol. Shibasaki-Kitakawa et al. (2007) found catalytic activity of anion-exchange resins correlated positively with cross- linking degree and particle size. The activity of acid ion-exchange resins for the esterification reaction is influenced by the accessibility of reactants to the matrix anchored sulfuric acid groups located at the surface or inside the resins. Mass-transfer restriction is another factor affecting catalytic activity. Internal diffusion was found to cause mass-transfer restriction and is rate-limiting for regular resins. Most of the active sites are embedded in the gel matrix, so the resins with macro-pores have high catalytic activity. Furthermore, the catalytic activity decreased when the cross-linking degree of polymeric matrix increased. Reusability is an important evaluation index for industrial applications of resins. Mechanical strength and thermo-stability are important for the large-scale applications of resins in biodiesel production. Ion exchange resins usually don't change catalytic property for long time operation at low temperature (< 100 °C).

Caetano et al. (2009) studied esterification of palmitic acid with methanol using poly(vinyl alcohol) cross-linked with sulfosuccinic acid (SSA) resin at 60°C, about 90% conversion rate was achieved after 2 h. Only about 5% sulfosuccinc acid was leached after 7 recycles. Activity of NKC-9 resin even slightly increased at the first 10 runs, due to breakdown of resin particles under mechanical agitation (Feng et al., 2010). Continuous production of biodiesel in a fixed-bed reactor packed with resins was successively operated (Shibasaki- Kitakawa et al., 2007; Liu et al., 2009; Feng et al., 2011). After 500 h, conversion yield of FFAs still kept over 98%. Amberlyst-15 performed high activity at 100 °C in the fixed-bed, and a 97.5% FAMEs yield was achieved (Son et al. 2011). Combination of fixed-bed reactor with supercritical CO_2 may develop a continuous process that is preferred for massive biodiesel production. Catalyst deactivation is caused by salt contaminants and water-swelling. Catalytic active sites on acidic resins can exchange with salt ions contained in oil. Traces of Na, K, Mg and Ca lead to a continuous activity loss (Russbueldt and Hoelderich, 2009). Deactivated catalyst can be recovered to its original activity by acid washing. It was found that temperature has negligible effect on water-swelling, but the water absorbed on the resin surface can be extracted by excessive methanol (Tesser et al., 2010)..

2.1.2 Zeolites

Zeolites are crystalline alumino-silicates with a three-dimensional porous structure. They can be synthesized with different crystal structures, definitive pore sizes, framework Si/Al ratios and adjustable acid centers to have some important catalytic properties. Aluminum atoms and ions in skeletons and porosity supply the acidic sites. Zeolites have extremely high internal surface area (600 m^2/s) and high thermal-stability (1000 °C), as the most popular solid catalysts. Acidic-shape selectivity is a significant feature of zeolite, derived from the influence of pore size and shape on a reaction. Zeolites were used in biodiesel production as a heterogeneous catalyst. Pérez-Pariente et al. (2003) studied the selective synthesis of fatty monoglycerides with Zeolites. Compared with reaction parameters, catalyst properties have more effects on monoglyceride yield.

Zeolite β is a high silica zeolite with both Lewis-acid sites and Brønsted-acid sites, containing an intersecting three-dimensional structure of 12-membered ring channels (Shu et al., 2007). Lewis-acid sites are mainly present in the micro-porous walls. On the contrast, Brønsted-acid sites are present on the internal and external surface. Zeolite β does not exhibit high activity in transesterificaiton, but it can be used for selective removal of FFAs in waste oil (Chung et al., 2008).

HY zeolite has a large number of weak acid sites. When one Si^{4+} is substituted by an Al^{3+}, the zeolite framework generates one Brønsted acid site. On the other hand, one Na^{+} cation neutralizes one acid site. F urthermore, hydroxyl groups formed by ion exchange of HY zeolite with ions, such as Ca^{2+}, Mg^{2+} and La^{3+}, can strengthen the acidic sites. Acidity of zeolite can also be adjusted by introducing protons with dilute hydrochloric acid. Sasidharan and Kumar (2004) found that large-pore zeolites such as Y, mordenite, and β showed higher activity (biodiesel yield 92%) than the medium-pore ZSM-5 and aluminum containing mesoporous MCM-41 (biodiesel yield < 30%). The high pore volume of large- pore zeolites favored reaction by rendering the active sites more accessible to the bulky triglyceride molecules. However, Hβ-Zeolite catalyzed transesterification of crude *Pongamia pinnata* oil gave low yield of 59% at a long reaction time (24 h) (Karmee and Chadha, 2005). Internal diffusion resistances are considered to limit reaction rate significantly. Thus, large- pore zeolites are active for the reaction with satisfactory reaction rate.

Most of zeolites exhibit not only acidic property, but they also provide high activity and selectivity in various acid catalysts as carrier. Bifunctional catalyst can usually be prepared by combining active catalytic sites on an acid zeolite. Shu et al. (2007) introduced La ion into zeolite β with $La(NO_3)_3$ as the ion exchange precursor. La/zeolite β resulted in higher conversion with higher stability than zeolite β because it has more external Brønsted acid sites available for the reactants. Triglyceride conversion yield of 48.9 wt% was obtained at 60 °C for reaction time of 4 h.

2.1.3 Heteropoly acids (HPAs)

A heteropoly acid is a class of acid made up of a particular combination of hydrogen and oxygen with certain metals (i.e., tungsten, molybdenum and vanadium) and non-metals (i.e., silicon, phosphorus). Heteropoly acid is frequently used as a re-usable acid catalyst in chemical reactions, but their long term stability and performances are not yet fully characterized. With inherent advantages of strong Brønsted acidity, stability and high proton mobility, HPAs are favorable as environmentally benign and economical solid catalysts. Owing to their unique physicochemical properties, HPAs are profitably used in homogeneous, biphasic and heterogeneous systems. There are many types of heteropolyacids, and the Keggin ($H_nXM_{12}O_{40}$) and Dawson ($H_nX_2M_{18}O_{62}$) structures are two of the better known groups (Kozhevnikov, 1998). HPAs (e.g., $H_3PW_{12}O_{40}$) are soluble in water and possess acidic strength as strong as sulfuric acid. HPAs solubility can be changed via alkali-exchange, and modified HPAs exhibiting significantly higher activity.

HPAs are excellent and environmentally benign acid catalyst for the production of biodiesel, which are tolerant to contaminations contained in oil resources such as FFAs and water. The Keggin HPA (i.e., $H_3PW_{12}O_{40}$) is soluble in methanol, and the use of Keggin heteropolyacids for triglyceride (trans)esterification has been reported. Alsalme et al. (2008) studied the intrinsic catalytic activity of Keggin HPAs, indicating activity of HPAs is significantly higher than that of the conventional acid catalysts in (trans)esterificaiton. Their acid strength in the descending order is as follows: $H_3PW_{12}O_{40}$ > $Cs_{2.5}H_{0.5}PW_{12}O_{40}$ > $H_4SiW_{12}O_{40}$ > 15%$H_3PW_{12}O_{40}/Nb_2O_5$, 15%$H_3PW_{12}O_{40}/ZrO_2$, 15%$H_3PW_{12}O_{40}/TiO_2$ > H_2SO_4 > HY, H-Beta > Amberlyst-15. HPA is able to efficiently promote the esterification with a similar performance to sulfuric acid. However, the recovery and reutilization of HPAs is difficult.

The main disadvantage of HPAs is their solubility in water and polar solvents. This problem can be overcome by converting it into its salt (e.g., ammonium salt) with decreases of acidity and catalytic activity. It is reported that partial exchange of ammonium salt in 12- tungstophosphoric acid with offers more acidic strength to the catalyst than the fully exchanged ammonium salt (Giri et al., 2005). Exchange of protons in HPA can help promote its activity in transesterification of triglycerides. The protons replacement has similar effects on activity as cations concentration increase. $Cs_xH_{3-x}PW_{12}O_{40}$ (x = 0.9-3), one kind of insoluble Keggin HPAs, offers excellent performance in (trans)esterification (Narasimharao et al., 2007). The catalytic activity of Cs-salts decreases as the content of Cs in HPW grows, due to the decrease of pH and the increase of conductivity of colloidal solutions in direct relation with the acidity of surface layers of primary particles. Furthermore, low-Cs loading on HPAs shows some dissolution of an active acid component after reflux in hot methanol, while high-Cs loading on HPAs is stable in hot methanol.

Immobilization of HPAs on carrier is also an efficiency method to obtain insoluble catalyst. Such supported solid acids performed high thermal-stability even under reaction conditions of 200 °C. Caetano et al. (2008) used tungstophosphoric acid, molibdophosphoric acid and tungstosilicic acid immobilized by sol-gel technique on silica to catalyze esterificaiton of palmitic acid with methanol. The higher heteropolyacids load on silica, the lower the catalytic activity is observed. Tungstophosphoric acid-silica (with 4.2 wt.%) showed the highest catalytic activity, 100% palmitic acid conversion was achieved after 30 h reaction time with methanol. Zięba et al. (2010) tested catalytic performance of Amberlyst-15, Nafion-SAC-13, polyaniline-sulfate, silver and cesium salts of HPAs in transesterifiaction of triglycerides with methanol. $Cs_2HPW_{12}O_{40}$ was the most active catalyst due to its highest strength of acid sites, but the great affinity toward glycerol led to its deactivation during recycling process.

2.1.4 Supported acid catalysts

Supports can provide higher surface area through the existence of pores where acidic sites can be anchored. Supports should be modified during preparation of catalysts to anchor catalytic species and obtain reusability. Furthermore, some amorphous carriers also showed good activity for (trans)esterification. Metal oxides are widely used as catalyst supports because of their thermal and mechanical stability, high specific surface area, and large pore size and pore volume. Because solid acids function the same as H^+ in sulfuric acid for (trans)esterification, sulphonated metal oxides, such as SO_4^{2-}/Al_2O_3, SO_4^{2-}/TiO_2, SO_4^{2-}/ZrO_2, SO_4^{2-}/SnO_2 and SO_4^{2-}/V_2O_5 (Garcia et al., 2008) can supply more acid species. Such solid acids are usually prepared by impregnating the hydroxides from ammonia precipitation of corresponding metal salt solutions with aqueous sulfuric acids followed by calcination. In addition to acid amount and acid strength adjustment, the catalysts are satisfactorily active in a heterogeneous liquid–solid system and are recoverable and reusable.

A cheap and high efficiency solid acid catalyst (SAC) derived from sulfonation of carbonized D-glucose or sucrose was reported, and used in transesterification of vegetable oil with alcohol (Shu et al., 2009; Toda et al., 2005; Zong et al., 2007). The catalyst was prepared from carbohydrates by carbonizing at 400 °C under N_2 atmosphere and then sulphonating at 150 °C. The solid acid catalyst can also be prepared by direct sulphonation of lignin consisting of polyethers and C-C linked phenylpropanes as shown in Fig. 1. The carbon carriers are amorphous, polycyclic aromatic carbon sheets containing SO_3H groups as active sites (Shu et al., 2009; Toda et al., 2005). The polycyclic carbon sheets can absorb long-chain hydrocarbon for reactants in solution to access SO_3H groups. Hydrolysis of cellulose to saccharides using such amorphous carbon bearing SO_3H, COOH, and OH function was studied (Suganuma et al., 2008). Phenolic OH groups bonded to the grapheme sheets can absorb β-1,4 glycosidic bonds and provide good access of reactants in solution to the SO_3H groups in the carbon material.

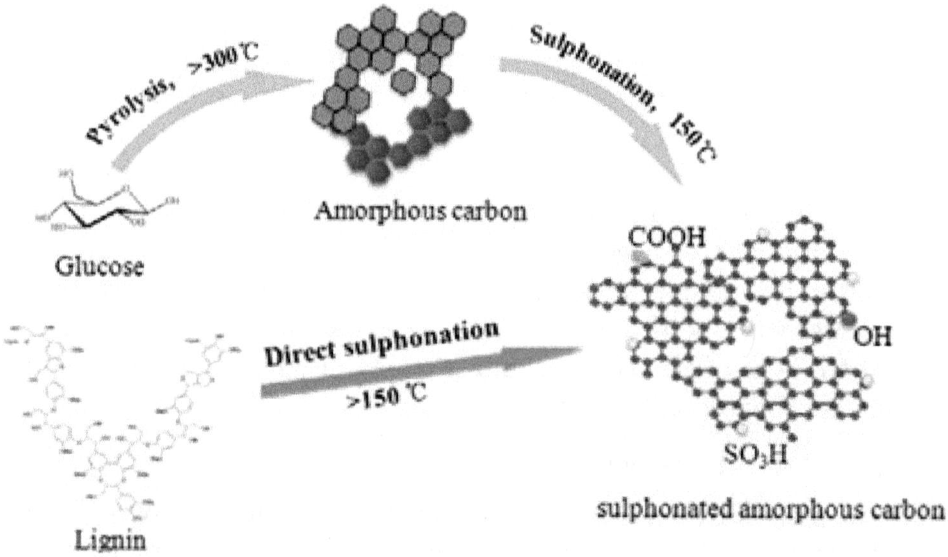

Fig. 1. Preparation process of sulphonated amorphous carbon from glucose and lignin.

Zong et al. (2007) utilized SAC as a solid acid catalyst for transesterification of waste oil (27.8% FFAs) with methanol. The reaction was carried out at 80 °C for 15 h, a high yield of above 90% obtained as compared with below 80% yield when sulfated ZrO_2, Amberlyst-15 and niobic acid were used. SAC was also used for other types of organics reactions, such as oxidations of organic compounds (e.g., sulfides, tertiary amines, aldehydes) with hydrogen peroxide (Shokrolahi et al., 2008). Specific surface area, pore size, pore volume and active site concentration on the surface of catalyst are effective factors on catalytic activity. Stability of the active sites is important for their industrial applications.

2.2 Heterogeneous base catalysts

The transesterification of vegetable oils or animal fats to biodiesel by chemical catalysts, especially in the presence of a strong basic solution, such as sodium hydroxide and potassium hydroxide, has been widely used in industrial production of biodiesel. Such basic solutions can transform triglycerides to their corresponding FAMEs with higher yield at lower temperature and shorter time than those by acid catalysts. However, separating the catalysts from products is technically difficult. Moreover, natural vegetable oils and animal fats usually contain small amounts of FFAs and water, which can have significant negative effects on the transesterification of glycerides with alcohols, and also hinder the separation of FAMEs and glycerol due to saponification of FFAs. Compared with basic solutions, solid base catalyst is preferred due to easy separation.

Heterogeneous base catalysis has a shorter history than that of heterogeneous acid catalysis. Solid bases refer mainly to solids with Brønsted basic and Lewis basic activity centers, that can supply electrons (or accept protons) for (or from) reactants. Heterogeneous base- catalyzed transesterification for biodiesel synthesis has been studied intensively over the last decade. Low-qualified oil or fat with FFAs and water can be used. However, the catalytic efficiency of conventional heterogeneous base catalysts is relative low and needs to be improved. Various types of catalytic materials have been studied to improve the transesterification of glycerides. Heterogeneous base catalysts, such as hydrotalcites, metal oxides, metallic salt, supported base catalyst and zeolites are introduced herein details.

2.2.1 Hydrotalcites

Hydrotalcites (HTs) are a class of anionic and basic clays known as layered double hydroxides (LDHs) with the formula $Mg_6Al_2(OH)_{16}CO_3 \cdot 4H_2O$. HTs consist of positively charged brucite-like layers and interstitial layers formed by CO_3^{2-} anions, and water molecules compensate the positive charge resulting from the substitution. LDHs have strong alkali sites and high stability with good adjustability of composition and structure. However, low surface area affected its catalytic activity. Mg/Al mole ratio and calcination temperature are the determining factors for the base-catalyzed activities. HTs with a 3:1 molar ratio of Mg to Al have the highest basicity and activity (Xie et al., 2006). Decomposition of HTs after calcination yields a high surface area Mg-Al mixed oxide, which presumably exposes strong Lewis basic sites. During calcination process, the interlayer water is lost first, followed by dehydroxylation and decomposition of interlayer carbonate to CO_2, which generate a porous structure and specific surface area ranging from 150 to

300 m^2/g (Lee et al., 2009). Furthermore, Mg^{2+} can be replaced by Zn^{2+}, Fe^{3+}, Co^{2+}, Ni^{2+}, and Al^{3+} by Cr^{3+}, Ga^{3+}, Fe^{3+}. HTs substituted with copper have a relatively uniform porous structure with decreased specific surface area. For iron substituted HTs, microporosity features developed, pore size decreased and specific surface area increased. The initial study by Cantrell et al. (2005) on biodiesel synthesis with HTs indicated magnesium content has obvious effects on catalytic activity. Increase of both magnesium content and electron- density enhances alkaline of HTs and finally increases biodiesel yield.

As classical solid base materials, calcined HTs were widely used as catalyst in the production of biodiesel (Brito et al., 2009; Deng et al., 2010). The basicity and surface area of HTs can be tuned by modifying chemical composition and preparation procedure. A co-precipitation method usually used to synthesize HTs with Mg/Al molar ratio of 3/1 using urea as precipitating agent. In previous work (Xie et al., 2006), transesterification process was carried out with reflux of methanol, methanol/soybean-oil molar ratio of 15/1, reaction time of 9 h and catalyst amount of 7.5%, and oil conversion rate was only 67%. In the work of Brito et al. (2009), waste oil as feedstock, biodiesel production was performed at temperatures ranging from 80 to 160 °C, methanol/oil molar ratio from 12/1 to 48/1 and catalyst concentration from 3 to 12%, respectively, and 90% biodiesel yield was achieved.

Fig. 2. SEM images of (A) hydrotalcite and (B) calcined hydrotalcite at 500 °C for 6 h.

It is known that HTs present lamellar structure thus not pose accessibility restrictions of vegetable oil molecules to catalyst sites. Improvement of specific surface area becomes necessary to obtain high catalytic activity. Deng et al. (2011) synthesized a series of nanosized HTs by a modified co-precipitation method. SEM images of HTs and calcined HTs were given in Fig. 2. Variables of temperature, solution pH and ageing time have a strong influence on the final basicity of the mixed oxides. Mg-Al ratio in the precursor HT depends on the basic properties of these sites. In the transesterification experiment using Mg-Al HT catalysts, 95% biodiesel yield was achieved from *Jatropha* oil in 1.5 h. Pre-mixture of HTs with methanol is essential to optimize catalyst activity to avoid lagging in reaction activity due to mass diffusion. Xi et al. (2008) tested influence of water on the activity and stability of activated Mg-Al HTs. In the presence of certain amount of interlayer water, Brønsted base sites were active. However, high degree of hydration caused rapid deactivation of the catalyst. Mg-Al HT shows relatively robust activity in the presence of water or FFAs tolerate, which is an attractive feature for biodiesel production.

2.2.2 Metal oxides

Metal oxides are composed of cations possessing Lewis acid and anions with Brønsted base. Metal oxides used in transesterification are classified as single metal oxides (e.g., MgO, CaO and SrO) and mixed metal oxides [A-B-O type metal oxides, where A is an alkaline-earth metal (Ca, Ba, Mg), alkaline metal (Li), or rare earth metal (La) and B is a transition metal (Ti, Mn, Fe, Zr, Ce)] (Kawashima et al., 2008; Liu et al., 2007; Liu et al., 2008; Montero et al., 2009). Early studies on heterogeneously catalyzed transesterification were focused on the catalysis by single metal oxides. The basicity of oxides (especially, basic sites) directly depends on reaction rate. A comparison of several metal oxides (MgO, CeO_2, La_2O_3 and ZnO) indicated that the most basic one is La_2O_3, followed by MgO, CeO_2 and ZnO (Bancquart et al., 2001).

The order of activity among alkaline earth oxide catalysts is BaO > SrO > CaO > MgO. CaO is the most frequently applied metal oxide catalyst for biodiesel preparation, due to its cheap price, relatively high basic strength and less environmental impacts. Reddy et al. (2006) used CaO as solid base catalyst in the transesterification of soybean oil, only 2% biodiesel yield achieved. However, the intrinsic basicity of nano-CaO is much higher, 99% biodiesel yield was obtained with nano-CaO. In addition to specific surface area, other variables such as temperature and molar ratio of methanol to oil also influenced the catalytic activity. A measured amount of water in oil is wonderful for promotion of catalytic activity. Study showed 95% biodiesel yield was obtained using CaO as catalyst in present of about 2 wt% water (Xi et al., 2008). SrO has basic sites stronger than H_0 = 26. It can catalyze many chemical reactions, such as oxidative coupling of methane, selective oxidation of propane, nitro-Aldol reactions and mixed Tishchenko reactions. Liu et al (2007) reported SrO performed high catalytic activity to convert soybean oil to biodiesel with a yield over 95% at temperature below 70 °C for 30 min. MgO has weak basic strength and low solubility in methanol. It is usually produced by direct heating of magnesium carbonate or magnesium hydroxide. There is a striking linear correspondence between the catalytic activity and surface basicity of MgO. High reaction temperature (e.g., 523 °C) and high pressure (e.g., 24 MPa) are usually needed for achieving a high biodiesel yield (Wang and Yang, 2007).

To increase the basic strength of a single metal oxide, mixed metal oxides are synthesized. Peterson et al. (1984) prepared CaO-MgO and found that it provided higher catalytic activity than CaO powders for transforming rapeseed oil to biodiesel. Catalytic activity tests were performed for CaO-TiO_2, CaO-MnO_2, $(CaO)_2$-Fe_2O_3, CaO-ZrO_2 and CaO-CeO_2 samples, approximately 90% biodiesel yield were obtained (Kawashima et al., 2008). The Ca catalysts were found to have higher basicity and activity. Such catalysts performed noticeably decreased activity in transesterificaiton when ethanol or branched alcohols was used, attributed to the steric effects on the catalytic activity of these catalysts. Furthermore, active sites of metal oxides are easily blocked by adsorbing intermediates (diglyceride, monoglyceride) or products. Deactivated catalysts can be recovered nearly to the initial value through calcination.

2.2.3 Metallic salts

Inorganic solid bases, such as sodium silicate (Guo et al., 2010), vanadyl phosphate (Serio et al., 2007), calcium zincate (Rubio-Caballero et al., 2009) and calcium methoxide (Liu et al., 2008), are low-cost and easy-to-use heterogeneous catalysts. Reports on metallic salts catalyzed conversion in biodiesel preparation are rare. Here, only sodium silicate, vanadyl phosphate (VOP) and calcium zincate are reviewed.

Sodium silicate was used as starting materials to synthesize γ-zeolite, NaY zeolite, and NaX zeolite. Guo et al. (2010) used sodium silicate to catalyze the transesterification reaction for the first time. It catalyzed soybean oil to biodiesel with a yield of almost 100% under the conditions: sodium silicate of 3.0 wt %, a molar ratio of methanol/oil of 7.5:1, reaction time of 60 min, reaction temperature of 60 °C, and stirring rate of 250 rpm. In addition to high catalyst activity, sodium silicate also has other similar characteristics to supported-solid base catalysts. Most of basic sites were in the interior of the solid catalyst due to low surface area and high density of the basic sites. The calcined sodium silicate could tolerate 4.0 wt% water or 2.5 wt% FFAs contained in soybean oil. The water tolerance is related to its special crystal and porous structure. In the presence of high amount of water, a sequential hydration will occur in three steps:

$$\begin{aligned} &\equiv Si-O-Na+H_2O \rightarrow \equiv Si-O-H+OH^- \\ &\equiv Si-O-Si\equiv +OH^- \rightarrow \equiv Si-O-H+\equiv Si-O^- \\ &\equiv Si-O^- +H_2O \rightarrow \equiv Si-O-H+OH^- \end{aligned} \quad (1)$$

As a result, Si-O-Si bridges would be hydrolyzed and H_4SiO_4 monomers are sequentially released. Such series of reactions not only produce OH^-, but also avoid the formation of soap. Furthermore, sodium silicate could also be used to catalyze dehydration of glycerol. Long et al. (2011) used sodium silicate as catalyst for transesterification of rapeseed oil for several recycles, and subsequently the used sodium silicate without any modification was catalyzed for the hydrothermal production of lactic acid from glycerol at 300 °C. A yield of 80.5% lactic acid and only minor amounts of formic, acetic acid and acrylic acid were produced.

Previous applications of VOP were mainly in hydrocarbon oxidation, dehydration and isomerization (Serio et al., 2006). Serio et al. (2007) confirmed that VOP-based catalysts were very active in the transesterification of vegetable oil with methanol despite their low specific surface area. VOP was deactivated due to a progressive reduction of vanadium (V) species from V^{5+} to V^{4+} and V^{3+} by methanol. Because the deactivation is reversible and catalyst activity can easily be restored by calcination. Rubio-Caballero et al. (2009) investigated the use of calcium zincate in the methanolysis of sunflower oil for biodiesel production. The activated calcium zincate at 400 °C is stable against lixiviation, attributed to its strong interaction with a much less soluble zinc oxide. But, calcium zincate is more sensitive to water (> 0.2 wt.%) rather than FFAs. Calcium methoxide has a moderate surface area, relative broad particle size distribution, narrow pore size distribution, strong basicity, long catalyst lifetime and better stability in organic solvent (Liu et al., 2008). It has tremendous potential to replace some homogeneous catalysts.

2.2.4 Supported base catalysts

Alkali metals (Li, Na, K) and alkaline earth metals (Mg, Ca, Ba) are the most common sources of super basicity, and selected as the active species of supported catalysts for biodiesel synthesis. They are frequently used in the metallic form or as various ionic forms of hydroxide, halide, carbonate and nitrate, such as K^+, Li^+ La^{3+}, KOH, NaOH, KF, K_2CO_3, KNO_3 (Shu et al., 2007; Sun et al., 2008; Vyas et al., 2009). Alumina, silica, zinc oxide, zirconium oxide and zeolite were used as supports for these catalysts. Surface basicity is the primary determinant of catalyst activity, then the specific surface area and pore volume (Sun et al., 2008). During the preparation of such catalysts, the mechanical intensity and surface area of carriers can be adjusted to obtain different basic intensities and activity sites.

Almost all supported base catalysts were synthesized via loading of active species on carriers by covalent bond, ionic bond or physical adsorption. Despite of formation of M-O-carrier (e.g., Al-O-K, Si-O-Na and Ca-O-K), other possible interactions of the alkali species with supports include formation of solid solutions and acid-base reactions. Hydroxyl groups introduced to the surface of solids play an important role in transesterification reaction (Xie et al. 2006). The hydroxyls with alkali species enhance the catalytic activity.

As a most popular carrier, Al_2O_3 has almost all noteworthy properties such as high temperature resistant, high surface area, high porosity, low density and transition crystalline phase existed in a wide temperature range. Furthermore, it serves as carrier with both solid acid and base. Most super basicity sources can be well dispersed on the Al_2O_3 support in the form of a monolayer at a low loading. Furthermore, alumina is more resistant than other supports (e.g., SiO_2, CaO and zeolite) for alkali species. Taking KNO_3/Al_2O_3 as an example, it is usually prepared by impregnation and subsequent calcination at 500 °C (Xie et al. 2006). K^+ ions replaced protons of isolated hydroxyl groups to form Al-O-K groups. The Al-O-K groups and K_2O derived from KNO_3 are active basic species. The base strength could be tentatively denoted as $15 < H_0 < 18.4$ by using Hammett indicator. Basic strength of KNO_3/Al_2O_3 was influenced by KNO_3 loading and temperature. The 35% KNO_3/Al_2O_3 sample calcined at 500 °C had the highest basicity. However, the sample prepared at 700 °C was most stable. Because part of potassium species are loss by a solid-solid reaction leading to formation of spinels or penetration into the subsurface.

Aends and Sheldon (2001) indicated that such kind of catalyst is unstable during reaction, mainly due to M-O-Al decomposed in present of methanol. Arzamendi et al (2007) confirmed that NaOH reacted with the support to form aluminates during preparation of $NaOH/Al_2O_3$. Leaching of sodium species from Al_2O_3 was also found. Furthermore, problems of high cost, difficult preparation and easy poisoning by absorption of H_2O and CO_2 should be solved. The supported solid base catalysts are excellent for transesterification of triglyceride, but a higher temperature is needed.

3. CATALYTIC MECHANISM

3.1 Heterogeneous solid acid-catalyzed esterification mechanism

Low-cost feedstocks need pretreatment (esterification) to remove FFAs before base- catalyzed transesterification reaction. The esterification path is relatively simple reversible reaction as follows:

(2)

In the reaction (2), FFA is converted to FAME. When homogenous acid (e.g., sulfonate acid, phosphorus acid and hydrochloric acid) was used, esterification reaction is a process that FFA supply hydroxide and methanol supply proton without intermediate process.

Different to homogeneous catalysis, heterogeneous catalytic process is known to follow a carbonium ion mechanism. The mechanism of solid acid-catalyzed esterification consists of following steps as shown in Fig. 3. Firstly, solid catalysts provided protons, and carbonyl carbon was protonated. Next, nucleophilic attack of CH_3OH on the carbonium ion formed a tetrahedral intermediate. Finally, FAME was produced after proton migrated and the intermediate broke down, and proton was reformed.

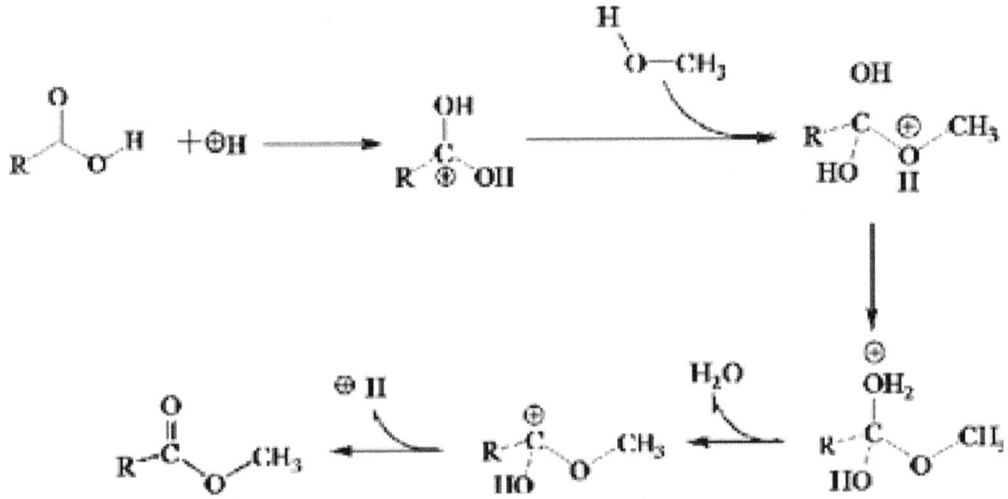

Fig. 3. Solid acid-catalyzed reaction mechanism of esterification.

The esterification reaction path is slightly different in various acidic species types. The whole reaction process is through proton-exchange. Tesser et al. (2005) proposed a kinetic model based on the following hypotheses: (1) major part of the active sites are occupied by methanol in a protonated form, and the rest part are also occupied; (2) fatty acid, water and methyl ester reach proton-exchange equilibrium with the protonated methanol; (3) inside the resin particles, an Eley-Rideal mechanism occurs between protonated fatty acid and the methanol. Deviate from the mechanism shown in Fig. 3, steps of protonation of carbonyl carbon, nucleophilic attack, proton migration and breakdown of intermediate are undergoing in a proton-exchange way.

3.2 Transesterification mechanism

The transesterification reaction involves catalytic reaction between triglyceride and alcohol (e.g., methanol, ethanol, propanol and butanol) to form biodiesel (FAMEs) and glycerol (Fig. 4). In the reaction, three consecutive reactions are required to complete the transesterification of a triglyceride molecule. In the presence of acid or base, a triglyceride molecule reacts with an alcohol molecule to produce a diglyceride and FAME. Then, a diglyceride reacts with alcohol to form a monoglyceride and FAME. Finally, an monoglyceride reacts with alcohol to form FAME and glycerol. Diglyceride and monoglyceride are the intermediates in this process.

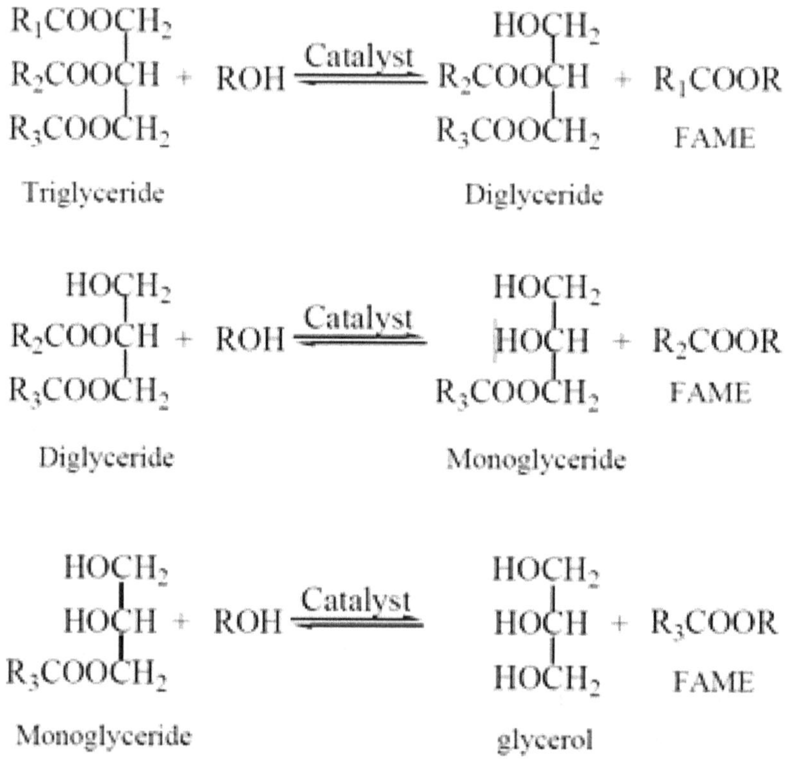

Fig. 4. Transesterification reactions of glycosides with alcohol.

3.2.1 Mechanism for heterogeneous acid-catalyzed transesterification

Acidic or basic functional groups in the active sites of solid catalysts catalyze the reaction by donating or accepting protons. Acid-catalyzed reaction mechanism for the transesterification of triglycerides is shown in Fig. 5. Firstly, triglycerides are protonated at the carbonyl group on the surface of solid acid. Then, a nucleophilic attack of the alcohol to carbocation forms a tetrahedral intermediate (hemiacetal species). Unstable tetrahedral intermediate leads to proton migration, followed by breakdown of the tetrahedral intermediate with assistance of solvent. After repeating twice, three new FAME as products were produced and the catalyst was regenerate. During the catalytic process, protonation of carbonyl group boosts the catalytic effect of solid acid catalyst by increasing the electrophilicity of the adjacent carbonyl carbon atom.

Different with Brønsted acids, Lewis acids [e.g., $Fe_2(SO_4)_3$, titanate complexes, carboxylic salts, divalent metal pyrone] act as electron-acceptors via the formation of a four-membered ring transition state (Abreu et al., 2004; Di Serio et al., 2005). The reactant triglyceride and metal form a Lewis complex, which assists solid Lewis acids during process of the carbonyl groups activating for a nucleophilic attack by the reactant alcohol. The triglyceride carbonyl coordinates at a vacant site in the catalytic active specie. Formation of a more electrophilic species is responsible for the catalytic activity. Stearate metals (Ca, Ba, Mg, Cd, Mn, Pb, Zn, Co and Ni) were tested as catalysts for methanolysis of soybean oil (2.0 g) with methanol (0.88 g) at 200 °C (Di Serio et al., 2005). A high FAMEs yield (96%) and a low final FFAs concentration (<1%) were obtained in a relatively short reaction time (200 min).

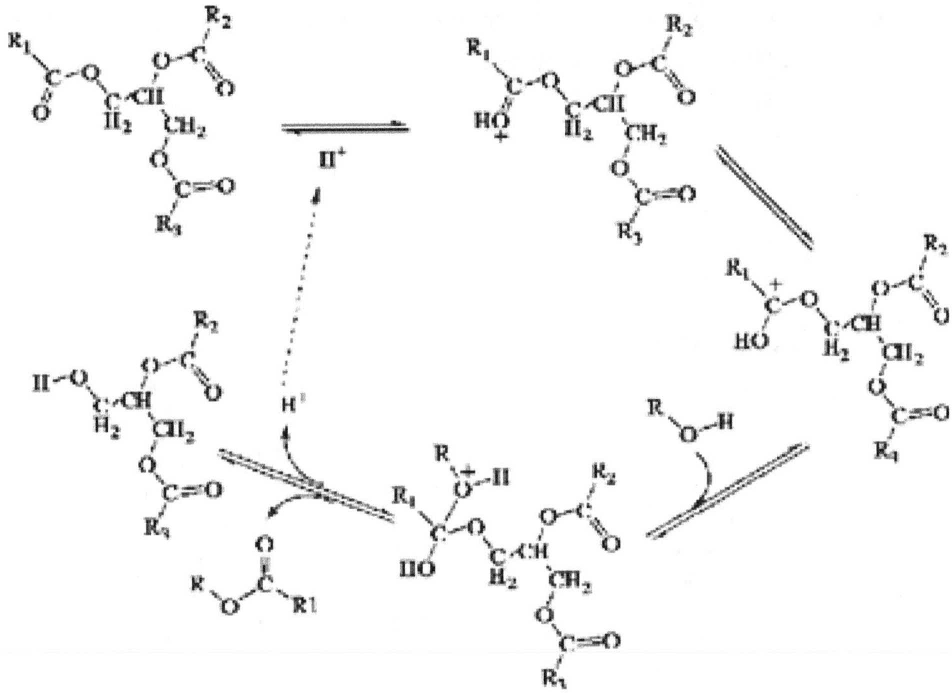

Fig. 5. Acid-catalyzed reaction mechanism of transesterification.

3.2.2 Mechanism for heterogeneous base-catalyzed transesterification

Base-catalyzed crude oil to biodiesel gets more studies than acid-catalyzed method. In base- catalyzed process, OH^- or CH_3O^- ions performed as active species. Catalytic reactions started on the surface of heterogeneous base (Fig. 6). The mechanistic pathway for solid base- catalyzed transesterification seems to follow a similar mechanism to that of a homogeneous base catalyst. First, ion-exchange proceeded after methanol absorbed on the surface of solid base, producing catalytic active specie (CH_3O^-) which is strongly basic and highly catalytic active. Secondly, nucleophilic attack of CH_3O^- on the carbonyl carbon of triglyceride formed a tetrahedral intermediate. Thirdly, rearrangement of the intermediate resulted in the formation of FAME. Finally, protons were converted to diglyceride ion to generate diglyceride. This sequence was then repeated twice to yield glycerol and biodiesel.

Formation of CH_3O^- is different according to solid base types. Taking CaO as an example, surface O^{2-} is the basic site, which can extract H^+ from H_2O to form

OH-, and OH- extracts H+ from methanol to generate CH_3O^- (Liu et al., 2008). It is interesting that CaO generates more methoxide anions in the presence of a little water (less than 2.8% by weight of crude oil), avoiding formation of soap. Surface oxides or hydroxide groups depend on the basicity and catalytic activities. The basic strengths of Na/CaO and K/CaO are slightly lower than that of Li/CaO (Ma and Hanna, 1999). The presence of the electron-deficient M^+ on the support enhances the basicity and activity of the catalysts towards the transesterification reaction.

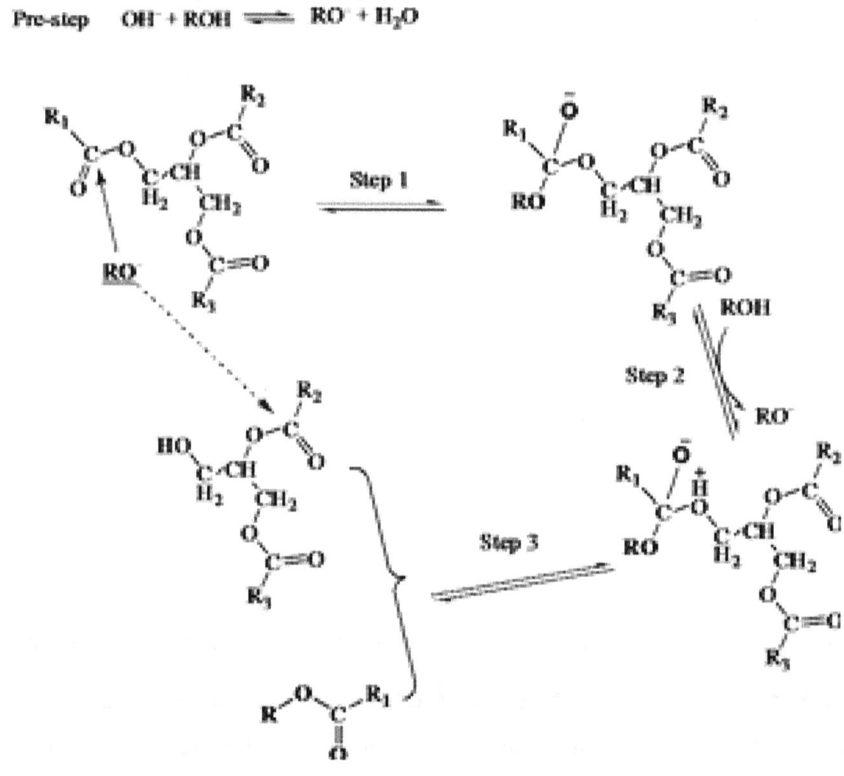

Fig. 6. Base-catalyzed reaction mechanism of transesterification.

4. OTHER METHODS OR TECHNOLOGIES

4.1 Microwave technology

Microwave heating has been widely used in many areas to affect chemical reaction pathways and accelerate chemical reaction rates. Microwave irradiation can accelerate the chemical reaction, and high product yield can be achieved in a short time. Microwave irradiation assisted biodiesel synthesis is a physicochemical process since both thermal and non-thermal effects are often involved, which activates the smallest degree of variance of polar molecules and ions such as alcohol with the continuously changing magnetic field. Upon microwave heating, rapid rising of temperature would result in interactions of changing electrical field with the molecular dipoles and charged ion, leading to a rapid generation of rotation and heat due to molecular friction. Dielectric properties are important in both the design calculations for high frequency and microwave heating equipment. Furthermore, dielectric constant depends on frequency, and is strongly influenced by temperature, mixed ratio and solvent type.

In Azcan and Danisman's work (2007), microwave heating effectively reduced reaction time from 30 min (for a conventional heating system) to 7 min. Ozturk et al. (2010) studied microwave assisted transesterification of maize oil, using a molar ratio alcohol/maize-oil of 10:1, and 1.5% w/w NaOH as catalyst. A 98.3% conversion rate is obtained using methanol for 5 min. Based on special heating manner, microwave irradiation performed well in transesterification of vegetable oil with heterogeneous base. Hsiao et al. (2011) introduced nano-powder calcium oxide as solid base in converting soybean oil to biodiesel. A 96.6% of conversion rate was obtained under conditions of methanol/oil molar ratio of 7:1, amount of catalyst of 3.0 wt.%, reaction temperature of 65 °C and reaction time of 60 min. While a biodiesel conversion rate exceeded 95% was achieved under conditions of 12:1 molar ratio of methanol to oil, 8 wt.% catalyst, 65 °C reaction temperature and 2.0% water content for 3 h (Xie et al., 2008). Microwave irradiation is also used for extraction of bioactive compounds for value-added products, including oil extraction systems. Microwave heating can be used for biodiesel production by in-situ simultaneous extraction and transesterification from oil seeds.

4.2 Ultrasonic technology

There are three primary effects on an object under ultrasound: (1) Mechanical effects; (2) Cavity effects; (3) Thermal effects. The above effects of ultrasound not only change the structure of the object, but also lead to chemical reactions. Ultrasonic radiation is a relative new technique that results in the formation and collapse of micro-scale bubbles in liquid to generate local high temperature and high pressure. So, it is used as alternative energy source to promote reactions. The cavitation in ultrasonic wavelength is the phenomenon of expansion and contraction of the transfer media bubbles. Ultrasonic energy is propagated into solution by the destruction of pressurized micro-bubbles into small droplets. Furthermore, ultrasonication device placed near the liquid–liquid interface in a two-phase reaction system benefited for producing large interfacial areas (Wu et al., 2007). Cavitation induced by ultrasound has significant effects on liquid phase reactions. When ultrasound irradiation increased from 30 to 70 W, the mean droplet size decreased from 156 nm to 146 nm. Nevertheless, effect of droplet size on biodiesel yield was not studied.

Ultrasound has a short wavelength, slow transfer rate, and high energy transmittance as the vibrating type energy. Irradiation of ultrasonic energy has been used for the (trans)esterification of vegetable oils to shorten reaction time and to increase product yield (Deng et al., 2010). A comparison study between conventional and ultrasonic preparation of beef tallow biodiesel was carried out (Teixeira et al., 2009). The results showed that conversion rate and biodiesel quality were similar. The use of ultrasonic irradiation decreased reaction time from 1 h to 70 s. In addition to the mentioned advantages, ultrasonic can promote the deposition of glycerol at the bottom of reactor. Stavarache et al. (2007) investigated a bench-scale continuous process for biodiesel synthesis from neat vegetable oils under high power, low frequency ultrasonic irradiation. Reaction time and alcohol-oil molar ratio were mainly variables affecting the transesterification. Their research confirmed that ultrasonic irradiation is suitable for large-scale processing of vegetable oils since relatively simple devices can be used to perform the reaction. In the process, however, real irradiation time decreased during increasing pulse interval for tuning temperature, leading to biodiesel yield decrease. To reduce the effect of irradiation time loss, reaction temperature should be kept constant.

Mass transfer resistance is one of the main reasons for poor catalytic performance of solid catalysts in (trans)esterification. Very fine ultrasonic emulsions greatly improve the interfacial area available for reaction, increase the effective local concentration of reactive species, and enhance the mass-transfer in interfacial region. Therefore it leads to a remarkable increase in reaction rate under phase-transfer conditions transesterification with solid catalyst. Ultrasonication could reduce the transesterification reaction time to around 10 min compared with over 6 h for conventional processing.

4.3 Ionic liquids

Ionic liquids (ILs) are defined as salts that are in the state of liquid at low temperatures (below 100 °C). They are composed solely of cations and anions, and were used as solvents/catalysts for reactions. ILs are nonvolatile and thermal stable, hence they are excellent alternatives to traditional solvents. Some ILs are Lewis and Franklin acids. Acidic ILs are new-type of catalysts with high-density active sites as liquid acids but non- volatilization as solid acids. Furthermore, cations and anions of ILs can be designed to bind a series of groups with specific properties, so as to achieve the purpose of regulating the acidity. Recently, they have been used to replace traditional liquid acids such as sulfuric acid and hydrochloric acid for biomass conversion (Qi et al., 2010).

ILs were originally used as solvents for biodiesel synthesis with high biodiesel yield in short reaction time, by forming an effective biphasic catalytic system for the transesterification reaction. Neto et al. (2007) introduced a complex [Sn(3-hydroxy-2-methyl-4-pyrone)$_2$(H_2O)$_2$] immobilized in BMI·$InCl_4$ with high price metal salts, and a maximum biodiesel yield of 83% was achieved. Later, biodiesel synthesis from vegetable oils using imidazolium-based ionic liquids under multiphase acidic and basic conditions was reported (Lapis et al., 2008). It is found that the acid is almost completely retained in ionic liquid phase, and ILs could be reused at least six times without any significant loss in the biodiesel yield or selectivity. However, the ILs is expensive and was only used for neutral vegetable oils. Brønsted acidic ILs were highly efficient catalysts for biodiesel synthesis from vegetable oils. Sulfuric acid groups in these ILs are the active sites for transesterification. Dicationic ILs exhibited better stability than the traditional ones. The acidic dicationic ILs with an alkane sulfuric acid group gave a superior catalytic performance in esterification reaction. Neto et al. (2007) assumed that the use of ILs with inherent Lewis acidity may constitute a more stable and robust catalytic system for the transesterification reaction. Guo et al. (2011) used 7 low-cost commercial ILs as both catalysts and solvents for the direct production of biodiesel from un-pretreated *Jatropha* oil. It was found that [BMIm][CH_3SO_3] had the highest catalytic activity with 93% of oleic acid being converted into ethyl oleate. When $FeCl_3$ was added to [BMIm][CH_3SO_3], a maximum biodiesel yield of 99.7% was achieved from un-pretreated *Jatropha* oil. However, it is complicated to synthesize these functional ILs and their cost is too high for industrial applications. Therefore, further investigation is necessary to synthesize inexpensive, stable and highly-active ILs.

5. CONCISE SUMMARY

At now, homogeneous catalysis is the primary technique for the transesterification reaction. Isolating the catalyst from a combination of reactants and products presents significant technical challenges. In comparison to liquid acid catalysts, solid acid catalysts has notable advantages in terms of recyclability, separation, and environmental sustainability. Solid acid catalysts can be readily isolated from the resulting mixture for subsequent usage following the reaction. Both Lewis acid-base sites and Brønsted acid-base sites possess the capability to catalyze the transesterification process of oil. In addition to specific surface area, pore size, and pore volume, the concentration of active sites and the kind of acidity are critical determinants of solid acid efficacy. Furthermore, the nature of active precursors significantly influences the catalytic activity of supported catalysts. Nevertheless, the concentration of the active site was determined to be the paramount element influencing the performance of solid catalysts. Solid acids with significant potential for biodiesel synthesis must possess a high density of Brønsted acid sites and have excellent thermal stability. An effective and robust catalyst with adequate catalytic properties.The integration of activity with suitable reactor architecture should enable the practical-scale generation of biodiesel..

Among solid catalysts introduced in this chapter, Solid acid (i.e. ion-exchange resins, HPAs and supported acid catalysts) and Solid base (i.e. hydrotalcites, metallic salts and supported base catalysts) are promising material for study. Low-cost catalysts that still retain the advantages of a supported base catalyst should be developed to simplify the preparation process. The development of solid catalysts with enhanced activity is a crucial advancement for the sustainable generation of biodiesel. Advancements and breakthroughs in the hydrolysis process are essential for the commercialization of solid acid catalysts. In the near future, the integration of green solvents, chemical processes, biotechnology, and catalysis is anticipated to result in the emergence of novel solid catalysts that will supplant the currently utilized homogeneous catalysts in biodiesel production.

REFERENCES

Abreu, F.R., Lima, D.G., Hamúa, E.H., Wolf, C., Suarez, P.A.Z. (2004) Utilization of metal complexes as catalysts in the transesterification of Brazilian vegetable oils with different alcohols. *Journal of Molecular Catalysis A: Chemical*, Vol.209, No.1-2, pp. 29- 33.

Aends, I.W.C.E., Sheldon, R.A. (2001) Activities and stabilities of heterogeneous catalysts in selective liquid phase oxidations: recent developments. *Applied catalysis A: General*, Vol.212, No.1-2, pp. 175-187.

Alsalme, A., Kozhevnikova, E.F., Kozhevnikov, I.V. (2008) Heteropoly acids as catalysts for liquid-phase esterification and transesterification. *Applied Catalysis A: General*, Vol.349, pp. 170-176.

Arzamendi, G., Campo, I., Arguínarena, E., Sánchez, M., Montes, M., Gandá, L.M. (2007) Synthesis of biodiesel with heterogeneous NaOH/alumina catalysts: Comparison with homogeneous NaOH. *Chemical Engineering Journal*, Vol.134, pp 123-130.

Azcan, N., Danisman, A. (2007) Alkali catalyzed transesterification of cottonseed oil by microwave irradiation. *Fuel*, Vol.86, pp. 2639-2644.

Bancquart,S.,Vanhove,C.,Pouilloux,Y.,Barrault,J.(2001)Glyceroltransesterification with methyl stearate over solid basic catalysts: I. Relationship between activity and basicity. *Applied Catalysis A: General*, Vol.218, pp 1-11.

Brito, A., Borges, M.E., Garín, M., Hernández, A. (2009) Biodiesel production from waste oil using Mg-Al layered doublehydrotalcite catalysts. Energy&Fuels, Vol.23, pp. 2952- 2958.

Caetano, C.S., Fonseca, I.M., Ramos, A.M., Vital, J., Castanheiro, J.E. (2008) Esterification of free fatty acids with methanol using heteropolyacids immobilized on silica. *Catalysis Communications*, Vol.9, pp. 1996-1999.

Caetano, C.S., Guerreiro, L., Fonseca, I.M., Ramos, A.M., Vital, J., Castanheiro, J.E. (2009) Esterification of fatty acids to biodiesel over polymers with sulfonic acid groups. *Applied Catalysis A: General*, Vol.359, No.1-2, pp. 41-46.

Cantrell, D.G., Gilie, L.J., Lee, A.F., Wilson, K. (2005) Structure-reactivity correlations in Mg- Al hydrotalcite catalysts for biodiesel synthesis. *Applied Catalysis A: General*, Vol.287, No. 2, pp. 183-190.

Chung, K.H., Chang, D.R., Park, B.G. (2008) Removal of free fatty acid in waste frying oil by esterification with methanol on zeolite catalysts. *Bioresource Technology*, Vol.99, pp. 7438-7443.

Deng, X., Fang, Z., Liu, Y.H. (2010) Ultrasonic transesterification of Jatropha curcas L. oil to biodiesel by a two-step process. *Energy Conversion and Management*, Vol.51, pp. 2802-2807.

Deng, X., Fang, Z., Liu, Y.H., Yu, C.L. (2011) Production of biodiesel from Jatropha oil catalyzed by nanosized solid basic catalyst. *Energy*, Vol.36, No.2, pp. 777-784.

Di Serio, M., Dimiccoli, M., Cammarota, F., Nastasi, M., Santacesaria, E. (2005) Synthesis of biodiesel via homogeneous Lewis acid catalyst. *Journal of Molecular Catalysis A: Chemical*, Vol.239, pp.111-115.

Feng, Y.H., He, B.Q., Cao, Y.H., Li, J.X., Liu, M., Yan, F., Liang, X.P. (2010) Biodiesel production using cation-exchange resin as heterogeneous catalyst. *Bioresource Technology*, Vol.101, pp. 1518-1521.

Feng, Y.H., Zhang, A.Q., Li, J.X., He, B.Q. (2011) A continuous process for biodiesel production in a fixed bed reactor packed with cation-exchange resin as heterogeneous catalyst. *Bioresource Technology*, Vol. 102, pp.3607-3609.

Garcia, C.M., Teixeira, S., Marciniuk, L.L., Schuchardt, U. (2008) Transesterification of soybean oil catalyzed by sulfated zirconia. *Bioresource Technology*, Vol. 99, pp. 6608- 6613.

Giri, B.Y., Rao, K.N., Devi, B.L.A.P., Lingaiah, N., Suryanarayana, I., Prasad, R.B.N., Prasad, P.S.S. (2005) Esterification of palmitic acid on the ammonium salt of 12- tungstophosphoric acid: The influence of partial proton exchange on the activity of the catalyst. *Catalysis Communications*, Vol. 6, pp. 788-792.

Guo, F., Fang, Z., Tian, X.F., Long, Y.D., Jiang, L.Q. (2011) One-step production of biodiesel from Jatropha oil with high-acid value in ionic liquids. *Bioresource Technology*, Vol.102, No.11, pp.6469-6472.

Guo, F., Peng, Z.G., Dai, J.Y., Xiu, Z.L. (2010) Calcined sodium silicate as solid base catalyst for biodiesel production. *Fuel Processing Technology*, Vol. 91,No.33, pp.322-328.

Hsiao, M.C., Lin, C.C., Chang, Y.H. (2011) Microwave irradiation-assisted transesterification of soybean oil to biodiesel catalyzed by nanopowder calcium oxide. *Fuel*, Vol.90, pp.1963-1967.

Karmee, K.S., Chadha, A. (2005) Preparation of biodiesel from crude oil of *Pongamia pinnata*. *Bioresource Technology*, Vol.96, pp.1425-1429.

Kawashima, A., Matsubara, K., Honda, K. (2008) Development of heterogeneous base catalysts for biodiesel production. *Bioresource Technology*, Vol.99, pp.3439-3443.

Kozhevnikov, I.V. (1998) Catalysis by heteropoly acids and multicomponent polyoxometalates in liquid-phase reactions. *Chemical Reviews*, 98: 171-198.

Lapis, A.A., Oliveira, L.F., Neto, B.A.D., Dupont, J. (2008) Ionic liquid supported acid/ base- catalyzed production of biodiesel. *ChemSusChem*, Vol.1, pp.759-762.

Lee, D.W., Park, Y.M., Lee, K.Y. (2009) Heterogeneous Base Catalysts for Transesterification in biodiesel synthesis. *Catal Surv Asia*, Vol.13, pp.63-77.

Liu, X., He, H., Wang, Y., Zhu, S.L. (2007) Transesterification of soybean oil to biodiesel using SrO as a solid base catalyst. *Catalysis Communications*, Vol.8, pp. 1107-1111.

Liu, X.J., He, H.Y., Wang, Y.J., Zhu, S.L., Piao, X. (2008) Transesterification of soybean oil to biodiesel using CaO as a solid base catalyst. *Fuel*, Vol.87, pp.216-221.

Liu, X.J., Piao, X.G., Wang, Y.J., Zhu, S.L., He, H.Y. (2008) Calcium methoxide as a solid base catalyst for the transesterification of soybean oil to biodiesel with methanol. *Fuel*, Vol.87, pp.1076-1082.

Liu, Y., Wang, L., Yan, Y. (2009) Biodiesel synthesis combining pre-esterification with alkali catalyzed process from rapeseed oil deodorizer distillate. *Fuel Processing Technology*, Vol.90, pp. 857-862.

Lotero, E., Liu, Y.J., Lopez, D.E. (2005) Suwannakarn K, Bruce DA, Goodwin JG. Synthesis of biodiesel via acid catalysis. *Industrial & Engineering Chemistry Research*, Vol.44, No.14, pp. 5353-5363.

Long, Y.D., Guo, F., Fang, Z., Tian, X.F., Jiang, L.Q., Deng, X., Zhang, F. (2011) Production of biodiesel and lactic acid from rapeseed oil using sodium silicate as catalyst. *Bioresource Technology*, Vol.102, pp.6884-6886.

Ma, F., Hanna, M.A. (1999) Biodiesel production: a review. *Bioresource Technology*, 70: 1-15.

Montero, J.M., Gai, P., Wilson, K., Lee, A.F. (2009) Structure-sensitive biodiesel synthesis over MgO nanocrystals. *Green Chemistry*, Vol.11, pp. 265-268.

Narasimharao, K., Brown, D.R., Lee, A.F., Newman, A.D., Siril, P.F., Tavener, S.J., Wilsona, K. (2007) Structure-activity relations in Cs-doped heteropolyacid catalysts for biodiesel production. *Journal of Catalysis*, Vol.248, pp.226-234.

Neto, B.A.D., Alves, M.B., Lapis, A.A.M., Nachtigall, F.M., Eberlin, M.N., Dupont, J., Suarez, P.A.Z. (2007) 1-n-Butyl-3-methylimidazolium tetrachloro-indate (BMI·$InCl_4$) as a media for the synthesis of biodiesel from vegetable oils. *Journal of Catalysis*, Vol.249, pp.154-161.

Noiroj, K., Intarapong, P., Luengnaruemitchai, A., Jai-In, S. (2009) A comparative study of KOH/Al_2O_3 and KOH/NaY catalysts for biodiesel production via transesterification from palm oil. *Renewable Energy*, Vol.34, pp.1145-1150.

Okuhara, T., Nishimura, T., Watanabea, H., Misono, M. (1992) Insoluble heteropoly compounds as highly active catalysts for liquid-phase reactions. *Journal of Molecular Catalysis*, Vol.74, No.1-3, pp. 247-256.

Ozturk, G., Kafadar, A.B., Duz, M.Z., Saydut, A., Hamamci, C. (2010) Microwave assisted transesterification of maize (*Zea mays L.*) oil as a biodiesel fuel. *Energy Exploration & Exploitation*, Vol.28, No.1, pp.47-57.

Peterson, G.R., Scarrach, W.P. (1984) Rapeseed oil transester- ification by heterogeneous catalysis. *Journal of the American Oil Chemists' Society*, Vol.61, pp.1593-1596.

Pérez-Pariente, J., Díaz, I., Mohino, F., Sastre, E. (2003) Selective synthesis of fatty monoglycerides by using functionalised mesoporous catalysts. *Applied Catalysis A: General*, Vol.254, pp.173-188.

Qi, X.H., Watanabe, M., Aida, TM., Smith Jr., R.L. (2010) Efficient one-pot production of 5- hydroxymethylfurfural from inulin in ionic liquids. *Green Chemistry*, Vol.12, pp.1855-1860.

Reddy, C.R.V., Oshel, R., Verkase, J.G. (2006) Room-temperature conversion of soybean oil and poultry fat to biodiesel catalyzed by nanocrystalline calcium oxides. *Energy & Fuels*, Vol.20, No.3, pp.1310-1314.

Rubio-Caballero, J.M., Santamaría-González, J., Moreno-Tost, R., Jiménez-López, A., Maireles-Torres, P. (2009) Calcium zincate as precursor of active catalysts for biodiesel production under mild conditions. *Applied Catalysis B: Environmental*, Vol.91, pp.339-346.

Russbueldt, B.M.E., Hoelderich, W.F. (2009) New sulfonic acid ion-exchange resins for the preesterification of different oils and fats with high content of free fatty acids. *Applied Catalysis A: General*, Vol.362, pp.47-57.

Sasidharan, M., Kumar, R. (2004) Transesterification over various zeolites under liquid- phase conditions. *Journal of Molecular Catalysis A: Chemical*, Vol.210, pp.93-98.

Serio, M.D., Cozzolino, M., Tesser, R., Patrono, P., Pinzarij Bonelli, B., Santacesaria, E. (2007) Vanadyl phosphate catalysts in biodiesel production. *Applied Catalysis A: General*, Vol.320, pp.1-7.

Serio, M.D., Ledda, M., Cozzolino, M., Minutillo, G., Tesser, R., Santacesaria, E. (2006) Transesterification of soybean oil to biodiesel by using heterogeneous basic catalysts. *Industrial & Engineering Chemistry Research*, Vol.45, No.9, pp.3009-3014.

Shibasaki-Kitakawa, N., Honda, H., Kuribayashi, H., Toda, T., Fukumura, T., Yonemoto, T. (2007) Biodiesel production using anionic ion-exchange resin as heterogeneous catalyst. *Bioresource Technology*, Vol.98, pp. 416-421.

Shokrolahi, A., Zali, A., Pouretedal, H.R., Mahdavi, M. (2008) Carbon-based solid acid catalyzed highly efficient oxidations of organic compounds with hydrogen peroxide. *Catalysis Communications*, Vol.9, pp. 859–863.

Shu, Q., Yang, B., Yuan, H., Qing, S., Zhu, G. (2007) Synthesis of biodiesel from soybean oil and methanol catalyzed by zeolite beta modified with La^{3+}. *Catalysis Communications,* Vol.8, pp.2159-2165.

Shu, Q., Zhang, Q., Xu, G.H., Nawaz, Z., Wang, D.Z., Wang, J.F. (2009) Synthesis of biodiesel from a model waste oil feedstock using a carbon-based solid acid catalyst: reaction and separation. *Fuel Processing Technology*, Vol.90, pp.1002-1008.

Son, S.M., Kimura, H., Kusakabe, K. (2011) Esterification of oleic acid in a three-phase, fixed- bed reactor packed with a cation exchange resin catalyst. *Bioresource Technology*, Vol.102, pp.2130-2132.

Stavarache, C., Vinatoru, M., Bandow, H. (2007) Ultrasonically driven continuous process for vegetable oil transesterification. *Ultrasonics Sonochemistry*, Vol.14, pp.413-417.

Suganuma, S., Nakajima, K., Kitano, M., Yamaguchi, D., Kato, H., Hayashi, S., Hara, M. (2008) Hydrolysis of cellulose by amorphous carbon bearing SO_3H, COOH, and OH groups. *JACS*, Vol.130, pp.12787-12793.

Sun, H., Hu, K., Lou, H., Zheng, X.M. (2008) Biodiesel production from transesterification of rapeseed oil using KF/Eu_2O_3 as a catalyst. *Energy Fuels*, Vol.22, No.4, pp.2756-2760.

Teixeira, L.S.D., Assis, J.C.R., Mendonça, D.R., Santos, I.T.V., Guimarães, P.R.B., Pontes, L.A.M., Teixeira, J.S.R. (2009) Comparison between conventional and ultrasonic preparation of beef tallow biodiesel. *Fuel Processing Technology*, Vol.90, pp.1164- 1166.

Tesser, R., Serio, M.D., Carotenuto, G., Santacesaria, E. (2010) Absorption of water/ methanol binary system on ion-exchange resins. *Canadian journal of chemical engineering*, Vol.88, No.6, pp.1044-1053.

Toda, M., Takagaki, A., Okamura, M., Kondo, J.N., Hayashi, S., Domen, K., Hara, M. (2005) Biodiesel made with sugar catalyst. *Nature*, Vol.438, pp.178.

Vyas, A.P., Subrahmanyam, N., Patel, P.A. (2009) Production of biodiesel through transesterification of Jatropha oil using KNO_3/Al_2O_3 solid catalyst. *Fuel*, Vol.88, No.4, pp. 625-628.

Wang, L., Yang, J. (2007) Transesterification of soybean oil with nano-MgO or not in supercritical and subcritical methanol. *Fuel*, Vol.86, pp. 328-333.

Wu, P., Yang, Y., Colucci, J.A., Grulke, E.A. (2007) Effect of ultrasonication on droplet size in biodiesel mixtures. *Journal of the American Oil Chemist's Society*, Vol.84, pp. 877-884.

Xi, Y.Z., Davis, R.J. (2008) Influence of water on the activity and stability of activated Mg–Al hydrotalcites for the transesterification of tributyrin with methanol. *Journal of Catalysis*, Vol.254, pp. 190-197.

Xie, W.L., Peng, H., Chen, L.G. (2006) Calcined Mg-Al hydrotalcites as solid base catalysts for methanolysis of soybean oil. *Journal of Molecular Catalysis A: Chemical*, Vol.246, pp.24-32.

Xie, W.L., Peng, H., Chen, L.G. (2006) Transesterification of soybean oil catalyzed by potassium loaded on alumina as a solid-base catalyst. *Applied Catalysis A: General*, Vol.300, pp.67-74.

Xu, L.L., Li, W., Hu, J.L., Yang, X., Guo, Y.H. (2009) Biodiesel production from soybean oil catalyzed by multifunctionalized Ta_2O_5/SiO_2-$[H_3PW_{12}O_{40}/R]$ (R = Me or Ph) hybrid catalyst. *Applied Catalysis B: Environmental*, Vol.90, pp.587-594.

Zięba, A., Drelinkiewicz, A., Chmielarz, P., Matachowski, L., Stejskal, J. (2010) Transesterification of triacetin with methanol on various solid acid catalysts: A role of catalyst properties. *Applied Catalysis A: General*, Vol.387, pp.13-25.

Zong, M.H., Duan, Z.Q., Lou, W.Y., Smith, T.J., Wu, H. (2007) Preparation of a sugar catalyst and its use for highly efficient production of biodiesel. *Green Chemistry*, Vol.7, pp.434-437.

17
Chapter

AN ECO-FRIENDLY ALTERNATIVE FOR CASTOR OIL BIODIESEL: UTILIZATION OF SOLID SUPPORTED ACIDIC SALT CATALYST

1. INTRODUCTION: AN OVERVIEW

In 1885, when Rudolf Diesel initially built the diesel engine, it was designed to operate on oil derived from vegetable sources. Over time, as fossil fuel reserves have diminished, this has become a contemporary necessity. Because it will play an important part in sustainable fuel and energy production solution for the future. Vegetable oil which remains in the form of triglyceride(1) of long chain fatty acid with carbon chain C16-C18 is not fit to use directly but needs certain transformation such as pyrolysis, microemulsion formation, transesterification etc. to suit it to use as diesel fuel

$$\begin{array}{l} CH_2OR \\ \;| \\ CHOR \\ \;| \\ CH_2OR| \end{array}$$

(1)

R= Long Chain Fatty acid moiety

Triglyceride

The transformed oil is termed as biodiesel due to its original biological source. Finite fossil fuel reserve, political, economic, biodegradability, low toxicity, health and environmental issues have led it to consider as the alternative and more importantly renewable and eco- friendly fuel. It has been found to show its ability to meet the world energy demand in transportation along with agricultural and other industrial sectors (Akoh et al. 2007). Since the source is plant, it is green as it does not have ash content, sulphur, aromatic ring compounds, renewable and so it has come out as superlative alternative and can be used in compression ignition engine with minor or no modification of the engine (Xu and Wu, 2005). A breakthrough in the process of converting vegetable oil into useful form promises a cheaper way to go green as it contributes mitigating global warming also. However the slow pace of progress in this direction in alternative fuel technologies has prevented the vision from materializing. On the other hand vegetable oil as such is expensive and direct use of it in diesel engine is not possible. Because, firstly the vegetable oils are very viscous. High viscosity in fuel causes transportation problems, carbon deposits in engine, suffering of engine liner, injection nozzle failure, and gum formation, lubricating oil thickening and high cloud and pour point. Secondly, the glyceride moiety in the triglyceride form of the vegetable oil during combustion could lead to formation of acrolein (2) and this in turn lead to formation of different aromatics (3) as polluting by-products. This is one of the reasons why fatty esters of vegetable oils are preferred over triglycerides.

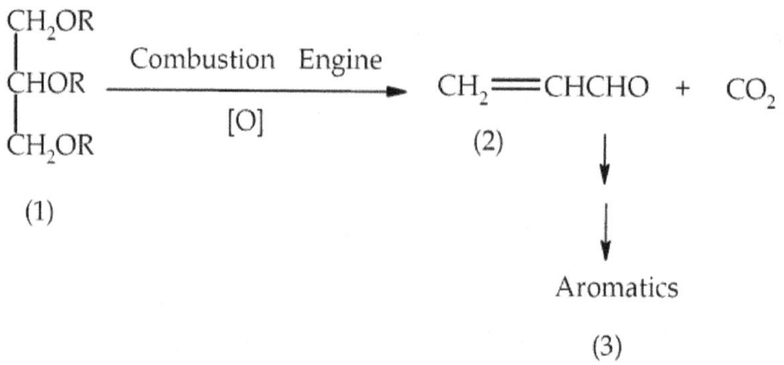

Scheme 1.

In 1970 it was discovered that reduction of viscosity of vegetable oils could be made by simple chemical process called transesterification by which the vegetable oil is treated with a low alkyl alcohol such as methanol or ethanol in presence of a suitable catalyst to form low alkyl esters whereby it could perform as petro diesel in modern engine. Glycerol that is produced during transesterification as by-product can be utilised in other industries. Thus by definition, biodiesel is low alkyl esters of long hydrocarbon chain fatty acids prepared from vegetable oils and animal fats through chemical or by biochemical process of transesterification.

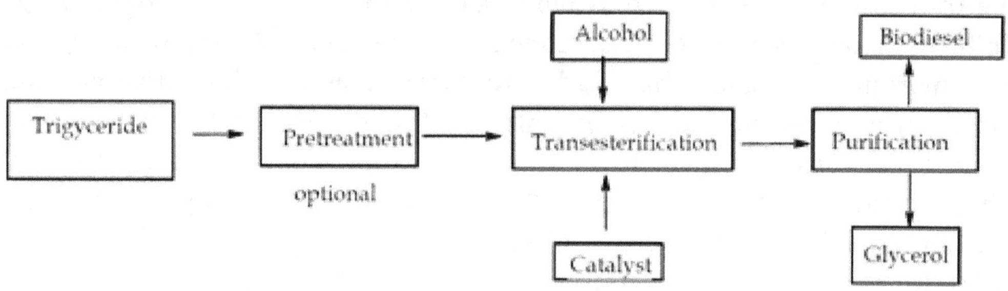

Scheme 2.

2. FEEDSTOCK OF BIODIESEL

Different feedstocks have been explored for extraction of vegetable oils in order to transform it to biodiesel. The feedstocks are animal fats, renewable plant resources basically from Euphorbiace family viz. Jatropha caracas, Soya, Sunflower, Castor seeds etc. besides waste cooking oils from different restaurants and food processing industries. Considering several aspects castor oil from castor seeds seem to be an alternative promising feedstock for commercial production of biodiesel particularly for cold climatic regions.

2.1 Why castor oil

Although from the economic point of view waste cooking oils from different sources is a better choice for biodiesel preparation compared to all other sources and vegetable oils, considering the multifarious advantages oil from castor seeds from *Ricinous Communis* (Palma christi)- a species from Euphorbeace family is believed to be a better option. Because castor oil is possibly the plant oil which is industry's most unappreciated asset that contains about 90% ricinolic acid as the major constituent. The plant originates in Africa but now is available in all the tropical and subtropical countries. The plant can stand long periods of drought. The oil has versatile utility such as cosmetics, lubricants, brake fluids, softener in tanning, solar cell, textile company, small components of PC, mobile phones, boots and shoe manufacturing etc. Presently India is the largest producer of castor oil in the world with China and Brazil being the next two. India exports about 15000 tonnes castor seeds per year and 1,00,000 tonnes of castor oil annually to European Union and the domain has been increasing rapidly. In the seed, the oil content is about 50% of the total weight. It is the only unique oil which has an unusual chemical composition of triglyceride of fatty acid. It is the only source of an 18-carbon hydroxylated fatty acid *viz* ricinoleic acid with one double bond. It is reported that fuels having fatty acids with 18 or more carbon atoms and one double bond have viscosity low, higher cetane number and lower cloud and pour point properties are better. From that point of view alkyl ester from castor oil satisfies most of the criteria with the exception of viscosity and cetane number to stand as promising biodiesel candidate. The chemical composition of castor oil triglyceride (castoroil. in - the Home of castor oil in Internet) is

1. Ricinolic acid- 89.5%
2. Linoleic acid- 4.2%
3. Oleic acid- 3%
4. Stearic acid- 1%
5. Palmitic acid- 1%
6. Dihydroxy stearic acid- 0.7%
7. Eicosanoic acid- 0.3%
8. Free fatty acid in refined castor oil- 8.45%

Although considerable researches have been done on palm oil, soya oil, sunflower oil, coconut oil, rapeseed oil, tung oil, jatropha oil etc not much informations are available on castor oil as biodiesel even though it is currently undergoing a phase of active research in several institutions.

Production of castor oil worldwide is 0.5 million tonnes per annum. Consumption of petrodiesel per day is approximately 10 million tonnes. If the entire petrodiesel is to be replaced by castor biodiesel it needs to produce 7000 times the castor oil that is being produced today. However, since it is one of the oldest traded goods mankind has been trading a few thousand years ago, it has a lot of industrial usages and therefore market is already in existences. Further, as the plantation of castor plant has been cultivated commercially, its biology is well understood and high yield hybrid is available. It can also be found in medium climate areas as an annual crop or in tropical area as a small tree. It gives faster oil yield and can be planted as marginal plant in unattended idle areas. The gestation period of harvesting the plant for oil is 4-6 months only.

2.2 Properties of biodiesel from castor oil

The biodiesel prepared from castor oil has certain properties that are attractive particularly for cold climate. It may be mentioned that it has flash point of 190.7^{o}C which is much higher than petrodiesel and other vegetable oil biodiesel. The oil is stable at low temperature and makes it an ideal combustible for region of extreme seasonal weather. From cost point of view although 100% biodiesel from castor oil (B_{100}) seems to be expensive its 10% (B_{10}) or 20% (B_{20}) blending with petrodiesel show good flow properties and further lowers the cloud and pour point. Further, due to its ability of displaying as a solvent, sedimentation does not occur which could otherwise potentially obstruct pipes and filters. However,the oil is sensitive to contamination by ferrous salts and rusts particles.Its higher cooling capacity is a key factor in the conservation of engine components. Considering the technical features, castor oil biodiesel is advisable taking into accounts its renewable resources. Because of its biodegradability and lower emissions, it presents a favourable impact on the environment. Moreover, it could be used as a crop substitution program turning it into a factor that promotes growth in many regions affected by several economic problems. Awareness is there in recent times for cultivation of castor plants boosting rural economy by government and private agencies by establishment of transesterification plant with million tonnes capacity per day, trial run using biodiesel from castor oil by Indian Railways, roadways, IOCL, HPLC etc. In addition to it a national mission on biodiesel has been proposed by the government of India with six micro missions to cover different aspects.

3. TRANSESTERIFICATION OF VEGETABLE OILS

Transesterification of vegetable oils has now come a long way for preparation of biodiesel. There are four basic methods for biodiesel production. These are acid catalysed, base catalysed, enzymatic/microbial transesterification and conversion of the oil to its fatty acids and then esterification to have ester as biodiesel.

3.1 Transesterification catalysts

The transesterification reactions require a catalyst in order to obtain a reasonable conversion rate and the nature of the catalyst must conform to the feedstock. Further, the reaction condition and post separation steps are predetermined by the nature of the catalyst.

Generally, transesterification of vegetable oil is done with methanol or ethanol in presence of a base catalyst such as NaOH, KOH, K_2CO_3, NaOMe, NaOEt, NaOPr, NaOBu etc. A minimum content of water and free fatty acid result in the saponification with consequent formation of soap. Presence of large content of water results in hydrolysis of the product formed. Theoretically 3 moles of methanol are required per mole of triglyceride. As the transesterification reaction of triglyceride is a reversible reaction, the excess of methanol shifts the equilibrium towards the direction of ester formation. Freedman et al (Freedman et al, 1984) suggested that 6:1 molar ratio of alcohol to oil is necessary to get the maximum ester yield thus minimising the concentration of tri, di and mono glycerides.

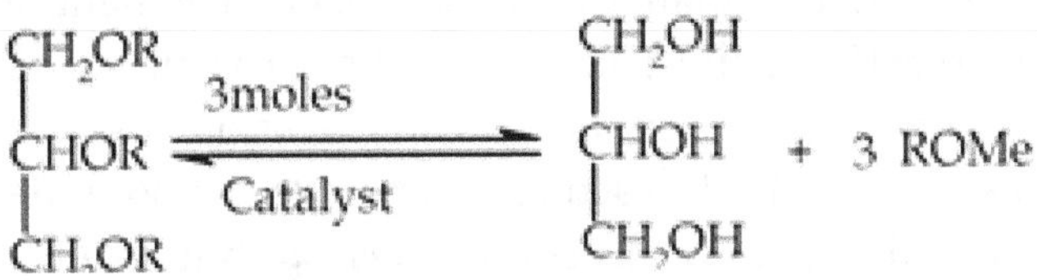

Scheme 3.

3.2 Solid catalysts for transesterification reactions:

3.2.1 Solid base catalysts

There are reports of many solid base catalysts to be active in transesterification reactions such as supported CaO catalysts (Yan et al, 2008), supported VO_2 catalysts (Kim et al, 2008) various other metal oxides such as BaO, SrO, MgO etc to have transesterified camaelina sativa oil as biodiesel with upto 80-89% yield(Patil and Deng, 2009). However, these solid base catalysts show much lower activity than traditional homogeneous catalysts.Potassium nitrate supported on alumina as solid base catalyst was reported by Vyas et al (Vyas et al,2009) for production of biodiesel from jatropha oil and has been successful in getting 84% yield. Certain of these catalysts are very much sensitive to trace amount of free fatty acid present.

Reports of lanthanum based (Kurian et al, 1998) strong basic catalysts have appeared for transesterification and esterification reaction.

3.2.2 Enzyme catalysis

Over the last few decades considerable research have been done on the use of enzyme in transesterification using lipase enzyme from filamentous fungi and recombinant bacteria under various condition. However not considerable attention has been received except in China where 20,000 tonnes of biodiesel per year(Du et al, 2008) is produced. But due to large reaction volume, time, higher conc. of catalyst, cost ($1000 per kg), loosening of catalyst activity on repeated use the process is not commercially viable although friendly to the environment.

3.2.3 Acid catalysts

Homogenous acid catalysts such as H_2SO_4, HCl, sulfonic acid etc. have the potentials to replace base catalysts since they do not show measurable susceptibility to free fatty acid (FFA) and can catalyse esterification and transesterification simultaneously (Kulkarni et al, 2006). However, separation problem, requirement of high temperature, high molar ratio of oil and alcohol, serious environmental and corrosion related problem make their use non practical for biodiesel production.

The demanding feedstock specification for base catalysed reactions have led researchers to seek catalytic process alternative that can ease this difficulty and lower production cost. To eliminate the corrosion, environment problem and time saving for multiple reaction, solid acid catalysts have recently replaced liquid acids for biodiesel production by simultaneous esterification and transesterification. Methodologies based on acid catalysed reaction have the potential to achieve this since acid catalysts did not show measurable susceptibility to FFAs. Compared to homogenous acid catalysts heterogeneous solid acid catalysts have great potential due to advantage in separation and corrosion related problems and such catalysts having large-pores, moderate to strong acid sites and a hydrophobic surface are idea for biodiesel production.

3.2.4 Solid acid catalysts

There have appeared in the literature several solid supported acid catalysts such as heteropolyacid, having Keggin structure viz-12-tungsto-phosphoric acid impregnated on various solid supports like hydrous zirconia (Kulkarni et al. 2006), silica, alumina, and activated carbonate using as solid acid catalyst for biodiesel preparation from different feedstock with achievement of more than 77% yield of biodiesel. Zeolites (Lotero E et al, 2005, Wang et al,2009) with large pore size have been used with success with fatty acid esterification but at higher temperature. Few other solid supported catalysts for esterification and transesterification of vegetable oils are zeolites with different pore size framework of Si/Al ratio and proton exchange level. These characteristics permit tailoring important catalytic properties such as acid strength. It was observed that zeolite catalysis in transesterification/esterification reaction using large molecules takes place on the external surface of the zeolite catalysts. However, it requires high temperature and the reaction rate is slow. The reactivity on such solid surface catalysts depends upon acid site strength and hydrophobicity of the surface. In fact, pore size, dimensionality of the catalyst channel system related to the diffusion of the reagents and products and aluminium content of zeolite framework strongly affect the zeolite catalytic activity for esterification. Related to zeolites, but with amorphous pore walls, silica molecular sieves such as MCM-41, mesoporous materials are generally not sufficiently acidic to catalyse esterification reaction due to pure silica structure. However introducing aluminium, zirconium, titanium or tin compounds into silica matrix of these solids can significantly improve their acidic properties. However, metal doped materials behave more like weak acids and can only be used for reactions that do not require a strong acid catalyst. It has also been reported that SO_4^{-2}/ZrO_2 has been shown to have applicability for several acid catalysed reactions. However the problem is that SO_4^{-2}/ZrO_2 deactivates in presence of water due to leaching of SO_4^{-2} either in the form of H_2SO_4 or HSO^{-}. Sulphated tin oxide (SO_4^{-2}/SnO_2) prepared from *m*-stannic acid has shown activity superior to that of SO_4^{-2}/ZnO_2 for esterification of octanoic acid by methanol at 150^{0}C due to superior acid strength (Furuta, S. et al 2004). The use of solid catalyst to produce biodiesel requires a better understanding of the factors that govern their reactivity. Thus, an ideal solid catalyst should show some underlying characteristics such as an interconnected system of large pores, moderate to high concentration of high acid sites and a hydrophobic surface. Large interconnected pores would minimise diffusion problem of molecules having long alkyl chain and strong acid sites are needed for the reaction to proceed at an acceptable rate. It is recently attracted considerable attention for solid acid catalyst such as Bronsted acid zeolites, ion exchange resin, metal oxides viz sulphated zirconia WO_3/ZrO_2, MoO_3/ZrO_2, sugar based catalyst (Zong et al, 2007). It has been noted that Bronsted acid catalysts are active mainly in esterification while Lewis acid catalysts are active in transesterification reaction. Therefore, preparation of such solid supported catalysts that contain both Bronsted acid and Lewis acid catalyst site having enhanced water tolerance and large pores, hydrophobic surface and low cost is still a challange. National Chemical Laboratory, Pune India has developed a novel solid double metal composition for transesterification of vegetable oils containing up to 18% FFA to biodiesel (Sree Prasanth et al, 2006). A series of layered alumino silicates with H_2SO_4 impregnation has been reported for transesterification. Activated montmorillonite KSF showed 100% conversion of transesterification within 4 hour at 200°C and 52 bar pressure. However problem encountered is leaching, for which reimpregnation of H_2SO_4 on the clay surface is required for reusability. Several other solid acid catalysts were reported but needed higher temperature (>200°C) for conversion. The use of age old polymer matrix Amberlyst-15 has also been reported but need mild condition to avoid degradation (Vicente et al,1998).

4. MATERIALS AND METHODS

In view of the above and having observed certain advantages of castor oil over others it was studied the transesterification of it using a simple, cheap and easily prepared solid supported acidic catalyst considering the positives of this clean catalysts.

$KHSO_4$, an acidic salt is ordinarily used as dehydrating agent of alcohol to olefinic compounds at high temperature. It was observed in our earlier study (Goswami et al, 2007) that treatment of it as such with an ester in presence of an alcohol, the ester undergoes partial transesterification very slowly. Dispersing this acidic salt on microporous surface silica gel uniformly triggers transesterification (Goswami et al, 2007) of esters in simple alcohol very satisfactorily giving the product yield more than 95%. The system behaves in a completely different manner on treatment with olefin (Das et al, 2010) leading to dimerization through C-C bond formation or addition product with alcohols depending upon the condition applied. Application of this system to castor oil triglyceride in methanol at its boiling point in 5-6 hours transform it into methyl ester of ricinoleic acid the primary constituent of castor oil along with other fatty acid methyl esters present in it with more than 95% yield.

4.1 Experimental condition

Instruments : The GC was recorded on Chemito 1000 GC using column OVIE+SP2401 (2mX10.635 cm, od) glass column and nitrogen as carrier gas. The textural properties were recorded on Quantachrome Automated Gas Sorption system. The FTIR was recorded on Perkin Elmer System-2000 and FT NMR was recorded on Bruker Avance-DPX-300MHz instrument. Reagents: Castor oil was obtained from local grocery shop (Dabur, 99%). methanol (99.8%) from Fisher Scientific, potassium bisulphate (98%) from Rankem and silica gel (60-100 mesh) were taken from Aldrich Chemicals. Methanol taken was made super dry following standard method:

4.1.1 Catalyst preparation

Potassium bisulphate ($KHSO_4$) 20 gm (144mmol) was dissolved in 100ml distilled water to have a clear saturated solution. The solution was soaked completely in microporous silica (40gm). The soaked mixture was thoroughly mixed and dried in a hot air oven at 150°C for 24 hours to have a free flowing powdery solid. The dried solid mixture was than kept in vacuum desiccator to use as a stock solid supported catalyst (A) in different reactions.

4.1.2 Experimental procedure

25 ml of refined castor oil containing 8.4% FFA was charged with 1 litre dry methanol in a 1.5 litres round-bottomed flask fitted with a condenser and fused calcium chloride guard tube on a preheated oil bath under vigorous stirring. To it was added 1.25gm (5%) catalyst A and stirred at 600 rpm under heating at 70°C (external) for 5 hours. Occasionally TLC was monitored to check the progress of the reaction. After completion, the reaction mixture was distilled to recover methanol. The product with the catalyst remained after separation of methanol was obtained with glycerol as a separate layer. Methyl ester of castor oil along with glycerol layer was decanted out from the solid catalyst surface. Glycerol separated as the bottom layer was taken out from the methyl ester of castor oil (CastMe) layer. The solid catalyst was washed several times with petroleum ether and dried at 150°C for 24 hours in a hot air oven for subsequent runs. The product isolated was found to have yield 95%. During the period of the reactions, samples were taken out at regular intervals and analysed on GC (Fig. 1) using carrier gas nitrogen at flow rate of 2.5kg/cm^2. Triglyceride, diglyceride, monoglyceride and methyl ester CastMe as transesterified product were quantified by comparing the peak areas of their corresponding standard.

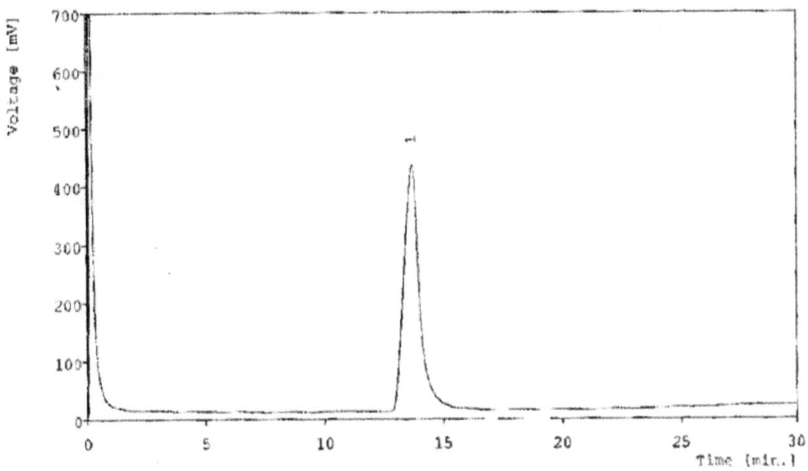

Fig. 1. GC of standard ricinoleic acid methyl ester

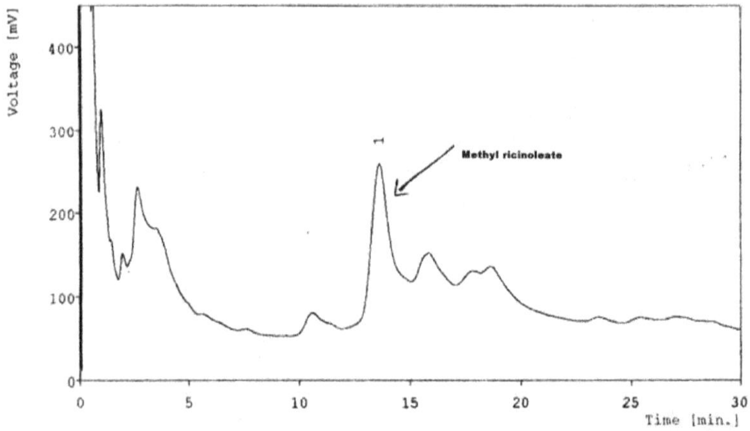

Fig. 2. GC of CastMe

4.1.3 Physical properties of CastMe determined by ASTM D6751 standard

The physical properties of CastMe viz. kinematic viscosity, density, pour point, and cloud point have been determined following standard ASTM D675 method and given in the Table 1 along with reported (Forero, C.L.B., 2004) values of corresponding castor oil, petrodiesel and methyl esters of few other vegetable oils. In Table 2 suggested ASTM standard for pure biodiesel (100%) were given. The properties of CastMe are comparable to those of petrodiesel and acceptable within what is specified for 100 % pure biodiesel as per ASTM standard except that of viscosity and cetane numbers which are the bottlenecks. However, 10% or 20% blended CastMe with petrodiesel that are known as B_{10} and B_{20} have their kinetic viscosity 4.54 & 4.97 mm^2/s and are within ASTM standard.

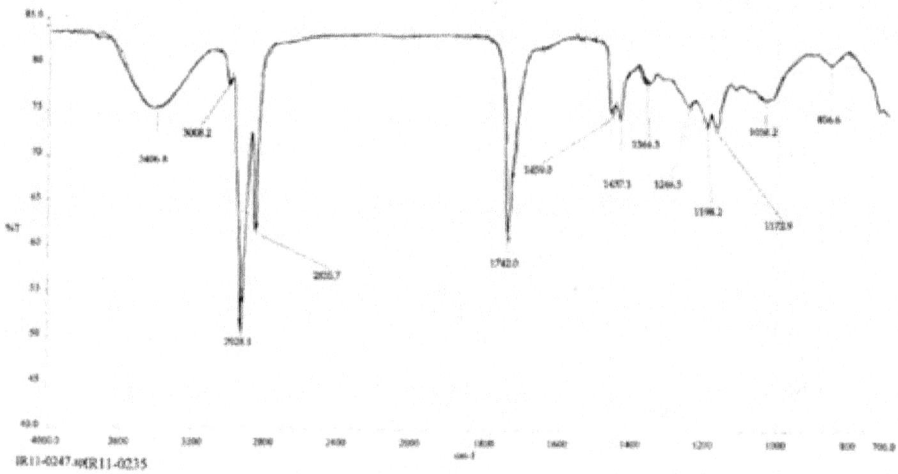

Fig. 3. FTIR of CastMe

The corresponding significant FT IR frequencies (Fig 3) and superimposable FTIR spectra of CastMe and standard methyl ricinoleate (Fig 4) and 300 MHz NMR spectral data (Fig5) of methyl ester of castor oil (CastMe) have been as given below. FT IR (Cm^{-1}, thin film): 1742 (COOMe str),2855 & 2928(CH str),3407(OH str) (Fig 3):

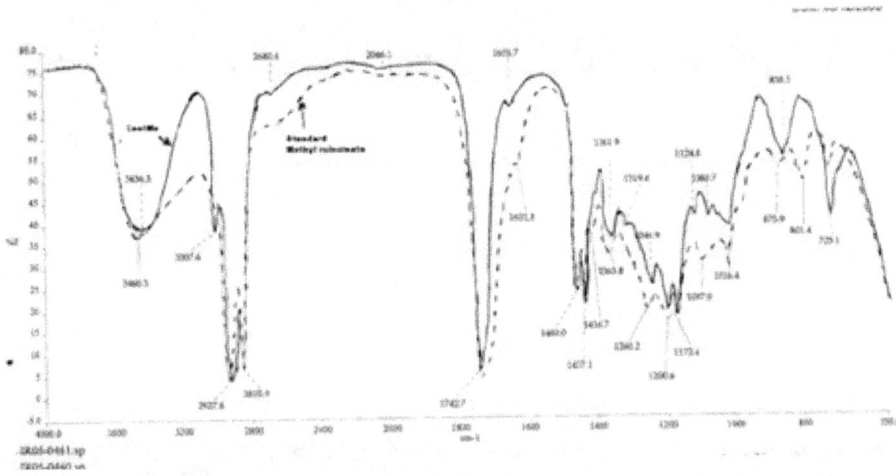

Fig. 4. Superimposable FTIR spectra of CastMe and standard methyl ricinoleate

^{1}HNMR(δ ppm,$CDCl_3$) : 1.96(s,1H,O<u>H</u>), 1.23-2.24(m, nH,$(C\underline{H}_2)_n\underline{Me}$), 3.55-3.57(m,1H,C<u>H</u>-OH), 3.57(s,3H,COO<u>Me</u>), 5.25-5.47(m, nH, olefinic protons),(n=different numbers of protons of fatty acids), the hydroxyl group present in ricinoleic acid, the major constituent of castor oil imparts unique properties. Because of the branching created by it causes the low cetane number and higher viscosity (Knothe et al, 2008). However the advantage of the present method is that unlike other acidic catalyst, this catalyst system does not facilitate anymethanololefinetherification (Goodwin et al,2002) even though the constituent of the oil do possess olefinic bonds .

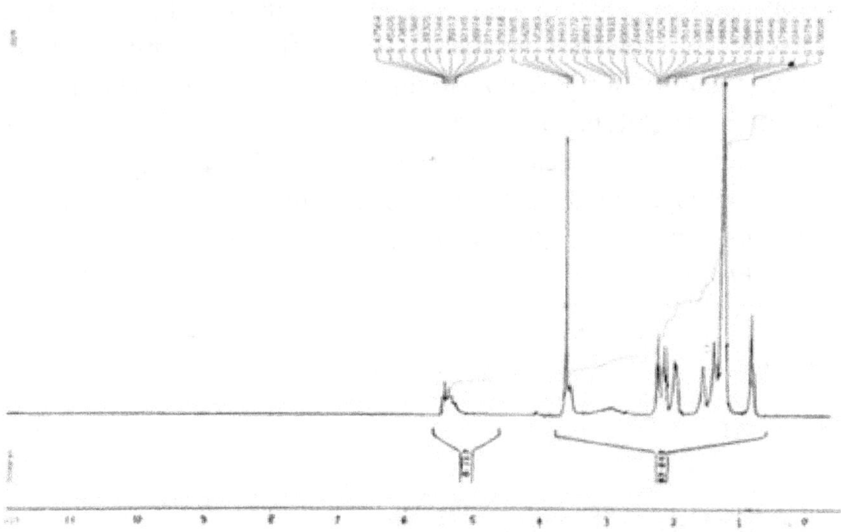

Fig. 5. ^{1}HNMR (300MHz) of CastMe

Table 1. Physical values of castor oil methyl ester (CastMe) determined along with values of other vegetable oil methyl esters

Item	Density (g/cc)	Kinematic Viscosity (CST)(380C)	Pour point (OC)	Cloud point (OC)	Flash point	Cetane no.
Castor oil	0.963	297	-32	-20	260	
CastMe	0.34	9.4	-45	-23	190.7	42
Petrodiesel	0.86-0.95	3.81	-6	-15	68.3	47
Soy ME	0.885	4.8-4.3	-3.8	-0.5	131	48
Rape ME	0.883	4.53	-10.8	-4.0	170	48
Tallow ME	0.876	51.15	9	13.9	117	35
Canola ME	0.88		-9	1	163	48

Table 2. Suggested standard for pure (100%) biodiesel as per ASTM.

Property	ASTM standard	limit	units
Flash point	93	100	^{0}C
Carbon residue	4530	0.050	wt%
Sulphated ash	874	0.020	wt%
Kinematic viscosity	445	1.9-6.0	mm^2/s
Sulphur	2622	0.05	wt%
cetane	613	40	^{0}C
Cloud point	2500	By customer	^{0}C
Free glycerol	GC	0	wt%

5. RESULTS AND DISCUSSION

Potassium bisulphate (PBS) impregnated microporous silica has been evaluated as solid acid catalyst for biodiesel production from refined castor oil containing 8.4% FFA compared to other support viz. alumina with 95% yield. The determination of surface area, pore volume and pore diameter and also FTIR spectra of $KHSO_4$ supported on microporous silica revealed that $KHSO_4$ is well dispersed very evenly generating Bronsted acid site that is responsible for its higher activity.The FT IR spectrum of pure $KHSO_4$,pure silica gel and $KHSO_4$ supported silica gel (Fig-6) have been depicted below.

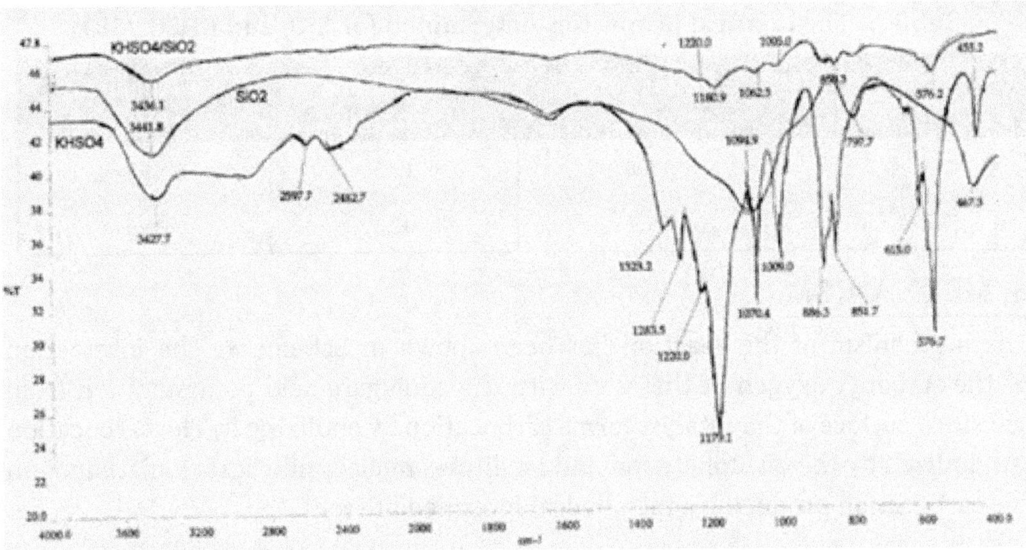

Fig. 6. FT IR spectra of pure $KHSO_4$,Pure SiO_2 and $KHSO_4$ supported on SiO_2

The pure silica FTIR spectra of $KHSO_4$ exhibited typically six major bands located at 577,852,886,1009,1070 and 1179 cm^{-1} which are stretching modes of oxygen bonded to sulphur and hydrogen. In supported $KHSO_4$ catalyst no clear bands were observed. These results indicated that $KHSO_4$ is highly dispersed on the surface of support SiO_2. A 40:1 alcohol to oil ratio at 70^0C (external) temperature and 5 wt% catalysts loading gave a maximum yield of CastMe up to 95%.

The textural properties (Kulkarni et al, 2006) of the catalyst were summarized in Table 3. The surface area of microporous silica of 60-100 mesh particle size has $300m^2/g$ and pore volume $1.15cm^2/g$ and its average pore diameter is 150 A^0. After loading 50 wt% of $KHSO_4$ the accessible surface area of silica gel left was only $55.45m^2/g$ and pore volume and average pore diameter were reduced to $0.13cm^2/g$ and 98.9 A^0. The reason is attributed to uniform dispersing of KHSO4 on the surface leaving only $55.45m^2/s$ surface and pore plugging of the support. The same reaction when carried out in a similar fashion supporting $KHSO_4$ on alumina surface, the reaction gives very poor or no yield at all. It may be due to too narrow micropores of alumina which cannot accommodate $KHSO_4$ molecule to disperse uniformly to enhance catalytic activity (Kulkarni et al, 2006) although its surface area is higher ($260m^2/g$). Even though alumina is an interesting support it is assumed that the surface basicity could bring about decomposition of $KHSO_4$. It means that particles of $KHSO_4$ conform to silica gel particles in order to disperse on its surface. Large pores can easily accommodate a bulky triglyceride molecule giving $KHSO_4/SiO_2$ large active site and surface area resulting in highest activity (Igarashi et al, 1979; Furuta, 2004 and Lecleroq et al, 2001).

Table 3. The textural properties determined for SiO_2 and $KHSO_4/SiO_2$

Solid support	Surface area (m2/g)	Pore volume (cm2/g)	Pore diameter (A0)
SiO_2	300	1.15	150
$KHSO_4/SiO_2$	55.45	0.13	98.9

6. MECHANISM

The mechanism of the reaction has been shown in Scheme 4. The interaction of the carbonyl oxygen of the ester with the conjugate acid potassium ion from the silica surface of the catalyst forms carbocation by enolizing it. The carbocation is stabilized by the bisulphate ion and facilitates nucleophilic attack methanol on the carbocation producing a tetrahedral intermediate (c).

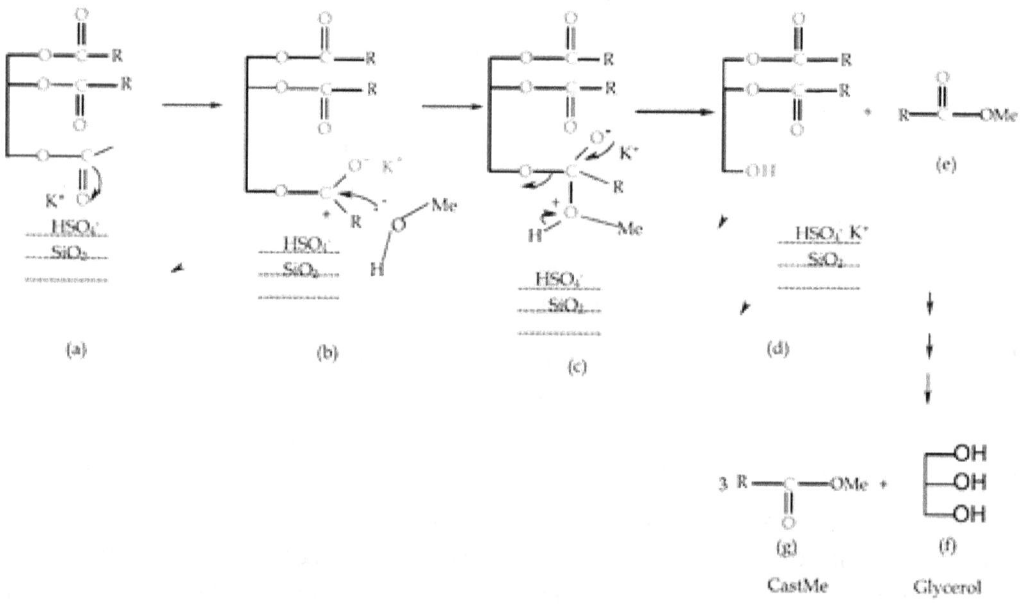

Scheme 4.

In the reaction sequence the triglyceride was converted stepwise to di and mono glyceride and finally to glycerol. The tetrahedral intermediate (c) formed during the reaction eliminate di, monoglyceride and glycerol when tri, di and monoglyceride came in contact with the acidic site respectively to give one mole of ester in each step. It has been reported (Freedman, B, 1986) in fact that the rate limiting step varied over time and in three stages in accordance with the observed reaction rate could categorize the overall reaction progress. In the first stage the reaction was characterized by a mass transfer controlled phase in which the low miscibility of the catalyst and the reagent or the non-polar oil was separated from the polar alcohol phase. The second phase is product formation stage whereby the product formed acts as an emulsifier. It is a kinetically controlled stage and is characterised by abrupt range of product formation. Finally the equilibrium is reached at the completion stage. It was found in castor oil transesterification with 40:1 alcohol to oil ratio acceptable reaction rate was achieved. Thus from this observation it can be stated that the forward reaction is pseudo first order kinetics while the backward or the reverse reaction is second order kinetics.

7. INFLUENCE OF REACTION PARAMETERS

The transesterification of castor oil in presence of $KHSO_4$ supported on silica gel in methanol is influenced by certain reaction parameters which have been studied thoroughly varying the conditions at different stages and the results have been appended below.

7.1 Reaction temperature

Initially the transesterification reaction was attempted at room temperature under stirring at 600 rpm for more than 48 hours. However the reaction rate at room temperature was found to be very slow and only 30-35% conversion was observed. It means that the rate of reaction is influenced by the reaction temperature. Gradually when the reaction temperature was raised by 10°C the reaction rate is increased with increase of product formation and at 70°C (external) temperature the formation of the product was found to be maximum of 95%. Beyond this temperature there was found to be no further increase of yield (Fig. 7).

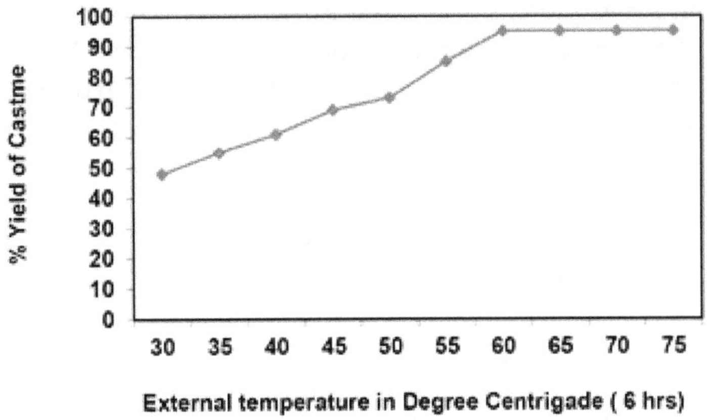

Fig. 7. Effect of external temperature on the reaction course

7.2 Effect of time

The effect of reaction time was studied and result was shown in Fig. 8. It was found that increasing the reaction time upto 5 hours enhanced the castor oil methyl ester yield and beyond it there found to be no further improvement. It means 5 hours time is optimum period required.

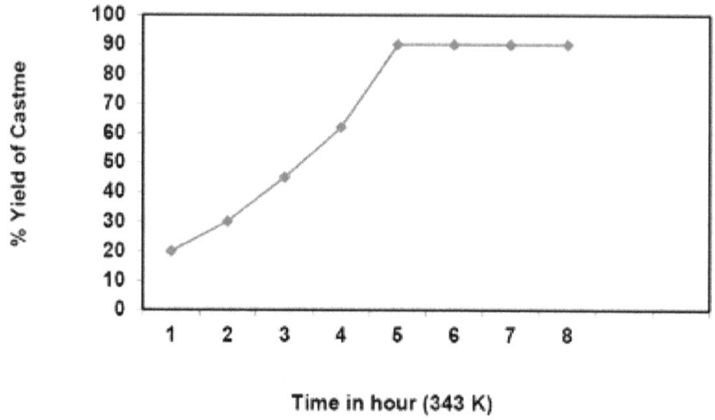

Fig. 8. Effect of time duration on the reaction course

7.3 Effect of alcohol to castor oil ratio

Methanol to castor oil weight ratio is one of the important parameters that affect the yield of methyl ester of castor oil. Theoretically the transesterification reaction requires 3 moles of methanol per mole of triglyceride (Lotero et al, 2005). Since the reaction is a reversible one, the excess methanol shifts the equilibrium towards the direction of ester formation (Cannkei et al, 1999). Generally heterogeneous acid catalytic of transesterification reaction is well known for slow reaction rate. In order to improve the rate of this reaction, use of excess alcohol is an option. It was reported (Xie et al, 2005) that increase of the ratio up to 275:1 of alcohol to oil improves the rate of transesterification reaction.In the present work with preoptomized reaction parameters the methanol to castor oil ratio was varied in the range 5:1 to 40:1 and its influence on the yield of CastMe was investigated at the end of 5 hours. It was clearly observed that at 70^{0}C (external) temperature with increase in ratio of alcohol to oil from 5:1 to 40:1 increased the yield of CastMe from 75% to 95%.

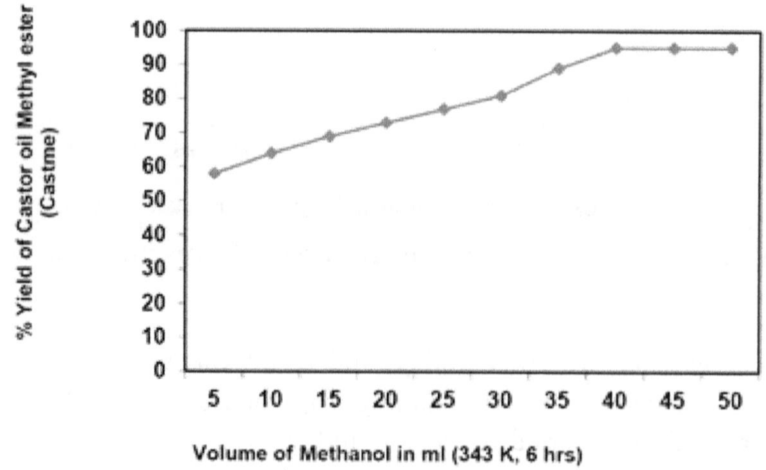

Fig. 9. Effect of volume of alcohol on the reaction course

Presence of 8.4% FFA in the refined castor oil did not affect the activity of the catalyst. Further increase of methanol did not show any significant improvement (Fig 9). The excess methanol can be recovered for reuse and low cost of methanol makes it the first choice for transesterification.

7.4 Effect of catalyst amount

The catalyst amount is also an important parameter that needs to be optimized for increasing the yield of castor oil methyl ester(CastMe). The effect of $KHSO_4/SiO_2$ wt/wt of castor oil on the reaction was studied. At low catalyst amount (< 5 wt %) there were not enough active site for reaction. The optimum amount of catalyst employed was found to be 5 wt% of castor oil to isolate a yield of 95% of the product (Fig. 10).

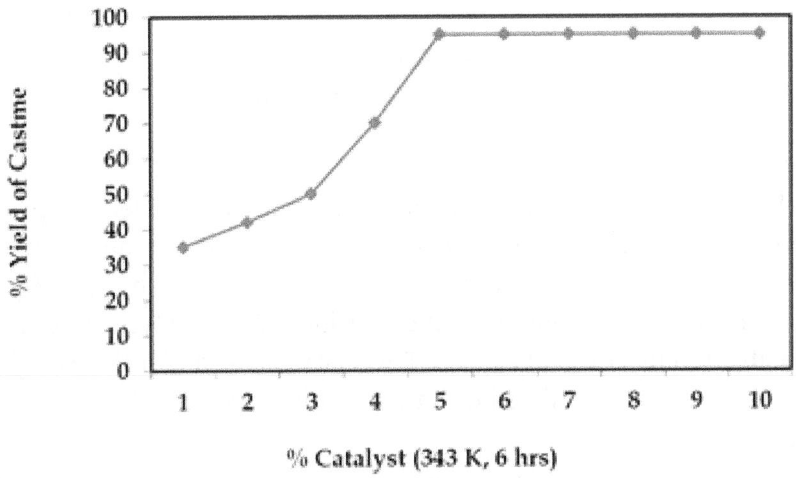

Fig. 10. Effect of catalyst amount

7.5 Catalyst recycling

The cost of a process depends upon the recyclability of a catalyst. It has been found that the dispersed catalyst $KHSO_4$ on silica gel surface after the transesterification reaction of castor oil in methanol, a certain amount gets leached out with methanol either in the form of H_2SO_4 or in HSO_4^-. However, after the completion of the reaction, methanol is distilled out completely and methyl ester of castor oil (CastMe) was extracted in dichloromethane whereby $KHSO_4$ is retained on the surface of silica. The catalyst was washed several times with petroleum ether and then dried completely at 150°C for 8-10 hours. On use of this catalyst for 5 runs with same amount of castor oil and methanol the yield of CastMe decreased was subtle even at fifth reuse (Fig. 11).

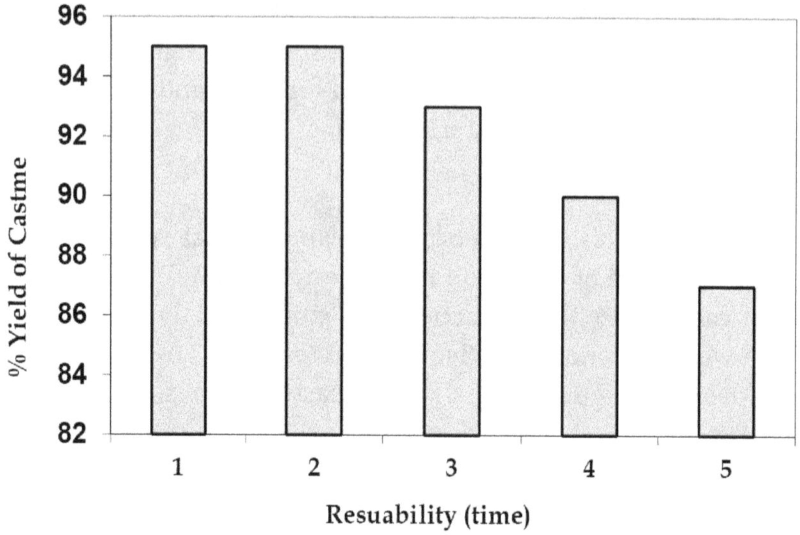

Fig. 11. Catalyst recycling

8. CONCISE SUMMARY

A silica gel-supported KHSO4 acidic catalyst has been developed for the biodiesel production from refined castor oil with 8.4% free fatty acid, demonstrating a simple, cost-effective, eco-friendly, and recyclable catalyst system that achieves a great yield of castor oil biodiesel under mild conditions. The catalytic system's activity remains unaffected by free fatty acids. The system is sufficiently straightforward that it necessitates no specialized design in comparison to other solid-supported acidic catalysts.

In this context, it is noteworthy that the foremost oil companies globally are seeking to exploit the business prospects of biodiesel. The procedure outlined in this chapter, when expanded to commercial levels by increasing numbers of oil firms, might represent a significant advancement in the development of an environmentally sustainable transportation fuel that is reasonably clean upon combustion and offers considerable cash to farmers.

REFERENCES

Akoh, C.C., Chang, S.W., Lee, G.C. and Shaw, J.F. (2007). Enzymatic approach to biodiesel production. *J. Agric. Food. Chem.*, 55, 8995-9005. DOI: 10, 1021/ jf071724y

Canakei,M. and Van,G.J. (1999). Biodiesel production via acid catalysis *Trans. ASAE*, 49,1203-1210. http:/www.castoroil.in-The Home of Castor oil in the Internet. (2010). Castor oil chemicals and derivatives: Chapter3 ,*Comprehensive Castor Oil Report:*,pp 10

Das, R.N., Sarma, K. and Goswami, A. (2010). Silica supported potassium bisulphate: an efficient system for activation of aromatic terminal olefins, *Synlett*,19,2908-2912

Du, W., Li,W., Sun,T.,Chen,X. And Liu, D. (2008). Perspectives for biotechnological production of biodiesel and impacts, *Applied Microbiol.Biotechnol*, 79,331-337, DOI: 10.1007/c00253-008-1448-8

Forero, C.L.B. (2004). Biodiesel from castor oil : a promising fuel for cold weather, *Report from Francisco de Paula Santander University*, Columbia, www.icrepq.com

Freedman, B; Butterfield, R.O. Pryde B.H.(1986). Transesterification kinetics of soyabean oil,*J. Am. Oil. Chem. Soc.* 63, 1375-1380.

Freedman, B., Pryde, E.H and Mounts, T.L. (1984). Variables affecting the yields of fatty esters from transesterifies vagatable oils. *J.Oil.Chem Society*. 61, 1638-1643, DOI: 10.1007/B IO 2541649, 2008.

Furuta, S, Matsuhashi,H, Arata K. (2004). Catalytic action of sulphated tin oxide for etherification and esterification in comparison with sulphated zirconia. *Appl-Catal A*. 269, 187-191.

Furuta, S., Matsuhashi ,H. and Arata ,K. (2004).Performance evaluation of biodiesel production by solid/liquid catalysts, *Catalysis Commun.*, 5,5721-5723

Ghaly, A.E., Dave , D., Brooks,M.S. and Budge, S. (2010). Production of biodiesel by Enzymatic transesterification: Review. *American Journal of Biochem. and Biotech* 6(2) 54-76.

Goodain, J.G.; Natesalehawat. S; Nikolopolous A.A Kim. S.Y. (2002). Etherification on zeolites: MTBE Synthesis, *Catal., Rev. Sci. Eng.* 44,287-320

Goswami,A., Das,R.N. and Borthakur,N. (2007). An efficient selective solid supported system for deprotection of alcohols from esters, *Ind,J Chem*,46B,1893-1895

Igarashi, S., Matsuda, T and Ogino, Y. (1979). Catalysis for vapour phase synthesis of methyl tertiary butyl ether,*J. Jpn Petrol.Ins.* 22, No, 331-336.

Kim, M., Yan,S.,Wang,H., Wilson, J.,Salley,S.O., Simon Ng, K.Y. (2008). Catalytic activities of vanadia loaded heterogenous catalysts on transesterification of soyabean oil with methanol, *99th AOCS Annual Meeting* in Seattle, Washington

Knothe, G. (2008). 'Designer' Biodiesel, optimozing fatty ester composition to improve fuel properties, *Energy & Fuels*, 22, 1358-1364.

Kulkarni, M.G, Gopinath, R., Meher, L.C. and Dalai A,K. (2006). Solid acid catalysed biodiesel production by simultaneous esterification and transesterification, *Green Chem*, 1056-1062.

Kurian J.V., Liang,Y. (1998). Transesterification process using lanthanum compound catalyst, *US Patent* 5840957

Lecleroq, E. Finiels, A. and Moreau,C.(2001). Transesterification of Sunflower oil over zeolite using different metal loading: A case of leaching and agglomeration studies,*J. Am Oil Chem Soc.*,11,1161.

Lotero, E., Liu,Y., Lopez, D.E., Suwannakarn K, Bruce, D. A and Goodwin, Jr. J.G. (2005). Synthesis of biodiesel via acid catalysis, *Ind.Eng. Chem. Rev*, 44, 5353-5363.

Lou, W.Y. Zong,M.H. and Duan Z.Q. (2008). Efficient production of biodiesel from high free fatty acid containing waste oils using various carbohydrate-derived solid acid catalysts, *Bioresource Technology*, 99(18), 8752-8758.

Patil, P.D. and Deng, S. (2009). Transesterification of camaelina sativa oil using heterogenous metal oxide catalysts, *Energy Fuels*,23,4619-4624

Robels-Medina, A., Gonzalez-Moreno,P.A., Esteban-Cerdan, L. and Molina-Grima, E. (2009). *Biocatalysis Towards Ever Greener Biodiesel Production. Biotechnol Adv.*, 27, 398- 408. DOI: 100.1016/j, biotechadv, 2008, 10.008.

Singh, A.K. and Fernando, S.D. (2008). Transesterification of soyabean oil using heterogenous catalysts, *Energy & Fuels*, 22,2067-2069

Sreeprasanth, P.S., Srivastava, R., Srinivas, D and Ratnaswami P. (2006). Hydrophobic solid acid catalysts for production of biofuel and bolubricantst *Appl. Catal A. General* 314, 148-159.

Vicente,G., Coteron, A., Martinez, M., Aracil, J. (1998). Application of the factorial design of experiments and response surface methodology to optimize biodiesel production, *Ind. Crop Products*, 8,29-35

Vyas, A.P., Subramanyam, N. and Patel, P.A. (2009). Production of biodiesel through transesterification of jatropha oil using potassium nitrate supported on alumina solid catalyst, *Fuel*, 88,4,625-628

Xie, W; Peng, H and Chen, L. (2006). Calcined Mg-Al hydrotalcites as solid base catalysts for methanolysis of soyabean oil *J. Mol. Catal A: Chem*, 246, 24-32.

Xu, G. And Wu, G.Y. (2003). The investigation of blending properties of biodiesel and no. 0 diesel fuel. J. *Jiangsu Polytechnique University*, 15; 16-18.

Wang, J.,Chem, Y, Wang, X and CaO, F. (2009). Aluminium dodecatungstophosphate ($Al_{0.9}H_{0.3}PW_{12}O_{40}$) nanotube as a solid acid catalyst one.pot production of biodiesel from waste cooking oil, *BioResources* ,4(11), 1477-1486.

Yan,S., Lu,H. and Liang, B.(2008). Supported calcium oxide catalysts used in the transesterification of rapeseed oil for the purpose of biodiesel production, *Energy and Fuel,* 22(1),646-651

Zong, M.H., Duan Z.Q., Lou, W.Y., Smith, T.J. and Wu, H. (2007). Preparation of a sugar catalyst and its use for highly efficient production of biodiesel, *Green Chem.* 9(5), 434-437.

18 Chapter

THE IMMOBILIZED LIPASES IN BIODIESEL PRODUCTION

1. INTRODUCTION: AN OVERVIEW

ASTM International, previously known as ASTM (American Society of Testing and Materials), describes biodiesel as a fuel consisting of mono-alkyl esters derived from long-chain fatty acids (ASTM D6751). It is often produced through the transesterification of triglycerides with methanol or ethanol in the presence of a catalyst, as illustrated by the following reaction:

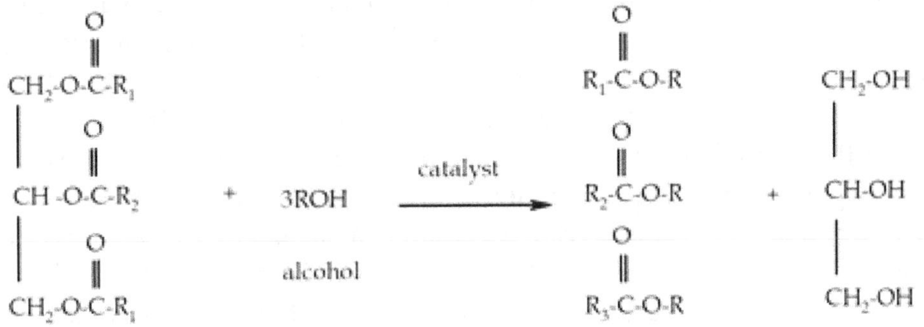

triglycerides (vegetable oil, animal fat) — mono-alkyl esters (biodiesel) — glycerol

Recently, biodiesel production by lipase catalyzed transesterification has been suggested as a promising alternative to the conventional chemical catalysis, in spite of the high conversion and reaction rates of the latter (Akoh et al., 2007; Bajaj et al., 2010; Bisen et al., 2010; Demirbas, 2009; Fjerbaek et al., 2009; Fukuda et al., 2001, 2009; Ghaly et al., 2010; Helwani et al., 2009; Jegannathan & Abang, 2008; Man Xi Ao et al., 2009; Marchetti et al., 2007; Ranganathan et al., 2008; Robles-Medina et al., 2009; Semwal et al., 2011). The enzymatic process enables eliminating the drawbacks of the alkali- or acid-catalyzed transesterification, namely: product contamination, wastewater release, free fatty acids and water interferences, and difficult glycerol recovery. Nevertheless, the commercialization of the lipase-catalyzed biodiesel synthesis remains problematic, because of the cost of the enzyme: approximately 1000 USD per kg of Novozym 435 lipase. Therefore, the implementation of strategies, such as enzyme immobilization, for the development of economic and effective enzyme based technologies for biodiesel production is of crucial importance. Enzyme immobilization ensures several issues: repetitive and continuous use of the enzyme and its stabilization, localization of the interaction, prevention of product contamination, reduction of effluent problems and material handling, and effective control of the reaction parameters (D'Souza, 1982). All these factors are manifested in the production cost.

This chapter aims to summarize the application of immobilized lipases in biodiesel synthesis, the methods employed for enzyme immobilization, and the factors influencing the process.

2. LIPASES MODE OF ACTION AND CLASSIFICATION

Lipases (EC 3.1.1.3 triacylglycerol acylhydrolase) represent a group of water soluble enzymes that originally catalyze the hydrolysis of ester bonds in water insoluble lipid substrates, acting at the interface between the aqueous and the organic phases. This unique heterogeneous reaction is feasible because of: (i) the specific lipases molecule 3D structure consisting of three domains: "contact domain", responsible for distinguishing of substrate surface, "hydrophobic" domain, responsible for extracting of one substrate molecule and its association with the "functional" domain, and "functional" domain, containing the catalytic triad Ser, Hys and Asp/Glu; (ii) the transition from closed to open conformation in the presence of the lipidic phase (Guncheva & Zhiryakova 2011; Panalotov & Verger, 2000). Enzymatic action of lipases on the substrate is a result of a nucleophilic attack on the carbonyl carbon atom from ester groups. Some lipases are also able to catalyze the processes of esterification, interesterification, transesterification, acidolysis, amynolysis and may show enantioselective properties (Hasan et al., 2009). The mechanism of the lipase catalyzed transesterification of triglycerides with an alcohol to produce biodiesel (Fjerbaek et al., 2009) could be presented by the following sequence of reactions:

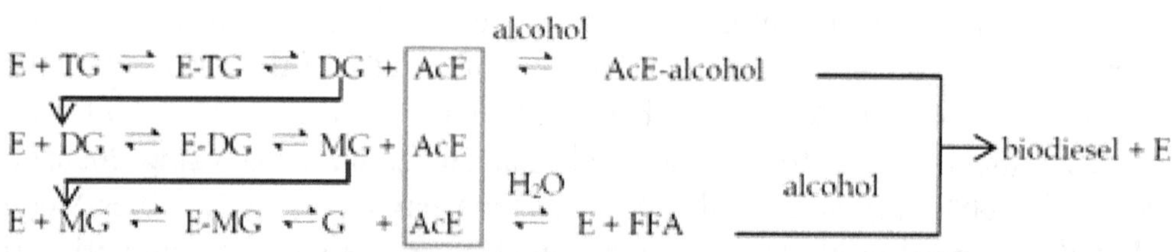

with: E-enzyme; TG-triglyceride; DT-diglyceride; MG-monoglyceride; G-glycerol; AcE- acylated enzyme; FFA-free fatty acid.

According to the origin lipases are plant, animal and microbial. The mostly used lipases in biodiesel production are of bacterial and fungal origin, such as: *Candida antarctica* (Novozym 435), *Candida Rugosa* (Lipase AY), *Pseudomonas cepacia* (Lipase PS), *Pseudomonas fluorescens* (lipase AK), *Pseudomonas aeruginosa*, and

Thermomyces lanuginose (Lipozime TL), among other. The catalytic properties and potential applications of *Bacillus* lipases are extensively reviewed by Guncheva & Zhiryakova (2011). Among the available lipase producing microorganisms, filamentous fungi belonging to various species of genera *Aspergillus* (Adinarayana et al., 2004; Karanam, & Medicherla, 2008), *Rhizopus* (Hiol et al., 2000; Shukla & Gupta, 2007), *Penicillium* (Chahinian et al, 2000; Lima et al., 2003; Vardanega et al., 2010), and *Trichoderma* (Kashmiri et al., 2006; Rajesh et al., 2010) are described as the most prospective lipase producers. Only microbial lipases are a matter of practical interest for biodiesel production, because only microbial lipases are produced in industrial scale. The application of the microbial lipases, all together with their immobilization which allows the regeneration and the reuse of the enzyme preparation in several working cycles reduces the production costs, and respectively the final cost of biodiesel. A review on microbial lipase production with emphasis on lipase engineering and use of mathematical models for process improvement and control is provided by Treichel et al. (2010).

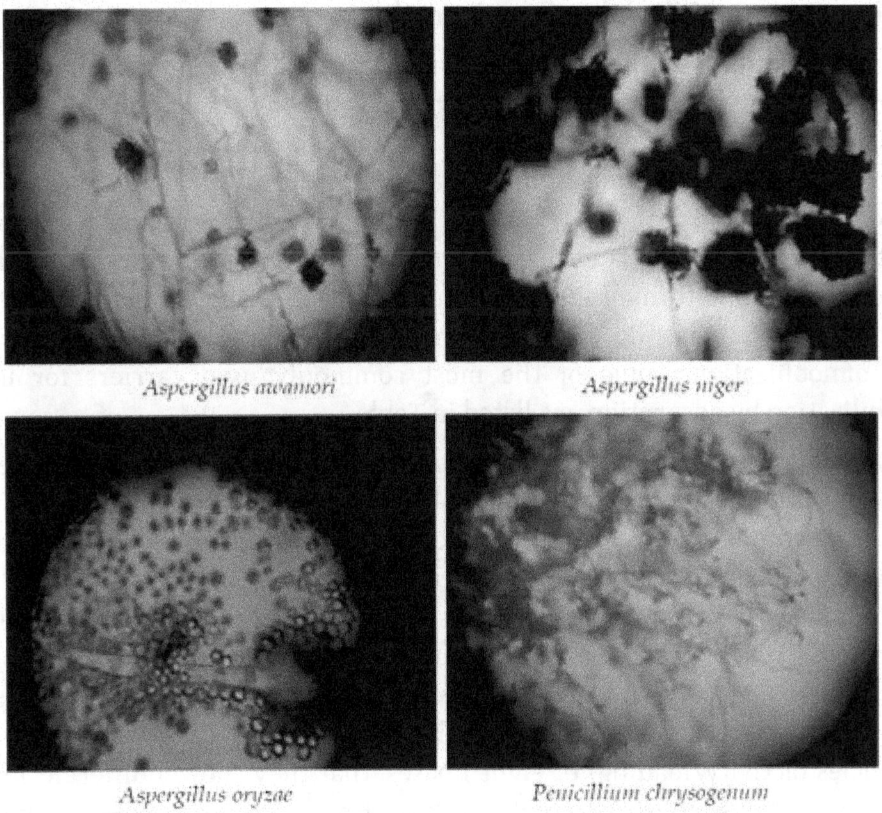

Fig. 1. Filamentous fungi belonging to various species of genera *Aspergillus* and *Penicillium*, considered as prospective lipase producers.

Some fungi cultured by the authors as powerful producers of lipase of use in biodiesel synthesis are shown in Fig. 1.

3. THE IMMOBILIZED LIPASES IN BIODIESEL PRODUCTION

The term "enzyme immobilization" was defined at the first Enzyme Engineering Conference held at Hennicker, NH, USA, in 1971. It describes "enzymes physically confined at or localized in a certain region of space with retention of their catalytic activity and which can be used repeatedly and continuously" (Powel, 1996). It is considered that lipase immobilization induces the enzyme conformational change required to enable the free access of substrate to the active centre. Especially, hydrophobic supports allow the adsorption of the open form of the lipases via interfacial activation, mimicking the lipophilic substrate (Ahn et al., 2010; Rodrigues & Fernandez-Lafuente, 2010; Salis et al., 2008; Séverac et al., 2011).

The revision of the literature covering the period 2005-2011 demonstrates that a large variety of matrices have been used for lipases immobilization, and that the main methods applied include adsorption, entrapment and/or encapsulation, and covalent attachment.

3.1 Lipases immobilization by adsorption

Physical adsorption is considered as the simplest method for enzyme immobilization. Enzyme fixation is performed through hydrogen bonds, salt linkages, and Van der Waal's forces. The process is carried out in mild conditions, without or with minimum support activation and clean up procedures application, and in the absence of additional reagents. Thus, it is economic and allows preserving enzyme activity and specificity. The chemical composition of the carrier, the molar ratio of hydrophilic to hydrophobic groups, as well as the particle size and the surface area determine the amount of enzyme bound and the enzyme behaviour after immobilization. Some of the most commonly used carriers for lipases immobilization by adsorption are listed in Table 1.

Data shown in Table 1 indicate that among the variety of lipase immobilization supports, Accurel has found a large application. Accurel is the trade name of a group of macroporous polymers. As carriers for lipase immobilization are used the polypropylene based hydrophobic Accurel MP (MP1000 with particle size below 1500 μm, Accurel MP1001 with particle size below 1000 μm and Accurel MP1004 with particle size below 400 μm), and Accurel EP-100. On the most hydrophobic support tested, Accurel MP1001, no glycerol adsorption was observed (Séverac et al., 2011). It has been demonstrated (Salis et al., 2009) comparing the catalytic efficiencies (activity/loading) of eight lipases, that they show a different level of adaptation to the support. Immobilized *Pseudomonas fluorescens* lipase is the most active biocatalyst, followed by immobilized *Pseudomonas cepacia* lipase. The other lipases tested (from *Rhizopus oryzae, Candida rugosa, Mucor javanicus, Penicillium roqueforti, Aspergillus niger, Penicillium camembertii*), are inactive toward biodiesel synthesis in the described conditions.

Enzyme immobilization on Accurel could be performed by direct contact between the lipase solution and the support (Cheirsilp et al., 2008). However, it has been confirmed that ethanol pre-treatment improves the immobilization process by inducing a better penetration of the enzyme solution inside the hydrophobic Accurel and by reducing the enzyme thermodynamic activity, thus forcing the adsorption process (Foresti, & Ferreira, 2004). Enzyme adsorption with previous ethanol treatment of the support is carried out via: (i) wetting the support with, sequentially: ethanol, aqueous ethanol solution and finally with water, with intermediary filtration, or (ii) a single wetting with ethanol and then a direct contact with the enzyme solution without removing ethanol.

Accurel, due to its hydrophobic properties, should stabilise the enzyme in its open (active) conformation. Thus, it is considered as efficient for lipase immobilization.

Other synthetic polymers used for lipases immobilization comprise: hydrophobic polystyrene macroporous resin (Li & Yan, 2010), electrospun polyacrylonitrile nanofibers with higher porosity and interconnectivity compared with other nanostructured carriers (Sakai et al., 2010), polymethacrylate (Salis et al., 2009), etc. The naturally occurring materials used as carriers for lipase immobilization include: activated carbon (Moreno- Parajàn & Giraldo, 2011; Naranjo et al., 2010) and carbon cloth (Naranjo et al., 2010), celite (Ji et al., 2010; Shah & Gupta, 2007); hydrotalcite (Yagiz et al., 2007; Zeng et al., 2009), zeolites (Yagiz et al., 2007), etc. The role of the nature of the support surface on the loading and the activity, as well as on the operational stability of the immobilized enzyme has been investigated in details.

Table 1. Carrier used for lipases immobilization by adsorption.

Carrier	Immobilized lipase origin	Reference
Accurel	Candida Antarctica Candida rugosa Pseudomonas cepacia Pseudomonas sp. *Pseudomonas fluorescens Pseudomonas fluorescens Mucor javanicus Penicillium roqueforti Penicillium camembertii Rhizopus oryzae Thermomyces lanuginosus*	Séverac et al., 2011; Tongboriboon et al., 2010 Salis et al., 2008 ; Tongboriboon et al., 2010 Salis et al., 2008 ; Tongboriboon et al., 2010 Cheirsilp et al., 2008 Salis et al., 2008, 2009 Tongboriboon et al., 2010 Salis et al., 2008 Salis et al., 2008 Salis et al., 2008 Salis et al., 2008 Tongboriboon et al., 2010
Activated carbon	Candida Antarctica *Candida rugosa*	Naranjo et al., 2010 Moreno-Parajàn & Giraldo, 2011

Celite	*Candida rugosa Pseudomonas cepacia Pseudomonas aeroginosa Pseudomonas fluorescens*	Shah & Gupta, 2007 Shah & Gupta, 2007 Ji et al., 2010 Shah & Gupta, 2007
Polystyrene	*Pseudomonas cepacia*	Li & Yan, 2010
Carbon cloth	*Pseudomonas cepacia*	23 Naranjo et al., 2010
Poly(acrylonitrile)	*Pseudomonas cepacia*	16 Sakai et al., 2010
Ceramics	*Pseudomonas cepacia*	Shah & Gupta, 2007
Pre-treated textile	*Candida sp.*	Chen et al., 2009; Li et al., 2010 ; Lu et al., 2007 and 2010
Hydrophilic resins	*Rhizomucor miehei*	De Paola et al., 2009
Silica	*Rhizomucor miehei Pseudomonas fluorescens*	Chen et al., 2009
Mg-Al hydrotalcites	*Saccharomyces cerevisiae*	Zeng et al., 2009
Resin D4020	*Penicillium expansum*	Li et al., 2009
Polymethacrylate	*Pseudomonas fluorescens*	Salis et al., 2009
Organosilicate	*Pseudomonas fluorescens*	Salis et al., 2009
Hydrotalcite	*Thermomyces lanuginosus*	Yagiz et al., 2007
Zeolites	*Thermomyces lanuginosus*	Yagiz et al., 2007

3.2 Lipases immobilization by entrapment and/or encapsulation

Entrapment involves capture of the enzyme within a matrix of a polymer, although enzyme encapsulation refers to the formation of a membrane-like physical barrier around the enzyme preparation (Cao, 2005). The matrix is usually formed during the process of the immobilization. The enzyme entrapped in a gel matrix can be further encapsulated. Both processes require simple equipment and relatively inexpensive reagents. It is supposed that enzymes immobilized by entrapment and/or encapsulation are more stable than the physically adsorbed ones. At the same time the immobilized enzymes maintain their activity and stability.

Table 2. Carrier used for lipases immobilization by entrapment and/or encapsulation.

Carrier	Immobilized lipase origin	References
κ-carrageenan	*Candida Antarctica Candida rugosa Burkholderia cepacia Pseudomonas fluorescens Aspergillus niger*	Jegannathan et al., 2010 Jegannathan et al., 2010 Jegannathan et al., 2009, 2010 Jegannathan et al., 2010 Jegannathan et al., 2010

Silica gel	*Thermomyces lanuginosus* *R. miehei* *Pseudomonas cepacia*	Khor et al., 2010 Macario et al., 2009 Noureddini et al., 2005
Celite supported sol-gel	*Candida Antarctica* Lipase NS44035	Meunier & Legge, 2010 Meunier & Legge, 2010
Silica aerogel	*Candida Antarctica* *Candida Antarctica* *Burkholderia cepacia*	Nassreddine et al., 2008 Orçaire et al., 2006 Orçaire et al., 2006

Numerous materials and techniques have been used for lipases entrapment and/ or encapsulation. Some of the immobilization matrices developed during the last years (2005- 2011) are enumerated in Table 2.

For instance, a simple technique for lipase encapsulation in κ-carrageenan by co-extrusion was suggested by Jegannathan et al. (2009, 2010). Carrageenan has been selected because of its availability, biodegradability, low cost, and lack of toxicity. It was found that at optimized reaction conditions a methyl ester conversion up to 100% could be achieved in transesterification of palm oil using the liquid core encapsulated lipase PS from *Burkholderia cepacia*. The immobilized lipase was stable and retained 82% relative transesterification activity after five cycles.

Another technique for lipase immobilization by entrapment and/or encapsulation, which has received a considerable attention in recent years, is the sol-gel process. The method involves an aqueous solution of the enzyme, a catalyst (NaOH, NaF, HCl), and an inorganic- organic matrix precursor (alkoxysilane). The hydrolysis and condensation of the precursor result in an amorphous silica matrix that covers the enzyme. The method has been applied for *R. miehei* lipase encapsulation within the micellar phase of a surfactant that is self- assembled with silica (Macario et al., 2009). It has been demonstrated that the enzyme preserves its mobility and activity. More over, because of the activation of the enzyme catalytic centre by the hydrophobic groups of the surfactant, the immobilized lipase was more active than its free form. In addition, the obtained ordered mesoporous structure improved the stability of the enzyme and decreased the rate of leaching.

Comprehensive characterization of sol–gel immobilized lipase has been performed by Noureddini et al. (2005). Lipase PS was entrapped within a sol–gel polymer matrix, prepared by polycondensation of hydrolyzed tetramethoxysilane and iso-butyltrimethoxysilane. The immobilized lipase was stable and more active than the free lipase toward the transesterification of soybean oil.

Various supports could be used to improve the stability of the entrapped encapsulated enzymes. Celite supported lipase sol–gels were investigated aiming such problems as activity, stability and reusability of the enzyme

(Meunier & Legge, 2010). The three types of Celite considered (R633, R632, and R647) were compared to unsupported lipase sol–gels. It has been established that sol–gel immobilized lipase supported on Celite R632 allowed achieving an average conversion of 60% per gram of material for 6 h, and exhibited an average initial lipase activity comparable to that of the unsupported sol–gel formulation.

Orçaire et al. (2006) report a technique for encapsulation of *Candida Antarctica* and *Burkholderia cepacia* lipases in silica aerogels reinforced with silica quartz fibre felt and dried by the CO_2 supercritical technique. The aerogel encapsulation permits maintaining the enzymes in a dispersion state similar to the dispersion prevailing in an aqueous solution, even in organic media, while agglomeration of the lipase occurs if it is used directly in the organic solvent. At present, sol-gel enzyme entrapment/encapsulation is considered to be the most successful immobilization technique for lipase immobilization.

3.3 Lipases immobilization by covalent attachment

Covalent attachment is a result of a chemical reaction between the active amino acid residues outside the active catalytic and binding site of the enzyme, and the active functionalities of the carrier (Cao, 2005). Although drastic and complicated, and strongly affected by the carriers' properties, covalent attachment is the most efficient technique for enzyme immobilization. Some carriers used for covalent lipase immobilization are displayed in Table 3.

Table 3. Carrier used for lipases immobilization by covalent attachment

Carrier	Immobilized lipase origin	References
Olive pomace	*Thermomyces lanuginosus*	Yücel, 2011
Resins	*Thermomyces lanuginosus* *Pseudomonas fluorescens*	Mendes et al., 2011 Mendes et al., 2011
Polymers	*Thermomyces lanuginosus*	Dizge et al., 2008, 2009a, 2009b
Polyurethane foam	*Thermomyces lanuginosus*	Dizge & Keskinler, 2008
Nb_2O_5 and SiO_2-PVA	*Burkholderia cepacia*	Da Rys et al., 2010
Chitosan	*Candida rugosa*	Shao et al., 2008 Ting et al., 2008
Lewatit	*Thermomyces lanuginosus*	Rodrigues et al., 2010
Silica	*Rhizopus orizae+Candida rugosa* *Rhizopus orizae+Candida rugosa Enterobacter aerogenes*	Lee and al., 2008 Kumari et al., 2009
Magnetic nanostructures	*Candida rugosa* *Thermomyces lanuginosus*	Dussan et al., 2007, 2010 Xie & Ma, 2010

Yücel (2011) reports a method for *Thermomyces lanuginosus* lipase covalent binding on polyglutaraldehyde-activated olive pomace powder. The technique is cost effective, because of the low price of the support and because of the strong covalent bond formed, leading to enzyme stabilization without loss of activity, allowing the multiple reuse of the enzyme. Immobilized lipase was stable for 10 batches of pomace oil transesterification retaining more than 80% residual activity.

Among the other naturally occurring materials, chitosan is considered as appropriate for enzyme binding. Its membrane forming and adhesion ability, high mechanical strength and facility of forming insoluble in water thermally and chemically inert films make it suitable for lipase immobilization. For instance, *Candida rugosa* type VII lipase was fixed onto chitosan beads using a binary method consisting in the follows: (i) lipase immobilization onto the hydroxyl groups of chitosan by activation with 1-ethyl-3-(3- dimethylaminopropyl) carbodiimide hydrochloride; (ii) immobilization of additional lipase molecules through their amino groups to chitosan by cross-linking with glutaraldehyde. The immobilized enzyme has been used to catalyse the hydrolysis of soybean oil. Then, the feedstock containing free fatty acids, mono-, di- and triglycerides was esterified with methanol in the presence of an acid catalyst to produce biodiesel. It has been demonstrated that the enzymatic/acid-catalyzed hybrid process uses milder reaction conditions and allows avoiding the inactivation of the immobilized enzyme by polar compounds and increase biodiesel yields.

A new method for the synthesis of hydrophobic microporous matrices for enzyme immobilization, namely styrene–divinylbenzene–polyglutaraldehyde and poly(styrene- divinylbenzene)-polyglutaraldehyde copolymers, applying High Internal Phase Emulsions (HIPE) technique has been developed by Dizge et al. (2008, 2009a, 2009b). *Thermomyces lanuginosus* lipase was successfully attached to the support by covalent binding. According to the authors, the copolymers could be prepared in a short time and in large amounts and shapes. The immobilization efficiency, defined as the ratio of the activity of the immobilized enzyme to the activity of the free enzyme was found to vary from 80% to 89%. The immobilized enzyme retained its activity during 10-15 repeated batch reactions.

Promising results in terms of enzyme thermal and operational stability improvement have been obtained using as supports for lipases immobilization silanized Nb_2O_5 and SiO_2-PVA (Da Rys et al., 2010), and glutaraldehyde or ethanolamine activated silica gels (Lee and al., 2008; Kumari et al., 2009). However, the stability of the immobilized enzyme depends not only on the chemical/physical nature of the carrier, but also on the binding mode, the binding number, and the position of the binding on the enzyme surface (Cao, 2005), among other. Mendes et al. (2011) and Rodrigues et al. (2010) demonstrate that the multipoint covalent attachment of *Thermomyces lanuginosus* lipase on Toyopearl AF-amino-650M resin and on aldehyde-Lewatit is an efficient strategy for enzyme stabilization. It has been demonstrated that *Thermomyces lanuginosus* lipase immobilized on glyoxyl-resin is between 27 and 31 times more stable than the soluble lipase (Mendes et al., 2011).

Another important issue provided by enzyme immobilization concerns the localization of the interaction in the zone where the maximum concentration of reagents is present and/or at the interface between the immiscible heterogeneous phases, regarding lipases. For this purpose, lipases (from *Candida rugosa* and *Thermomyces lanuginosus*) have been immobilized on magnetic nanostructures (Fe_3O_4) and localized by application of a magnetic field. In addition, the method favours the simple and fast separation of the enzyme from the reaction mixture. Thus, it allows the intensification of the process and the reduction of the production costs.

3.4 Cells immobilization

The technological and economic advantages of immobilized cells over immobilized enzymes are well known (D'Souza, 1982): higher operational stability, higher yields of enzyme activity after immobilization, greater resistance to environmental perturbations, greater potential for multistep processes, and lower effective enzyme cost (enzyme purification and extraction are avoided). Despite of these benefits, only few investigations on the use of immobilized cells in biodiesel synthesis are reported until now (Fukuda et al., 2009; Hama et al., 2006, 2007; Li et al., 2007; Oda et al., 2005; Tamalampudi et al., 2008). The research efforts were focused on the immobilization no more than of *Rhysopus oryzae* within porous biomass support particles. The fixation was achieved spontaneously during batch cultivation. The applied technique (Atkinson et al., 1979) offers numerous advantages over other methods: the particles are reusable and mechanically resistant; additional reagents, aseptic handling of particles, and preproduction of cells are not necessary. It has been demonstrated that *Rhysopus oryzae* cells immobilized within biomass support particles can be used as low cost catalyst for biodiesel production.

4. CONCISE SUMMARY

The enzymatic method for biodiesel generation presents numerous benefits compared to the chemical catalysis now utilized. It is more efficient due to enzyme specificity and selectivity, entails less energy consumption owing to moderate reaction conditions, and is environmentally sustainable due to minimal emission of byproducts or trash. Catalyst immobilization offers several advantages, including the repeated utilization of enzymes, improved thermal and operational stability, localization of interactions, and effective regulation of reaction parameters, thereby decreasing production costs and rendering enzyme-based biodiesel synthesis a compelling alternative to other technologies. This article offers an overview of the approaches utilized for lipase immobilization.

REFERENCES

Adinarayana, K.; Bapi Raju, K. V. V. S. N.; Zargar, M. I. ; Devi, R. B.; Lakshmi, P. J. & Ellaiah, P. (2004). Optimization of process parameters for production of lipase in solid-state fermentation by newly isolated *Aspergillus* species, *Ind. J. Biotechnol.*, 3, 65-69

Ahn, K.; Ye, S.; Chun, W.; Rah, H. & Kim, S. (2010). Yield and component distribution of biodiesel by methanolysis of soybean oil with lipase-immobilized mesoporous silica. *Microporous and Mesoporous Materials* doi:10.1016/j.micromeso.2010.11.008 (in press)

Akoh, C.; Chang, Shu-Wei; Lee, Guan-Chiun & Shaw, Jei-Fu (2007). Enzymatic approach to biodiesel production. J. *Agric. Food Chem.* 55, 8995-9005

ASTM D6751: American Society for Testing and Materials standard specification for biodiesel fuel (B 100) blend stock for distillate fuels

Atkinson, B.; Black, G. M.; Lewis, P. J. S. & Pinches, A. (1979). Biological particles of given size, shape, and density for use in biological reactors. *Biotechnol. Bioeng.*, 21, 193-200

Bajaj, A.; Lohan, P.; Jha, P. & Mehrotra, R. (2010). Biodiesel production through lipase catalyzed transesterification: An overview. *J. Mol. Catal. B: Enzymatic*, 62, 9-14

Bisen, P.; Sanodiya, B.; Thakur, G.; Baghel, R. & Prasad, G. B. K. S. (2010). Biodiesel production with special emphasis on lipase-catalyzed transesterification. *Biotechnol. Lett.*, 32, 1019–1030

Cao, L. (2005). *Carrier-bound immobilized enzymes. Principles, application and design.* Wiley- VCH, ISBN-10: 3-527-31232-3, Weinheim

Chahinian, H.; Vanot, G.; Ibrik, A.; Rugani, N.; Sarda, L. & Comeau, L. C. (2000). Production of extracellular lipases by *Penicillium cyclopium* purification and characterization of a partial acylglycerol lipase. *Biosci. Biotechnol. Biochem.*, 64, 215-222

Cheirsilp, B.; H-Kittikuna, A. & Limkatanyu, S. (2008). Impact of transesterification mechanisms on the kinetic modeling of biodiesel production by immobilized lipase. *Biochem. Eng. J.*, 42, 261–269

Chen, Y.; Xiao, B.; Chang, J.; Fu, Y.; Lv, P. & Wang, X. (2009). Synthesis of biodiesel from waste cooking oil using immobilized lipase in fixed bed reactor. *Energy Conversion and Management*, 50, 668–673

Da Rys, P.; Silva, G.; Mendes, A.; Santos, J. & Castro, H. (2010). Evaluation of the catalytic properties of *Burkholderia cepacia* lipase immobilized on non-commercial matrices to be used in biodiesel synthesis from different feedstocks. *Bioresource Technology*, 101, 5508–5516

De Paola, M. G.; Ricca, E.; Calabrò, V.; Curcio, S. & Iorio, G. (2009). Factor analysis of transesterification reaction of waste oil for biodiesel production. *Bioresource Technology*, 100, 5126–513

Demirbas, A. (2009). Progress and recent trends in biodiesel fuels. *Energy Conversion and Management*, 50, 14–34

Dizge, N. & Keskinler, B. (2008). Enzymatic production of biodiesel from canola oil using immobilized lipase. *Biomass and Bioenergy*, 32, 1274–1278

Dizge, N.; Keskinler, B. & Tanriseven, A. (2008). Covalent attachment of microbial lipase onto microporous styrene-divinylbenzene copolymer by means of poly glutaraldehyde. *Colloids and Surfaces B: Biointerfaces*, 66, 34–38

Dizge, N.; Keskinler, B. & Tanriseven, A. (2009a). Biodiesel production from canola oil by using lipase immobilized onto hydrophobic microporous styrene-divinylbenzene copolymer. *Biochem. Eng. J.*, 44, 220–225

Dizge, N.; Aydiner, C.; Imer, D.; Bayramoglu, M.; Tanriseven, A. & Keskinler, B. (2009b). Biodiesel production from sunflower, soybean, and waste cooking oils by transesterification using lipase immobilized onto a novel microporous polymer. *Bioresource Technology*, 100, 1983–1991

D'Souza, S. F. (1989). Immobilized cells: techniques and applications. *Indian J. Microbiol.*, 29, 83-117

Dussán, K.; Giraldo, O. & Cardona, C. (2007). Application of magnetic nanostructures in biotechnological processes: Biodiesel production using lipase immobilized on magnetic carriers. *Proceedings of European Congress of Chemical Engineering* (ECCE-6), Copenhagen, 16-20 September 2007

Dussan, K.; Cardona, C.; Giraldo, O.; Gutiérrez, L. & Pérez, V. (2010). Analysis of a reactive extraction process for biodiesel production using a lipase immobilized on magnetic nanostructures. *Bioresource Technology*, 101, 9542–9549

Fjerbaek, L.; Christensen, K. & Norddahl, B. (2009). A Review of the Current State of Biodiesel Production Using Enzymatic Transesterification. *Biotechnol Bioeng.*, 102, 5, 1298-1315

Foresti, M. L. & Ferreira M. L. (2004). Ethanol pretreatment effect and particle diameter issues on the adsorption of *Candida rugosa* lipase onto polypropylene powder. *Appl. Surf. Sci.*, 238, 86–90

Fukuda, H.; Kondo, A. & Noda, H. (2001). Biodiesel fuel production by transesterification of oils. *J. Bioscience Bioeng.*, 5, 405-416

Fukuda, H.; Kondo, A. & Tamalampudi, S. (2009). Bioenergy: Sustainable fuels from biomass by yeast and fungal whole-cell biocatalysts. *Biochem. Eng. J.*, 44, 2–12

Ghaly, A. E.; Dave, D.; Brooks, M. S. & Budge, S. (2010). Production of biodiesel by enzymatic transesterification: Review. *Am. J. Biochem. Biotechnol.*, 6, 54-76

Guncheva, M. & Zhiryakova, D. (2011). Catalytic properties and potential applications of Bacillus lipases. *J. Mol. Catalysis B: Enzymatic*, 68, 1-21

Hama, S.; Tamalampudi, S.; Fukumizu, T.; Miura, K.; Yamaji, H.; Kondo, A. & Fukuda, H. (2006). Lipase localization in *Rhizopus oryzae* cells immobilized within biomass support particles for use as whole-cell biocatalysts in biodiesel-fuel production. *J. Bioscience Bioeng.*, 101, 328-333

Hama, S.; Yamaji, H.; Fukumizu, T.; Numata, T.; Tamalampudi, S.; Kondo, A. Noda, H. & Fukuda, H. (2007). Biodiesel-fuel production in a packed-bed reactor using lipase- producing *Rhizopus oryzae* cells immobilized within biomass support particles. *Biochem. Eng. J.*, 34, 273-278

Hasan, F.; Shah, A. & Hameed, A. (2009). Methods for detection and characterization of lipases: A comprehensive review. *Biotechnol. Adv.*, 27, 782-798

Helwani, Z.; Othman, M. R.; Aziz, N.; Fernando, W. J. N. & Kim, J. (2009). Technologies for production of biodiesel focusing on green catalytic techniques: A review. *Fuel Processing Technology*, 90, 1502–1514

Hiol, A.; Jonzo, M. D.; Rugani, N.; Druet, D.; Sarda, L. & Comeau, L. C. (2000). Purification and characterization of an extracellular lipase from a thermophilic *Rhizopus oryzae* strain isolated from palm fruit. *Enzyme Microb. Technol.*, 26, 421–30

Jegannathan, K.; Jun-Yee, L.; Chan, E. & Ravindra, P. (2009). Design an immobilized lipase enzyme for biodiesel production. *J. Renewable and Sustainable Energy*, 1, 063101-1 - 063101-8

Jegannathan K. Abang, S. (2008). Production of biodiesel using immobilized lipase-A critical review. *Critical Reviews in Biotechnology*, 28, 253–264

Jegannathan, K.; Jun-Yee, L.; Chan, E. & Ravindra, P. (2010). Production of biodiesel from palm oil using liquid core lipase encapsulated in κ-carrageenan. *Fuel*, 89, 2272–2277

Ji, Q.; Xiao, S.; He, B. & Liu, X. (2010). Purification and characterization of an organic solvent-tolerant lipase from *Pseudomonas aeruginosa* LX1 and its application for biodiesel production. *J. Mol. Catal. B: Enzymatic*, 66, 264–269

Karanam, S. K. & Medicherla, N. R. (2008). Enhanced lipase production by mutation induced *Aspergillus japonicus*. *African J. Biotechnol.*, 7, 2064-2067

Kashmiri, M. A.; Adnan, A.; Butt, B. W. (2006). Production, purification and partial characterization of lipase from *Trichoderma viride*. *African J. Biotechnol.*, 5, 878-882

Khor, G.; Sim, J.; Kamaruddin, A. & Uzir, M. (2010). Thermodynamics and inhibition studies of lipozyme TL IM in biodiesel production via enzymatic transesterification. *Bioresource Technology*, 101, 6558–6561

Kumari, A.; Mahapatra, P.; Garlapati, V. & Banerjee, R. (2009). Enzymatic transesterification of Jatropha oil. *Biotechnology for Biofuels*, 2:1, doi:10.1186/1754-6834-2-1

Lee Jong Ho; Dong Hwan Lee; Jung Soo Lim; Byung-Hwan Um; Chulhwan Park; Seong Woo Kang & Seung Wook Kim (2008). Optimization of the process for biodiesel production using a mixture of immobilized *Rhizopus oryzae* and *Candida rugosa* lipases. *J. Microbiol. Biotechnol.*, 18, 1927–1931

Li, N.; Zong, M.; Wu, H. (2009). Highly efficient transformation of waste oil to biodiesel by immobilized lipase from *Penicillium expansum*. *Process Biochemistry*, 44, 685–688

Li, Q. & Yan, Y. (2010). Production of biodiesel catalyzed by immobilized *Pseudomonas cepacia* lipase from Sapium sebiferum oil in micro-aqueous phase. *Applied Energy*, 87, 3148–3154

Li, W.; Du, W. & Liu, D. (2007). Rhizopus oryzae IFO 4697 whole cell catalyzed methanolysis of crude and acidified rapeseed oils for biodiesel production in tert-butanol system. *Process Biochemistry*, 42, 1481–1485

Li, Z.; Deng, L.; Lu, J.; Guo, X.; Yang, Z. & Tan, T. (2010). Enzymatic synthesis of fatty acid methyl esters from crude rice bran oil with immobilized *Candida sp.* 99-125. *Chinese Journal of Chemical Engineering*, 18, 870-875

Lima, V. M. G.; Krieger, N.; Sarquis, M. I. M.; Mitchell, D. A.; Ramos, L. P. & Fontana, J. D. (2003). Effect of nitrogen and carbon sources on lipase production by *Penicillium aurantiogriseum. Food Technol. Biotechnol.,* 41, 105–110

Lu, J.; Nie, K.; Xie, F.; Wang, F. & Tan, T. (2007). Enzymatic synthesis of fatty acid methyl esters from lard with immobilized *Candida sp.* 99-125. *Process Biochemistry*, 42, 1367– 1370

Lu, J.; Deng, L.; Zhao, R.; Zhang, R.; Wang, F. & Tan, T. (2010). Pretreatment of immobilized *Candida sp.* 99-125 lipase to improve its methanol tolerance for biodiesel production. *J. Mol. Catal. B: Enzymatic*, 62, 15–18

Macario, A.; Moliner, M; Corma, A. & Giordano, G. (2009). Increasing stability and productivity of lipase enzyme by encapsulation in a porous organic–inorganic system. *Microporous and Mesoporous Materials*, 118, 334–340

Man Xi Ao, Sini Mathew & Obbard, J. (2009). Biodiesel fuel production via transesterification of oils using lipase biocatalyst. *GCB Bioenergy*, 1, 115–125

Marchetti, J. M.; Miguel, V. U. & Errazu, A.F. (2007). Possible methods for biodiesel production. *Renewable and Sustainable Energy Reviews*, 11, 1300-1311

Mendes, A.; Giordano, R. C.; Giordano, R. L. C. & Castro, H. (2011). Immobilization and stabilization of microbial lipases by multipoint covalent attachment on aldehyde- resin affinity: Application of the biocatalysts in biodiesel synthesis. *J. Mol. Catal. B: Enzymatic*, 68, 109–115

Meunier, S. & Legge, R. (2010). Evaluation of diatomaceous earth as a support for sol–gel immobilized lipase for transesterification. *J. Mol. Catal. B: Enzymatic*, 62, 54–58

Moreno-Parajàn, J. C. & Giraldo, L. (2011). Study of immobilized candida rugosa lipase for biodiesel fuel production from palm oil by flow microcalorimetry. *Arabian Journal of Chemistry*, 4, 55–62

Naranjo, J.; Córdoba, A.; Giraldo, L.; García, V. & Moreno-Parajàn, J. C. (2010). Lipase supported on granular activated carbon and activated carbon cloth as a catalyst in the synthesis of biodiesel fuel. *J. Mol. Catal. B: Enzymatic*, 66, 166–171

Nassreddine, S.; Karout, A.; Christ, M. & Pierre, A. (2008). Transesterification of a vegetal oil with methanol catalyzed by a silica fibre reinforced aerogel encapsulated lipase. *Applied Catalysis A: General*, 344, 70–77

Oda, M.; Kaieda, M.; Hama, S.; Yamaji, H.; Kondo, A.; Izumoto, E. & Fukuda, H. (2005). Facilitatory effect of immobilized lipase-producing *Rhizopus oryzae* cells on acyl migration in biodiesel-fuel production. *Biochem. Eng. J.*, 23, 45–51

Orçaire, O.; Buisson, P. & Pierre, A. (2006). Application of silica aerogel encapsulated lipases in the synthesis of biodiesel by transesterification reactions. *J. Mol. Catal. B: Enzymatic*, 42, 106–113

Noureddini, H.; Gao, X. & Philkana, R. S. (2005). Immobilized Pseudomonas cepacia lipase for biodiesel fuel production from soybean oil. *Bioresource Technology*, 96, 769–777

Panalotov, I. & Verger, R. (2000). *Physical Chemistry of Biological Interfaces.* Marcel Dekker Inc., New York

Powel, L. W. (1996). In *Industrial Enzymology*, T. Godfrey and S. West, Editors, 2nd ed. Stockton Press, NY

Rajesh, E. M.; Arthe, R.; Rajendran, R.; Balakumar, C.; Pradeepa, N. & Anitha, S. (2010). Investigation of lipase production by *Trichoderma reesei* and optimization of production parameters. *EJEAFChe*, 9, 1177-1189

Ranganathan, S.; Narasimhan, S. & Muthukumar, K. (2008). An overview of enzymatic production of biodiesel. *Bioresource Technology*, 99, 3975–3981

Robles-Medina, A.; González-Moreno, P.A.; Esteban-Cerdán, L. & Molina-Grima, E. (2009). Biocatalysis: Towards ever greener biodiesel production. *Biotechnology Advances*, 27, 398–408

Rodrigues, R. & Fernandez-Lafuente, F. (2010). Lipase from Rhizomucor miehei as an industrial biocatalyst in chemical process. *J. Mol. Catal. B: Enzymatic*, 64, 1–22

Rodrigues, R.; Pessela, B.; Volpato, G.; Fernandez-Lafuente, R.; Guisan, J. & Ayub, M. (2010). Two step ethanolysis: A simple and efficient way to improve the enzymatic biodiesel synthesis catalyzed by an immobilized–stabilized lipase from *Thermomyces lanuginosus*. *Process Biochemistry*, 45, 1268–1273

Sakai, S.; Yuping Liu, Y.; Yamaguchi, T.; Watanabe, R.; Kawabe, M. & Kawakami, K. (2010). Production of butyl-biodiesel using lipase physically-adsorbed onto electrospun polyacrylonitrile fibers. *Bioresource Technology*, 101, 7344–7349

Salis, A.; Pinna, M.; Monduzzi, M. & Solinas, V. (2008). Comparison among immobilised lipases on macroporous polypropylene toward biodiesel synthesis. *J. Mol. Catal. B: Enzymatic*, 54, 19–26

Salis, A.; Bhattacharyya, M.; Monduzzi, M. & Solinas, V. (2009). Role of the support surface on the loading and the activity of *Pseudomonas fluorescens* lipase used for biodiesel synthesis. *J. Mol. Catal. B: Enzymatic*, 57, 262–269

Semwal, S.; Arora, A.; Badoni, R. & Tuli. D. (2011). Biodiesel production using heterogeneous catalysts. *Bioresour. Technol.*, 102, 2151-2161

Séverac, E.; Galy, O.; Turon, F.; Pantele, C.; Condoret, J-S.; Monsan, P. & Marty, A. (2011). Selection of CalB immobilization method to be used in continuous oil transesterification: Analysis of the economical impact. *Enzyme and Microbial Technology*, 48, 61–70

Shah, S. & Gupta, M. (2007). Lipase catalyzed preparation of biodiesel from Jatropha oil in a solvent free system. *Process Biochemistry*, 42, 409–414

Shao, P.; Meng, X.; He, J. & Sun, P. (2008). Analysis of immobilized *Candida rugosa* lipase catalyzed preparation of biodiesel from rapeseed soapstock. *Food and Bioproducts Processing*, 8 6, 283–289

Shukla, P. & Gupta, K. (2007). Ecological screening for lipolytic molds and process optimization for lipase production from *Rhizopus oryzae* KG-5. *J Appl. Sci. Environ. Sanit.*, 2, 35–42

Tamalampudi, S.; Talukder, M.; Hama, S.; Numata, T.; Kondo, A. & Fukuda, H. (2008). Enzymatic production of biodiesel from Jatropha oil: A comparative study of immobilized-whole cell and commercial lipases as a biocatalyst. *Biochem. Eng. J.*, 39, 185–189

Tan, T.; Lu, J.; Nie, K.; Deng, L. & Wang, F. (2010). Biodiesel production with immobilized lipase: A review. *Biotechnology Advances*, 28, 628–634

Ting, W.; Huang, C.; Giridhar, N. & Wu, W. (2008). An enzymatic/acid-catalyzed hybrid process for biodiesel production from soybean oil. J. *Chinese Institute of Chemical Engineers*, 39, 203–210

Tongboriboon, K.; Cheirsilp, P. & H-Kittikun, A. (2010). Mixed lipases for efficient enzymatic synthesis of biodiesel from used palm oil and ethanol in a solvent-free system. *J. Mol. Catal. B: Enzymatic*, 67, 52–59

Treichel, H.; Oliveira, D.; Mazutti, M.; Di Luccio, M. & Oliveira, J. (2010). A Review on Microbial Lipases Production. *Food Bioprocess Technol.*, 3, 182-196

Vardanega, R.; Remonatto, D.; Arbter, F.; Polloni, A.; Rigo, E.; Ninow, L. N.; Treichel, H.; Oliveira, D. & Luccio, M. (2010). A Systematic study on extraction of lipase obtained by solid-state fermentation of soybean meal by a newly isolated strain of *Penicillium* sp. *Food Process. Technol.*, 3, 461-465

Xie, W. & Ma, N. (2010). Enzymatic transesterification of soybean oil by using immobilized lipase on magnetic nano-particles. *Biomass and Bioenergy*, 34, 890-896

Yagiz, F; Kazan, D. & Nilgun Akin, A. (2007). Biodiesel production from waste oils by using lipase immobilized on hydrotalcite and zeolites. *Chem. Eng. J.*, 134, 262–267

Yücel, Y. (2011). Biodiesel production from pomace oil by using lipase immobilized onto olive pomace. *Bioresource Technology*, 102, 3977–3980

Zeng, H.; Liao, K.; Deng, X.; Jiang, H. & Zhang, F. (2009). Characterization of the lipase immobilized on Mg–Al hydrotalcite for biodiesel. *Process Biochemistry*, 44, 791–798

Index

A

Acetic, 258

Acetone, 255-256

Activity, 402-405, 407, 409-411, 419, 452, 455, 457

Adsorption, 452-453

Adults, 31

Agronomic, 18

Alcohols, 34, 255-256, 324, 327, 330

Ammonia, 116

Ammonium, 267-268, 405

Aqueous, 249, 267

Aromatic, 52

Azeotropes, 350

B

Bacteria, 261

Basicity, 409, 411

Bilayer, 92

Biobutanol, 261, 269

Biodiesel, 1-3, 5, 8, 10, 18-23, 25-27, 34, 42-45, 49-53, 58, 60-61, 65-66, 71-81, 83, 86, 99-100, 110-111, 117, 127, 133-134, 136-140, 142, 144-147, 149-150, 152, 154-156, 159-163, 168, 170-172, 178, 181-192, 204-210, 215-216, 226-229, 237, 239, 241, 245, 248, 252, 269, 273, 275, 284-293, 301, 308, 313, 315-321, 323, 329, 332, 334, 336-337, 339-348, 355, 372-373, 375-376, 378, 380, 382, 396-398, 402-403, 407-410, 417-425, 429-432, 444-446, 449-451, 458-465

Bioenergy, 156

Biofuels, 126, 136, 143, 145, 156, 195-196, 208-209

Biogas, 243, 269

Biomass, 7, 96-97, 100, 111-113, 117, 123-124, 128, 147-148, 150, 172, 175-177, 182-183, 190-191, 196-197, 200, 202-204, 208, 213, 222, 226-230, 232, 240, 243-245, 265, 269, 273

Brassica, 3-4

Butanol, 252-254, 260-262, 264, 330

Butyric, 258

C

Calcium, 251, 410-411

Candida, 85, 103, 462

Castor, 5, 360, 373, 429-430, 434-435, 441-444

Catalysis, 77, 79, 420-423, 425

Catalysts, 12-13, 34-36, 52, 66, 253, 311-312, 344, 354, 401-402, 404, 407, 410-411, 418-420, 422-423, 431-433, 446

Cerevisiae, 93

Cetane, 205-207

Chlorella, 167, 184, 186, 189

Chloroform, 139

Corrosion, 432

Cottonseed, 292, 361, 364, 373-375

Covalent, 456-457

D

Density, 11, 45, 118, 224-225

Department, 50

Design, 39-40, 217, 353

Diameter, 353

Dielectric, 325

Diesel, 1, 25, 52, 110, 136-137, 153, 205, 227, 252, 273, 286-287, 323, 387, 390-391, 393-396, 398-399, 427

E

Emission, 137, 394

Endosulfan, 31

Energy, 1, 5, 51, 54, 73-75, 78, 111, 135-137, 154-156, 159, 183, 185, 192, 195, 199, 210, 212, 217-218, 220, 225-226, 228-229, 234-235, 239, 241, 243-245, 269-270, 273, 289, 304, 307, 315-316, 319, 324-326, 335, 340, 342-347, 396, 398-399, 417, 423

Enzyme, 432, 450, 452-458

Ethanol, 34, 109, 245, 252, 260-261, 306, 453

G

Generation, 196, 208

Glucose, 98

Glycerol, 63, 99-100, 114, 178, 248, 250, 254-256, 258-259, 313, 320, 355, 435

Greenhouse, 2

Growth, 54, 56, 94-98, 112-113, 115-116, 140, 148, 163, 172-173, 189, 200-201, 220-223, 230, 233, 237

H

Heptanal, 262-263

Heteropoly, 404

Hexane, 311

Hydrogen, 128, 249, 302

Hydrolysis, 65, 117, 299, 318

Hydrolytic, 120

I

In-situ, 276, 281, 286

Intensity, 173

Internal, 277

Inulin, 94

Iodine, 10

J

Journal, 76-77, 104, 107, 129-130, 184, 186-187, 190, 193, 236-238, 241, 270-272, 289, 291-292, 317-320, 423, 425

Juncea, 4, 6-7, 18

L

Larvae, 31

Linoleic, 8, 207

Linolenic, 207

Lipase, 450-453, 455-465

Lipases, 450-452, 456

Lipids, 83-86, 92-94, 98, 112, 114, 126, 131, 148, 150, 164, 172-174, 181, 215-216

Lipolytica, 84, 92-94, 101, 105-106

M

Malate, 89

Manure, 7

Metabolism, 93

Methanol, 14, 17, 34, 77, 139, 246-251, 253-255, 275-276, 278-281, 284, 288, 290, 298-299, 305-306, 308-310, 312, 315-318, 320-321, 334, 337, 356, 362-363, 369-371, 373-374, 405, 408-409, 413, 417, 425, 431, 434-435, 442-443

Methoxide, 35

Microalgae, 112-113, 115-118, 121-122, 127, 133, 139-140, 142-143, 146, 148,

150, 154, 156, 162-163, 168, 170, 172, 182, 184, 186, 196-198, 201, 204, 208-210, 212, 214-215, 236-241

N

Nematodes, 3

Nitrogen, 87, 95-97, 116, 148, 150, 167, 173, 233, 244

Nucleic, 87

Nutrients, 151-152, 221

O

Oilseed, 3

Oliveira, 465

Oryzae, 129, 461

Osmotic, 177

Oxidation, 207, 255

Oxygen, 175, 218, 255

P

Pellets, 122

Petroleum, 51, 137, 347

Phosphorus, 116, 148

Phycology, 236-238, 240-241

Potassium, 35, 249, 265, 267-268

Poultry, 60, 72, 80

Proteins, 87

Proton, 413

Pyruvate, 89

Q

Quality, 45, 205

R

Radiation, 199, 201, 324

Reductase, 179

Resins, 402

Rhizopus, 461

S

Salina, 211

ativa, 3

Sodium, 35, 410

Soxhlet, 288

Soybean, 253, 297, 305, 315, 321, 422, 425, 465

Spectrum, 200

Starch, 181

Strains, 114, 151

Sulfuric, 249, 251, 265-267

Sunflower, 6, 32

Systems, 118, 196, 198, 217-219

T

Thermal, 300-301, 392

Tobacco, 3, 5, 18

U

Ultrasonic, 177, 335-336, 417-418

Ultrasound, 335-337, 340, 417

V

Vegetable, 59, 137, 246, 250-251, 253-254, 289, 297, 323, 329, 360, 377, 398, 418, 428, 431

Viscosity, 11, 34, 207, 323, 436

W

Wastewater, 116, 151, 236

Weevil, 31

Y

Yeasts, 84-88, 94, 98

Z

Zeolites, 403-404, 433

Zincate, 411

www.ingramcontent.com/pod-product-compliance
Lightning Source LLC
LaVergne TN
LVHW081942200926
848484LV00009B/2495
* 9 7 8 1 7 8 7 1 5 6 0 2 9 *